D0845800

Category Theory for Computing Science

Prentice Hall International Series in Computer Science

C. A. R. Hoare, Series Editor

BACKHOUSE, R. C., *Program Construction and Verification*
BACKHOUSE, R. C., *Syntax of Programming Languages: Theory and practice*
DE BAKKER, J. W., *Mathematical Theory of Program Correctness*
BEN-ARI, M., *Principles of Concurrent and Distributed Programming (2nd edn)*
BIRD, R. and WADLER, P., *Introduction to Functional Programming*
BJÖRNER, D. and JONES, C. B., *Formal Specification and Software Development*
BORNAT, R., *Programming from First Principles*
BUSTARD, D., ELDER, J. and WELSH, J., *Concurrent Program Structures*
CLARK, K. L. and McCABE, F. G., *micro-Prolog: Programming in logic*
CROOKES, D., *Introduction to Programming in Prolog*
DROMEY, R. G., *How to Solve it by Computer*
DUNCAN, F., *Microprocessor Programming and Software Development*
ELDER, J., *Construction of Data Processing Software*
ELLIOTT, R. J. and HOARE, C. A. R., (eds.), *Scientific Applications of Multiprocessors*
GOLDSCHLAGER, L. and LISTER, A., *Computer Science: A modern introduction (2nd edn)*
GORDON, M. J. C., *Programming Language Theory and its Implementation*
HAYES, I. (ed.), *Specification Case Studies*
HEHNER, E. C. R., *The Logic of Programming*
HENDERSON, P., *Functional Programming: Application and implementation*
HOARE, C. A. R., *Communicating Sequential Processes*
HOARE, C. A. R. and JONES, C. B. (ed.), *Essays in Computing Science*
HOARE, C. A. R. and SHEPHERDSON, J. C. (eds.), *Mathematical Logic and Programming Languages*
HUGHES, J. G., *Database Technology: A software engineering approach*
INMOS LTD, *occam Programming Manual*
INMOS LTD, *occam 2 Reference Manual*
JACKSON, M. A., *System Development*
JOHNSTON, H., *Learning to Program*
JONES, C. B., *Systematic Software Development using VDM (2nd edn)*
JONES, C. B. and SHAW, R. C. F. (eds.), *Case Studies in Systematic Software Development*
JONES, G., *Programming in occam*
JONES, G. and GOLDSMITH, M., *Programming in occam 2*
JOSEPH, M., PRASAD, V. R., and NATARAJAN, N., *A Multiprocessor Operating System*
LEW, A., *Computer Science: A mathematical introduction*
MacCALLUM, I., *Pascal for the Apple*
MacCALLUM, I., *UCSD Pascal for the IBM PC*
MARTIN, J. J., *Data Types and Data Structures*
MEYER, B., *Introduction to the Theory of Programming Languages*
MEYER, B., *Object-oriented Software Construction*
MILNER, R., *Communication and Concurrency*
MORGAN, C., *Programming from Specifications*
PEYTON JONES, S. L., *The Implementation of Functional Programming Languages*
POMBERGER, G., *Software Engineering and Modula-2*
REYNOLDS, J. C., *The Craft of Programming*
RYDEHEARD, D. E. and BURSTALL, R. M., *Computational Category Theory*
SLOMAN, M. and KRAMER, J., *Distributed Systems and Computer Networks*
SPIVEY, J. M., *The Z Notation: A reference manual*
TENNENT, R. D., *Principles of Programming Languages*
WATT, D. A., *Programming Language Concepts and Paradigms*
WATT, D. A., WICHMANN, B. A., and FINDLAY, W., *ADA: Language and methodology*
WELSH, J., and ELDER, J., *Introduction to Modula-2*
WELSH, J., and ELDER, J., *Introduction to Pascal (3rd edn)*
WELSH, J., ELDER, J., and BUSTARD, D., *Sequential Program Structures*
WELSH, J., and HAY, A., *A Model Implementation of Standard Pascal*
WELSH, J., and McKEAG, M., *Structured System Programming*
WIKSTRÖM, Å., *Functional Programming using Standard ML*

Category Theory for Computing Science

Michael Barr

McGill University, Montreal, Canada

Charles Wells

Case Western Reserve University, Cleveland, Ohio, USA

Prentice Hall

New York London Toronto Sydney Tokyo Singapore

First published 1990 by
Prentice Hall International (UK)
66 Wood Lane End, Hemel Hempstead
Hertfordshire HP2 4RG
A division of
Simon & Schuster International Group

Printed and bound in Great Britain by
BPCC Wheatons Ltd, Exeter

Library of Congress Cataloging-in-Publication Data

Barr, Michael
Category theory for computing science / Michael Barr, Charles Wells.
p. cm. — (Prentice Hall International series in computer science)
"June 15, 1988."
Bibliography: p.
Includes index.
ISBN 0-13-120486-6 : $40.00
1. Electronic data processing—Mathematics. 2. Categories (Mathematics) I. Wells, Charles. II. Title. III. Series.
QA76.9.M35B37 1989
512′55—dc19 89-3450
CIP

British Library Cataloguing in Publication Data

Barr, M. (Michael), *1937-*
Category theory for computing science. — (Prentice Hall international series in computer science)
1. Computer systems. Applications of category theory
I. Title II. Wells, C. (Charles), *1937–*
004′.01′51255

ISBN 0-13-120486-6

2 3 4 5 94 93 92 91

Contents

Preface

This book is a textbook in basic category theory, written specifically to be read by researchers and students in computing science. We expound the constructions we feel are basic to category theory in the context of examples and applications to computing science. Some categorical ideas and constructions are already used heavily in computing science and we describe many of these uses. Other ideas, in particular the concept of adjoint, have not appeared as widely in the computing science literature. We give here an elementary exposition of those ideas we believe to be basic categorical tools, with pointers to possible applications when we are aware of them.

In addition, this text advocates a specific idea: the use of sketches as a systematic way to turn finite descriptions into mathematical objects. This aspect of the book gives it a particular point of view. We have, however, taken pains to keep much of the material on sketches in separate sections, which it is not necessary to read to learn many of the topics covered by the book.

As a way of showing how you can use categorical constructions in the context of computing science, we describe several examples of modeling linguistic or computational phenomena categorically. These are not intended as the final word on how categories should be used in computing science; indeed, they hardly constitute the initial word on how to do that! We are mathematicians, and it is for those in computing science, not us, to determine which is the best model for a given application.

The emphasis in this book is on understanding the concepts we have introduced, rather than on giving formal proofs of the theorems. We include proofs of theorems only if they are enlightening in their own right. We have attempted to point the reader to the literature for proofs and further development for each topic.

In line with our emphasis on understanding, we frequently recommend one or another way of thinking about a concept. It is typical of most of the useful concepts in mathematics that there is more than one way of perceiving or understanding them. It is simply not true that everything about a mathematical concept is contained in its definition. Of course it is true that in some sense all the *theorems* are inherent in its definition, but not what makes it useful to mathematicians or to scientists who use mathematics. We believe that the more ways you have of perceiving an idea, the more likely you are to recognize situations in your own work where the idea is useful.

We have acted on the belief just outlined with many sentences beginning with phrases such as 'This concept may be thought of as ...'. We have been warned that doing this may present difficulties for a non-mathematician who has only just mastered *one* way of thinking about something, but we feel it is part of learning about a mathematical topic to understand the contextual associations it has for those who use it.

About categories

Categories originally arose in mathematics out of the need of a formalism to describe the passage from one type of mathematical structure to another. A category in this way represents a kind of mathematics, and may be described as **category as mathematical workspace.**

A category is also a mathematical structure. As such, it is a common generalization of both ordered sets and monoids, and questions motivated by those topics often have interesting answers for categories. This is **category as mathematical structure.**

A third point of view is emphasized in this book. A category can be seen as a structure that formalizes a mathematician's description of a type of structure. This is the role of **category as theory**. Formal descriptions in mathematical logic are traditionally given as formal languages with rules for forming terms, axioms and equations. Algebraists long ago invented a formalism based on tuples, the method of signatures and equations, to describe algebraic structures. In this book, we advocate categories in their role as formal theories as being in many ways superior to the others just mentioned. Continuing along the same path, we advocate sketches as finite specifications for the theories.

Topics

The first eight chapters explain the topics they cover in considerable detail. The remaining chapters cover a large number of topics in a more superficial way, with pointers to the literature.

Chapter 1 contains preliminary material on graphs, sets and functions. The reader who has taken a discrete mathematics course may wish to skip this chapter. However, some fine points concerning sets and functions are discussed that may be worth looking at.

Chapter 2 introduces categories and gives many examples. We also give certain simple constructions on categories and describe elementary properties of objects and arrows.

Chapter 3 introduces functors, which are the mappings that preserve the structure of categories. We make certain constructions here that will be needed in later chapters.

Chapter 4 deals with three related topics: diagrams, natural transformations and sketches. Probably the first thing noncategorists notice about category theory is the proliferation of diagrams: here we begin the heavy use of diagrams in this book. We discuss representable functors, universal objects and the Yoneda embedding, which are fundamental tools for the categorist. We also introduce in this chapter a very weak version of sketch called a linear sketch.

Chapter 5 introduces products and sums. This allows one to use categories, in their role as theories, to specify functions of several variables and to specify alternatives. In programming languages these appear as record structures and variant records.

Chapter 6 is an introduction to cartesian closed categories, which have been a major source of interest to computer scientists because they are equivalent in theoretical power to typed lambda calculus. In this chapter, we outline briefly the process of translating between typed lambda calculus and cartesian closed categories.

Normally, in learning a new language, one should plunge right in speaking it instead of translating. However, it may be helpful for the suspicious reader to see that translation is possible. We outline two translation processes in this book: the one just mentioned and another in Chapter 7, where the way to translate between a type of sketch called FP sketch and signatures and equations is given. Except for those two places, category theory is everywhere presented in its own terms.

Chapter 7 is the first of three chapters on sketches; the others are Chapters 9 and 10. Chapter 7 introduces sketches with the expressive power of universal algebra, and a more general kind that allows union types. The sketches in Chapter 9 are more powerful, having more expressive power than universal Horn theories. The constructions in Chapter 10 (mappings between sketches) are applied to the description of parametrized data types, and sketches are shown to be institutions in the sense of Goguen and Burstall.

Chapter 8 introduces the general concepts of limit and colimit of which the constructions in Chapter 5 were a special case. These are basic constructions in category theory that allow the formation of equationally defined subtypes and quotients. We describe unification in terms of coequalizers of free models of a certain kind of theory (free theory).

Chapter 11 describes fibrations and the Grothendieck construction, which have applications to programming language semantics. We also consider wreath products, a type of fibration which has been used in the study of automata.

Chapter 12 discusses the concept of adjointness, which is one of the grand unifying ideas of category theory. It is closely related to two other ideas: representable functors and the Yoneda Lemma. Many of the constructions in preceding chapters are examples of adjoints.

We discuss representable functors, universal objects and the Yoneda Lemma in Chapter 4, but we have deliberately postponed adjoints until we have several examples of them in applications. The concept of adjoint appears difficult and unmotivated if introduced too early. Nevertheless, most of Chapter 12 can be read after having finished Section 4.5. Some of the material in the chapter requires the concepts of limit and colimit from Chapters 5 and 8. One major example of an adjoint, cartesian closed categories, requires 5.1 and 6.1.

Chapter 13 contains a miscellany of topics centered around the idea of the algebra for a functor. We use this to define fixed points for a functor and also to introduce the notion of a triple (monad).

The book ends with Chapter 14, which introduces toposes. A topos is a kind of generalized set theory in which the logic is intuitionistic instead of classical. Toposes (or computable subcategories thereof) have often been thought the correct arena for programming language semantics. Categories of fuzzy sets are recognized as almost toposes, and modest sets, which are thought by many to be the best semantic model of polymorphic lambda calculus, live in a specific topos.

Most sections have exercises which provide additional examples of the concepts and pursue certain topics further. Many exercises can be solved by carefully keeping track of the definitions of the terms involved. A few exercises are harder and are marked with a dagger. Some of those so marked require a certain amount of ingenuity (although we do not expect the reader to agree in every case with our judgment on this!). Others require familiarity with some particular type of mathematical structure. For example, although we define monoids in the text, a problem asking for an example of a monoid with certain behavior can be difficult for someone who has never thought about them before reading this book.

We provide solutions to all the exercises. The solutions to the easy exercises, especially in the early chapters, go into considerable detail. The solutions to the harder exercises often omit routine verifications.

Other categorical literature

Nearly all of the topics in category theory in this book are developed further in the authors' monograph [Barr and Wells, 1985]. Indeed, the present text could be used as an introduction to that monograph. Most of the topics, except sketches, are also developed further in [Mac Lane, 1971] and [Arbib and Manes, 1975]. The text by Rydeheard and Burstall [1988] treats category theory from a computational point of view. A certain amount of category theory in the context of computer science is developed in [Manes and Arbib, 1986]. Various aspects of the close relationship between logic and categories (in their role as theories) are treated in [Makkai and Reyes, 1977], [Lambek and Scott, 1986] and [Bell, 1988].

Recent collections of papers in computer science which have many applications of category theory are [Pitt, Abramsky, Poigné and Rydeheard, 1986], [Pitt, Poigné and Rydeheard, 1987], [Main, Melton, Mislove and Schmidt, 1988] and [Gray and Scedrov, 1989].

Since this is an expository text, we make no effort to describe the history of the concepts we introduce or to discover the earliest references to theorems we state. In no case does our statement of a theorem constitute a claim that the theorem is original with us. In the few cases where it is original, we have announced the theorem in a separate research article.

We do give an extensive bibliography; however the main criteria for inclusion of a work in the bibliography are its utility and availability, not the creation of a historical record.

Prerequisites

This text assumes some familiarity with abstract mathematical thinking, and some specific knowledge of the basic language of mathematics and computing science of the sort taught in an introductory discrete mathematics course.

Terminology

In most scientific disciplines, notation and terminology are standardized, often by an international nomenclature committee. (Would you recognize Einstein's equation if it said $p = HU^2$?) We must warn the non-mathematician reader that such is not the case in mathematics. There is no standardization body and terminology and notation are individual and often idiosyncratic. We will introduce and stick to a fixed notation in this book, but any reader who looks in another source must expect to find different notation and even different names for the same concept – or what is worse, the same name used for a different concept. We have tried to give warnings when this happens, with the terminology at least.

Acknowledgments

In the preparation of this book, the first author was assisted by a grant from the NSERC of Canada and the second by NSF grant CCR-8702425. The authors would also like to thank McGill University and Case Western Reserve University, respectively, for sabbatical leaves, and the University of Pennsylvania for a very congenial setting in which to spend those leaves.

We learned much from discussions with Adam Barr, Robin Cockett, George Ernst, C. A. R. Hoare, Colin McLarty, Pribhakar Mateti, William F. Ogden, Robert Paré and John Power. Bob Harper, C. A. R. Hoare, and an anonymous referee made many helpful corrections and suggestions. We would especially like to thank Benjamin Pierce, who has read the book from beginning to end and found scores of errors, typographical and otherwise.

Chapter Dependency Chart

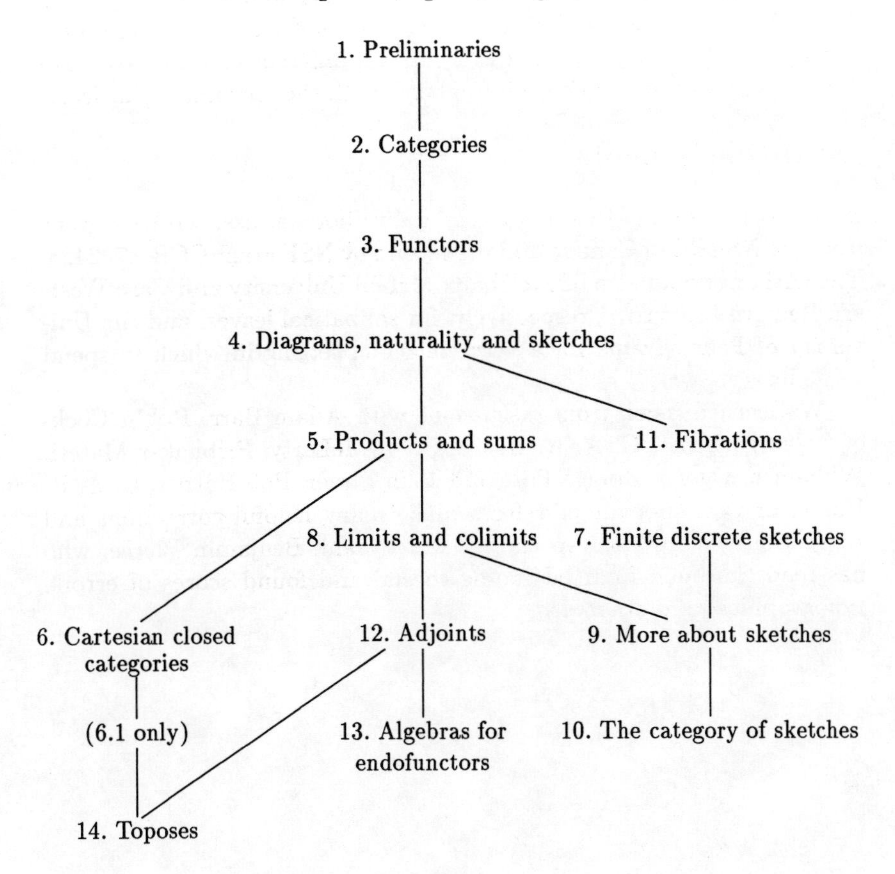

Notes: Section 8.7 needs Section 7.1, Section 13.6 needs Chapter 6, and Section 14.7 needs Chapter 9 and Section 11.2. The introductions to many chapters describe the status of individual sections (in particular where they are used later) in more detail. The Preface discusses the prerequisites for Chapter 12.

Chapter Dependency Chart

1

Preliminaries

This chapter is concerned with some preliminary ideas needed for the introduction to categories that begins in Chapter 2. The main topics are (i) an introduction to our notation and terminology for sets and functions and a discussion of some fine points that can cause trouble if not addressed early, and (ii) a discussion of graphs, by which we mean a specific type of directed graph. Graphs are a basis for the definition of category and an essential part of the definition of both commutative diagrams and sketches.

1.1 Sets

The concept of **set** is usually taken as known in mathematics. Instead of attempting a definition, we will give a *specification* for sets and another one for functions that is adequate for our purposes.

1.1.1 Specification A **set** is a mathematical entity that is distinct from, but completely determined by, its elements (if any). For every entity x and set S, the statement $x \in S$ is a proposition with a well defined truth value.

A finite set can be defined by listing its elements within braces; for example, $\{1, 3, 5\}$ is a set completely determined by the fact that its only elements are the numbers 1, 3 and 5. In particular, the set $\{\ \}$ with no elements is called the **empty set** and is denoted $\emptyset$.

The **setbuilder notation** defines a set as the collection of entities that satisfy a predicate; if x is a variable ranging over a specific type of data and $P(x)$ is a predicate about that type of data, then the notation $\{x \mid P(x)\}$ denotes the set of things of type x about which $P(x)$ is true. Thus $\{x \in \mathbf{R} \mid x > 7\}$ is the set of real numbers greater than 7. The set $\{x \mid P(x)\}$ is called the **extension** of the predicate P.

1.1.2 Russell's paradox The setbuilder notation (which implicitly supposes that every predicate determines a set) has a bug occasioned by **Russell's paradox**, which uses setbuilder notation to define something that cannot be a set:

$$\{S \mid S \text{ is a set and } S \notin S\}$$

This purports to be the set of all those sets that are not elements of themselves. If this were indeed a set T, then $T \in T$ implies by definition that $T \notin T$, whereas $T \notin T$ implies by definition that $T \in T$. This contradiction shows that there is no such set T.

A simple way to avoid this paradox is to restrict x to range over a particular type of data (such as one of the various number systems – real, integers, etc.) that already forms a set. This prophylaxis guarantees safe sets.

If you find it difficult to comprehend how a set could be an element of itself, rest assured that most approaches to set theory rule that out. Following those approaches, the (impossible) set T consists of *all* sets, so that we now know that there is no set whose elements are all the sets that exist – there is no 'set of all sets'.

1.1.3 Definition If S and T are sets, the **cartesian product** of S and T is the set $S \times T$ of ordered pairs (s, t) with $s \in S$ and $t \in T$. Thus in setbuilder notation, $S \times T = \{(s, t) \mid s \in S \text{ and } t \in T\}$.

The ordered pair (s, t) is determined uniquely by the fact that its first coordinate is s and its second coordinate is t. That is essentially a specification of ordered pairs. The formal categorical definition of product is based on this.

More generally, an **ordered n-tuple** is a sequence $(a_1, \ldots, a_n)$ determined uniquely by the fact that for $i = 1, \ldots, n$, the ith coordinate of $(a_1, \ldots, a_n)$ is a_i. Then the cartesian product $S_1 \times S_2 \times \cdots \times S_n$ is the set of all n-tuples $(a_1, \ldots, a_n)$ with $a_i \in S_i$ for $i = 1, \ldots, n$.

1.1.4 Notation We denote the set of natural numbers (nonnegative integers) by $\mathbf{N}$, the set of all integers by $\mathbf{Z}$, the set of all rational numbers by $\mathbf{Q}$, and the set of all real numbers by $\mathbf{R}$.

1.2 Functions

1.2.1 Specification A **function** f is a mathematical entity with the following properties:

F–1 f has a **domain** and a **codomain**, each of which must be a set.

F–2 For every element x of the domain, f has a **value** at x, which is an element of the codomain and is denoted $f(x)$.

F–3 The domain, the codomain, and the value $f(x)$ for each x in the domain are all determined completely by the function.

F–4 Conversely, the data consisting of the domain, the codomain, and the value $f(x)$ for each element x of the domain completely determine the function f.

The domain and codomain are often called the **source** and **target** of f, respectively.

The notation $f : S \to T$ is a succinct way of saying that the function f has domain S and codomain T. It is used both as a verb, as in 'Let $f : S \to T$', which would be read as 'Let f be a function from S to T', and as a noun, as in 'Any function $f : S \to T$ satisfies ...', in which the expression would be read as 'f from S to T'. One also says that f is of type 'S arrow T'.

We will use the **barred arrow notation** to provide an anonymous notation for functions. For example, if $\mathbf{R}$ denotes the set of real numbers, the function from $\mathbf{R}$ to $\mathbf{R}$ that squares its input can be denoted $x \mapsto x^2 : \mathbf{R} \to \mathbf{R}$. The barred arrow goes from *datum to datum* and the straight arrow goes from *domain to codomain*. The barred arrow notation serves the same purpose as the logicians' lambda notation, $\lambda x.x^2$, which we do not use except in the discussion of λ-calculus in Chapter 6. The barred arrow notation, like the lambda notation, is used mainly for functions defined by a formula.

1.2.2 The significance of F–3 is that a function is not merely a rule, but a rule together with its domain and codomain. This is the point of view taken by category theorists concerning functions, but is not shared by all mathematicians. Thus category theorists insist on a very strong form of typing. For example, these functions

(i) $x \mapsto x^2 : \mathbf{R} \to \mathbf{R}^+$

(ii) $x \mapsto x^2 : \mathbf{R} \to \mathbf{R}$

(iii) $x \mapsto x^2 : \mathbf{R}^+ \to \mathbf{R}^+$

(iv) $x \mapsto x^2 : \mathbf{R}^+ \to \mathbf{R}$

(where $\mathbf{R}^+$ is the set of nonnegative reals) are four different functions. This distinction is not normally made in college mathematics courses, and indeed there is no reason to make the distinction there, but it turns out to be necessary to make it in category theory and some other branches of abstract mathematics.

It can be useful to make the distinction even at an elementary level. For example, every set S has an **identity function** $\mathrm{id}_S : S \to S$ for which $\mathrm{id}_S(x) = x$ for all $x \in S$. If S is a subset of a set T, then there is an **inclusion function** $i : S \to T$ for which $i(x) = x$ for all $x \in S$. The functions id_S and i are different functions because they have different codomains, even though their value at each element of their (common) domain is the same.

1.2.3 Definition The **graph of a function** $f : S \to T$ is the set of ordered pairs: $\{(x, f(x)) \mid x \in S\}$.

The graph of a function has the **functional property** that for all $s \in S$, there is one and only one $t \in T$ such that (s, t) is in the graph. Many texts, but not this one, define a function to be a relation with the functional property, and some use the word **mapping** for what we call a function (where the domain and codomain are part of the definition).

1.2.4 Definition The **image** of a function (also called its range) is its set of values; that is, the image of $f : S \to T$ is $\{t \in T \mid \exists s \in S$ for which $f(s) = t\}$. The image of each of the squaring functions mentioned in 1.2.2 is of course the set of nonnegative reals.

1.2.5 Definition A function $f : S \to T$ is **injective** if whenever $s \neq s'$ in S, then $f(s) \neq f(s')$ in T.

Another name for injective is **one to one**. The identity and inclusion functions described previously are injective. The function $x \mapsto x^2 : \mathbf{R} \to \mathbf{R}$ is not injective since it takes 2 and -2 to the same value, namely 4. On the other hand, $x \mapsto x^3$ is injective.

There is a unique function $e : \emptyset \to T$ for any set T. It has no values. It is vacuously injective.

Do not confuse the definition of injective with the property that all functions have that if $s = s'$ then $f(s) = f(s')$. Another way of saying that a function is injective is via the contrapositive of the definition of injective: if $f(s) = f(s')$, then $s = s'$.

1.2.6 Definition A function $f : S \to T$ is **surjective** if its image is T.

The identity function on a set is surjective, but no other inclusion function is surjective.

Observe that the definition of surjective depends on the specified codomain; for example, of the four squaring functions listed in 1.2.2, only (i) and (iii) are surjective. A surjective function is often said to be 'onto'.

A function is **bijective** if it is injective and surjective. A bijective function is also called a **one to one correspondence**.

1.2.7 Functions and cartesian products If S and T are sets, the cartesian product $S \times T$ is equipped with two **coordinate** or **projection** functions $\mathrm{proj}_1 : S \times T \to S$ and $\mathrm{proj}_2 : S \times T \to T$. The coordinate functions are surjective if S and T are both nonempty. Coordinate functions for products of more than two sets are defined analogously.

There are two additional notational devices connected with the cartesian product.

1.2.8 Definition If X, S and T are sets and $f : X \to S$ and $g : X \to T$ are functions, then the function $\langle f, g \rangle : X \to S \times T$ is defined by $\langle f, g \rangle(x) = (f(x), g(x))$ for all $x \in X$.

1.2.9 Definition If X, Y, S and T are sets and $f : X \to S, g : Y \to T$ are functions, then $f \times g : X \times Y \to S \times T$ is the function defined by $(f \times g)(x, y) = (f(x), g(y))$. It is called the **cartesian product** of the functions f and g.

These functions are discussed further in Chapter 5.

1.2.10 Definition If $f : S \to T$ and $g : T \to U$, then the **composite function** $g \circ f : S \to U$ is defined to be the unique function with domain S and codomain U for which $(g \circ f)(x) = g(f(x))$ for all $x \in S$. In the computing science literature, $f; g$ is often used for $g \circ f$.

Category theory is based on composition as a fundamental operation in much the same way that classical set theory is based on the 'element of' or membership relation.

In categorical treatments, it is necessary to insist, as we have here, that the *codomain of f be the domain of g* for the composite $g \circ f$ to be defined. Many texts in some other branches of mathematics require only that the image of f be included in the domain of g.

1.2.11 Definition If $f : S \to T$ and $A \subseteq S$, then the **restriction** of f to A is the composite $f \circ i$, where $i : A \to S$ is the inclusion function. Thus the squaring function in 1.2.2(iii) is the restriction to $\mathbf{R}^+$ of the squaring function in 1.2.2(i).

Similarly, if $T \subseteq B$, f is called the **corestriction** of the function $j \circ f : S \to B$ to T, where j is the inclusion of T in B. Thus, in 1.2.2, the function in (i) is the corestriction to $\mathbf{R}^+$ of the function in (ii).

1.2.12 Functions in theory and practice The concept of function can be explicitly defined in terms of its domain, codomain and graph. Precisely, a function $f : S \to T$ could be defined as an ordered triple (S, Γ, T) with the property that Γ is a subset of the cartesian product $S \times T$ with the functional property (Γ is the graph of f). Then for $x \in S$,

$f(x)$ is the unique element $y \in T$ for which $(x, y) \in \Gamma$. Such a definition clearly satisfies specification 1.2.1.

The description of functions in 1.2.1 is closer to the way a mathematician thinks of a function than the definition in 1.2.12. For a mathematician, a function has a domain and a codomain, and if x is in the domain, then there is a well defined value $f(x)$ in the codomain. It is wrong to think that a function is *actually* an ordered triple as described in the preceding paragraph in the same sense that it is wrong for a programmer writing in a high level language to think of the numbers he deals with as being expressed in binary notation. The possible definition of function in the preceding paragraph is an *implementation* of the specification for function, and just as with program specifications the expectation is that one normally works with the specification, not the implementation, in mind. We make a similar point in 5.1.1 when we discuss ordered pairs in the context of categorical products.

In understanding the difference between a specification of something and an implementation of it, it may be instructive to read the discussion of this point in [Halmos, 1960], Section 6, who gives the definition of an ordered pair. The usual definition is rather unnatural and serves only to demonstrate that a construction with the required property exists.

1.2.13 Exercises

Most introductory texts in discrete mathematics provide dozens of exercises concerning sets, functions and their properties, and operations such as union, intersection, and so on. We regard our discussion as establishing notation, not as providing a detailed introduction to these concepts, and so do not give such exercises here. The exercises we do provide here allow a preliminary look at some categorical constructions that will appear in detail later in the book.

1. Let S and T be sets, and let $\mathrm{Hom}(S, T)$ denote the set of all functions with domain S and codomain T. (This fits with the standard notation we introduce in Chapter 2.) Let $f : T \to V$ be a function; define the function

$$\mathrm{Hom}(S, f) : \mathrm{Hom}(S, T) \to \mathrm{Hom}(S, V)$$

by

$$\mathrm{Hom}(S, f)(g) = f \circ g$$

Note that $\mathrm{Hom}(S, x)$ is overloaded notation: when x is a set, $\mathrm{Hom}(S, x)$ is a set of functions, but when x is a function, so is $\mathrm{Hom}(S, x)$. Show that if S is not the empty set, then f is injective if and only if $\mathrm{Hom}(S, f)$ is injective.

2. In the notation of Exercise 1, let $h : W \to S$ be a function and define $\mathrm{Hom}(h,T) : \mathrm{Hom}(S,T) \to \mathrm{Hom}(W,T)$ by $\mathrm{Hom}(h,T)(g) = g \circ h$. Show that if T has at least two elements, then h is *surjective* if and only if $\mathrm{Hom}(h,T)$ is *injective*.

3. a. Using the notation of Exercise 1, show that the mapping that takes a pair $(f : X \to S, g : X \to T)$ of functions to the function $\langle f, g\rangle : X \to S \times T$ defined in Definition 1.2.8 is a bijection from $\mathrm{Hom}(X,S) \times \mathrm{Hom}(X,T)$ to $\mathrm{Hom}(X, S \times T)$.

b. If you set $X = S \times T$ in (a), what does $\mathrm{id}_{S\times T}$ correspond to under the bijection?

1.3 Graphs

The type of graph that we discuss in this section is a specific version of directed graph, one that is well adapted to category theory, namely what is often called a directed multigraph with loops. A graph is a constituent of a sketch, which we introduce in Chapter 4, and is an essential ingredient in the definition of commutative diagram, which is the categorist's way of expressing equations (and is also a constituent of sketches). The concept of graph is also a precursor to the concept of category itself: a category is, roughly speaking, a graph in which paths can be composed.

1.3.1 Definition and notation Formally, to specify a **graph**, you must specify its **nodes** (or **objects**) and its **arrows**. Each arrow must have a specific **source** (or **domain**) node and **target** (or **codomain**) node. If the graph is small enough, it may be drawn with its nodes indicated by dots or labels and each arrow by an actual arrow drawn from its source to its target.

There may be one or more arrows – or none at all – with given nodes as source and target. Moreover, the source and target of a given arrow need not be distinct; thus graphs in this book may have loops. The notation '$f : a \to b$' means that f is an arrow and a and b are its source and target, respectively. A loop, that is an arrow with the same source and target node, will be called an **endoarrow** or **endomorphism** of that node.

We will systematically denote the collection of nodes of a graph $\mathcal{G}$ by G_0 and the collection of arrows by G_1: the nodes form the zero-dimensional part of the graph and the arrows the one-dimensional part.

1.3.2 Definition A graph is called **discrete** if it has no arrows.

In particular, the empty graph, with no nodes and no arrows, is discrete. A discrete graph is essentially a set; discrete graphs and sets are usefully regarded as the same thing for most purposes.

1.3.3 Definition A graph is **finite** if the number of nodes *and* arrows is finite.

1.3.4 Example It is often convenient to picture a relation on a set as a graph. Here we take the common definition that a **relation** α from a set A to a set B is an arbitrary subset of the set $A \times B$ of ordered pairs with first coordinates in A and second coordinates in B. For example, let $A = \{1,2,3\}$, $B = \{2,3,4\}$ and $\alpha = \{(1,2),(2,2),(2,3),(1,4)\}$. Then α can be pictured as

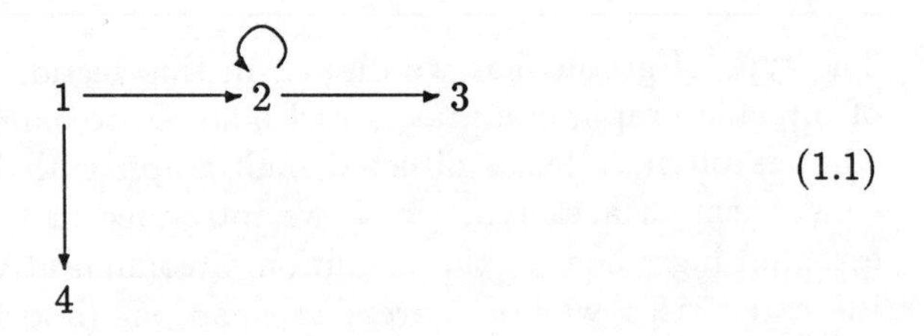

Of course, graphs that arise this way never have more than one arrow with the same source and target. Such graphs are called **simple graphs**.

Note that the graph of a function, as defined in 1.2.1, is a relation, and so corresponds to a graph in the sense just described. The resulting picture has an arrow from each element x of the domain to $f(x)$ so it is not the graph of the function in the sense of analytic geometry.

1.3.5 Example A data structure can sometimes be represented by a graph. This graph represents the set **N** of natural numbers in terms of zero and the successor function (adding 1):

$$1 \xrightarrow{\;0\;} n \quad \circlearrowleft \text{succ} \tag{1.2}$$

The name '1' for the left node is the conventional notation to require that the node denote a **singleton set**, that is, a set with exactly one element. In 4.7.6 we provide a formal mathematical meaning to the idea that this graph generates the natural numbers. Right now, this is just a graph with nodes named '1' and 'n'.

This informal idea of a graph representing a data type will become the basis of the formal theory of sketches in Chapter 4.

1.3.6 Example In the same spirit as the example above, let us see what data type is represented by the graph

$$a \underset{t}{\overset{s}{\rightrightarrows}} n \tag{1.3}$$

The data type has a signature consisting of two objects, call them a and n, and two arrows, let us call them (temporarily) $s, t : a \to n$. But if we interpret a as arrows, n as nodes and s and t as source and target, this is exactly what we have defined a graph to be: two sets and two functions from one of them to the other. For this reason, this graph is called the graph of graphs. (See [Lawvere, 1989].)

1.3.7 Example The **graph of sets and functions** has all sets as nodes and all functions between sets as arrows. The source of a function is its domain, and its target is its codomain.

In this example, unlike the previous ones, the nodes do not form a set. (See 1.1.2.) This fact will not cause trouble in reading this book, and will not usually cause trouble in applications. We use some standard terminology for this distinction.

1.3.8 Definition A graph that has a *set* of nodes and arrows is a **small graph**; otherwise, it is a **large** graph.

Thus the graph of sets and functions is a large graph. More generally we refer to any kind of mathematical structure as 'small' if the collection(s) it is built on form sets, and 'large' otherwise.

1.3.9 Exercises

1. The graphs in this section have labeled nodes; for example, the two nodes in (1.2) are labeled '1' and 'n'. Produce a graph analogous to (1.3) that expresses the concept of 'graph with nodes labeled from a set L'.

2. Let $\mathcal{G}$ be a graph with set of nodes N and set of arrows A. Show that $\mathcal{G}$ is simple if and only if the function $\langle\text{source}, \text{target}\rangle : A \to N \times N$ is injective. (This uses the pair notation for functions to products as described in 1.2.8.)

1.4 Homomorphisms of graphs

A homomorphism of graphs should preserve the abstract shape of the graph. A reasonable translation of this vague requirement into a precise mathematical definition is as follows.

1.4.1 Definition A **homomorphism** ϕ from a graph $\mathcal{G}$ to a graph $\mathcal{H}$, denoted $\phi : \mathcal{G} \to \mathcal{H}$, is a pair of functions $\phi_0 : G_0 \to H_0$ and $\phi_1 : G_1 \to H_1$ with the property that if $u : m \to n$ is an arrow of $\mathcal{G}$, then $\phi_1(u) : \phi_0(m) \to \phi_0(n)$ in $\mathcal{H}$.

It is instructive to restate this definition using the source and target mappings from 1.3.6: let $\text{source}_{\mathcal{G}} : G_1 \to G_0$ be the source map that takes an arrow (element of G_1) to its source, and similarly define $\text{target}_{\mathcal{G}}$, $\text{source}_{\mathcal{H}}$ and $\text{target}_{\mathcal{H}}$. Then the pair of maps $\phi_0 : G_0 \to H_0$ and $\phi_1 : G_1 \to H_1$ is a graph homomorphism if and only if

$$\text{source}_{\mathcal{H}} \circ \phi_1 = \phi_0 \circ \text{source}_{\mathcal{G}}$$

and

$$\text{target}_{\mathcal{H}} \circ \phi_1 = \phi_0 \circ \text{target}_{\mathcal{G}}$$

1.4.2 Notation of the form $a : B \to C$ is overloaded in several ways. It can denote a set-theoretic function, a graph homomorphism or an arrow in a graph. In fact, all three are instances of the third since there is a large graph whose nodes are sets and arrows are functions and another whose nodes are (small) graphs and arrows are graph homomorphisms.

Another form of overloading is that if $\phi : \mathcal{G} \to \mathcal{H}$ is a graph homomorphism, ϕ actually stands for a pair of functions we here call $\phi_0 : G_0 \to H_0$ and $\phi_1 : G_1 \to H_1$. In fact, it is customary to omit the subscripts and use ϕ for all three (the graph homomorphism as well as its components ϕ_0 and ϕ_1).

This does not lead to ambiguity in practice; in reading about graphs you are nearly always aware of whether the author is talking about nodes or arrows. We will keep the subscripts in this section and drop them thereafter.

1.4.3 Example If $\mathcal{G}$ is any graph, the **identity homomorphism** $\text{id}_{\mathcal{G}} : \mathcal{G} \to \mathcal{G}$ is defined by $(\text{id}_{\mathcal{G}})_0 = \text{id}_{G_0}$ (the identity function on the set of nodes of $\mathcal{G}$) and $(\text{id}_{\mathcal{G}})_1 = \text{id}_{G_1}$.

1.4.4 Example If $\mathcal{G}$ is the graph

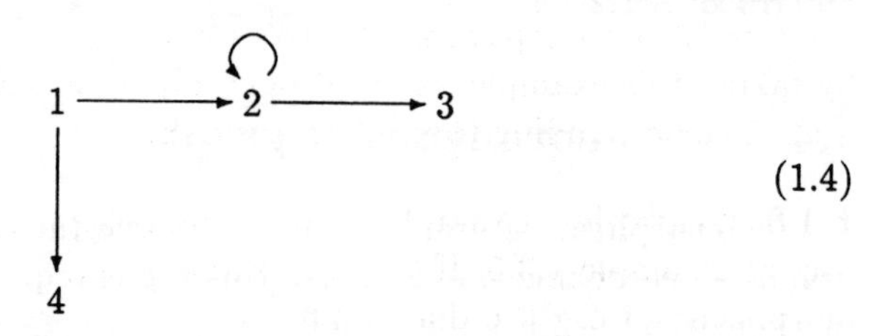

and $\mathcal{H}$ is this graph,

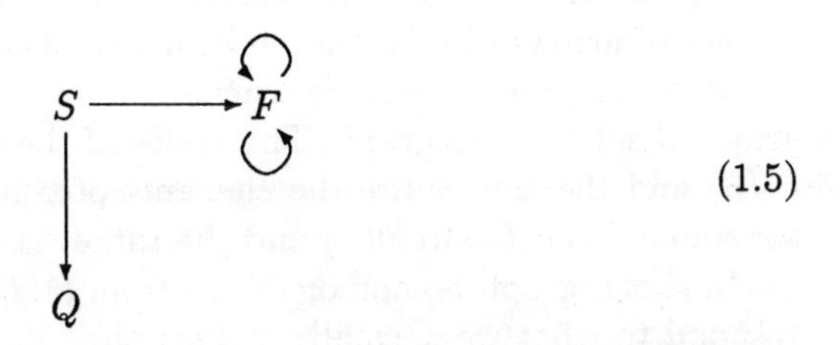

(1.5)

then there is a homomorphism $\phi : \mathcal{G} \to \mathcal{H}$ for which $\phi_0(1) = S$, $\phi_0(2) = \phi_0(3) = F$ and $\phi_0(4) = Q$, and ϕ_1 takes the loop on 2 and the arrow from 2 to 3 both to the upper loop on F; what ϕ_1 does to the other two arrows is forced by the definition of homomorphism. Because there are two loops on F there are actually four possibilities for ϕ_1 on arrows (while keeping ϕ_0 fixed).

1.4.5 Example If $\mathcal{H}$ is any graph with a node n and a loop $u : n \to n$, then there is a homomorphism from *any* graph $\mathcal{G}$ to $\mathcal{H}$ that takes every node of $\mathcal{G}$ to n and every arrow to u. This construction gives two other homomorphisms from $\mathcal{G}$ to $\mathcal{H}$ in Example 1.4.4 besides the four mentioned there. (There are still others.)

1.4.6 Example There is a homomorphism σ from Example 1.3.5 to the graph of sets that takes the node called 1 to a one-element set, which in contexts like this we will denote $\{*\}$, and that takes the node n to the set N of **natural numbers**. (Following the practice in computing science rather than mathematics, we start our natural numbers at 0.) The homomorphism σ takes the arrow $1 \to n$ to the function $* \mapsto 0$ that picks out the natural number 0, and $\sigma(\text{succ})$, naturally, is the function that adds 1. This is an example of a model of a sketch, which we discuss in 4.7.7. This homomorphism is a semantics for the sketch constituted by the abstract graph of 1.3.5.

1.4.7 Example The homomorphism in Example 1.4.6 is not the only homomorphism from Example 1.3.5 to sets. One can let n go to the set of integers $(\text{mod}\, k)$ for a fixed k and let succ be the function that adds one $(\text{mod}\, k)$ (it wraps around). You can also get other homomorphisms by taking this example (or 1.4.6) and adjoining some extra elements to the set corresponding to n which are their own successors.

1.4.8 Example Example 1.3.6 can be given a semantics in the same way as Example 1.3.5. If $\mathcal{G}$ is *any* small graph, there is a graph homomorphism ϕ from the diagram in (1.3) to the graph of sets for which $\phi_0(n)$ is the set of nodes of $\mathcal{G}$, $\phi_0(a)$ is the set of arrows, and ϕ_1 takes the arrows labeled source and target to the corresponding functions from the set of arrows of $\mathcal{G}$ to the set of nodes of $\mathcal{G}$.

Moreover, the converse is true: any homomorphism from (1.3) to the graph of sets gives a graph. The nodes of the graph are the elements of $\phi_0(n)$ and the arrows are the elements of $\phi_0(a)$. If $f \in \phi_1(a)$, then the source of f is $\phi_1(\text{source})(f)$ and the target is $\phi_1(\text{target})(f)$.

In short, graph homomorphisms from (1.3) to the graph of sets correspond to what we normally call graphs.

1.4.9 Notation In an expression like '$\phi_1(\text{source})(f)$', ϕ_1 is a function whose value at 'source' is a function that applies to an arrow f. As this illustrates, the application operation associates to the left.

1.4.10 Exercises

1. Show that if the codomain $\mathcal{H}$ of a graph homomorphism ϕ is a simple graph, then ϕ_1 is determined uniquely by ϕ_0.

2. Show that the composite of graph homomorphisms is a graph homomorphism. Precisely: if $\phi : \mathcal{G} \to \mathcal{H}$ and $\psi : \mathcal{H} \to \mathcal{K}$ are graph homomorphisms, then define the composite $\psi \circ \phi$ by requiring that $(\psi \circ \phi)_0 = \psi_0 \circ \phi_0$ and $(\psi \circ \phi)_1 = \psi_1 \circ \phi_1$. Then prove that $\psi \circ \phi$ is a graph homomorphism.

3. Let ϕ be a homomorphism $\mathcal{G} \to \mathcal{H}$ for which both ϕ_0 and ϕ_1 are bijective. Define $\psi : \mathcal{H} \to \mathcal{G}$ by $\psi_0 = (\phi_0)^{-1}$ and $\psi_1 = (\phi_1)^{-1}$.

a. Show that ψ is a graph homomorphism from $\mathcal{H}$ to $\mathcal{G}$.

b. Using the definition of composite in the preceding exercise, show that $\psi \circ \phi = \text{id}_{\mathcal{G}}$ and $\phi \circ \psi = \text{id}_{\mathcal{H}}$.

2

Categories

A category is a graph with a rule for composing arrows head to tail to give another arrow. This rule is subject to certain conditions, which we will give precisely in Section 2.1. The connection between functional programming languages and categories is described in Section 2.2. Some special types of categories are given in Section 2.3. Sections 2.4 and 2.5 are devoted to a class of examples of the kind that originally motivated category theory. The reader may wish to read through these examples rapidly rather than trying to understand every detail.

Constructions that can be made with categories are described in Section 2.6. Sections 2.7, 2.8 and 2.9 describe certain properties that an arrow of a category may have.

2.1 Basic definitions

Before we define categories, we need a preliminary definition.

2.1.1 Definition In a graph $\mathcal{G}$, a **path** from a node x to a node y of length k is a sequence $(f_1, f_2, \ldots, f_k)$ of (not necessarily distinct) arrows for which

(i) $\text{source}(f_k) = x$,

(ii) $\text{target}(f_i) = \text{source}(f_{i-1})$ for $i = 2, \ldots, k$, and

(iii) $\text{target}(f_1) = y$.

Observe that if you draw a path as follows:

$$\cdot \xrightarrow{f_k} \cdot \xrightarrow{f_{k-1}} \cdots \xrightarrow{f_2} \cdot \xrightarrow{f_1} \cdot$$

with the arrows going from left to right, f_k will be on the left and the subscripts will go down from left to right. We do it this way for consistency with composition (compare 2.1.3, C–1).

For each node x of the graph, we have one path of length 0, denoted (), with source and target x. It is called the **empty path** at x. For any arrow f, (f) is a path of length 1. As an example, in Diagram (1.2) on page 8, there is just one path of each length k from n to n, namely (), (succ), (succ, succ), and so on.

2.1.2 Definition The set of paths of length k in a graph $\mathcal{G}$ is denoted G_k.

In particular, G_2, which will be used in the definition of category, is the set of pairs of arrows (g, f) for which the target of f is the source of g. These are called **composable pairs** of arrows.

We have now assigned two meanings to G_0 and G_1. This will cause no conflict as G_1 refers indifferently either to the collection of arrows of $\mathcal{G}$ or to the collection of paths of length 1, which is essentially the same thing. Similarly, we use G_0 to represent either the collection of nodes of $\mathcal{G}$ or the collection of empty paths, of which there is one for each node. In each case we are using the same name for two collections that are not the same but are in a natural one to one correspondence. Compare the use of '2' to denote either the integer or the real number. As this last remark suggests, one might want to keep the two meanings of G_1 separate for purposes of implementing a graph as a data structure.

The one to one correspondences mentioned in the preceding paragraph were called 'natural'. The word is used informally here, but in fact these correspondences are natural in the technical sense (Exercise 10 of Section 4.3).

2.1.3 Categories A **category** is a graph together with two functions $c : G_2 \to G_1$ and $u : G_0 \to G_1$ with properties C–1 through C–4 below. The function c is called **composition**, and if (g, f) is a composable pair, $c(g, f)$ is written $g \circ f$ and is called the **composite** of g and f. If A is an object, $u(A)$ is denoted id_A, which is called the **identity** of the object A.

C–1 The source of $g \circ f$ is the source of f and the target of $g \circ f$ is the target of g.

C–2 $(h \circ g) \circ f = h \circ (g \circ f)$ whenever either side is defined.

C–3 The source and target of id_A are both A.

C–4 If $f : A \to B$, then $f \circ \mathrm{id}_A = \mathrm{id}_B \circ f = f$.

In the category theory literature, id_A is often written just A.

Note that the significance of the fact that the composite c is defined on G_2 is that $g \circ f$ is defined if and only if the source of g is the target of f. This means that composition is a function whose domain is an equationally defined subset of $G_1 \times G_1$: the equation requires that the source of g equal the target of f. It follows from this and C–1 that in C–2, one side of the equation is defined if and only if the other side is defined.

2.1.4 Terminology Although a category is a graph, the terminology for some of the data differs. In particular, for categories the words 'object', 'morphism', 'domain' and 'codomain' are more common than 'node', 'arrow', 'source' and 'target'. In this book we normally use 'object' for 'node', but stay with the words 'arrow', 'source' and 'target' even for categories. We will normally denote objects of categories by capital letters but nodes of graphs (except when we think of a category as a graph) by lower case letters. Arrows are always lower case.

In the computing science literature, the composite $g \circ f$ is sometimes written $f; g$, a notation suggested by the perception of a typed functional programming language as a category (see 2.2.1).

We have presented the concept of category as a two-sorted data structure; the sorts are the objects and the arrows. Categories are sometimes presented as one-sorted – arrows only. The objects can be recovered from the fact that C–3 and C–4 together characterize id_A (Exercise 3), so that there is a one to one correspondence between the objects and the identity arrows id_A.

2.1.5 Definition A category is **small** if its objects and arrows constitute sets; otherwise it is **large** (see the discussion of Russell's paradox, 1.1.2.) The category of sets and functions defined in 2.1.11 below is an example of a large category. Although one must in principle be wary in dealing with large classes, it is not in practice a problem; category theorists have rarely, if ever, run into set-theoretic difficulties.

2.1.6 Definition If A and B are two objects of a category $\mathcal{C}$, then the set of all arrows of $\mathcal{C}$ that have source A and target B is denoted $\mathrm{Hom}_{\mathcal{C}}(A, B)$, or just $\mathrm{Hom}(A, B)$ if the category is clear from context.

Thus for each triple A, B, C of objects, composition induces a function

$$\mathrm{Hom}(B, C) \times \mathrm{Hom}(A, B) \to \mathrm{Hom}(A, C)$$

A set of the form $\mathrm{Hom}(A, B)$ is called a **hom set**. Other common notations for $\mathrm{Hom}(A, B)$ are $\mathcal{C}(A, B)$ and $\mathcal{C}(AB)$.

2.1.7 The reference to the *set* of all arrows from A to B constitutes an assumption that they do indeed form a set. A category with the property that $\mathrm{Hom}(A, B)$ is a set for all objects A and B is called **locally small**. All categories in this book are locally small.

2.1.8 Definition For any path $(f_1, f_2, \ldots, f_n)$ in a category $\mathcal{C}$, define $f_1 \circ f_2 \circ \cdots \circ f_n$ recursively by

$$f_1 \circ f_2 \circ \cdots \circ f_n = (f_1 \circ f_2 \circ \cdots \circ f_{n-1}) \circ f_n, \quad n > 2$$

2.1.9 Proposition **The general associative law.** *For any path*

$$(f_1, f_2, \ldots, f_n)$$

in a category C and any integer k with $1 < k < n$,

$$(f_1 \circ \cdots \circ f_k) \circ (f_{k+1} \circ \cdots \circ f_n) = f_1 \circ \cdots \circ f_n$$

In other words, you can unambiguously drop the parentheses.

In this proposition, the notation $f_{k+1} \circ \cdots \circ f_n$ when $k = n-1$ means simply f_n.

This is a standard fact for associative binary operations (see [Jacobson, 1974], Section 1.4) and can be proved in exactly the same way for categories.

2.1.10 Little categories The smallest category has no objects and (of course) no arrows. The next smallest category has one object and one arrow, which must be the identity arrow. This category may be denoted **1**. Other categories that will be occasionally referred to are the categories **1** + **1** and **2** illustrated below (the loops are identities). In both cases the choice of the composites is forced.

2.1.11 Categories of sets The **category of sets** is the category whose objects are sets and whose arrows are functions (see 1.2.1) with composition of functions for c and the identity function from S to S for id_S. The statement that this is a category amounts to the statements that composition of functions is associative and that each identity function $\text{id}_S : S \to S$ satisfies $f \circ \text{id}_S = f$ and $\text{id}_S \circ g = g$ for all f with source S and all g with target S.

In this text, the category of sets is denoted **Set**. There are other categories whose objects are sets, as follows.

2.1.12 Definition The **category of finite sets**, denoted Fin, is the category whose objects are finite sets and whose arrows are all the functions between finite sets.

2.1.13 Definition A **partial function** from a set S to a set T is a function with domain S_0 and codomain T, where S_0 is some subset of S. The **category of sets and partial functions** has all sets as objects and all partial functions as arrows. If $f : S \to T$ and $g : T \to V$ are partial functions with f defined on $S_0 \subseteq S$ and g defined on $T_0 \subseteq T$, the composite $g \circ f : S \to V$ is the partial function from S to V defined on the subset $\{x \in S_0 \mid f(x) \in T_0\}$ of S by the requirement $(g \circ f)(x) = g(f(x))$.

Other examples of categories whose objects are sets are the category of sets and injective functions and the category of sets and surjective functions (Exercises 1 and 2).

Categories also arise in computing science in an intrinsic way. Three examples of this concern functional programming languages (2.2.1), deductive systems (Section 5.6) and automata with typed states (3.2.6). In Sections 2.3, 2.4 and 2.5, we discuss some of the ways in which categories arise in mathematics.

2.1.14 Exercises

1. Prove that sets (as objects) and injective functions (as arrows) form a category with functional composition as the composition operation c.

2. Do the same as Exercise 1 for sets and surjective functions.

3. Prove the following for any arrow $u : A \to A$ of a category $\mathcal{C}$. It follows from these facts that C–3 and C–4 of 2.1.3 characterize the identity arrows of a category.

a. If $g \circ u = g$ for every object B of $\mathcal{C}$ and arrow $g : A \to B$, then $u = \mathrm{id}_A$.

b. If $u \circ h = h$ for every object C of $\mathcal{C}$ and arrow $h : C \to A$, then $u = \mathrm{id}_A$.

2.2 Functional programming languages as categories

The intense interest in category theory among researchers in computing science in recent years is due in part to the recognition that the constructions in functional programming languages make a functional programming language look very much like a category. The fact that deduction systems are essentially categories has also been noticed in computing science.

In this section we describe the similarities between functional programming languages and categories informally, and discuss some of the technical issues involved in making them precise. Deduction systems are discussed in Section 5.6.

This discussion is incomplete, since at this point we have no way to describe n-ary operations for $n > 1$, nor do we have a way of specifying the flow of control. The first will be remedied in Section 5.3.12. One approach to the second question is given in Section 13.2. See also [Wagner, 1986a, 1986b].

2.2.1 Functional programming languages A functional programming language may be described roughly as one that gives the user some primitive typed operations and some constructors from which one can produce more complicated types and operations.

What a pure functional programming language in this sense does *not* have is variables or assignment statements. One writes a program by applying constructors to the types, constants and functions. 'Running' a program consists of applying such an operator to constants of the input type to obtain a value.

This is functional programming in the sense of the 'function-level programming' of Backus [1981a, 1981b]. (See also [Williams, 1982]). Another widely held point of view is that functional programming means no assignment statements: variables may appear but are not assigned to. This is true of the lambda calculus, for example.

We will discuss Backus-style functional programming languages here. The lambda calculus, with variables, is discussed in Chapter 6; see particularly 6.4.3. Wagner [1986b, 1986c] discusses assignment statements from a categorical point of view.

2.2.2 The category corresponding to a functional programming language A functional programming language has:

FPL–1 Primitive data types, given in the language.

FPL–2 Constants of each type.

FPL–3 Operations, which are functions between the types.

FPL–4 Constructors, which can be applied to data types and operations to produce derived data types and operations of the language.

The language consists of the set of all operations and types derivable from the primitive data types and primitive operations. The word 'primitive' means given in the definition of the language rather than constructed by a constructor. Some authors use the word 'constructor' for the primitive operations.

2.2.3 If we make two assumptions about a functional programming language and one innocuous change, we can see directly that a functional programming language L corresponds in a canonical way to a category $C(L)$.

A–1 We must assume that there is a do-nothing operation id_A for each type A (primitive and constructed). When applied, it does nothing to the data.

A–2 We add to the language an additional type called 1, which has the property that from every type A there is a unique operation to 1. We interpret each constant c of type A as an arrow $c : 1 \to A$. This incorporates the constants into the set of operations; they no longer appear as separate data.

A–3 We assume the language has a composition constructor: take an operation f that takes something of type A as input and produces something of type B, and another operation g that has input of type B and output of type C; then doing one after the other is a derived operation (or program) typically denoted $f;g$, which has input of type A and output of type C.

Functional programming languages generally have do-nothing operations and composition constructors, so A–1 and A–3 fit the concept as it appears in the literature. The language resulting from the change in A–2 is operationally equivalent to the original language.

This composition must be associative in the sense that, if either of $(f;g);h$ or $f;(g;h)$ is defined, then so is the other and they are the same operation. We must also require, for $f : A \to B$, that $f;\mathrm{id}_B$ and $\mathrm{id}_A;f$ are defined and are the same operation as f. That is, we impose the **equations** $f;\mathrm{id}_B = f$ and $\mathrm{id}_A;f = f$ on the language. Both these requirements are reasonable in that in any implementation, the two operations required to be the same would surely do the same thing.

2.2.4 Under those conditions, a functional programming language L has a category structure $C(L)$ for which:

FPC–1 The types of L are the objects of $C(L)$.

FPC–2 The operations (primitive and derived) of L are the arrows of $C(L)$.

FPC–3 The source and target of an arrow are the input and output types of the corresponding operation.

FPC–4 Composition is given by the composition constructor, written in the reverse order.

FPC–5 The identity arrows are the do-nothing operations.

The reader may wish to compare the discussion in [Pitt, 1986].

Observe that $C(L)$ is a *model* of the language, not the language itself. For example, in the category $f; \mathrm{id}_B = f$, but in the language f and $f; \mathrm{id}_B$ are different source programs. This is in contrast to the treatment of languages using context-free grammars: a context-free grammar generates the actual language.

2.2.5 Example As a concrete example, we will suppose we have a simple such language with three data types, **NAT** (natural numbers), **BOOLEAN** (true or false) and **CHAR** (characters). We give a description of its operations in categorical style.

(i) **NAT** should have a constant $0 : 1 \to \mathbf{NAT}$ and an operation $\mathbf{succ} : \mathbf{NAT} \to \mathbf{NAT}$.

(ii) There should be two constants **true**, **false** $: 1 \to \mathbf{BOOLEAN}$ and an operation $\neg$ subject to the equations $\neg \circ \mathbf{true} = \mathbf{false}$ and $\neg \circ \mathbf{false} = \mathbf{true}$.

(iii) **CHAR** should have one constant ‘c’: $1 \to \mathbf{CHAR}$ for each desired character ‘c’.

(iv) There should be two type conversion operations $\mathbf{ord} : \mathbf{CHAR} \to \mathbf{NAT}$ and $\mathbf{chr} : \mathbf{NAT} \to \mathbf{CHAR}$. These are subject to the equation $\mathbf{chr} \circ \mathbf{ord} = \mathrm{id}_{\mathbf{CHAR}}$. (You can think of **chr** as operating modulo the number of characters, so that it is defined on all natural numbers.)

An example program is the arrow ‘next’ defined to be the composite $\mathbf{chr} \circ \mathbf{succ} \circ \mathbf{ord} : \mathbf{CHAR} \to \mathbf{CHAR}$. It calculates the next character in order. This arrow ‘next’ is an arrow in the category representing the language, and so is any other composite of a sequence of operations.

2.2.6 The objects of the category $C(L)$ of this language are the types **NAT**, **BOOLEAN**, **CHAR** and 1. Observe that typing is a natural part of the syntax in this approach.

The arrows of $C(L)$ consist of all programs, with two programs being identified if they must be the same because of the equations. For example, the arrow

$$\mathbf{chr} \circ \mathbf{succ} \circ \mathbf{ord} : \mathbf{CHAR} \to \mathbf{CHAR}$$

just mentioned and the arrow

$$\mathbf{chr} \circ \mathbf{succ} \circ \mathbf{ord} \circ \mathbf{chr} \circ \mathbf{ord} : \mathbf{CHAR} \to \mathbf{CHAR}$$

must be the same because of the equation in (iv).

Observe that **NAT** has constants **succ** ∘ **succ** ∘ ... ∘ **succ** ∘ 0 where **succ** occurs zero or more times. In the exercises, n is the constant defined by induction by $1 =$ **succ** ∘ 0 and $n + 1 =$ **succ** ∘ n.

Composition of the category is composition of programs. Note that for composition to be well defined, if two composites of primitive operations are equal, then their composites with any other program must be equal. For example, we must have

$$\mathbf{ord} \circ (\mathbf{chr} \circ \mathbf{succ} \circ \mathbf{ord}) = \mathbf{ord} \circ (\mathbf{chr} \circ \mathbf{succ} \circ \mathbf{ord} \circ \mathbf{chr} \circ \mathbf{ord})$$

as arrows from **CHAR** to **NAT**. This is handled systematically in 3.5.8 using the quotient construction.

Other aspects of functional programming languages are considered in 5.3.12 and 5.4.5.

2.2.7 Exercise

1. Describe how to add a predicate '**nonzero**' to the language of this section. When applied to a constant of **NAT** it should give **true** if and only if the constant is not zero.

2.3 Mathematical structures as categories

Certain common mathematical structures can be perceived as special types of categories.

2.3.1 Preordered and ordered sets If S is a set, a subset $\alpha \subseteq S \times S$ is called a **binary relation** on S. It is often convenient to write $x\alpha y$ as shorthand for $(x, y) \in \alpha$. We say that α is **reflexive** if $x\alpha x$ for all $x \in S$ and **transitive** if $x\alpha y$ and $y\alpha z$ implies $x\alpha z$ for all x, y, $z \in S$.

A set S with a reflexive, transitive relation α on it is a structure (S, α) called a **preordered set**. This structure determines a category $C(S, \alpha)$ defined as follows.

CO–1 The objects of $C(S, \alpha)$ are the elements of S.

CO–2 If $x, y \in S$ and $x\alpha y$, then $C(S, \alpha)$ has exactly one arrow from x to y, denoted (y, x).

CO–3 If x is not related by α to y there is no arrow from x to y.

The identity arrows of $C(S, \alpha)$ are those of the form (x, x); they belong to α because it is reflexive. The transitive property of α is needed to ensure the existence of the composite described in 2.1.3, so that $(z, y) \circ (y, x) = (z, x)$.

A preordered set (S, α) for which α is antisymmetric (that is $x\alpha y$ and $y\alpha x$ imply $x = y$) is called an **ordered set** or **poset** (for 'partially ordered set'), a type of structure that is widely used in mathematics. Two examples of posets are $(\mathbf{R}, \leq)$, the real numbers with the usual ordering, and for any set S, the poset $(\mathcal{P}(S), \subseteq)$, the set of subsets of S with inclusion as ordering.

It is often quite useful and suggestive to think of a category as a generalized ordered set, and we will refer to this example to illuminate constructions we make later.

2.3.2 Semigroups A **semigroup** is a set S together with an associative binary operation $m : S \times S \to S$. The set S is called the **underlying set** of the semigroup.

Normally for s and t in S, $m(s, t)$ is written 'st' and called 'multiplication', but note that it does not have to satisfy the commutative law; that is, we may have $st \neq ts$. A **commutative semigroup** is a semigroup whose multiplication is commutative.

It is standard practice to talk about 'the semigroup S', naming the semigroup by naming its underlying set. This will be done for other mathematical structures such as posets as well. Mathematicians call this practice 'abuse of notation'. It is occasionally necessary to be more precise; that happens in this text in Section 12.1.

If k is any positive integer, we set $s^1 = s$ and $s^k = ss^{k-1}$. Such powers of an element obey the laws $s^k s^n = s^{k+n}$ and $(s^k)^n = s^{kn}$ (for *positive* k and n). On the other hand, the law $(st)^k = s^k t^k$ requires commutativity.

We specifically allow the **empty semigroup**, which consists of the empty set and the empty function from the empty set to itself. (Note that the cartesian product of the empty set with itself *is* the empty set.) This is not done in most of the non-category theory literature; it will become evident later why we should include the empty semigroup.

2.3.3 Definition An **identity element** e for a semigroup S is an element of S that satisfies the equation $se = es = s$ for all $s \in S$. There can be at most one identity element in a semigroup (Exercise 3).

2.3.4 Definition A **monoid** is a semigroup with an identity element. It is **commutative** if its binary operation is commutative.

It follows from the definition that a monoid is *not* allowed to be empty: it must contain an identity element.

2.3.5 Examples One example of a semigroup is the set of positive integers with addition as the operation; this is a semigroup but not a monoid. If you include 0 you get a monoid.

The **Kleene closure** A^* of a set A is the set of strings (or lists) of finite length of elements of A. The string which would be written 'abda' in much of the computer science literature will be written (a, b, d, a) here.

The operation of concatenation makes the Kleene closure a monoid $F(A)$, called the **free monoid** determined by A. We write concatenation as juxtaposition; thus

$$(a, b, d, a)(c, a, b) = (a, b, d, a, c, a, b)$$

Note that the underlying set of the free monoid is A^*, not A. The identity element is the empty string $()$. In the literature, A is usually assumed finite, but the Kleene closure is defined for any set A. The elements of A^* are strings of *finite* length in any case. When A is nonempty, A^* is an infinite set.

The concept of freeness is a general concept applied to many kinds of structures. It is treated systematically in Chapter 12.

2.3.6 Definition A **submonoid** of a monoid M is a subset S of M with the properties:

SM–1 The identity element of M is in S.

SM–2 If $m, n \in M$ then $mn \in M$. (One says that S is **closed** under the operation.)

2.3.7 Examples The natural numbers with addition form a submonoid of the integers with addition. For another example, consider the integers with multiplication as the operation, so that 1 is the identity element. Again the natural numbers form a submonoid, and so does the set of *positive* natural numbers, since the product of two positive numbers is another one. Finally, the singleton set $\{0\}$ is a subset of the integers that is closed under multiplication, and it is a monoid, but it is not a submonoid of the integers on multiplication because it does not contain the identity element 1.

2.3.8 Monoid as category Any monoid M determines a category $C(M)$.

CM–1 $C(M)$ has one object, which we will denote $*$; $*$ can be chosen arbitrarily. A simple uniform choice is to take $* = M$.

CM–2 The arrows of $C(M)$ are the elements of M with $*$ as source and target.

CM–3 Composition is the binary operation on M.

(This construction is revisited in Section 3.4.)

Thus a category can be regarded as a generalized monoid, or a 'monoid with many objects'. This point of view has not been as fruitful in mathematics as the perception of a category as a generalized poset. However, for computing science, we believe that the monoid metaphor is worth considering. It is explored in this book primarily in Chapter 11.

2.3.9 Exercises

1. For which sets A is $F(A)$ a commutative monoid?

2. Prove that for each object A in a category $\mathcal{C}$, $\mathrm{Hom}(A, A)$ is a monoid with composition of arrows as the operation.

3. Prove that a semigroup has at most one identity element. (Compare Exercise 3 of Section 2.1.)

2.4 Categories of sets with structure

The typical use of categories has been to consider categories whose objects are sets with mathematical structure and whose arrows are functions that preserve that structure. The definition of category is an abstraction of basic properties of such systems. Typical examples have included categories whose objects are spaces of some type and whose arrows are continuous (or differentiable) functions between the spaces, and categories whose objects are algebraic structures of some specific type and whose arrows are homomorphisms between them.

In this section we describe various categories of sets with structure. The following section considers categories of semigroups and monoids.

Note the contrast with Section 2.3, where we discussed certain mathematical structures as categories. Here, we discuss categories whose *objects* are mathematical structures.

2.4.1 Definition The **category of graphs** has graphs as objects and homomorphisms of graphs (see 1.4.1) as arrows. It is denoted **GRF**. The category of small graphs (see 1.3.8) and homomorphisms between them is denoted **Grf**.

Let us check that the composite of graph homomorphisms is a graph homomorphism (identities are easy). Suppose $\phi : \mathcal{G} \to \mathcal{H}$ and $\psi : \mathcal{H} \to \mathcal{K}$ are graph homomorphisms, and suppose that $u : m \to n$ in $\mathcal{G}$. Then by definition $\phi_1(u) : \phi_0(m) \to \phi_0(n)$ in $\mathcal{H}$, and so by definition

$$\psi_1(\phi_1(u)) : \psi_0(\phi_0(m)) \to \psi_0(\phi_0(n)) \text{ in } \mathcal{K}$$

Hence $\psi \circ \phi$ is a graph homomorphism.

The identity homomorphism $\mathrm{id}_{\mathcal{G}}$ is the identity function for both nodes and arrows.

2.4.2 The category of posets If (S, α) and (T, β) are posets, a function $f : S \to T$ is **monotone** if whenever $x\alpha y$ in S, $f(x)\beta f(y)$ in T.

The identity function on a poset is clearly monotone, and the composite of two monotone functions is easily seen to be monotone, so that posets with monotone functions form a category. A variation on this is to consider only **strictly monotone** functions, which are functions f with the property that if $x\alpha y$ and $x \neq y$ then $f(x)\beta f(y)$ and $f(x) \neq f(y)$.

In 2.3.1, we saw how a single poset is a category. Now we are considering the *category* of *posets*.

We must give a few words of warning on terminology. The usual word in mathematical texts for what we have called 'monotone' is 'increasing' or 'monotonically increasing'. The word 'monotone' is used for a function that either preserves *or reverses* the order relation. That is, in mathematical texts a function $f : (X, \alpha) \to (T, \beta)$ is also called monotone if whenever $x\alpha y$ in S, $f(y)\beta f(x)$ in T.

2.4.3 ω-complete partial orders We now describe a special type of poset that has been a candidate for programming language semantics. Actually, these days more interest has been shown in various special cases of this kind of poset, but the discussion here shows the approach taken.

Let $(S, \leq)$ be a poset. An **ω-chain** in S is an infinite sequence $s_0, s_1, s_2, \ldots$ of elements of S for which $s_i \leq s_{i+1}$ for all natural numbers i. Note that repetitions are allowed. In particular, for any two elements s and t, if $s \leq t$, there is a chain $s \leq t \leq t \leq t \leq \ldots$.

A **supremum** or **least upper bound** of a subset T of a poset $(S, \leq)$ is an element $v \in S$ with the following two properties:

SUP–1 For every $t \in T$, $t \leq v$.

SUP–2 If $w \in S$ has the property that $t \leq w$ for every $t \in T$, then $v \leq w$.

The supremum of a subset T is unique: if v and v' are suprema of T, then $v \leq v'$ because v' satisfies SUP–1 and v satisfies SUP–2, whereas $v' \leq v$ because v satisfies SUP–1 and v' satisfies SUP–2. Then $v = v'$ by antisymmetry.

If v satisfies SUP–1, it is called an **upper bound** of T.

2.4.4 Definition A poset $(S, \leq)$ is an **ω-complete partial order**, or **ω-CPO**, if every chain has a supremum. If the poset also has a minimum element, it is called a **strict ω-CPO**. In this context, the minimum element is usually denoted $\perp$ and called 'bottom'.

Note: this usage of the word 'complete' follows the customary usage in computing science. However, you should be warned that this use of 'complete' conflicts with standard usage in category theory, where 'complete' refers to limits, not colimits, so that ω-complete means closed under infimums of all *descending* chains. This concept is discussed in Chapter 8.

For example, every powerset of a set is a strict ω-CPO with respect to inclusion (Exercise 4).

2.4.5 Example A more interesting example from the point of view of computing science is the set $\mathcal{P}$ of partial functions from some set S to itself as defined in 2.1.13. A partial function on S can be described as a set f of ordered pairs of elements of S with the property that if $(s, t) \in f$ and $(s, t') \in f$, then $t = t'$. (Compare 1.2.3.)

We give the set $\mathcal{P}$ a poset structure by defining $f \leq g$ to mean $f \subseteq g$ as a set of ordered pairs. It follows that if f and g are partial functions on S, then $f \leq g$ if and only if the domain of f is included in the domain of g and for every x in the domain of f, $f(x) = g(x)$.

2.4.6 Proposition *$\mathcal{P}$ is a strict ω-CPO.*

Proof. Let $\mathcal{T}$ be a chain in $\mathcal{P}$. The supremum t of $\mathcal{T}$ is simply the union of all the sets of ordered pairs in $\mathcal{T}$. This set t is clearly the supremum if indeed it defines a partial function. So suppose (x, y) and (x, z) are elements of t. Then there are partial functions u and v in $\mathcal{P}$ with $(x, y) \in u$ and $(x, z) \in v$. Since $\mathcal{T}$ is a chain and the ordering of $\mathcal{P}$ is inclusion, there is a partial function w in $\mathcal{T}$ containing both (x, y) and (x, z), for example whichever of u and v is higher in the chain. Since w is a partial function, $y = z$ as required.

The bottom element of $\mathcal{P}$ is the empty function. □

2.4.7 Definition A function $f : S \to T$ between ω-CPOs is **continuous** if whenever s is the supremum of a chain $\mathcal{C} = (c_0, c_1, c_2, \ldots)$ in S, then $f(s)$ is the supremum of the image $f(\mathcal{C}) = \{f(c_i) \mid i \in \mathbf{N}\}$ in T. A continuous function between strict ω-CPOs is **strict** if it preserves the bottom element.

A continuous function is monotone, as can be seen by applying it to a chain $(s, t, t, t, \ldots)$ with $s \leq t$. Thus it follows that the image $f(\mathcal{C})$

of the chain $\mathcal{C}$ in the definition is itself a chain. See [Barendregt, 1984], pp.10ff, for a more detailed discussion of continuous functions.

The ω-complete partial orders and the continuous maps between them form a category. Strict ω-complete partial orders and strict continuous functions also form a category. In fact the latter is a subcategory in the sense to be defined in 2.6.1.

2.4.8 Functions as fixed points Let $f : S \to S$ be a set function. An element $x \in S$ is a **fixed point** of f if $f(x) = x$. Fixed points of functions are of interest in computing science because they provide a way of solving recursion equations. Complete partial orders provide a natural setting for expressing this idea.

As an example, consider the ω-CPO $\mathcal{P}$ of partial functions from the set $\mathbf{N}$ of natural numbers to itself. There is a function $\phi : \mathcal{P} \to \mathcal{P}$ that takes a partial function $h : \mathbf{N} \to \mathbf{N}$ in $\mathcal{P}$ to a partial function k defined this way:

(i) $k(0) = 1$;

(ii) for $n > 0$, $k(n)$ is defined if and only if $h(n-1)$ is defined, and $k(n) = nh(n-1)$.

ϕ is a continuous function from $\mathcal{P}$ to itself. Furthermore, the factorial function f is the unique fixed point of ϕ (Exercise 6).

The following proposition, applied to the poset $\mathcal{P}$ of partial functions, warrants the general recursive construction of functions.

2.4.9 Proposition *Let $(S, \leq)$ be a strict ω-CPO and $f : S \to S$ a continuous function. Then f has a least fixed point, that is an element $p \in S$ with the property that $f(p) = p$ and for any $q \in S$, if $f(q) = q$ then $p \leq q$.*

Proof. Form the chain

$$\mathcal{C} = (\bot, f(\bot), f \circ f(\bot), \ldots, f^k(\bot), \ldots)$$

Note that indeed $f^k(\bot) \leq f^{k+1}(\bot)$ for $k = 0, 1, \ldots$ by induction: $\bot \leq f(\bot)$, and if $f^{k-1}(\bot) \leq f^k(\bot)$, then $f^k(\bot) \leq f^{k+1}(\bot)$ because f is continuous.

The required least fixed point p is the supremum of the chain $\mathcal{C}$ (Exercise 8). □

This construction will be made in a wider context in Section 6.5 and Section 13.1.

2.4.10 Exercises

1. Let (S, α) and (T, β) be sets with relations on them. A **homomorphism** from (S, α) to (T, β) is a function $f : S \to T$ with the property that if $x\alpha y$ in S then $f(x)\beta f(y)$ in T.

a. Show that sets with relations and homomorphisms between them form a category.

b. Show that if (S, α) and (T, β) are both posets, then $f : S \to T$ is a homomorphism if and only if it is a monotone map.

2. Show that (strict) ω-complete partial orders and (strict) continuous functions form a category.

3. Let $\mathbf{R}^+$ be the set of nonnegative real numbers. Show that the poset $(\mathbf{R}^+, \leq)$ is not an ω-CPO.

4. Show that for every set S, the poset $(\mathcal{P}(S), \subseteq)$ is a strict ω-CPO.

5.† Give an example of ω-CPOs with a monotone map between them that is not continuous. (Hint: Adjoin two elements to $(\mathbf{Z}, \leq)$ that are greater than any integer.)

6.† Show that the function ϕ defined in 2.4.8 is continuous and that the factorial function is the unique fixed point of ϕ.

7. Exhibit the Fibonacci function as the least fixed point of a continuous function from an ω-CPO to itself, in the way we treated the factorial function in 2.4.8.

8.† Complete the proof of Proposition 2.4.9.

2.5 Categories of algebraic structures

In this section, we discuss categories whose objects are semigroups or monoids. These are typical of categories of algebraic structures; we have concentrated on semigroups and monoids because of their familiarity in computing science. The material in this section will come up primarily in examples later, and need not be thoroughly understood in order to read the rest of the book.

2.5.1 Homomorphisms of semigroups and monoids If S and T are semigroups, a function $h : S \to T$ is a **homomorphism** if for all $s, s' \in S$, $h(ss') = h(s)h(s')$.

A **homomorphism of monoids** is a semigroup homomorphism between monoids that preserves the identity elements: if e is the identity element of the domain, $h(e)$ must be the identity element of the codomain.

2.5.2 Examples The identity function on any monoid is a monoid homomorphism. If M is a monoid and S is a submonoid (see 2.3.6), the inclusion function from S to M is a monoid homomorphism. Another example is the function that takes an even integer to 0 and an odd integer to 1. This is a monoid homomorphism from the monoid of integers on multiplication to the set $\{0, 1\}$ on multiplication.

Since identity functions are homomorphisms and homomorphisms compose to give homomorphisms (see Exercise 2), we have two categories: **Sem** is the category of semigroups and semigroup homomorphisms, and **Mon** is the category of monoids and monoid homomorphisms.

2.5.3 Example Let S be a semigroup with element s. Let $\mathbf{N}^+$ denote the semigroup of positive integers with addition as operation. There is a homomorphism $p : \mathbf{N}^+ \to S$ for which $p(k) = s^k$. That this is a homomorphism is just the statement $s^{k+n} = s^k s^n$.

2.5.4 A *semigroup* homomorphism between monoids need not preserve the identities. An example of this involves the trivial monoid E with only one element e (which is perforce the identity element) and the monoid of all integers with multiplication as the operation, which is a monoid with identity 1. The function that takes the one element of E to 0 is a semigroup homomorphism that is not a monoid homomorphism. And, by the way, even though $\{0\}$ is a subsemigroup of the integers with multiplication and even though it is actually a monoid, it is *not* a submonoid.

2.5.5 Isomorphisms of semigroups If f is a bijective homomorphism, then it can be shown (Exercise 1) that the inverse function is also a homomorphism. In that case, we say that the homomorphism is an **isomorphism**. The two semigroups in question have the same abstract structure and are said to be **isomorphic**. It is important to understand that there may in general be many different isomorphisms between isomorphic semigroups (Exercise 5). As we will see later, the property of possessing an inverse is taken to define the categorical notion of isomorphism (2.7.2).

We note two important types of examples of monoid homomorphisms that will reappear later in the book.

2.5.6 Kleene closure induces homomorphisms Let A be a set, thought of as an alphabet. The set A^* of lists of elements of A is a monoid $F(A)$ (see 2.3.5). Let $f : A \to B$ be any set function. We define $f^* : A^* \to B^*$ by $f^*\Big((a_1, a_2, \ldots, a_k)\Big) = (f(a_1), f(a_2), \ldots, f(a_k))$. In particular, $f^*() = ()$ and for any $a \in A$, $f^*(a) = f(a)$. Then f^* is a homomorphism of monoids, a requirement that, in this case, means it preserves concatenation. Thus any set function between sets induces a monoid homomorphism between the corresponding free monoids. f^* is called αf in [Backus, 1981a] and is often called 'apply to all' or 'maplist' in the computing science literature.

2.5.7 The remainder function The set $\mathbf{Z}$ of all integers forms a monoid with respect to either addition or multiplication. If k is any positive integer, the set $\mathbf{Z}_k = \{0, 1, \ldots, k-1\}$ of **remainders** of k is also a monoid with respect to addition or multiplication $(\text{mod}\, k)$. Here are more precise definitions.

2.5.8 Definition Let k be a positive integer and n any integer. Then $n \bmod k$ is the unique integer $r \in \mathbf{Z}_k$ for which there is an integer q such that $n = qk + r$ and $0 \leq r < k$. It is not difficult to see that there is indeed a unique integer r with these properties.

Define an operation '$+_k$' of addition $(\text{mod}\, k)$ by requiring that

$$r +_k s = (r + s \bmod k)$$

Then the following proposition holds.

2.5.9 Proposition *$(\mathbf{Z}_k, +_k)$ is a monoid with identity 0.*

We also have the following.

2.5.10 Proposition *The function $n \mapsto (n \bmod k)$ is a monoid homomorphism from $(\mathbf{Z}, +)$ to $(\mathbf{Z}_k, +_k)$.*

A similar definition and proposition can be given for multiplication.

2.5.11 Exercises

1. Show that the inverse of a bijective semigroup homomorphism is also a semigroup homomorphism.

2. Show that the composite of semigroup (respectively monoid) homomorphisms is a semigroup (respectively monoid) homomorphism.

3. Prove Proposition 2.5.9.

4. Prove Proposition 2.5.10.

5.† Exhibit two distinct isomorphisms between the monoid with underlying set $\{0, 1, 2, 3\}$ and addition $(\text{mod}\, 4)$ as operation and the monoid with underlying set $\{1, 2, 3, 4\}$ and multiplication $(\text{mod}\, 5)$ as operation.

6. Show that if A is a set and $f : A \to M$ is a set function to a monoid M, then there is a unique monoid homomorphism $\widehat{f} : F(A) \to M$ for which for any element $a \in A$, $\widehat{f}(a) = f(a)$. (This property is called the **universal property** of the free monoid F(A). Note that in the statement, the input to $\widehat{f}$ is a string of length 1, whereas the input to f is an element of A.)

7. Using the terminology of 2.5.6, with $f : A \to B$ any function, show that f^* preserves concatenation.

8. Using the terminology of 2.5.6, show that if f is an isomorphism then so is f^*.

2.6 Constructions on categories

If you are familiar with some branch of abstract algebra (for example the theory of semigroups, groups or rings) then you know that given two structures of a given type (e.g., two semigroups), you can construct a 'direct product' structure, defining the operations coordinatewise. Also, a structure may have substructures, which are subsets closed under the operations, and quotient structures, formed from equivalence classes modulo a congruence relation. Another construction that is possible in many cases is the formation of a 'free' structure of the given type for a given set.

All these constructions can be performed for categories. We will outline the constructions here, except for quotients, which will be described in Section 3.5. We will also describe the construction of the slice category, which does not quite correspond to anything in abstract algebra (although it is akin to the adjunction of a constant to a logical theory). You do not need to be familiar with the constructions in other branches of abstract algebra, since they are all defined from scratch here.

2.6.1 Definition A **subcategory** $\mathcal{D}$ of a category $\mathcal{C}$ is a category for which:

S–1 All the objects of $\mathcal{D}$ are objects of $\mathcal{C}$ and all the arrows of $\mathcal{D}$ are arrows of $\mathcal{C}$ (in other words, $\mathcal{C}_0 \subseteq \mathcal{D}_0$ and $\mathcal{C}_1 \subseteq \mathcal{D}_1$).

S–2 The source and target of an arrow of $\mathcal{D}$ are the same as its source and target in $\mathcal{C}$ (in other words, the source and target maps for $\mathcal{D}$ are the restrictions of those for $\mathcal{C}$). It follows that for any objects A and B of $\mathcal{D}$, $\mathrm{Hom}_{\mathcal{D}}(A, B) \subseteq \mathrm{Hom}_{\mathcal{C}}(A, B)$.

S–3 If A is an object of $\mathcal{D}$ then its identity arrow id_A in $\mathcal{C}$ is in $\mathcal{D}$.

S–4 If $f : A \to B$ and $g : B \to C$ in $\mathcal{D}$, then the composite (in $\mathcal{C}$) $g \circ f$ is in $\mathcal{D}$ and is the composite in $\mathcal{D}$.

2.6.2 Examples As an example, the category Fin of finite sets and all functions between them is a subcategory of **Set**, and in turn **Set** is a subcategory of the category of sets and *partial* functions between sets (see 2.1.12 and 2.1.13). These examples illustrate two phenomena:

(i) If A and B are finite sets, then $\mathrm{Hom}_{\mathrm{Fin}}(A, B) = \mathrm{Hom}_{\mathbf{Set}}(A, B)$. In other words, every arrow of **Set** between objects of Fin is an arrow of Fin.

(ii) The category of sets and the category of sets and partial functions, on the other hand, have exactly the same *objects*. The phenomenon of (i) does not occur here: there are generally *many* more partial functions between two sets than there are full functions.

Example (i) motivates the following definition.

2.6.3 Definition If $\mathcal{D}$ is a subcategory of $\mathcal{C}$ and for every pair of objects A, B of $\mathcal{D}$, $\mathrm{Hom}_{\mathcal{D}}(A, B) = \mathrm{Hom}_{\mathcal{C}}(A, B)$, then $\mathcal{D}$ is a **full subcategory** of $\mathcal{C}$.

Thus Fin is a full subcategory of **Set** but **Set** is not a full subcategory of the category of sets and partial functions.

Example 2.6.2(ii) also motivates a (less useful) definition, as follows.

2.6.4 Definition If $\mathcal{D}$ is a subcategory of $\mathcal{C}$ with the same objects, then $\mathcal{D}$ is a **wide** subcategory of $\mathcal{C}$. In this case only the arrows are different. **Set** is a wide subcategory of the category of sets and partial functions.

2.6.5 Example Among all the objects of the category of semigroups are the monoids, and among all the semigroup homomorphisms between two monoids are those that preserve the identity. Thus the category of monoids is a subcategory of the category of semigroups that is neither wide nor full (for the latter, see 2.5.4).

As it stands, being a subcategory requires the objects and arrows of the subcategory to be identical with some of the objects and arrows of the category containing it. This requires an uncategorical emphasis on what something *is* instead of on the specification it satisfies. We will return to this example in 3.1.6, where it will be reformulated to avoid this defect.

2.6.6 Definition If $\mathcal{C}$ and $\mathcal{D}$ are categories, the **product** $\mathcal{C} \times \mathcal{D}$ is the category whose objects are all ordered pairs (C, D) with C an object of $\mathcal{C}$ and D an object of $\mathcal{D}$, and in which an arrow $(f, g) : (C, D) \to (C', D')$ is a pair of arrows $f : C \to C'$ in $\mathcal{C}$ and $g : D \to D'$ in $\mathcal{D}$. The identity of (C, D) is $(\mathrm{id}_C, \mathrm{id}_D)$. If $(f', g') : (C', D') \to (C'', D'')$ is another arrow, then the composite is defined by

$$(f', g') \circ (f, g) = (f' \circ f, g' \circ g) : (C, D) \to (C'', D'')$$

2.6.7 The dual of a category Given any category $\mathcal{C}$, you can construct another category denoted $\mathcal{C}^{\mathrm{op}}$ by reversing all the arrows. The **dual** or **opposite** $\mathcal{C}^{\mathrm{op}}$ of a category $\mathcal{C}$ is defined by:

D–1 The objects and arrows of $\mathcal{C}^{\mathrm{op}}$ are the objects and arrows of $\mathcal{C}$.

D–2 If $f : A \to B$ in $\mathcal{C}$, then $f : B \to A$ in $\mathcal{C}^{\mathrm{op}}$.

D–3 If $h = g \circ f$ in $\mathcal{C}$, then $h = f \circ g$ in $\mathcal{C}^{\mathrm{op}}$.

The meaning of D–2 is that source and target have been reversed. It is easy to see that the identity arrows have to be the same in the two categories $\mathcal{C}$ and $\mathcal{C}^{\mathrm{op}}$ and that C–1 through C–4 of Section 2.1 hold, so that $\mathcal{C}^{\mathrm{op}}$ is a category.

If M is a monoid, then the opposite of the category $C(M)$ is the category determined by a monoid M^{op}; if $xy = z$ in M, then $yx = z$ in M^{op}. Similar remarks may be made about the opposite of the category $C(P)$ determined by a poset P. The opposite of the poset $(\mathbf{Z}, \leq)$, for example, is $(\mathbf{Z}, \geq)$.

2.6.8 Both the construction of the product of two categories and the construction of the dual of a category are purely formal constructions. Even though the original categories may have, for example, structure-preserving functions of some kind as arrows, the arrows in the product category are simply pairs of arrows of the original categories.

Consider **Set**, for example. Let A be the set of letters of the English alphabet. The function $v : A \to \{0, 1\}$ that takes consonants to 0 and vowels to 1 is an arrow of **Set**. Then the arrow $(\mathrm{id}_A, v) : (A, A) \to (A, \{0, 1\})$ of **Set** $\times$ **Set** is not a function, not even a function of two variables; it is merely the arrow of a product category and as such is an ordered pair of functions.

A similar remark applies to duals. In $\mathbf{Set}^{\mathrm{op}}$, v is an arrow from $\{0, 1\}$ to A. And that is all it is. It is in particular *not* a function from $\{0, 1\}$ to A.

Nevertheless, it is possible in some cases to prove that the dual of a familiar category is essentially the same as some other familiar category. One such category is **Set**: see 3.4.4.

The product of categories is a formal way to make constructions dependent on more than one variable. The major use we make of the concept of dual is that many of the definitions we make have another meaning when applied to the dual of a category that is often of independent interest. The phrase **dual concept** or **dual notion** is often used to refer to a concept or notion applied in the dual category.

2.6.9 Slice categories If $\mathcal{C}$ is a category and A any object of $\mathcal{C}$, the **slice category** $\mathcal{C}/A$ is described this way:

SC–1 An object of $\mathcal{C}/A$ is an arrow $f : C \to A$ of $\mathcal{C}$ for some object C.

SC–2 An arrow of $\mathcal{C}/A$ from $f : C \to A$ to $f' : C' \to A$ is an arrow $h : C \to C'$ with the property that $f = f' \circ h$.

SC–3 The composite of $h : f \to f'$ and $h' : f' \to f''$ is $h' \circ h$.

It is necessary to show that $h' \circ h$ satisfies the requirements of being an arrow from f to f'' (Exercise 5).

The usual notation for arrows in $\mathcal{C}/A$ is deficient: the same arrow h can satisfy $f = f' \circ h$ and $g = g' \circ h$ with $f \neq f'$ or $g \neq g'$ (or both). Then $h : f \to f'$ and $h : g \to g'$ are *different* arrows of $\mathcal{C}/A$.

The importance of slice categories comes in part with their connection with indexing. An S**-indexed set** is a set X together with a function $\tau : X \to S$. If $x \in X$ and $\tau(x) = s$ then we say x is of **type** s, and we also refer to X as a **typed set**.

The set $G = G_0 \cup G_1$ of objects and arrows of a graph $\mathcal{G}$ is an example of a typed set, typed by the function $\tau : \mathcal{G} \to \{0, 1\}$ that takes a node to 0 and an arrow to 1. Note that this depends on the fact that a node is not an arrow: G_0 and G_1 are disjoint.

A function from a set X typed by S to a set X' typed by the same set S that preserves the typing (takes an element of type s to an element of type s) is exactly an arrow of the slice category **Set**$/S$. Such a function is called a **indexed function** or **typed function**. It has been fruitful for category theorists to pursue this analogy by thinking of objects of any slice category $\mathcal{C}/A$ as objects of $\mathcal{C}$ indexed by A.

2.6.10 Example Let (P, α) be a poset and let $C(P)$ be the corresponding category as in 2.3.1. For an element $x \in P$, the slice category $C(P)/x$ is the category corresponding to the set of elements greater than or equal to x. The dual notion of coslice gives the set of elements less than or equal to a given element.

2.6.11 The free category generated by a graph For any given graph $\mathcal{G}$ there is a category $F(\mathcal{G})$ whose objects are the nodes of $\mathcal{G}$ and whose arrows are the paths in $\mathcal{G}$. Composition is defined by the formula

$$(f_1, f_2, \ldots, f_k) \circ (f_{k+1}, \ldots, f_n) = (f_1, f_2, \ldots, f_n)$$

This composition is associative, and for each object A, id_A is the empty path from A to A. The category $F(\mathcal{G})$ is called the **free category generated by the graph $\mathcal{G}$**. It is also called the **path category** of $\mathcal{G}$.

The free category generated by the graph with one node and no arrows is the category with one object and only the identity arrow, which is the empty path. The free category generated by the graph with one node and one loop on the node is the free monoid with one generator (Kleene closure of a one-letter alphabet); this is isomorphic with the nonnegative integers with + as operation.

It is useful to regard the free category generated by *any* graph as analogous to Kleene closure (free monoid) generated by a set (see 2.3.5). The paths in the free category correspond to the strings in the Kleene closure. The difference is that you can concatenate any symbols together to get a string, but arrows can be strung together only head to tail, thus taking into account the typing.

In 12.2.3 we give a precise technical meaning to the word 'free'.

2.6.12 Exercises

1. Let M be a monoid. Show that the opposite of the category $C(M)$ determined by M is also the category determined by a monoid, called M^{op}.

2. Do the same as the preceding exercise for posets.

3. Give examples of posets P, Q and R for which $C(P)$ (the category determined by the poset) is a wide but not full subcategory of $C(Q)$ and $C(P)$ is a full but not wide subcategory of $C(R)$.

4.† Show by example that the requirement S–3 in 2.6.1 does not follow from the other requirements.

5. Show that definition SC–3 of 2.6.9 does indeed give an arrow from f to f''.

6. Describe explicitly the free categories generated by the graphs in Examples 1.3.5 and 1.3.6.

2.7 Properties of objects and arrows in a category

The data in the definition of category can be used to define properties that the objects and arrows of the category may have. A property that is defined strictly in terms of the role the object or arrow has in the category, rather than in terms of what it really is in any sense, is called a **categorical definition**. Such definitions are *abstract* in the sense that a property a thing can have is defined entirely in terms of the external interactions of that thing with other entities.

The examples of categorical definitions in this section and the next are of several simple concepts that can be expressed directly in terms of the data used in the definition of category. Other concepts, such as limit, naturality and adjunction, require deeper ideas that will be the subject of succeeding chapters.

2.7.1 Isomorphisms In general, the word 'isomorphic' is used in a mathematical context to mean indistinguishable in form. We have already used it in this way in 2.5.5. It turns out that it is possible to translate this into categorical language in a completely satisfactory way.

2.7.2 Definition Suppose that $\mathcal{C}$ is a category and that A and B are two objects of $\mathcal{C}$. An arrow $f : A \to B$ is said to be an **isomorphism** if there is an arrow $g : B \to A$ such that $f \circ g$ is the identity arrow of B and $g \circ f$ is the identity arrow of A. In that case, f is the **inverse** of g and g is the inverse of f. If such a pair of arrows exists between A and B, then we say that A is **isomorphic to** B, written $A \cong B$.

In a monoid, an element which is an isomorphism in the corresponding category is usually called **invertible**.

2.7.3 Definition An arrow $f : A \to A$ in a category (with the source and target the same) is called an **endomorphism**. If it is invertible, it is called an **automorphism**.

2.7.4 Examples Any identity arrow in any category is an isomorphism (hence an automorphism). In the category determined by a partially ordered set, the *only* isomorphisms are the identity arrows. If, in the category determined by a monoid, every arrow is an isomorphism the monoid is called a **group**. Because of this, a category in which every arrow is an isomorphism is called a **groupoid**.

2.7.5 Definition A property that objects of a category may have is **preserved by isomorphisms** if for any object A with the property, any object isomorphic to A must also have the property.

2.7.6 To show that two objects are isomorphic, it is sufficient to exhibit an isomorphism between them. To show they are not isomorphic is often more difficult, since the definition requires checking every arrow between the two objects. In practice it is almost always done by finding some property that is preserved by isomorphisms and is possessed by one of the objects and not possessed by the other. See 2.7.9.

2.7.7 Proposition *A function in* **Set** *is an isomorphism if and only if it is a bijection.*

Proof. Suppose first that $f : S \to T$ is an isomorphism; it therefore has an inverse $g : T \to S$. Then (i) f is injective, for if $f(x) = f(y)$, then $x = g(f(x)) = g(f(y)) = y$. Also (ii) f is surjective, for if $t \in T$, then $f(g(t)) = t$.

Conversely, suppose that $f : S \to T$ is bijective. Define $g : T \to S$ by saying that $g(t)$ is the unique element $x \in S$ for which $f(x) = t$. There is such an element x because f is surjective, and x is unique because f is injective. The definition itself says that $f(g(t)) = t$ for *any* $t \in T$, so in particular $f(g(f(x))) = f(x)$; since f is known to be injective, it follows that $g(f(x)) = x$, as required. □

It follows that two finite sets are isomorphic (in the category of sets) if and only if they have the same number of elements.

2.7.8 Example A graph homomorphism $\phi : \mathcal{G} \to \mathcal{H}$ is an isomorphism if and only if both ϕ_0 and ϕ_1 are bijections. This follows immediately from Exercise 3 of Section 1.4.

2.7.9 Example For ordered sets, the situation is different. For example, the following partially ordered sets A and B containing three elements each are not isomorphic in the category of partially ordered sets and monotone functions: A consists of three elements $a < b < c$ and B consists of three elements $x < y$, $x < z$, but no relation holds between y and z.

It seems clear that A and B are not isomorphic, but it might seem hard to say why. One way is simply to observe that A is totally ordered (for any two elements u and v, either $u < v$ or $v < u$) and B is not totally ordered. Since being totally ordered is preserved by isomorphisms, the two posets cannot be isomorphic.

An alternative proof is possible in the case of the preceding example: exhaustively consider all 6 bijections between the two posets and show that none of those that are monotone have inverses that are monotone. This approach clearly is unacceptably time-consuming for any interesting problem.

It is often quite hard to show that two structures are not isomorphic. One often approaches this problem by trying to find numerical invariants. In the case at hand, a simple invariant is the **depth**, that is the length of the longest totally ordered subset: the depth of A is 3 and the depth of B is 2. A set of invariants is called **complete** if it is sufficient to decide the isomorphism class. Complete sets of invariants rarely exist. Depth is sufficient in the case of 2.7.9 but is certainly not in general.

2.7.10 In many cases of categories of sets with structure and structure-preserving functions, a structure-preserving function that is a bijection is automatically an isomorphism. This is not always the case, as is illustrated by the poset example in 2.7.9. In fact, the bijection from A to B that takes x to a, y to b and z to c preserves the order relation. Of course, it also takes a pair of incomparable elements to comparable ones, but the point is it preserves the order, insofar as it exists.

2.7.11 Terminal and initial objects An object T of a category $\mathcal{C}$ is called **terminal** if there is *exactly* one arrow $A \to T$ for each object A of $\mathcal{C}$. We usually denote the terminal object by 1 and the unique arrow $A \to 1$ by $\langle\rangle$.

The dual notion (see 2.6.7), an object of a category that has a unique arrow *to* each object (including itself), is called an **initial object** and is often denoted 0.

2.7.12 Examples In the category of sets, the empty set is initial and any one-element set is terminal. Thus the category of sets has a *unique* initial object but many terminal objects. The one-element monoid is both initial and terminal in the category of monoids. In the category determined by a poset, an initial object is an absolute minimum for the poset, and a terminal object is an absolute maximum. Since there is no largest or smallest whole number, the category determined by the set of integers with its natural order gives an example of a category without initial or terminal object.

2.7.13 Proposition *Any two terminal (respectively initial) objects in a category are isomorphic.*

Proof. Suppose T and T' are terminal objects. Since T is terminal, there is an arrow $f : T' \to T$. Similarly, there is an arrow $g : T \to T'$. The arrow $f \circ g : T \to T$ is an arrow with target T. Since T is a terminal object of the category, there can be only one arrow from T to T. Thus it must be that $f \circ g$ is the identity of T. An analogous proof shows that $g \circ f$ is the identity of T'. □

2.7.14 Constants In **Set**, an element x of a set A is the image of a function from a singleton set to A that takes the unique element of the singleton to x. Thus if we pick a specific singleton $\{*\}$ and call it 1, the elements of the set A are in one to one correspondence with $\mathrm{Hom}(1, A)$, which is the set of functions from the terminal object to A. Moreover, if $f : A \to B$ is a set function and x is an element of A determining the function $x : 1 \to A$, then the element $f(x)$ of B is just the composite $f \circ x : 1 \to B$. Because of this, the categorist typically thinks of an element $x \in A$ as *being* the constant $x : 1 \to A$.

An arrow $1 \to A$ in a category, where 1 is the terminal object, is called a **constant of type** A. Thus each element of a set is a constant in **Set**. On the other hand, each monoid M has just one constant $1 \to M$ in the category of monoids, since monoid homomorphisms must preserve the identity. The name 'constant' is explained by Exercise 9.

The more common name in the categorical literature for a constant is **global element** of A, a name that comes from sheaf theory (see Section 14.5).

A terminal object is an object with exactly one arrow $\langle\rangle : A \to 1$ to it from each object A. So the arrows *to* 1 are not interesting. Global elements are arrows *from* the terminal object. There may be none or many, so *they* are interesting.

2.7.15 If 1 and $1'$ are two terminal objects in a category and $x : 1 \to A$ and $x' : 1' \to A$ are two constants with the property that $x' \circ \langle\rangle = x$ (where $\langle\rangle$ is the unique isomorphism from 1 to $1'$), then we regard x and x' as the *same* constant. Think about this comment as it applies to elements in the category of sets, with two different choices of terminal object, and you will see why.

2.7.16 Exercises

1. Show that if an arrow in a category has an inverse, it has only one.

2. Show that if $f : A \to B$ and $g : B \to C$ are isomorphisms in a category with inverses $f^{-1} : B \to A$ and $g^{-1} : C \to B$, then $g \circ f$ is an isomorphism with inverse $f^{-1} \circ g^{-1}$. (This is sometimes called the 'Shoe–Sock Theorem': to undo the act of putting on your socks, then your shoes, you have to take off your *shoes*, then your *socks*.)

3.† Give examples of a monoid M for which $M = M^{\mathrm{op}}$; one for which $M \neq M^{\mathrm{op}}$ (meaning the binary operations are different) but M and M^{op} are isomorphic; and one for which M and M^{op} are not isomorphic. (See Exercise 1 of Section 2.6.)

4. Do the same as the preceding exercise for posets. (See Exercise 2 of Section 2.6.)

5. Show that a poset isomorphic to a totally ordered set must be totally ordered.

6. Let $(P, \leq)$ be a poset. Show that in the corresponding category $C(P, \leq)$ (see 2.3.1), no two distinct objects are isomorphic.

7. Show that in the category of semigroups (respectively monoids), the isomorphisms are exactly the bijective homomorphisms. (This is not true for ordered sets, as we saw in 2.7.9. It is true in any variety of universal algebras, and it is not true in most interesting categories of topological spaces.)

8. a. Show that in the category of semigroups, the empty semigroup (see 2.3.2) is the initial object and any one-element semigroup is a terminal object.

b.† Show that the category of nonempty semigroups does not have an initial object. (Warning: this is *not* a trivial consequence of part (a) of this problem!) Some sources define 'semigroup' in a way that requires it to be nonempty.

9. Call a set function $f : A \to B$ constant if it factors as $f = k \circ \langle\rangle$ where $k : 1 \to B$ (hence is a constant in the sense of 2.7.14) and $\langle\rangle : A \to 1$ is the unique map given by the definition of terminal object. Show that if $B \neq \emptyset$, then a function $f : A \to B$ in the category of sets is constant in the sense that $f(x) = f(y)$ for all $x, y \in A$ if and only if f is constant.

10. Show that in the category of graphs and graph homomorphisms, the graph with one node and one arrow is the terminal object.

11. An arrow $f : A \to A$ in a category is **idempotent** if $f \circ f = f$. Show that in **Set** a function is idempotent if and only if its image is the same as its set of fixed points. (For example, applying a specific sorting method to a set of files is idempotent, since if you sort an already sorted file you leave it the same.)

12. An idempotent (see the preceding problem) $f : A \to A$ in a category is **split** if there is an object B and functions $g : A \to B$ and $h : B \to A$ for which $h \circ g = f$ and $g \circ h = \mathrm{id}_B$.

a. Show that every idempotent in **Set** is split.

b.† Give an example of a category with a non-split idempotent.

13. A category in which every arrow is an isomorphism is a **groupoid**. A category in which every arrow is an identity arrow is called **discrete**. Prove or disprove:

(i) Any two objects in a groupoid are isomorphic.

(ii) A groupoid in which no two distinct objects are isomorphic is discrete.

(iii) A poset P for which $C(P)$ is a groupoid is discrete.

2.8 Monomorphisms and subobjects

2.8.1 Monomorphisms A monomorphism is a type of arrow in a category which generalizes the concept of injective function; in particular, a monomorphism in the category of sets is exactly an injective function. A function $f : A \to B$ in **Set** is injective if for any $x, y \in A$, if $x \neq y$, then $f(x) \neq f(y)$. If f is an arrow in an arbitrary category, we use the same definition, except for one change required because the concept of 'element' no longer makes sense.

2.8.2 Definition $f : A \to B$ is a **monomorphism** if for any object T of the category and any arrows $x, y : T \to A$, if $x \neq y$, then $f \circ x \neq f \circ y$.

In this definition and many like it, what replaces the concept of element of A is an arbitrary arrow into A. In this context, an arbitrary arrow $a : T \to A$ is called a **variable element** of A, parametrized by T. When a is treated as a variable element and f has source A one may write $f(a)$ for $f \circ a$. Using this notation, f is a monomorphism if for any variable elements $x, y : T \to A$, if $x \neq y$, then $f(x) \neq f(y)$.

We often write $f : A \rightarrowtail B$ to indicate that f is a monomorphism and say that f is **monic** or that f is **mono**.

The following theorem validates the claim that 'monomorphism' is the categorical version of 'injective'.

2.8.3 Theorem *In the category of sets, a function is injective if and only if it is a monomorphism.*

Proof. Suppose $f : A \to B$ is injective, and let $a, a' : T \to A$ be variable elements of A. If $a \neq a'$ then there is an (ordinary) element $t \in T$ for which $a(t) \neq a'(t)$. Then $f(a(t)) \neq f(a'(t))$, so $f \circ a \neq f \circ a'$. Hence f is monic.

Conversely, suppose f is monic. Since global elements are elements, this says that for any global elements $x, y : 1 \to A$ with $x \neq y$, $f \circ x \neq f \circ y$, i.e., $f(x) \neq f(y)$, which means that f is injective. □

2.8.4 Examples In most familiar categories of sets with structure and structure-preserving functions, the monomorphisms are exactly the injective functions. In particular, the monomorphisms in **Mon** are the injective homomorphisms. Some other examples are in the exercises. This is evidence that Definition 2.8.2 is the correct categorical definition generalizing the set-theoretic concept of injectivity.

In the category determined by a poset, every arrow is monic. A monic element of the category determined by a monoid is generally called **left cancellable.**

2.8.5 Subobjects The concept of subobject is intended to generalize the concept of subset of a set, submonoid of a monoid, subcategory of a category, and so on. This idea cannot be translated exactly into categorical terms, since the usual concept of subset violates the strict typing rules of category theory: to go from a subset to a set requires a change of type, so there is no feasible way to say that the same element x is in both a set and a subset of the set.

Because of this, any categorical definition of subobject will not exactly give the concept of subset when applied to the category of sets. However, the usual definition of subobject (which we give here in Definition 2.8.8) produces, in **Set**, a concept that is naturally equivalent to the concept of subset in a strong sense that we will describe in 2.8.9. The definition, when applied to sets, defines subset in terms of the inclusion function.

2.8.6 We need a preliminary idea. If $f : A \to B$ is an arrow in a category, and for some arrow $g : C \to B$ there is an arrow $h : A \to C$ for which $f = g \circ h$, we say f **factors through** g. This is because the equation $g \circ h = f$ can be solved for h.

The use of the word 'factor' shows the explicit intention of categorists to work with functions in an algebraic manner: a category is an algebra of functions.

Suppose $f_0 : C_0 \to C$ and $f_1 : C_1 \to C$ are monomorphisms in a category. Let us say that $f_0 \sim f_1$ if each factors through the other.

2.8.7 Proposition *Let $f_0 \sim f_1$. Then the factors implied by the definition of $\sim$ are unique and are inverse isomorphisms. Moreover, the relation $\sim$ is an equivalence relation.*

Proof. The definition implies the existence of arrows $g : C_0 \to C_1$ and $h : C_1 \to C_0$ such that $f_1 \circ g = f_0$ and $f_0 \circ h = f_1$. The arrows g and h are unique because f_0 and f_1 are monomorphisms. Moreover, $f_1 \circ g \circ$

$h = f_0 \circ h = f_1 = f_1 \circ \mathrm{id}$; since f_1 is a monomorphism, we conclude that $g \circ h = \mathrm{id}$. Similarly, $h \circ g = \mathrm{id}$.

This result allows us to conclude that $\sim$ is symmetric, for if $f_1 \circ g = f_0$, then g is an isomorphism with inverse h; then $f_0 \circ h = f_1 \circ g \circ h = f_1 \circ \mathrm{id} = f_1$, so f_1 factors through f_0.

That $\sim$ is reflexive is obvious (the factor is the identity arrow). For transitivity, you get the required factor by composing the given factors; we leave the details to you. □

2.8.8 Definition In a category $\mathcal{C}$, a **subobject** of an object C is an equivalence class of monomorphisms under $\sim$.

2.8.9 Subobjects in the category of sets In **Set**, a monomorphism is an injection, so a subobject is an equivalence class of injections. The following sequence of statements are each easy to prove and together form a precise description of the connection between subobjects and subsets in the category of sets. Similar remarks can be made about other categories of sets with structure, such as semigroups, monoids or posets. In these statements, S is a set.

(a) Let $\mathcal{O}$ be a subobject of S.

- (i) Any two injections $m : A \to S$ and $n : B \to S$ in $\mathcal{O}$ have the same image; call the image I.
- (ii) The inclusion $i : I \to S$ is equivalent to any injection in $\mathcal{O}$, hence is an element of $\mathcal{O}$.
- (iii) If $j : J \to S$ is an inclusion of a subset J into S that is in $\mathcal{O}$, then $I = J$ and $i = j$.
- (iv) Hence every subobject of S contains exactly one inclusion of a subset of S into S, and that subset is the image of any element of $\mathcal{O}$.

(b) Let $i : T \to S$ be the inclusion of a subset T of S into S.

- (i) Since i is injective, it is an element of a subobject of S.
- (ii) Since the subobjects are equivalence classes of an equivalence relation, they are disjoint, so i is not in two subobjects.
- (iii) Hence the subsets of S form a complete set of class representatives for the subobjects of S.

Thus subobjects, given by a categorical definition, are not the same as subsets, but each subset determines and is determined by a unique subobject.

Because of this close relationship, one frequently says, of objects A and B in a category, 'Let A be a subobject of B', meaning that one has in mind a certain equivalence class of monomorphisms that in particular contains a monomorphism $A \rightarrowtail B$. You should be aware that there may be many other monomorphisms from A to B that are not in the equivalence class, just as from any subset A of a set B there are generally many injective functions from A to B other than the inclusion.

2.8.10 As a consequence of the properties of the subobject construction, categorists take a different attitude toward substructures such as subsets and submonoids, as compared to many other mathematicians. For them, A is a subobject or substructure of B if there is a monomorphism from A to B, and the subobject is the equivalence class determined by that monomorphism. For example, let $\mathbf{Z}$ denote the set of integers and $\mathbf{R}$ the set of real numbers. In calculus classes, $\mathbf{Z}$ is a subset of $\mathbf{R}$; an integer actually is a real number. For the categorist, it suffices that there be a monic (injective) map from $\mathbf{Z}$ to $\mathbf{R}$.

That monic map is a kind of *type conversion.* (See Reynolds [1980] for a more general view.) An integer need not actually be thought of as a real number, but there is a standard or canonical way (translate this statement as 'a monic map') to regard an integer as a real number. This mapping is regarded by the categorist as an inclusion, even though in fact it may change what the integer really is.

In a computer language, converting an integer to a real may increase the storage allotted to it and change its representation. Something similar happens in many approaches to the foundations of mathematics, where real numbers are constructed from integers by a complicated process (Dedekind cuts or Cauchy sequences), which results in an embedding of the integers in the real numbers. Just as for computer languages, this embedding changes the form of an integer: instead of whatever it was before, it is now a Dedekind cut (or Cauchy sequence).

In traditional texts on foundations, this construction had to be modified to replace the image of each integer by the actual integer, so that the integers were actually inside the real numbers. From the categorical point of view, this is an unnecessary complication. This change results in replacing a monomorphism $\mathbf{Z} \rightarrow \mathbf{R}$ by an equivalent monomorphism (one that determines the same subobject). From an *operational* point of view, the integers behave the same way whether this change is made or not.

2.8.11 Categories and typing In category theory, the inclusion map is usually made explicit. From the computing science point of view, category theory is a very strongly typed language, more strongly typed than any computer language. For example, the strict categorist will refer explicitly to the inclusion map from the nonzero real numbers to the set of all real numbers when talking of division. In a computer language this would correspond to having two different types, `REAL` and `NONZERO_REAL`, set up in such a way that you can divide a `REAL` only by a `NONZERO_REAL`. To multiply a `REAL` by a `NONZERO_REAL`, the strong typing would require you to convert the latter to a `REAL` first.

To be sure, categorists themselves are not always so strict; but when they are not strict they are aware of it. (Compare the comments in 1.2.2. Nor is this discussion meant to imply that computer languages should have such strict typing: rather, the intention is to illustrate the way category theory handles types.)

2.8.12 Exercises

1. Show that in any category an isomorphism $f : A \to B$ is a monomorphism.

2.† Show that a monomorphism in the category of semigroups (respectively monoids) is an injective homomorphism, and conversely.

3. a. Show that if an arrow is a monomorphism in a category, it is a monomorphism in any subcategory it happens to be in.

b. Give an example showing that a monomorphism in a subcategory need not be a monomorphism in the category containing the subcategory. (Hint: Look at small finite categories.)

4. Prove the statements in 2.8.9.

5. Show that if A is a subobject of a terminal object and B is any object then there is at most one arrow from B to A. Conclude that any arrow from A is monic.

6. Find all the subobjects of the terminal object in each category:

a. **Set**.

b. The category of graphs and graph homomorphisms.

c. The category of monoids and monoid homomorphisms.

2.9 Other types of arrow

2.9.1 Epimorphisms Epimorphisms in a category are the same as monomorphisms in the dual category. So $f : S \to T$ is an epimorphism if for any arrows $g, h : T \to X$, $g \circ f = h \circ f$ implies $g = h$. An epimorphism is said to be **epic** or an **epi**.

2.9.2 Proposition *A set function is an epimorphism in* **Set** *if and only if it is surjective.*

Proof. Suppose $f : S \to T$ is surjective, and $g, h : T \to X$ are two functions. If $g \neq h$, then there is some particular element $t \in T$ for which $g(t) \neq h(t)$. Since f is surjective, there is an element $s \in S$ for which $f(s) = t$. Then $g(f(s)) \neq h(f(s))$, so that $g \circ f \neq h \circ f$.

Conversely, suppose f is *not* surjective. Then there is some $t \in T$ for which there is no $s \in S$ such that $f(s) = t$. Now define two functions $g : T \to \{0, 1\}$ and $h : T \to \{0, 1\}$ as follows:

(i) $g(x) = h(x) = 0$ for all $x \in T$, $x \neq t$.

(ii) $g(t) = 0$.

(iii) $h(t) = 1$.

Then $g \neq h$ but $g \circ f = h \circ f$, so f is not an epimorphism. □

2.9.3 In contrast to the situation with monomorphisms, epimorphisms in categories of sets with structure are commonly not surjective. For example the nonnegative integers and the integers are both monoids under addition, and the inclusion function i is a homomorphism which is certainly not surjective. However, it is an epimorphism.

Here is the proof: any homomorphism h whose domain is the integers is determined completely by its value $h(1)$. For positive m, $m = 1 + 1 + \cdots + 1$, so

$$h(m) = h(1 + 1 + \cdots + 1) = h(1)h(1) \cdots h(1)$$

where we write the operation in the codomain as juxtaposition. Also, $h(-1)$ is the inverse of $h(1)$, since

$$h(1)h(-1) = h(-1)h(1) = h(-1 + 1) = h(0)$$

which must be the identity of the codomain. Since an element of a monoid can have only one inverse, this means $h(-1)$ is uniquely determined by $h(1)$. Then since every negative integer is a sum of -1's, the value of h at every negative integer is also determined by its value at 1.

Now suppose that g and h are two homomorphisms from the monoid of integers into the same codomain. Then g and h are both determined by their value at 1. Since 1 is a positive integer, this means that if $g \circ i = h \circ i$, then $g = h$. Thus i is an epimorphism.

From the variable element point of view, an epimorphism f is a variable element with the property that any two different arrows out of its target must have different values at f. Thus in some sense it is a variable element with a lot of variation.

This is not particularly familiar behavior for elements, and is an example of a situation where the variable element point of view is not very suggestive. Perhaps thirty years from now the variable element idea will be much more pervasive and this will be the natural way to look at epimorphisms.

2.9.4 An arrow $f : A \to B$ in a category is an isomorphism if it has an inverse $g : B \to A$ which must satisfy both the equations $g \circ f = \mathrm{id}_A$ and $f \circ g = \mathrm{id}_B$. If it only satisfies the second equation, $f \circ g = \mathrm{id}_B$, then f is a **left inverse** of g and (naturally) g is a **right inverse** of f.

2.9.5 Definition An arrow f with a right inverse is called a **split epimorphism.**

A split epimorphism is indeed an epimorphism: if $h \circ f = k \circ f$ and f has a right inverse g, then $h = h \circ f \circ g = k \circ f \circ g = k$, which is what is required for f to be an epimorphism.

(This comment is for those who know about the Axiom of Choice.) The statement that every epimorphism in the category of sets is split is equivalent to the Axiom of Choice: if $f : A \to B$ is surjective, a right inverse is exactly a choice function for the sets of the form $f^{-1}(b)$ as b ranges over B.

In most categories, epimorphisms need not be split. The function that includes the monoid of nonnegative integers on addition in the monoid of all the integers on addition, which we mentioned in 2.9.3, certainly does not have a right inverse in the category of monoids, since it does not have a right inverse in the category of sets. There are plenty of examples of epimorphisms of monoids which *are* surjective which have no right inverse in the category of monoids, although of course they do in the category of sets (Exercise 2).

2.9.6 Definition An arrow $f : A \to B$ with a *left* inverse is called a **split monomorphism.**

The left inverse is necessarily monic, dual to the case of split epis. Unlike epis, which always split in the category of sets, monics in **Set** do not always split. Every arrow out of the empty set is monic and, save for the identity of $\emptyset$ to itself, is not split. On the other hand, every monic with nonempty source does split. We leave the details to you.

2.9.7 Hom sets The elementary categorical definitions given in the last section and this one can all be phrased in terms of hom sets. In any category, $\mathrm{Hom}(A, B)$ is the set of arrows with source A and target B.

Thus a terminal object 1 satisfies the requirement that $\mathrm{Hom}(A, 1)$ is a singleton set for every object A, and an initial object 0 satisfies the dual requirement that $\mathrm{Hom}(0, A)$ is always a singleton. And $\mathrm{Hom}(1, A)$ is the set of constants (global elements) of A.

2.9.8 If $f : B \to C$, f induces a set function

$$\mathrm{Hom}(A, f) : \mathrm{Hom}(A, B) \to \mathrm{Hom}(A, C)$$

defined by composing by f on the left: for any $g \in \mathrm{Hom}(A, B)$, that is, for any $g : A \to B$, $\mathrm{Hom}(A, f)(g) = f \circ g$, which does indeed go from A to C. (Compare Exercise 1 of Section 1.2.)

Similarly, for any object D, $f : B \to C$ induces a set function

$$\mathrm{Hom}(f, D) : \mathrm{Hom}(C, D) \to \mathrm{Hom}(B, D)$$

(note the reversal) by requiring that $\mathrm{Hom}(f, D)(h) = h \circ f$ for any $h \in \mathrm{Hom}(C, D)$.

In terms of these functions, we can state this proposition, which we leave to you to prove.

2.9.9 Proposition *An arrow $f : B \to C$ in a category*

(i) *is a monomorphism if and only if* $\mathrm{Hom}(A, f)$ *is injective for every object A;*

(ii) *is an epimorphism if and only if* $\mathrm{Hom}(f, D)$ *is injective (!) for every object D;*

(iii) *is a split monomorphism if and only if* $\mathrm{Hom}(f, D)$ *is surjective for every object D;*

(iv) *is a split epimorphism if and only if* $\mathrm{Hom}(A, f)$ *is surjective for every object A.*

(v) *is an isomorphism if and only if any one of the following equivalent conditions holds:*

  (a) *it is both a split epi and a mono;*
  (b) *it is both an epi and a split mono;*
  (c) $\mathrm{Hom}(A, f)$ *is bijective for every object A;*
  (d) $\mathrm{Hom}(f, A)$ *is bijective for every object A.*

Although many categorical definitions can be given in terms of hom sets, no categorical definition *must* be; in fact, some mathematicians consider category theory to be a serious alternative to set theory as a foundation for mathematics (see many works of Lawvere, including [1963] and [1966], as well as [McLarty, 1989]), and for that purpose (which is not our purpose, of course), definition in terms of hom sets or any other sets must be avoided.

2.9.10 Discussion Categorical definitions, as illustrated in the simple ideas of Sections 2.7, 2.8 and 2.9, provide a method of abstract specification which has proved very useful in mathematics. They have, in particular, clarified concepts in many disparate branches of mathematics and provided as well a powerful unification of concepts across these branches.

The method of categorical definition is close in spirit to the modern attitude of computing science that programs and data types should be specified abstractly before being implemented and that the specification should be kept conceptually distinct from the implementation. We believe that the method of categorical definition is a type of abstract specification which is suitable for use in many areas of theoretical computing science. This is one of the major themes of this book.

When a category $\mathcal{C}$ is a category of sets with structure, with the arrows being functions which preserve the structure, a categorical definition of a particular property does not involve the elements (in the standard sense of set theory) of the structure. Such definitions are said to be **element-free**, and that has been regarded as a great advantage of category theory. Nevertheless, as we have seen, some definitions can be phrased in terms of *variable elements*, thereby allowing the option of retaining for general categories the way of thinking in terms of elements familiar from the category of sets. It is not clear whether this way of thinking with elements is a transitional phase of category theory, to be dropped with maturity, or will remain a permanent feature of the subject.

2.9.11 Exercises

1. Show that a surjective monoid homomorphism is an epimorphism.

2. Let $\mathbf{Z}_n$ denote the monoid of integers $(\bmod n)$ with addition $(\bmod n)$ as operation. Show that the map $\phi : \mathbf{Z}_4 \to \mathbf{Z}_2$ that takes 0 and 2 to 0 and 1 and 3 to 1 is a surjective monoid epimorphism and is not split.

3. Let M be a monoid.

a. Show that if M is finite then an element is a monomorphism in $C(M)$ if and only if it is an epimorphism in $C(M)$ if and only if it is an isomorphism in $C(M)$.

b. Give an example showing that the assumption of finiteness in (a) cannot be relaxed.

4. Show that in the category $C(P)$ determined by a poset P, the only split epis or split monos are the identity arrows.

5. Give an example of an arrow in a category that is both mono and epi but not an isomorphism.

6. Show that if h has a left inverse g, then $h \circ g$ is a split idempotent (see Exercise 12 of Section 2.7.)

7. Prove Proposition 2.9.9. (For (iii), set $D = B$.)

8.† Show that in the category of graphs and graph homomorphism, a homomorphism $f : \mathcal{G} \to \mathcal{H}$ has any of the following properties if and only if both $f_0 : G_0 \to H_0$ and $f_1 : G_1 \to H_1$ (which are set functions) have the property in **Set**:

(i) epic;
(ii) monic;
(iii) an isomorphism.

3
Functors

A functor F from a category $\mathcal{C}$ to a category $\mathcal{D}$ is a graph homomorphism which preserves identities and composition. It plays the same role as monoid homomorphisms for monoids and monotone maps for posets: it preserves the structure that a category has. Functors have another significance, however: since one sort of thing a category can be is a mathematical workspace (see Preface), many of the most useful functors used by mathematicians are *transformations from one type of mathematics to another.*

Less obvious, but perhaps more important, is the fact that many categories that are mathematically interesting appear as categories whose objects are a natural class of functors into the category of sets. This point of view will be explored in the chapters on sketches.

The first three sections define functors, give examples and describe some properties functors may have. Section 3.4 defines the concept of equivalence of categories, which captures the idea that two categories are the same from the categorical point of view. The last section concerns quotients of categories, which have quotients of monoids as special cases. This concept is used only in the chapters on sketches (and in 4.1.12, which itself is used only for sketches).

3.1 Functors

A functor is a structure-preserving map between categories, in the same way that a homomorphism is a structure-preserving map between graphs or monoids. Here is the formal definition.

3.1.1 Definition A **functor** $F : \mathcal{C} \to \mathcal{D}$ is a pair of functions $F_0 : \mathcal{C}_0 \to \mathcal{D}_0$ and $F_1 : \mathcal{C}_1 \to \mathcal{D}_1$ for which

F–1 If $f : A \to B$ in $\mathcal{C}$, then $F_1(f) : F_0(A) \to F_0(B)$ in $\mathcal{D}$.

F–2 For any object A of $\mathcal{C}$, $F_1(\mathrm{id}_A) = \mathrm{id}_{F_0(A)}$.

F–3 If $g \circ f$ is defined in $\mathcal{C}$, then $F_1(g) \circ F_1(f)$ is defined in $\mathcal{D}$ and $F_1(g \circ f) = F_1(g) \circ F_1(f)$.

By F–1, a functor is in particular a homomorphism of graphs. Following the practice for graph homomorphisms, the notation is customarily overloaded (see 1.4.2): if A is an object, $F(A) = F_0(A)$ is an object, and if f is an arrow, $F(f) = F_1(f)$ is an arrow. The notation for the constituents $F_0 : \mathcal{C}_0 \to \mathcal{D}_0$ and $F_1 : \mathcal{C}_1 \to \mathcal{D}_1$ is not standard, and we will use it only for emphasis.

3.1.2 Example It is easy to see that a monoid homomorphism $f : M \to N$ determines a functor from $C(M)$ to $C(N)$. On objects, a homomorphism f must take the single object to M to the single object of N, and F–1 is trivially verified since all arrows in M have the same domain and codomain, and similarly for N. Then F–2 and F–3 say precisely that f is a monoid homomorphism. Conversely, every functor is determined in this way by a monoid homomorphism.

3.1.3 Example Let us see what a functor from $C(S, \alpha)$ to $C(T, \beta)$ must be when (S, α) and (T, β) are posets as in 2.3.1. It is suggestive to write both relations α and β as '$\leq$' and the posets simply as S and T. Then there is exactly one arrow from x to y in S (or in T) if and only if $x \leq y$; otherwise there are no arrows from x to y.

Let $f : S \to T$ be the functor. F–1 says if there is an arrow from x to y, then there is an arrow from $f(x)$ to $f(y)$; in other words,

$$\text{if } x \leq y \text{ then } f(x) \leq f(y)$$

Thus f is an monotone map (see 2.4.2). F–2 and F–3 impose no additional conditions on f because they each assert the equality of two specified arrows between two specified objects and in a poset as category all arrows between two objects are equal.

3.1.4 Example If $\mathcal{C}$ is a category, the functor

$$P_1 : \mathcal{C} \times \mathcal{C} \to \mathcal{C}$$

(see 2.6.6) which takes an object (C, D) to C and an arrow $(f, g) : (C, D) \to (C', D')$ to f is called the **first projection**. There is an analogous second projection functor P_2 taking an object or arrow to its second coordinate.

3.1.5 The category of categories The category **Cat** has all small categories as objects and all functors between such categories as arrows. The composite of functors is their composite as graph homomorphisms: if $F : \mathcal{C} \to \mathcal{D}$ and $G : \mathcal{D} \to \mathcal{E}$, then $G \circ F : \mathcal{C} \to \mathcal{E}$ satisfies $G \circ F(C) = G(F(C))$ for any object C of $\mathcal{C}$, and $G \circ F(f) = G(F(f))$ for any arrow f of $\mathcal{C}$. Thus $(G \circ F)_i = G_i \circ F_i$ for $i = 0, 1$.

We note that the composition circle is usually omitted when composing functors so that we write $GF(C) = G(F(C))$.

It is sometimes convenient to refer to a category **CAT** which has all small categories and ordinary large categories as objects, and functors between them. Since trying to have **CAT** be an object of itself would raise delicate foundational questions, we do not attempt here a formal definition of **CAT**.

3.1.6 Example The inclusion map of a subcategory is a functor. As we pointed out in 2.8.10, the categorical point of view does not require that the object and arrows of a subcategory actually be objects and arrows of the bigger category, only that there be a monomorphism from the subcategory to the category.

This approach has the strange result that two different categories can each be regarded as subcategories of the other one (Exercise 9).

3.1.7 Underlying functors Forgetting some of the structure in a category of structures and structure-preserving functions gives a functor called an **underlying functor** or **forgetful functor**. The functor U : **Mon** $\rightarrow$ **Sem** which embeds the category of monoids into the category of semigroups by forgetting that a monoid has an identity is an example of an underlying functor.

Another example is the functor which forgets *all* the structure of a semigroup. This is a functor U : **Sem** $\rightarrow$ **Set**. There are lots of semigroups with the same set of elements; for example, the set $\{0, 1, 2\}$ is a semigroup on addition (mod 3) and also a different semigroup on multiplication (mod 3). The functor U applied to these two different semigroups gives the same set, so U is not injective on objects, in contrast to the forgetful functor from monoids to semigroups.

We will not give a formal definition of underlying functor. It is reasonable to expect any underlying functor U to be faithful (see 3.3.2 below) and that if f is an isomorphism and $U(f)$ is an identity arrow then f is an identity arrow.

3.1.8 Example If you forget you can compose arrows in a category and you forget which arrows are the identities, then you have remembered only that the category is a graph. This gives an underlying functor U : **Cat** $\rightarrow$ **Grf**, since every functor is a graph homomorphism although not vice versa.

3.1.9 Example A small graph has *two* underlying sets: its set of nodes and its set of arrows. Thus there is an underlying set functor U : **Grf** $\rightarrow$ **Set** $\times$ **Set** for which for a graph $\mathcal{G}$, $U(\mathcal{G}) = (A, N)$; an arrowset functor A : **Grf** $\rightarrow$ **Set** which takes a graph to its set of arrows and a graph homomorphism to the corresponding function from arrows to arrows; and a similarly defined nodeset functor $N : \mathbf{Grf} \to \mathbf{Set}$ which takes a graph to its set of nodes.

3.1.10 Example In 2.6.9, we described the notion of a slice category $\mathcal{C}/A$ based on a category $\mathcal{C}$ and an object A. An object is an arrow $B \to A$ and an arrow from $f : B \to A$ to $g : C \to A$ is an arrow $h : B \to C$ for which

$$h \circ g = f$$

There is a functor $U : \mathcal{C}/A \to \mathcal{C}$ that takes the object $f : B \to A$ to B and the arrow h from $B \to A$ to $C \to A$ to $h : B \to C$. This is called the underlying functor of the slice. In the case that $\mathcal{C} = \mathbf{Set}$, an object $B \to A$ of $\mathbf{Set}/A$ for some set A is an A-indexed object, and the effect of the underlying functor is to forget the indexing.

3.1.11 Example Since a category is a graph with extra structure, **Cat** has object and arrow functors $O : \mathbf{Cat} \to \mathbf{Set}$ and $A : \mathbf{Cat} \to \mathbf{Set}$ which take a category to its set of objects and set of arrows respectively, and a functor to the appropriate set map.

3.1.12 Free functors The **free monoid functor** from **Set** to the category of monoids takes a set A to the free monoid $F(A)$, which is the Kleene closure A^* with concatenation as operation (see 2.3.5), and a function $f : A \to B$ to the function $F(f) = f^* : F(A) \to F(B)$ defined in 2.5.6.

To see that the free monoid functor is indeed a functor it is necessary to show that if $f : A \to B$ and $g : B \to C$, then $F(g \circ f) : F(A) \to F(C)$ is the same as $F(g) \circ F(f)$, which is immediate from the definition, and that it preserves identity arrows, which is also immediate.

3.1.13 Example The free category on a graph is also the object part of a functor $F : \mathbf{Grf} \to \mathbf{Cat}$. What it does to a graph is described in 2.6.11. Suppose $\phi : \mathcal{G} \to \mathcal{H}$ is a graph homomorphism. The objects of the free category on a graph are the nodes of the graph, so we are forced to define $F(\phi)_0 = \phi_0$. Now suppose $(f_n, f_{n-1}, \ldots, f_1)$ is a path, that is, an arrow, in $F(\mathcal{G})$. Since functors preserve domain and codomain, we can define $F(\phi)_1(f_n, f_{n-1}, \ldots, f_1)$ to be $(\phi_1(f_n), \phi_1(f_{n-1}), \ldots, \phi_1(f_1))$ and know we get a path in $F(\mathcal{H})$. That F preserves composition of paths is also clear.

3.1.14 The arrow-lifting property The free category functor $F : \mathbf{Grf} \to \mathbf{Cat}$ and also other free functors, such as the free monoid functor (3.1.12), have an arrow-lifting property which will be seen in Section 12.2 as the defining property of freeness. We will describe the property for free categories since we use it later. For the free monoid, note Exercise 6 of Section 2.5.

Let $\mathcal{G}$ be a graph and $F(\mathcal{G})$ the free category generated by $\mathcal{G}$. There is a graph homomorphism with the special name $\eta G : \mathcal{G} \to U(F(\mathcal{G}))$ which includes a graph $\mathcal{G}$ into $U(F(\mathcal{G}))$, the underlying graph of the free category $F(\mathcal{G})$. The map $(\eta G)_0$ is the identity, since the objects of $F(\mathcal{G})$ are the nodes of $\mathcal{G}$. For an arrow f of $\mathcal{G}$, $(\eta G)_1(f)$ is the path (f) of length one. This is an inclusion arrow in the generalized categorical sense of 2.8.10, since f and (f) are really two distinct entities.

3.1.15 Proposition *Let $\mathcal{G}$ be a graph and $\mathcal{C}$ a category. Then for every graph homomorphism $h : \mathcal{G} \to U(\mathcal{C})$, there is a unique functor $\widehat{h} : F(\mathcal{G}) \to \mathcal{C}$ with the property that $U(\widehat{h}) \circ \eta G = h$.*

Proof. We define, for an object a of $F(\mathcal{G})$ (that is, node of $\mathcal{G}$), $\widehat{h}(a) = h(a)$. The value of $\widehat{h}$ on paths can be readily computed by a program in which $p = (f_1, f_2, \ldots, f_k)$ is a path from a to b:

```
Proc ĥ(p)
Begin
    If k = 0 Then
        Output id_a
    Else Output(ĥ(f_1, f_2, ..., f_{k-1}) ∘ h(f_k))
    EndIf
EndProc
```

□

As noted in 2.1.1, there is a unique empty path for each node a of $\mathcal{G}$. Composing the empty path at a with any path p from a to b gives p again, and similarly on the other side. That is why the program returns id_a for the empty path at a.

3.1.16 Powerset functors Any set S has a powerset $\mathcal{P}S$, the set of all subsets of S. There are three different functors F for which F_0 takes a set to its powerset; they differ on what they do to arrows. One of them is fundamental in topos theory; that one we single out to be called the powerset functor.

If $f : A \to B$ is any set function and C is a subset of B, then the **inverse image** of C, denoted $f^{-1}(C)$, is the set of elements of A which f takes into C: $f^{-1}(C) = \{a \in A \mid f(a) \in C\}$. Thus f^{-1} is a function from $\mathcal{P}B$ to $\mathcal{P}A$.

Note that for a bijection f, the symbol f^{-1} is also used to denote the inverse function. Context makes it clear which is meant, since the input to the inverse image function must be a subset of the codomain of f, whereas the input to the actual inverse of a bijection must be an *element* of the codomain.

3.1.17 Definition The **powerset functor** $\mathcal{P} : \mathbf{Set}^{\mathrm{op}} \to \mathbf{Set}$ takes a set S to the powerset $\mathcal{P}S$, and a set function $f : A \to B$ (that is, an arrow from B to A in $\mathbf{Set}^{\mathrm{op}}$) to the inverse image function $f^{-1} : \mathcal{P}B \to \mathcal{P}A$.

To check that $\mathcal{P}$ is a functor requires showing that $\mathrm{id}_A^{-1} = \mathrm{id}_{\mathcal{P}A}$ and that if $g : C \to C$, then $(g \circ f)^{-1} = f^{-1} \circ g^{-1}$, where both compositions take place in **Set**.

Although we will continue to use the notation f^{-1}, it is denoted f^* in much of the categorical literature.

3.1.18 A functor $F : \mathcal{C}^{\mathrm{op}} \to \mathcal{D}$ is also called a **contravariant functor** from $\mathcal{C}$ to $\mathcal{D}$. As illustrated in the preceding definition, the functor is often defined in terms of arrows of $\mathcal{C}$ rather than of arrows of $\mathcal{C}^{\mathrm{op}}$. The use of opposite categories is most commonly used to provide a way of talking about contravariant functors as ordinary (**covariant**) functors: the opposite category in this situation is a purely formal construction of no independent interest (see 2.6.8).

3.1.19 The other two functors which take a set to its powerset are both covariant. The **direct** or **existential image** functor takes $f : A \to B$ to the function $f_* : \mathcal{P}A \to \mathcal{P}B$, where $f_*(A_0) = \{f(x) \mid x \in A_0\}$, the set of values of f on A_0. The **universal image** functor takes A_0 to those values of f which come *only* from A_0: formally, it takes $f : A \to B$ to $f_! : \mathcal{P}A \to \mathcal{P}B$, with

$$f_!(A_0) = \{y \in B \mid f(x) = y \text{ implies } x \in A_0\} = \{y \in B \mid f^{-1}(\{y\}) \subseteq A_0\}$$

3.1.20 Hom functors Let $\mathcal{C}$ be a category with object C and arrow $f : A \to B$. In 2.9.8, we defined the function $\mathrm{Hom}(C, f) : \mathrm{Hom}(C, A) \to \mathrm{Hom}(C, B)$ by

$$\mathrm{Hom}(C, f)(g) = f \circ g$$

for every $g \in \mathrm{Hom}(C, A)$, that is for $g : C \to A$. We use this function to define the **covariant hom functor** $\mathrm{Hom}(C, -) : \mathcal{C} \to \mathbf{Set}$ as follows:

HF–1 $\mathrm{Hom}(C, -)(A) = \mathrm{Hom}(C, A)$ for each object A of $\mathcal{C}$;

HF–2 $\mathrm{Hom}(C, -)(f) = \mathrm{Hom}(C, f) : \mathrm{Hom}(C, A) \to \mathrm{Hom}(C, B)$ for $f : A \to B$.

The following calculations show that $\mathrm{Hom}(C, -)$ is a functor. For an object A, $\mathrm{Hom}(C, \mathrm{id}_A) : \mathrm{Hom}(C, A) \to \mathrm{Hom}(C, A)$ takes an arrow

$f: A \to A$ to $\mathrm{id}_A \circ f = f$; hence $\mathrm{Hom}(C, \mathrm{id}_A) = \mathrm{id}_{\mathrm{Hom}(C,A)}$. Now suppose $f: A \to B$ and $g: B \to C$. Then for any arrow $k: C \to A$,

$$\begin{aligned}\Big(\mathrm{Hom}(C,g) \circ \mathrm{Hom}(C,f)\Big)(k) &= \mathrm{Hom}(C,g)\Big(\mathrm{Hom}(C,f)(k)\Big)\\ &= \mathrm{Hom}(C,g)(f \circ k)\\ &= g \circ (f \circ k)\\ &= (g \circ f) \circ k\\ &= \mathrm{Hom}(C, g \circ f)(k)\end{aligned}$$

In terms of variable elements, $\mathrm{Hom}(C, f)$ takes the variable elements of A with parameter set C to the variable elements of B with parameter set C.

There is a distinct covariant hom functor $\mathrm{Hom}(C, -)$ for each object C. In this expression, C is a parameter for a family of functors. The argument of each of these functors is indicated by the dash. An analogous definition in calculus would be to define the function which raises a real number to the nth power as $f(-) = (-)^n$ (here n is the parameter). One difference in the hom functor case is that the hom functor is overloaded and so has to be defined on two different kinds of things: objects and arrows.

3.1.21 Definition For a given object D, the **contravariant hom functor**

$$\mathrm{Hom}(-, D) : \mathcal{C}^{\mathrm{op}} \to \mathbf{Set}$$

is defined for each object A by

$$\mathrm{Hom}(-, D)(A) = \mathrm{Hom}(A, D)$$

and for each arrow $f: A \to B$,

$$\mathrm{Hom}(-, D)(f) = \mathrm{Hom}(f, D) : \mathrm{Hom}(B, D) \to \mathrm{Hom}(A, D)$$

Thus if $g: B \to D$, $\mathrm{Hom}(f, D)(g) = g \circ f$.

3.1.22 Definition The **two-variable hom functor**

$$\mathrm{Hom}(-, -) : \mathcal{C}^{\mathrm{op}} \times \mathcal{C} \to \mathbf{Set}$$

takes a pair (C, D) of objects of $\mathcal{C}$ to $\mathrm{Hom}(C, D)$, and a pair (f, g) of arrows with $f: C \to A$ and $g: B \to D$ to

$$\mathrm{Hom}(f, g) : \mathrm{Hom}(A, B) \to \mathrm{Hom}(C, D)$$

where for $h : A \to B$,

$$\mathrm{Hom}(f, g)(h) = g \circ h \circ f$$

which is indeed an arrow from C to D.

In this case we also use the product of categories as a formal construction to express functors of more than one variable. From the categorical point of view, a functor always has one variable, which as in the present case might well be an object in a product category (an ordered pair).

3.1.23 Exercises

1. Show that in the definition of functor, the clause '$F_1(g) \circ F_1(f)$ is defined in $\mathcal{D}$' can be omitted.

2. Describe the initial and terminal objects in the category of categories and functors.

3. Prove that the existential and universal image functors of 3.1.19 are functors.

4. Give an example of a functor $F : \mathcal{C} \to \mathcal{D}$ with the property that the image of F is not a subcategory of $\mathcal{D}$. (Hint: it will be necessary for F to take two distinct objects to the same object.)

5. a. Prove that a functor is a monomorphism in the category of categories if and only if it is injective on both objects and arrows. (Compare Exercise 8(ii) of Section 2.9.)

b. Prove that the functor $U : \mathbf{Mon} \to \mathbf{Sem}$ described in 3.1.7 is a monomorphism.

6. Given a semigroup S, construct a monoid $M = S \cup \{e\}$, using a new element e not in S and different for each semigroup S. For example, you could take $e = \{S\}$. The multiplication in M is defined this way:

(i) xy is the product in S if both x and y are in S.

(ii) $xe = ex = x$ for all $x \in M$.

M is denoted S^1 in the semigroup literature. Show that

a. S^1 is a monoid (note that if S were already a monoid, S^1 is too but with a new identity element);

b. there is a functor $F : \mathbf{Sem} \to \mathbf{Mon}$ which takes each semigroup S to S^1 and each semigroup homomorphism $f : S \to T$ to a monoid homomorphism $f^1 : S^1 \to T^1$ which is the same as f on S and which takes the added element to the added element;

c. F is a monomorphism in $\mathbf{Cat}$.

7. Let $U : \mathbf{Mon} \to \mathbf{Set}$ be the underlying set functor, and $F : \mathbf{Set} \to \mathbf{Mon}$ the free monoid functor. For every set A and monoid M, construct a function

$$\beta : \mathrm{Hom}_{\mathbf{Set}}(A, U(M)) \to \mathrm{Hom}_{\mathbf{Mon}}(F(A), M)$$

by defining

$$\beta(f)(a_1 a_2 \cdots a_n) = f(a_1)f(a_2)\cdots f(a_n)$$

(the right side is the product in M) for $f : A \to U(M)$. Show that β is a bijection.

8. Let $U : \mathbf{Mon} \to \mathbf{Sem}$ be the functor of 3.1.6 and $F : \mathbf{Sem} \to \mathbf{Mon}$ the functor of Exercise 6. Define a function

$$\gamma : \mathrm{Hom}_{\mathbf{Sem}}(S, U(M)) \to \mathrm{Hom}_{\mathbf{Mon}}(F(S), M)$$

by $\gamma(h)(s) = h(s)$ if $s \in S$ and $\gamma(h)(e_S) = 1$ if e_S is the new element added to S to construct $F(S)$, where 1 is the identity element of M. Show that γ has the claimed codomain and is a bijection. (Compare Exercise 7.)

9. a. Show that each of **Mon** and **Sem** is a subcategory of the other.

b. Show that each of the following categories is a subcategory of the other: **Set** and the category of sets and partial functions. (Hint: to construct a monic functor from the category of sets and partial functions to **Set**, take each set A to the set $A \cup \{A\}$, and a partial function $f : A \to B$ to the full function f' for which $f'(x) = f(x)$ if $f(x)$ is defined, and $f'(x) = B$ otherwise.)

10.† Let $\mathcal{A}$ be a category. Show that $\mathcal{A}$ is discrete (see Exercise 13 of Section 2.7) if and only if every set function $F : \mathcal{A}_0 \to \mathcal{B}_0$, where $\mathcal{B}$ is any *category*, is the object part of a unique functor from $\mathcal{A}$ to $\mathcal{B}$.

11.† A category $\mathcal{B}$ is **indiscrete** if every set function $F : \mathcal{A}_0 \to \mathcal{B}_0$, where $\mathcal{A}$ is any category, is the object part of a unique functor from $\mathcal{A}$ to $\mathcal{B}$. Give a definition of 'indiscrete' in terms of the objects and arrows of $\mathcal{B}$.

3.2 Actions

In this section, we discuss set-valued functors as a natural generalization of finite state machines. This section is referred to only in Chapter 11 and in a few scattered examples. Set-valued functors also have theoretical importance in category theory because of the Yoneda Lemma (Section 4.5).

3.2.1 Monoid actions Let M be a monoid with identity 1 and let S be a set. An **action** of M on S is a function $\alpha : M \times S \to S$ for which

A–1 $\alpha(1, s) = s$ for all $s \in S$.

A–2 $\alpha(mn, s) = \alpha(m, \alpha(n, s))$ for all $m, n \in M$ and $s \in S$.

It is customary in mathematics to write ms for $\alpha(m, s)$; then the preceding requirements become

A′–1 $1s = s$ for all $s \in S$.

A′–2 $(mn)s = m(ns)$ for all $m, n \in M$ and $s \in S$.

When actions are written this way, S is also called an M-**set**. The same syntax ms for $m \in M$ and $s \in S$ is used even when different actions are involved. This notation is analogous to (and presumably suggested by) the notation $c\mathbf{v}$ for scalar multiplication, where c is a scalar and $\mathbf{v}$ is a vector.

It is useful to think of the set S as a **state space** and the elements of M as acting to induce **transitions** from one state to another.

3.2.2 Definition Let M be a monoid with actions on sets S and T. An **equivariant map** from S to T is a function $\phi : S \to T$ with the property that $m\phi(s) = \phi(ms)$ for all $m \in M$ and $s \in S$. The identity function is an equivariant map and the composite of two equivariant maps is equivariant. This means that for each monoid M, monoid actions and equivariant maps form a category M–**Act**.

3.2.3 Actions as functors Let α be an action of a monoid M on a set S. Let $C(M)$ denote the category determined by M as in 2.3.8. The action α determines a functor $F_\alpha : C(M) \to \mathbf{Set}$ defined by:

AF–1 $F_\alpha(*) = S$.

AF–2 $F_\alpha(m) = s \mapsto \alpha(m, s)$ for $m \in M$ and $s \in S$.

This observation will allow us to generalize actions to categories in 3.2.6.

3.2.4 Example One major type of action by a monoid is the case when the state space is a vector space and M is a collection of linear transformations closed under multiplication. However, in that case the linear structure (the fact that states can be added and multiplied by scalars) is extra structure which the definition above does not require. Our definition also does not require that there be any concept of continuity of transitions. Thus, the definition is very general and can be regarded as a *nonlinear, discrete* approach to state transition systems.

Less structure means, as always, that fewer theorems are true and fewer useful tools are available. On the other hand, less structure means that more situations fit the axioms, so that the theorems that are true and the tools that do exist work for more applications.

3.2.5 Example A particularly important example of a monoid action occurs in the study of finite state machines. Let A be a finite set, the **alphabet** of the machine, whose elements may be thought of as characters or tokens, and let S be another finite set whose elements are to be thought of as states. We assume there is a distinguished state $s_0 \in S$ called the **start state**, and a function $\phi : A \times S \to S$ defining a transition to a state for each token in A and each state in S. Such a system $\mathcal{M} = (A, S, s_0, \phi)$ is a **finite state machine**. Note that there is no question of imposing axioms such as A–1 and A–2 because A is not a monoid.

Any string w in A^* induces a sequence of transitions in the machine $\mathcal{M}$ starting at the state s_0 and ending in some state s. Precisely, we define a function $\phi^* : A^* \times S \to S$ by:

FA–1 $\phi^*((), s) = s$ for $s \in S$.

FA–2 $\phi^*((a)w, s) = \phi(a, \phi^*(w, s))$ for any $s \in S$, $w \in A^*$ and $a \in A$.

A^* *is* a monoid, under concatenation, and the function ϕ^* thus defined is an action of the free monoid $F(A)$ on S (Exercise 2).

Finite state machines in the literature often have added structure. The state space may have a subset F of **acceptor states** (or **final states**). The subset L of A^* of strings which drive the machine from the start state to an acceptor state is then the set of strings, or language, which is **recognized** by the machine $\mathcal{M}$. This is the machine as **recognizer**. A compiler typically uses a finite state machine to recognize identifiers in the input file.

Another approach is to assume that the machine outputs a string of symbols (not necessarily in the same alphabet) for each state it enters or each transition it undergoes. This is the machine as **transducer**.

Two important texts which use algebraic methods to study finite state machines (primarily as recognizers) are those by Eilenberg [1976] and Lallement [1979]. The latter book has many other applications of semigroup theory as well.

3.2.6 Set-valued functors as actions Suppose we wanted to extend the idea of an action by introducing typing. What would the result be?

To begin with, we would suppose that in addition to the state space S, there was a type set T and a function type : $S \to T$ that assigned to each element $s \in S$ an element type$(s) \in T$.

In describing the elements of M, one must say, for an $m \in M$ and $s \in S$, what is type(ms). Moreover, it seems that one might well want to restrict the types of the inputs on which a given m acts. In fact, although it might not be strictly necessary in every case, it seems clear that we can, without loss of generality, suppose that each $m \in M$ acts on only one kind of input and produces only one kind of output. For if m acted on two types of output, we could replace m by two elements, one for each type. Thus we can imagine that there are two functions we will call input and output from M to T for which input(m) is the type of element that m acts on and output(m) is the type of $m(s)$ for an element s of type input(m).

In the untyped case, we had that M was a monoid, but here it is clearly appropriate to suppose that $m_1 * m_2$ is defined only when $\text{output}(m_2) = \text{input}(m_1)$. It is reasonable to suppose that for each type t, there is an operation $1_t \in M$ whose input and output types are t and such that for any $m \in M$ of input type t, we have $m * 1_t = m$ and for any $m \in M$ of output type t, we have $1_t * m = m$.

As for the action, we will evidently wish to suppose that when $s \in S$ has type t and $m, m' \in M$ have input types t, t', respectively, and output types t', t'', respectively, then $m'(m(s)) = (m' * m)(s)$ and $1_t(s) = s$.

Now it will not have escaped the reader at this point that M and T together constitute a category $\mathcal{C}$ whose objects are the elements of T and arrows are the elements of M. The input and output functions are just the source and target arrows and the 1_t are the identities.

M and S make up exactly the data of a set-valued functor on $\mathcal{C}$. Define a functor $F : \mathcal{C} \to \mathbf{Set}$ by letting $F(t) = \{s \in S \mid \text{type}(s) = t\}$. If m is an arrow of $\mathcal{C}$, that is an element of M, let its input and output types be t and t', respectively. Then for F to be a functor, we require a function $F(m) : F(t) \to F(t')$. Naturally, we define $F(m)(s) = ms$, which indeed has type t'. The facts that F preserves composition and identities are an easy consequence of the properties listed above.

This construction can be reversed. Let $\mathcal{C}$ be a small category and suppose we have a functor $F : \mathcal{C} \to \mathbf{Set}$ for which $F(C)$ and $F(D)$ are disjoint whenever C and D are distinct objects of $\mathcal{C}$ (this disjointness requirement is necessary to have a category, but can be forced by a simple modification of F – see Exercise 5 of Section 4.5). Then we can let T be the set of objects of $\mathcal{C}$, M the set of arrows and $S = \bigcup_{t \in T} F(t)$. The rest of the definitions are evident and we leave them to the reader.

Thus if $\mathcal{C}$ is a small category, a functor $F : \mathcal{C} \to \mathbf{Set}$ is an action which generalizes the concept of monoid acting on a set.

3.2.7 Example For any given object C of a category $\mathcal{C}$, the hom functor $\mathrm{Hom}(C, -)$ (see 3.1.20) is a particular example of a set-valued functor. When the category $\mathcal{C}$ is a monoid, it is the action by left multiplication familiar in semigroup theory. A theorem generalizing the Cayley theorem for groups is true, too (see 4.5.2).

3.2.8 Variable sets It may be useful to think of a set-valued functor $F : \mathcal{C} \to \mathbf{Set}$ as an action, not on a typed set, but on a single *variable* set. The objects of $\mathcal{C}$ form a parameter space for the variation of the set being acted upon. Another way of saying this is that each object of $\mathcal{C}$ is a *point of view*, that the set being acted upon looks different from different points of view, and the arrows of $\mathcal{C}$ are changes in point of view (as well as inducing transitions). See [Barr, McLarty and Wells, 1985].

3.2.9 Machines with typed actions The concept generalizing finite state machines is based on the perception that words in a typed alphabet are paths in a graph whose nodes are the types.

Formally, a **typed finite state machine** consists of a graph $\mathcal{G}$ and a graph homomorphism ϕ from $\mathcal{G}$ to the category of finite sets. Thus for each node n of $\mathcal{G}$ there is a set $\phi(n)$, and for each arrow $f : m \to n$ of $\mathcal{G}$ there is a function $\phi(f) : \phi(m) \to \phi(n)$.

What corresponds to the action of the free monoid on the states in the case of ordinary finite state machines is the action of the free category $F(\mathcal{G})$ generated by $\mathcal{G}$. The words in the free monoid are now paths in the free category. The action ϕ generates an action $\phi^* : F(\mathcal{G}) \to \mathbf{Set}$ according to this recursive definition, which is a precise generalization of Section 3.2.5.

FA′–1 $\phi^*(()_C)(x) = x$ for $x \in F(C)$, where $()_C$ denotes the empty path from C to C.

FA′–2 For $x \in \mathrm{dom}(f_1)$,

$$\phi^*(f_n, f_{n-1}, \ldots, f_1)(x) = \phi^*(f_n, \ldots, f_2)[\phi(f_1)(x)]$$

This in other notation is a special case of the functor $F(\phi)$ defined in 3.1.13.

3.2.10 Exercises

1. Let S be a set. The **full transformation monoid** on S, denoted $FT(S)$, is the set of all functions from S to S with composition as the operation. Show that the following is equivalent to the definition in 3.2.1 of monoid action: an action by a monoid M on S is a monoid homomorphism from M to $TF(S)$.

2. Show that if ϕ^* is defined by FA–1 and FA–2 in 3.2.5, then it is an action of A^* on S in the sense of 3.2.1.

3.3 Types of functors

Since **Cat** is a category, we already know about some types of functors. Thus a functor $F : \mathcal{C} \to \mathcal{D}$ is an isomorphism if there is a functor $G : \mathcal{D} \to \mathcal{C}$ which is inverse to F. This implies that F is bijective on objects and arrows and conversely a functor which is bijective on objects and arrows is an isomorphism.

We have already pointed out (Exercise 5 of Section 3.1) that a functor is a monomorphism in **Cat** if and only if it is injective. Epimorphisms in **Cat** need not be surjective, since the example in 2.9.3 is actually an epimorphism in **Cat** between the categories determined by the monoids (Exercise 6).

3.3.1 Full and faithful We will now consider properties of functors which are more intrinsic to **Cat** than the examples just given.

Any functor $F : \mathcal{C} \to \mathcal{D}$ induces a set mapping

$$\mathrm{Hom}_{\mathcal{C}}(A, B) \to \mathrm{Hom}_{\mathcal{D}}(F(A), F(B))$$

for each pair of objects A and B of $\mathcal{C}$. This mapping takes an arrow $f : A \to B$ to $F(f) : F(A) \to F(B)$.

3.3.2 Definition A functor $F : \mathcal{C} \to \mathcal{D}$ is **faithful** if the induced mapping is injective on every hom set.

Thus if $f : A \to B$ and $g : A \to B$ are different arrows, then $F(f) \neq G(f)$. However, it is allowed that $f : A \to B$ and $g : C \to D$ may be different arrows, with $F(A) = F(C)$, $F(B) = F(D)$ and $F(f) = F(g)$, provided that either $A \neq C$ or $B \neq D$.

3.3.3 Example Underlying functors are typically faithful. Two different monoid homomorphisms between the same two monoids must be different as set functions.

On the other hand, consider the set $\{0, 1, 2\}$. It has two different monoid structures via addition and multiplication (mod 3) (and many other monoid structures, too), but the two corresponding identity homomorphisms are the same as set functions (have the same underlying function). Thus underlying functors need not be injective.

3.3.4 Definition A functor $F : \mathcal{C} \to \mathcal{D}$ is **full** if the induced mapping is surjective for every hom set.

A full functor need not be surjective on either objects or arrows. A full subcategory (2.6.3) is exactly one whose embedding is a full and faithful functor.

That the underlying functor from the category of semigroups to the category of sets is not full says exactly that not every set function between semigroups is a semigroup homomorphism. Note that this functor is surjective on objects, since every set can be made into a semigroup by letting $xy = x$ for every pair x and y of elements.

3.3.5 Example The functor $F : \mathcal{C} \to \mathcal{D}$ which takes A and B to C and X and Y to Z (and so is forced on arrows) in the picture below (which omits identity arrows) is *not* full. That is because $\mathrm{Hom}(A, B)$ is empty, but $\mathrm{Hom}(F(A), F(B)) = \mathrm{Hom}(C, C)$ has an arrow in it – the identity arrow. This functor is faithful even though not injective, since two arrows between the same two objects do not get identified.

$$\begin{array}{ccccc} A & & B & \qquad & C \\ \downarrow\uparrow & & \downarrow\uparrow & & \downarrow\uparrow \\ X & & Y & & Z \\ & \mathcal{C} & & & \mathcal{D} \end{array} \qquad (3.1)$$

3.3.6 Preservation of properties A functor $F : \mathcal{C} \to \mathcal{D}$ **preserves** a property P of arrows if whenever f has property P, so does $F(f)$.

For example, the statement that a monomorphism in the category of monoids must be injective can be worded as saying that the underlying functor preserves monomorphisms (since an injective function in **Set** is a monomorphism). The statement that an epimorphism in **Mon** need not be surjective is the same as saying that the underlying functor does not preserve epimorphisms.

3.3.7 Proposition *Every functor preserves isomorphisms.*

Proof. This is because the concept of isomorphism is defined in terms of equations involving composition and identity. If $f : A \to B$ is an isomorphism with inverse g, then $F(g)$ is the inverse of $F(f)$. One of the two calculations necessary to prove this is that $F(g) \circ F(f) = F(g \circ f) = F(\mathrm{id}_A) = \mathrm{id}_{F(A)}$; the other calculation is analogous. □

3.3.8 Definition A functor $F : \mathcal{C} \to \mathcal{D}$ **reflects** a property P of arrows if whenever $F(f)$ has property P then so does f (for *any* arrow that F takes to $F(f)$).

The statement that a bijective monoid homomorphism must be an isomorphism is the same as saying that the underlying functor from **Mon** to **Set** reflects isomorphisms. The underlying functor from the category of posets and monotone maps does not reflect isomorphisms (see 2.7.9).

3.3.9 You can also talk about a functor preserving or reflecting a property of objects. For example, since a terminal object in **Mon** is a one-element monoid and a one-element set is a terminal object, the underlying functor from **Mon** to **Set** preserves terminal objects. It also reflects terminal objects. It does not preserve initial objects, but it does reflect initial objects although vacuously: the empty set is the only initial object in **Set** and the underlying set of a monoid cannot be empty since it must have an identity element. We leave the details to you.

3.3.10 There is a third degree after preserving and reflecting. The functor $F : \mathcal{C} \to \mathcal{D}$ is to said to **create** a property (of an arrow or object) if given an arrow or object in $\mathcal{D}$ there is a unique arrow or object in $\mathcal{C}$ which has that property and is taken by F to the given arrow or object of $\mathcal{D}$. In the example of monoids in the last paragraph, the functor reflects both terminal and initial objects, but creates only the former. That is given the terminal set – the one with one element – there is a unique monoid structure on it that makes it the terminal monoid. There is no such structure on the empty set; a monoid cannot be empty.

All three concepts – preserve, reflect, create – are not really about properties of arrows and objects, but rather diagrams (which will be considered in the next chapter), of which arrows and objects are special cases.

3.3.11 Proposition *A full and faithful functor creates isomorphisms.*

Proof. Suppose $F : \mathcal{C} \to \mathcal{D}$ is full and faithful and suppose $u : F(A) \to F(B)$ is an isomorphism. Since F is full there must be arrows $f : A \to B$ and $g : B \to A$ for which $F(f) = u$ and $F(g) = u^{-1}$. Then

$$F(g \circ f) = F(g) \circ F(f) = u^{-1} \circ u = \mathrm{id}_{F(A)} = F(\mathrm{id}_A)$$

But F is faithful, so $g \circ f = \mathrm{id}_A$. A similar argument shows that $f \circ g = \mathrm{id}_B$, so that g is the inverse of f and $A \cong B$. □

3.3.12 Corollary *Let $F : \mathcal{C} \to \mathcal{D}$ be a full and faithful functor. If $F(A) = F(B)$ for two objects A and B of $\mathcal{C}$, then A and B are isomorphic.*

Compare Example 3.3.5.

Proof. Since F is full, there is an arrow $u : A \to B$ for which $F(u) = \mathrm{id}_{F(A)}$. Since $\mathrm{id}_{F(A)}$ is an isomorphism, u must be too, by Proposition 3.3.11. □

3.3.13 Exercises

1. What does it mean for a functor to be faithful if
 a. it is between the categories determined by monoids?
 b. it is between the categories determined by posets?

2. Same question as 1 for 'full'.

3. Is the forgetful functor from **Mon** to **Sem** full?

4. Is the free monoid functor faithful? Full?

5. Show that the powerset functor is faithful but not full.

6. Show that the example in 2.9.3 is an epimorphism in **Cat** when the monoids involved are regarded as categories; hence epimorphisms in **Cat** need not be surjective.

7. Give an example of a functor which does not preserve monomorphisms. (Hint: use an inclusion functor defined on a subcategory so small that some arrow is a monomorphism in the subcategory but not in the whole category.)

8. Prove that every functor preserves split monos and split epis.

9. a. Does the underlying functor from **Sem** to **Set** preserve, reflect or create initial objects? What about terminal objects?
 b. Same questions for the underlying functor from **Cat** to **Grf** (3.1.8).

3.4 Equivalences

In this section we define what it means for two categories to be equivalent. The correct concept turns out to be weaker than requiring that they be isomorphic – that is, that there is a functor from one to the other which has an inverse in **Cat**. In order to understand the issues involved, we first take a close look at the construction of the category corresponding to a monoid in Section 2.3.8. It turns out to be a functor.

3.4.1 Monoids and one-object categories For each monoid M we constructed a small category $C(M)$ in 2.3.8. We make the choice mentioned there that the one element of $C(M)$ is M. Note that although an element of $C(M)$ is now an arrow from M to M, it is *not* a set function.

For each monoid homomorphism $h : M \to N$, construct a functor $C(h) : C(M) \to C(N)$ as follows:

CF–1 On objects, $C(M) = N$.

CF–2 $C(h)$ must be exactly the same as h on arrows (elements of M).

It is straightforward to see that $C(h)$ is a functor and that this construction makes C a functor from **Mon** to the full subcategory of **Cat** of categories with exactly one object. We will denote this full subcategory as **Ooc**.

There is also a functor $U : \mathbf{Ooc} \to \mathbf{Mon}$ going the other way.

UO–1 For a category $\mathcal{C}$ with one object, $U(\mathcal{C})$ is the monoid whose elements are the arrows of $\mathcal{C}$ and whose binary operation is the composition of $\mathcal{C}$.

UO–2 If $F : \mathcal{C} \to \mathcal{D}$ is a functor between one-object categories, $U(F) = F_1$, that is, the functor F on arrows.

The functors U and C are not inverse to each other, and it is worthwhile to see in detail why.

The construction of C is in part arbitrary. We needed to regard each monoid as a category with one object. The choice of the elements of M to be the arrows of the category is obvious, but what should be the one object? We chose M itself, but we could have chosen some other thing, such as the set $\{e\}$, where e is the identity of M. The only real requirement is that it not be an element of M (such as its identity) in order to avoid set-theoretic problems caused by the category being an element of itself. The consequence is that we have given a functor $C : \mathbf{Mon} \to \mathbf{Ooc}$ in a way which required arbitrary choices.

The arbitrary choice of one object for $C(M)$ means that if we begin with a one-object category $\mathcal{C}$, construct $M = U(\mathcal{C})$, and then construct $C(M)$, the result will not be the same as C unless it happens that the one object of $\mathcal{C}$ is M. Thus $C \circ U \neq \mathrm{id}_{\mathbf{Ooc}}$, so that U is not the inverse of C. (In this case $U \circ C$ is indeed $\mathrm{id}_{\mathbf{Mon}}$.)

C is not surjective on objects, since not every small category with one object is in the image of C; in fact a category $\mathcal{D}$ is $C(M)$ for some monoid M only if the single object of $\mathcal{D}$ is actually a monoid and the arrows of $\mathcal{D}$ are actually the arrows of that monoid. This is entirely contrary to the spirit of category theory: we are talking about specific elements rather than specifying behavior. Indeed, in terms of specifying behavior, the category of monoids and the category of small categories with one object ought to be essentially the same thing.

The fact that C is not an isomorphism of categories is a signal that isomorphism is the wrong idea for capturing the concept that two categories are essentially the same.

However, every small category with one object is *isomorphic* to one of those constructed as $C(M)$ for some monoid M. This is the starting point for the definition of equivalence.

3.4.2 Definition A functor $F : \mathcal{C} \rightarrow \mathcal{D}$ is an **equivalence of categories** if there are:

E–1 A functor $G : \mathcal{D} \rightarrow \mathcal{C}$.

E–2 A family $u_C : C \rightarrow G(F(C))$ of isomorphisms of $\mathcal{C}$ indexed by the objects of $\mathcal{C}$ with the property that for every arrow $f : C \rightarrow C'$ of $\mathcal{C}$, $G(F(f)) = u_{C'} \circ f \circ u_C^{-1}$.

E–3 A family $v_D : D \rightarrow F(G(D))$ of isomorphisms of $\mathcal{D}$ indexed by the objects of $\mathcal{D}$, with the property that for every arrow $g : D \rightarrow D'$ of $\mathcal{D}$, $F(G(g)) = v_{D'} \circ g \circ v_D^{-1}$.

If F is an equivalence of categories, the functor G of E–1 is called a **pseudo-inverse** of F. That the functor C of 3.4.1 is an equivalence (with pseudo-inverse U) is left as an exercise.

The idea behind the definition is that not only is every object of $\mathcal{D}$ isomorphic to an object in the image of F, but the isomorphisms are compatible with the arrows of $\mathcal{D}$; and similarly for $\mathcal{C}$. (See Exercise 4 of Section 4.2.)

3.4.3 Theorem *Let $F : \mathcal{C} \rightarrow \mathcal{D}$ be an equivalence of categories and $G : \mathcal{D} \rightarrow \mathcal{C}$ a pseudo-inverse to F. Then F and G are full and faithful.*

Proof. Actually, something more is true: if F and G are functors for which E–2 is true, then F is faithful. For suppose $f, f' : C \rightarrow C'$ in $\mathcal{C}$ and $F(f) = F(f')$ in $\mathcal{D}$. Then $G(F(f)) = G(F(f'))$ in $\mathcal{C}$, so that

$$f = u_{C'}^{-1} \circ G(F(f)) \circ u_C = u_{C'}^{-1} \circ G(F(f')) \circ u_C = f'$$

Thus F is faithful. A symmetric argument shows that if E–3 is true then G is faithful.

Now suppose $F : \mathcal{C} \rightarrow \mathcal{D}$ is an equivalence of categories and $G : \mathcal{D} \rightarrow \mathcal{C}$ is a pseudo-inverse to F. We now know that F and G are faithful. To show that F is full, suppose that $g : F(C) \rightarrow F(C')$ in $\mathcal{D}$. We must find $f : C \rightarrow C'$ in $\mathcal{C}$ for which $F(f) = g$. Let $f = u_{C'}^{-1} \circ G(g) \circ u_C$. Then a calculation using E–2 shows that $G(F(f)) = G(g)$. Since G is faithful, $F(f) = g$. □

Proposition 3.3.11 implies that an equivalence of categories does not take nonisomorphic objects to isomorphic ones.

An alternative definition of equivalence sometimes given in the literature is that a functor $F : \mathcal{C} \to \mathcal{D}$ is an equivalence if it is full and faithful, and every object of $\mathcal{D}$ is isomorphic to an object in the image of F. This definition can be proved equivalent to ours.

3.4.4 Example The category of finite sets and functions between them is equivalent to the opposite of the category of finite Boolean algebras and homomorphisms between them. We sketch the construction here, omitting the many necessary verifications. (Boolean algebras are defined in 6.1.10.) A homomorphism $h : B \to B'$ is a monotone function which preserves meets, joins, $\top$, $\bot$ and complements (these requirements are redundant).

Let **FSet** denote the category of finite sets and functions and **FBool** the category of finite Boolean algebras and homomorphisms. Let $F : \mathbf{FSet}^{\mathrm{op}} \to \mathbf{FBool}$ take a finite set to its powerset, which is a Boolean algebra with inclusion for the ordering. If $f : S \to T$ is a function, $F(f) : \mathcal{P}T \to \mathcal{P}S$ takes a subset of T to its inverse image under f; this function $F(f)$ is a homomorphism of Boolean algebras.

To construct the pseudo-inverse, we need a definition. An **atom** in a finite Boolean algebra B is an element a for which there are no elements $b \in B$ such that $\bot < b < a$. It is a fact that any element $b \in B$ is the join of the set of atoms beneath it (this may be *false* in infinite Boolean algebras). It is also true that if $h : B \to B'$ is a homomorphism of Boolean algebras and A is the set of atoms of B, then the join of all the elements $h(a)$ for $a \in A$ is $\top$ in B', and for any two atoms a_1, a_2 of B, $h(a_1) \wedge h(a_2) = \bot$. It follows from this that if b' is an atom of B', then there is a unique atom a of B for which $b' \leq h(a)$.

Now define a functor $G : \mathbf{FBool}^{\mathrm{op}} \to \mathbf{FSet}$ as follows. Let B be a finite Boolean algebra. Then $G(B)$ is the set of atoms of B. If $h : B \to B'$ is a homomorphism, then $G(h) : G(B') \to G(B)$ takes an atom a' of B' to the unique atom a of B for which $a' \leq h(a)$. This makes G a functor.

3.4.5 Proposition *F is an equivalence with pseudo-inverse G.*

The component of the required natural isomorphism from a finite set S to $G(F(S))$ takes an element $x \in S$ to the singleton $\{x\}$. The component of the natural isomorphism from a finite Boolean algebra B to $F(G(B))$ (the latter is the set of all subsets of all the atoms of B) takes an element $b \in B$ to the set of atoms under b. We omit the details.

3.4.6 Exercises

1. Prove that a functor which is an isomorphism in **Cat** is an equivalence.

2. Give an example showing that an equivalence of categories can have two different pseudo-inverses.

3.† Let $\mathcal{PF}$ denote the category of sets and partial functions defined in 2.1.13. Let $\mathcal{PS}$ denote the category whose objects are sets with a distinguished element (called **pointed sets**) and whose arrows are functions which preserve the distinguished element. In other words, if S is a set with distinguished element s and T is a set with distinguished element t, then an arrow of $\mathcal{PS}$ is a function $f : S \to T$ for which $f(s) = t$. Show that $\mathcal{PF}$ and $\mathcal{PS}$ are equivalent categories. (Hint: the functor from $\mathcal{PF}$ adds a new element to each set and completes each partial function to a total function by assigning the new element to each element where it was formerly undefined. The functor in the other direction takes a pointed set to the set without the point and if a function has the distinguished point as a value for some input, it becomes undefined at that input.)

4. Show that the category of preordered sets and increasing maps is equivalent to the full category of those small categories with the property that Hom(A, B) never has more than one element.

5.† (For the reader conversant with vector spaces and linear mappings.) Let $\mathcal{L}$ denote the category of finite dimensional vector spaces and linear maps. Let $\mathcal{M}$ be the category whose objects are the natural numbers, and for which an arrow $M : m \to n$ is an $n \times m$ matrix. When $n = 0$ or $m = 0$ or both, there is just one arrow called 0. Composition is matrix multiplication. Any composite involving 0 gives the 0 arrow. Show that $\mathcal{L}$ and $\mathcal{M}$ are equivalent categories.

6. a. Prove that the functor C : **Mon** $\to$ **Ooc** constructed in 3.4.1 is an equivalence of categories.

b. Prove directly, without using Theorem 3.4.3, that it is full and faithful.

3.5 Quotient categories

A quotient of a category by a congruence relation on the arrows is very similar to the concept of the quotient of a monoid by a congruence relation. We will describe the construction from scratch; you need not know about congruence relations to understand it. The construction we describe is not the most general possible: it merges arrows, but not objects.

The constructions of this section are used in 4.1.12 and in the chapters on sketches.

3.5.1 Definition An equivalence relation $\sim$ on the arrows of a category $\mathcal{C}$ is a **congruence relation** if:

CR–1 Whenever $f \sim g$, then f and g have the same domain and the same codomain.

CR–2 In this diagram,

$$A \xrightarrow{h} B \underset{g}{\overset{f}{\rightrightarrows}} C \xrightarrow{k} D$$

if $f \sim g$, then $f \circ h \sim g \circ h$ and $k \circ f \sim k \circ g$.

We denote the congruence class containing the arrow f by $[f]$.

3.5.2 Definition Let $\sim$ be a congruence relation on the arrows of $\mathcal{C}$. Define the **quotient category** $\mathcal{C}/\sim$ as follows.

QC–1 The objects of $\mathcal{C}/\sim$ are the objects of $\mathcal{C}$.

QC–2 The arrows of $\mathcal{C}/\sim$ are the congruence classes of arrows of $\mathcal{C}$.

QC–3 If $f : A \to B$ in $\mathcal{C}$, then $[f] : A \to B$ in $\mathcal{C}/\sim$.

QC–4 If $f : A \to B$ and $g : B \to C$ in $\mathcal{C}$, then $[g] \circ [f] = [g \circ f] : A \to C$ in $\mathcal{C}/\sim$.

It follows from Exercise 1 that QC–4 is well defined and that the result is indeed a category.

3.5.3 Definition Let $\mathcal{C}/\sim$ be the quotient of a category $\mathcal{C}$ by a congruence relation $\sim$. Define $Q : \mathcal{C} \to \mathcal{C}/\sim$ by $QA = A$ for an object A and $Qf = [f]$ for an arrow f of $\mathcal{C}$.

It is immediate from QC–4 that Q is a functor.

3.5.4 Proposition *Let $\sim$ be a congruence relation on a category $\mathcal{C}$. Let $F : \mathcal{C} \to \mathcal{D}$ be any functor with the property that if $f \sim g$ then $F(f) = F(g)$. Then there is a unique functor $F_0 : \mathcal{C}/\sim \to \mathcal{D}$ for which $F_0 \circ Q = F$.*

The proposition says that every way of passing from $\mathcal{C}$ to some other category which merges congruent arrows factors through Q uniquely. This can be perceived in another way: $\mathcal{C}/\sim$ is the category constructed from $\mathcal{C}$ by making the fewest identifications consistent with forcing two congruent arrows in $\mathcal{C}$ to be the same arrow.

3.5.5 Factorization of functors Every functor factors through a faithful one in the following precise sense. Let $F : \mathcal{C} \rightarrow \mathcal{D}$ be a functor. Define a relation $\sim$ between two arrows $f, g : C \rightarrow D$ with the same domain and codomain by saying that $f \sim g$ if and only if $F(f) = F(g)$.

3.5.6 Proposition *The relation $\sim$ induced by F is a congruence relation on $\mathcal{C}$, and the functor $F_0 : \mathcal{C}/{\sim} \rightarrow \mathcal{D}$ induced by Proposition 3.5.4 is faithful.*

The proof is contained in Exercise 4 below.

A faithful set-valued functor (see 3.3.2) is one for which two different arrows act differently on at least one state. In the special case of monoid actions this is precisely the definition of 'faithful' used in the literature (not a coincidence), and the preceding proposition is a well-known fact about monoid actions.

3.5.7 The intersection of any set of congruence relations on a category is also a congruence relation (Exercise 2). This means that if α is any relation on $\mathcal{C}$ with the property that if $f\alpha g$ then f and g have the same domain and the same codomain, then there is a unique smallest congruence relation generated by α.

Thus in particular given two arrows $f : A \rightarrow B$ and $g : A \rightarrow B$ in a category $\mathcal{C}$, there is a quotient category by the congruence relation generated by requiring that $f = g$. This is called **imposing the relation** $f = g$. Two arrows in $\mathcal{C}$ are merged in the quotient category if requiring that $f = g$ forces them to be merged.

3.5.8 The category of a programming language We described in 2.2.6 the category $C(L)$ corresponding to a simple functional programming language L defined there. We can now say precisely what $C(L)$ is.

The definition of L in 2.2.5 gives the primitive types and operations of the language. The types are the nodes and the operations are the arrows of a graph. This graph generates a free category $F(L)$, and the equations imposed in 2.2.5(ii) and (iv) (each of which says that two arrows of $C(L)$ must be equal) generate a congruence relation as just described. The resulting quotient category is precisely $C(L)$.

When one adds constructors such as record types to the language, the quotient construction is no longer enough. Then it must be done using sketches. See [Wells, 1989]. The construction just given is in fact a special case of a model of a sketch (see Section 4.6).

3.5.9 Functorial semantics Functors provide a way to give a meaning to the constructs of the language L just mentioned. This is done by giving a functor from $C(L)$ to some category suitable for programming language semantics, such as those discussed in 2.4.3.

We illustrate this idea here using a functor to **Set** as the semantics for the language described in 2.2.5. **Set** is for many reasons unsuitable for programming language semantics, but it is the natural category for expressing our intuitive understanding of what programming language constructs mean.

Following the discussion in 2.2.5, we define a semantics functor $\Sigma : C(L) \to \mathbf{Set}$. To do this, first we define a function F on the primitive types and operations of the language.

(i) $F(\mathbf{NAT})$ is the set of natural numbers. The constant 0 is the number 0 and **succ** is the function which adds 1.

(ii) $F(\mathbf{BOOLEAN})$ is the set $\{\mathbf{true}, \mathbf{false}\}$. The constants **true** and **false** are the elements of the same name, and $F(\neg)$ is the function which switches true and false.

(iii) $F(\mathbf{CHAR})$ is the set of 128 ASCII symbols, and each symbol is a constant.

(iv) $F(\mathbf{ord})$ takes a character to its ASCII value, and $F(\mathbf{chr})$ takes a number n to the character with ASCII code n modulo 128.

Let $F(L)$ be the free category generated by the graph of types and operations, as in 2.6.11. By Proposition 3.1.15, there is a functor $\widehat{F} : F(L) \to \mathbf{Set}$ which has the effect of F on the primitive types and operations.

This functor $\widehat{F}$ has the property required by Proposition 3.5.4 that if $\sim$ is the congruence relation on $F(L)$ generated by the equations of 2.2.5(ii) and (iv), then $f \sim g$ implies that $\widehat{F}(f) = \widehat{F}(g)$ (Exercise 5). This means that there is a functor $\Sigma : C(L) \to \mathbf{Set}$ (called F_0 in Proposition 3.5.4) with the property that if x is any primitive type or operation, then $\Sigma(x) = F(x)$.

The fact that Σ is a functor means that it preserves the meaning of programs; for example the program (path of arrows) $\mathbf{chr} \circ \mathbf{succ} \circ \mathbf{ord}$ ought to produce the next character in order, and in fact

$$\Sigma(\mathbf{chr} \circ \mathbf{succ} \circ \mathbf{ord})$$

does just that, as you can check. Thus it is reasonable to refer to Σ as a possible semantics of the language L.

We will return to this example in Section 4.3.11. More general approaches to these questions are in Lair [1987] and Wells [1989]. The construction of $C(L)$ and Σ are instances of the construction of the theory of a sketch in Section 7.5.

3.5.10 Exercises

1. Show that an equivalence relation $\sim$ satisfying CR–1 is a congruence relation if and only if, for all arrows f_1, f_2, g_1, g_2 as in this diagram,

$$A \underset{f_2}{\overset{f_1}{\rightrightarrows}} B \underset{g_2}{\overset{g_1}{\rightrightarrows}} C$$

if $f_1 \sim f_2$ and $g_1 \sim g_2$, then $g_1 \circ f_1 \sim g_2 \circ f_2$.

2. Show that the intersection of congruence relations is a congruence relation.

3. Show that the quotient functor in 3.5.3 is full. (Warning: this exercise would be incorrect if we allowed the more general definition of quotient, which allows merging objects as well as arrows.)

4. Let $F : \mathcal{C} \to \mathcal{D}$ be a functor. Define a relation $\sim$ between two arrows $f, g : c \to d$ of $\mathcal{C}$ with the same domain and codomain by saying that $f \sim g$ if and only if $F(f) = F(g)$.

a. Show that the relation $\sim$ induced by F is a congruence relation.

b. Show that the induced functor $F_0 : \mathcal{C}/{\sim} \to \mathcal{D}$ is faithful.

c. Conclude from this and the preceding exercise that every functor $F : \mathcal{C} \to \mathcal{D}$ factors as a full functor followed by a faithful functor.

5. Let $\widehat{F}$ and $\sim$ be defined as in 3.5.9. Prove that $f \sim g$ implies that $\widehat{F}(f) = \widehat{F}(g)$.

6. Let M be a monoid. A **congruence** on M is an equivalence relation $\sim$ with the property that it is a congruence relation for the category $C(M)$ determined by M.

a. Show that an equivalence relation $\sim$ on M is a congruence relation if and only if for all elements m, n, n' of M, if $n \sim n'$ then $mn \sim mn'$ and $nm \sim n'm$.

b. Let K be the subset $\{(m, n) \mid m \sim n\}$ of the monoid $M \times M$. Show that K is a submonoid of $M \times M$ if and only if $\sim$ is a congruence relation. ($M \times M$ is the monoid whose elements are all ordered pairs of elements of M with multiplication $(m, n)(m', n') = (mm', nn')$.)

4

Diagrams, naturality and sketches

Commutative diagrams are the categorist's way of expressing equations. Natural transformations are maps between functors; one way to think of them is as a deformation of one construction (construed as a functor) into another. A sketch is a graph with imposed commutativity and other conditions; it is a way of expressing structure. Models of the structure are given by functors, and homomorphisms between them by natural transformations.

All this will become clearer as the chapter is read. It turns out that the concepts just mentioned are all very closely related to each other. Indeed, there is a sense in which diagrams, functors and models of sketches are all different aspects of the same idea: they are all types of graph homomorphisms in which some or all of the graphs are categories.

The first three sections introduce diagrams, commutative diagrams and natural transformations, three basic ideas in category theory. These concepts are used heavily in the rest of the book. Section 4.4 gives the Godement rules, which form the basis of the algebra of functors and natural transformations.

Section 4.5 introduces the concepts of representable functor, the Yoneda embedding and universal elements. Working through the details of this presentation is an excellent way of learning to work with natural transformations.

We also recommend studying the introduction to linear sketches and linear sketches with constants in the last two sections as an excellent way to familiarize yourself with both commutative diagrams and natural transformations. However, the last two sections may be skipped unless you are going to read Chapters 7 and 9.

4.1 Diagrams

We begin with diagrams in a graph and discuss commutativity later.

4.1.1 Definition Let $\mathcal{I}$ and $\mathcal{G}$ be graphs. A **diagram** in $\mathcal{G}$ of **shape** $\mathcal{I}$ is a homomorphism $D : \mathcal{I} \to \mathcal{G}$ of graphs. $\mathcal{I}$ is called the **shape graph** of the diagram D.

We have thus given a new name to a concept which was already defined (not uncommon in mathematics). A diagram is a graph homomorphism from a different point of view.

In much of the categorical literature, a diagram in $\mathcal{C}$ is a functor $D : \mathcal{E} \to \mathcal{C}$ where $\mathcal{E}$ is a category. Because of Proposition 3.1.15, a graph homomorphism into a category extends uniquely to a functor based on the free category generated by the graph, so that diagrams in our sense generate diagrams in the functorial sense. On the other hand, any functor is a graph homomorphism on the underlying graph of its domain (although not conversely!), so that every diagram in the sense of functor is a diagram in the sense of graph homomorphism.

4.1.2 Example Here is an example illustrating some subtleties involving the concept of diagram. Let $\mathcal{G}$ be a graph with objects A, B and C (and maybe others) and arrows $f : A \to B$, $g : B \to C$ and $h : B \to B$. Consider these two diagrams, where here we use the word 'diagram' informally:

$$A \xrightarrow{f} B \xrightarrow{g} C \qquad\qquad A \xrightarrow{f} B \circlearrowleft h \tag{4.1}$$

(a) (b)

These are clearly of different shapes (again using the word 'shape' informally). But the diagram

$$A \xrightarrow{f} B \xrightarrow{h} B \tag{4.2}$$

is the same shape as (4.1)(a) even though as a graph it is the same as (4.1)(b).

To capture the difference thus illustrated between a graph and a diagram, we introduce two shape graphs

$$1 \xrightarrow{u} 2 \xrightarrow{v} 3 \qquad\qquad 1 \xrightarrow{u} 2 \circlearrowleft w \tag{4.3}$$

$\mathcal{I}$ $\mathcal{J}$

(where, as will be customary, we use numbers for the nodes of shape graphs). Now diagram (4.1)(a) is seen to be the diagram $D : \mathcal{I} \to \mathcal{G}$ with $D(1) = A$, $D(2) = B$, $D(3) = C$, $D(u) = f$ and $D(v) = g$; whereas diagram (4.1)(b) is $E : \mathcal{J} \to \mathcal{G}$ with $E(1) = A$, $E(2) = B$, $E(u) = f$ and $E(w) = h$. Moreover, Diagram (4.2) is just like D (has the same shape), except that v goes to h and 3 goes to B.

4.1.3 Our definition in 4.1.1 of a diagram as a graph homomorphism, with the domain graph being the shape, captures both the following ideas:

(i) A diagram can have repeated labels on its nodes and (although the examples did not show it) on its arrows, and

(ii) Two diagrams can have the same labels on their nodes and arrows but be of different shapes: Diagrams (4.1)(b) and (4.2) are *different diagrams* because they have different shapes.

4.1.4 Commutative diagrams When the target graph of a diagram is the underlying graph of a category some new possibilities arise, in particular the concept of commutative diagram, which is the categorist's way of expressing equations.

In this situation, we will not distinguish in notation between the category and its underlying graph: if $\mathcal{I}$ is a graph and $\mathcal{C}$ is a category we will refer to a diagram $D : \mathcal{I} \to \mathcal{C}$.

We say that D is **commutative** (or **commutes**) provided for any nodes i and j of $\mathcal{I}$ and two paths

$$\begin{array}{ccccccccc} & & k_1 & \xrightarrow{s_2} & k_2 \to \cdots \to & k_{n-2} & \xrightarrow{s_{n-1}} & k_{n-1} & \\ & {\scriptstyle s_1}\nearrow & & & & & & & \searrow{\scriptstyle s_n} \\ i & & & & & & & & j \\ & {\scriptstyle t_1}\searrow & & & & & & & \nearrow{\scriptstyle t_m} \\ & & l_1 & \xrightarrow[t_2]{} & l_2 \to \cdots \to & l_{m-2} & \xrightarrow[t_{m-1}]{} & l_{m-1} & \end{array} \tag{4.4}$$

from i to j in $\mathcal{I}$, the two paths

$$\begin{array}{ccccccccc} & & Dk_1 & \xrightarrow{Ds_2} & Dk_2 \to \cdots \to & Dk_{n-2} & \xrightarrow{Ds_{n-1}} & Dk_{n-1} & \\ & {\scriptstyle Ds_1}\nearrow & & & & & & & \searrow{\scriptstyle Ds_n} \\ Di & & & & & & & & Dj \\ & {\scriptstyle Dt_1}\searrow & & & & & & & \nearrow{\scriptstyle Dt_m} \\ & & Dl_1 & \xrightarrow[Dt_2]{} & Dl_2 \to \cdots \to & Dl_{m-2} & \xrightarrow[Dt_{m-1}]{} & Dl_{m-1} & \end{array} \tag{4.5}$$

are the same in $\mathcal{C}$. This means that

$$Ds_n \circ Ds_{n-1} \circ \ldots \circ Ds_1 = Dt_m \circ Dt_{m-1} \circ \ldots \circ Dt_1$$

4.1.5 Much ado about nothing There is one subtlety to the definition of commutative diagram: what happens if one of the numbers m or n in the diagram (4.5) should happen to be 0? If, say, $m = 0$, then we interpret the above equation to be meaningful only if the nodes i and j are the same (you go nowhere on an empty path) and the meaning in this case is that

$$Dt_n \circ Dt_{n-1} \circ \ldots \circ Dt_1 = \mathrm{id}_i$$

(you do nothing on an empty path). In particular, a diagram D based on the graph

$$\overset{\circlearrowleft e}{i}$$

commutes if and only if $D(e)$ is the identity arrow from $D(i)$ to $D(i)$.

Note, and note well, that both shape graphs

$$\overset{\circlearrowleft e}{i} \qquad\qquad i \xrightarrow{\ d\ } j$$

(a) (b)

have models that one might think to represent by the diagram

$$\overset{\circlearrowleft f}{A}$$

but the diagram based on (a) commutes if and only if $f = \mathrm{id}_A$, while the diagram based on (b) commutes automatically (no two nodes have more than one path between them so the commutativity condition is vacuous). Although it is not really wrong to represent the diagram based on (b) by this shape it is certainly clearer to picture it as

$$A \xrightarrow{\ f\ } A$$

4.1.6 Examples of commutative diagrams – and others The prototypical commutative diagram is the triangle

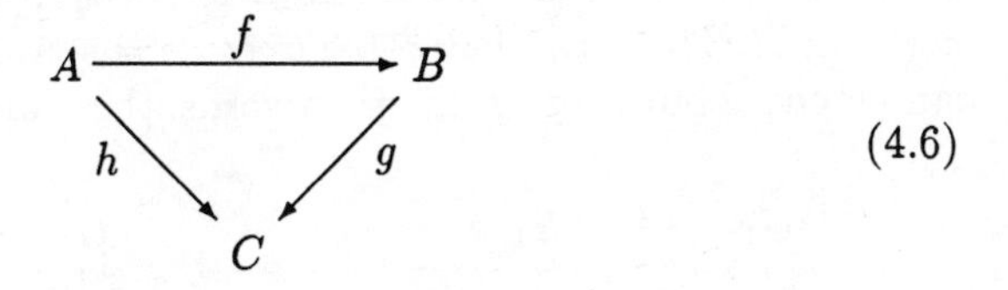

(4.6)

that commutes if and only if h is the composite $g \circ f$. The reason this is prototypical is that any commutative diagram – unless it involves an empty path – can be replaced by a set of commutative triangles. This fact is easy to show and not particularly enlightening, so we are content to give an example. The diagram

$$\begin{array}{ccc} A & \xrightarrow{h} & B \\ {\scriptstyle f}\downarrow & & \downarrow{\scriptstyle g} \\ C & \xrightarrow[k]{} & D \end{array} \tag{4.7}$$

commutes if and only if the two diagrams

$$\begin{array}{ccc} A & & \\ {\scriptstyle f}\downarrow & \searrow^{g \circ h} & \\ C & \xrightarrow[k]{} & D \end{array} \qquad \begin{array}{ccc} A & \xrightarrow{h} & B \\ & \searrow^{k \circ f} & \downarrow{\scriptstyle g} \\ & & D \end{array} \tag{4.8}$$

commute (in fact if and only if either one does).

4.1.7 Example An arrow $f : A \to B$ is an isomorphism with inverse $g : B \to A$ if and only if

$$A \underset{g}{\overset{f}{\rightleftarrows}} B \tag{4.9}$$

commutes. The reason is that for this diagram to commute, the empty path at A must be the same as $g \circ f$ and the empty path at B must be the same as $f \circ g$.

4.1.8 Graph homomorphisms by commutative diagrams The definition of graph homomorphism in 1.4.1 can be expressed by a commutative diagram. Let $\phi = (\phi_0, \phi_1)$ be a graph homomorphism from $\mathcal{G}$ to $\mathcal{H}$. For any arrow $u : m \to n$ in $\mathcal{G}$, 1.4.1 requires that $\phi_1(u) : \phi_0(m) \to \phi_0(n)$ in $\mathcal{H}$. This says that $\phi_0(\text{source}(u)) = \text{source}(\phi_1(u))$, and a similar statement about targets. In other words, these diagrams must commute:

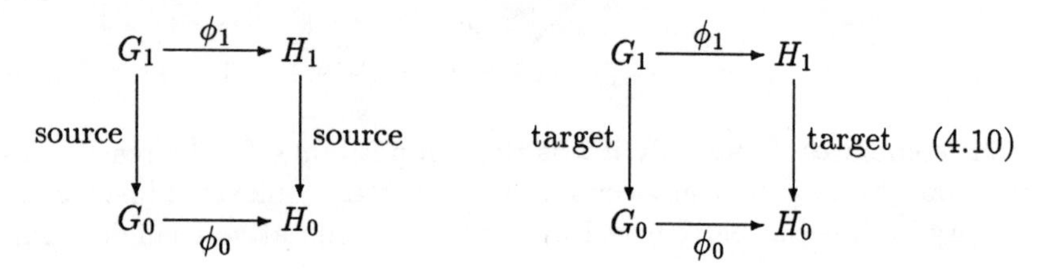

In these two diagrams the two arrows labeled 'source' are of course different functions; one is the source function for $\mathcal{G}$ and the other for $\mathcal{H}$. A similar remark is true of 'target'.

4.1.9 This point of view provides a pictorial proof that the composite of two graph homomorphisms is a graph homomorphism (see 2.4.1). If $\phi : \mathcal{G} \to \mathcal{H}$ and $\psi : \mathcal{H} \to \mathcal{K}$ are graph homomorphisms, then to see that $\psi \circ \phi$ is a graph homomorphism requires checking that the outside rectangle below commutes, and similarly with target in place of source:

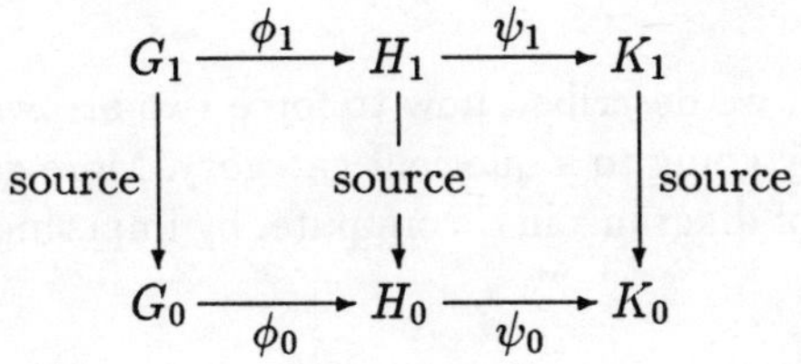

$$\begin{array}{ccccc} G_1 & \xrightarrow{\phi_1} & H_1 & \xrightarrow{\psi_1} & K_1 \\ {\scriptstyle \text{source}}\downarrow & & \downarrow{\scriptstyle \text{source}} & & \downarrow{\scriptstyle \text{source}} \\ G_0 & \xrightarrow[\phi_0]{} & H_0 & \xrightarrow[\psi_0]{} & K_0 \end{array} \qquad (4.11)$$

4.1.10 Associativity by commutative diagrams The fact that the multiplication in a monoid or semigroup is associative can be expressed as the assertion that a certain diagram in **Set** commutes.

Let S be a semigroup. Define the following functions, using the cartesian product notation for functions of 1.2.9:

(i) $\text{mult} : S \times S \to S$ satisfies $\text{mult}(x, y) = xy$.

(ii) $S \times \text{mult} : S \times S \times S \to S \times S$ satisfies

$$(S \times \text{mult})(x, y, z) = (x, yz)$$

(iii) $\text{mult} \times S : S \times S \times S \to S \times S$ satisfies

$$(\text{mult} \times S)(x, y, z) = (xy, z)$$

That the following diagram commutes is exactly the associative law.

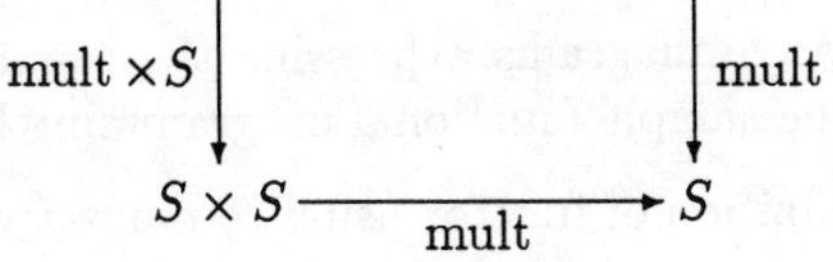

$$\begin{array}{ccc} S \times S \times S & \xrightarrow{S \times \text{mult}} & S \times S \\ {\scriptstyle \text{mult} \times S}\downarrow & & \downarrow{\scriptstyle \text{mult}} \\ S \times S & \xrightarrow[\text{mult}]{} & S \end{array} \qquad (4.12)$$

4.1.11 Normally, associativity is expressed by the equation $x(yz) = (xy)z$ for all x, y, z in the semigroup. The commutative diagram expresses this same fact *without the use of variables.* Of course, we did use variables in defining the functions involved, but we remedy that deficiency in Chapter 5 when we give a categorical definition of products.

Another advantage of using diagrams to express equations is that diagrams show the source and target of the functions involved. This is not particularly compelling here but in other situations the two-dimensional picture of the compositions involved makes it much easier to follow the discussion.

4.1.12 In 3.5.7, we described how to force two arrows in a category $\mathcal{C}$ to be the same by going to a quotient category. More generally, you can make any set $\mathcal{D}$ of diagrams in $\mathcal{C}$ commute, by imposing all the relations of the form

$$Ds_n \circ Ds_{n-1} \circ \ldots \circ Ds_1 \sim Dt_m \circ Dt_{m-1} \circ \ldots \circ Dt_1$$

where

$$\begin{array}{ccccc}
 & Dk_1 \xrightarrow{Ds_2} Dk_2 \to \cdots \to Dk_{n-2} \xrightarrow{Ds_{n-1}} Dk_{n-1} & \\
Ds_1 \nearrow & & \searrow Ds_n \\
Di & & Dj \\
Dt_1 \searrow & & \nearrow Dt_m \\
 & Dl_1 \xrightarrow[Dt_2]{} Dl_2 \to \cdots \to Dl_{m-2} \xrightarrow[Dt_{m-1}]{} Dl_{m-1} &
\end{array} \qquad (4.13)$$

are two paths in any diagram $D \in \mathcal{D}$. As before, if one of these paths is the empty path the other must be an identity arrow in order for the diagram to commute.

4.1.13 Exercises

1. Draw a commutative diagram expressing the fact that an arrow $f : A \to B$ factors through an arrow $g : C \to B$. (See 2.8.6.)

2. Draw a commutative diagram to express the fact that addition of real numbers is commutative.

3. Draw commutative diagrams expressing the equations occurring in the definition of the sample functional programming language in 2.2.5.

4. Express the definition of functor using commutative diagrams.

4.2 Natural transformations

4.2.1 Unary operations In Section 4.1 we saw that diagrams in a category are graph homomorphisms to the category from a different point of view. Now we introduce a third way to look at graph homomorphisms to a category, namely as models. To give an example, we need a definition.

4.2.2 Definition A **unary operation** on a set S is a function $u : S \to S$.

This definition is by analogy with the concept of binary operation on a set. A set with a unary operation is a (very simple) algebraic structure, which we call a **u-structure** (there is no standard name as far as we know).

4.2.3 A homomorphism of u-structures should be a function which preserves the structure. There is really only one definition that is reasonable for this idea: if (S, u) and (T, v) are u-structures, $f : S \to T$ is a **homomorphism of u-structures** if $f(u(s)) = v(f(s))$ for all $s \in S$. Thus this diagram must commute:

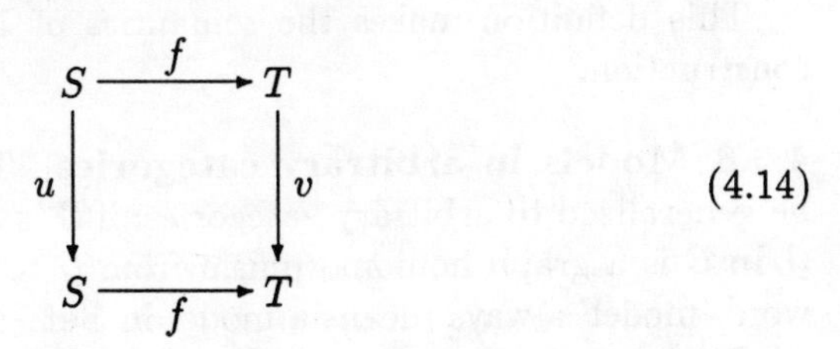

(4.14)

4.2.4 Models of graphs We now use the concept of u-structure to motivate the third way of looking at graph homomorphisms to a category.

4.2.5 Let $\mathcal{U}$ be the graph with one node u_0 and one arrow e. Then a u-structure is essentially the same as a diagram in **Set** of shape U: to be precise, if $D : \mathcal{U} \to$ **Set** is a diagram, then $D(u_0)$ is a set and $D(e)$ is a unary operation on $D(u_0)$, so that $(D(u_0), D(e))$ is a u-structure. Conversely, if (S, u) is a u-structure, it corresponds to the diagram D defined by $D(u_0) = S$ and $D(e) = u$. This suggests the following definition.

4.2.6 Definition A **model** M of a graph $\mathcal{G}$ is a graph homomorphism $M : \mathcal{G} \to$ **Set**.

We will see how to define a monoid as a model involving a graph homomorphism (and other ingredients) in Chapter 7. We had to introduce u-structures here to have an example for which we had the requisite techniques. The technique we are missing is the concept of product in a category, which allows the definition of operations of arity greater than one.

4.2.7 Example As another example, consider this graph (see 1.3.6):

$$g_1 \underset{\text{target}}{\overset{\text{source}}{\rightrightarrows}} g_0 \qquad (4.15)$$

A model M of this graph consists of sets $G_0 = M(g_0)$ and $G_1 = M(g_1)$ together with functions source $= M(\text{source}) : G_1 \to G_0$ and target $= M(\text{target}) : G_1 \to G_0$. To understand what this structure is, imagine a picture in which there is a dot corresponding to each element of G_0 and an arrow corresponding to each element $a \in G_1$ which goes from the dot corresponding to source(a) to the one corresponding to target(a). It should be clear that the picture so described is a graph and thus the graph (4.15) is a graph whose models are graphs!

This definition makes the semantics of 1.3.6 into a mathematical construction.

4.2.8 Models in arbitrary categories The concept of model can be generalized to arbitrary categories: if $\mathcal{C}$ is any category, a **model of** $\mathcal{G}$ **in** $\mathcal{C}$ is a graph homomorphism from $\mathcal{G}$ to $\mathcal{C}$. In this book, the bare word 'model' always means a model in **Set**.

For example, a model of the graph for u-structures in the category of posets and monotone maps is a poset and a monotone map from the poset to itself.

4.2.9 Natural transformations between models of a graph In a category, there is a natural notion of an arrow from one model of a graph to another. This usually turns out to coincide with the standard definition of homomorphism for that kind of structure.

4.2.10 Definition Let $D, E : \mathcal{G} \to \mathcal{C}$ be two models of the same graph in a category. A **natural transformation** $\alpha : D \to E$ is given by a family of arrows αa of $\mathcal{C}$ indexed by the nodes of $\mathcal{G}$ such that:

NT–1 $\alpha a : Da \to Ea$ for each node a of $\mathcal{G}$.

NT–2 For any arrow $s : a \to b$ in $\mathcal{G}$, the diagram

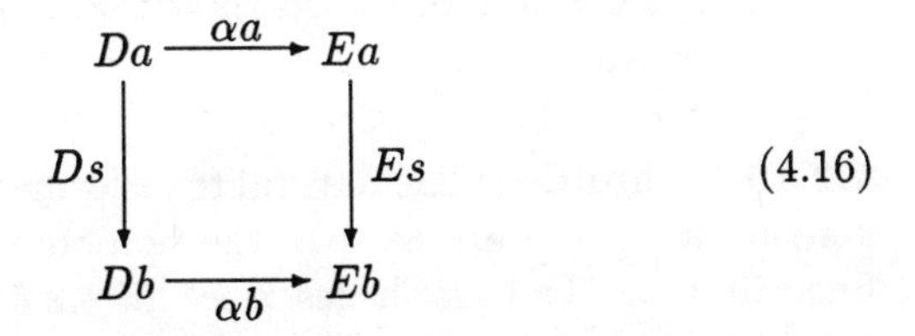

(4.16)

commutes.

The commutativity of the diagram in NT–2 is referred to as the **naturality condition** on α. The arrow αa for an object a is the **component** of the natural transformation α at a.

Note that you talk about a natural transformation from D to E only if they have the same domain as well as the same codomain and if, moreover, the codomain is a category.

4.2.11 Definition Let D, E and F be models of $\mathcal{G}$ in $\mathcal{C}$, and $\alpha : D \to E$ and $\beta : E \to F$ natural transformations. The **composite** $\beta \circ \alpha : D \to F$ is defined componentwise: $(\beta \circ \alpha)a = \beta a \circ \alpha a$.

4.2.12 Proposition *The composite of two natural transformations is also a natural transformation.*

Proof. The diagram that has to be shown commutative is the outer rectangle of

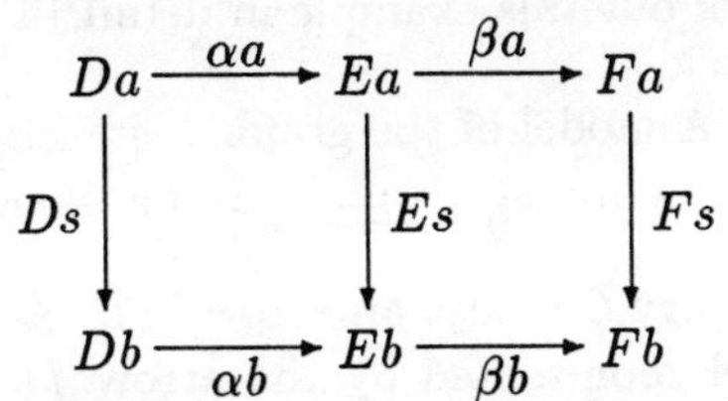

for each arrow $s : a \to b$ in $\mathcal{G}$. The rectangle commutes because the two squares do; the squares commute as a consequence of the naturality of α and β. □

It is interesting that categorists began using modes of reasoning like that in the preceding proof because objects of categories generally lacked elements; now one appreciates them for their own sake *because* they allow element-free (and thus variable-free) arguments.

4.2.13 It is even easier to show that there is an identity natural transformation between any model D and itself, defined by $(\mathrm{id}_D)a = \mathrm{id}_{Da}$. We then have the following proposition, whose proof is straightforward.

4.2.14 Proposition *The models of a given graph $\mathcal{G}$ in a given category $\mathcal{C}$, and the natural transformations between them, form a category, denoted* $\mathbf{Mod}(\mathcal{G},\mathcal{C})$.

4.2.15 Example The natural transformations between models of the u-structure graph are exactly the homomorphisms of u-structures defined in 4.2.3. The graph described in 4.2.5 has one object u_0 and one arrow e, so that a natural transformation from a model D to a model E has only one component which is a function from $D(u_0)$ to $E(u_0)$. If we set $S = D(u_0)$, $u = D(e)$, $T = E(u_0)$, $v = E(e)$, and we define $\alpha u_0 = f$, this is the single component of a natural transformation from D to E. Condition NT–2 in 4.2.10 coincides in this case with the diagram in 4.2.3: the naturality condition is the same as the definition of homomorphism of u-structures.

4.2.16 Example A homomorphism of graphs is a natural transformation between models of the graph

$$g_1 \underset{\text{target}}{\overset{\text{source}}{\rightrightarrows}} g_0$$

The two graphs in (4.10) are the two necessary instances (one for the source and the other for the target) of (4.16). In a similar way, the proof that the composite of two natural transformations is a natural transformation reduces in this case to the commutativity of Diagram (4.11). It is instructive to work out this example in detail.

4.2.17 Example A model of the graph

$$0 \longrightarrow 1$$

in an arbitrary category $\mathcal{C}$ is just an arrow in $\mathcal{C}$. A natural transformation from the model represented by the arrow $f : A \to B$ to the one represented by $g : C \to D$ is a pair of arrows $h : A \to C$ and $k : B \to D$ making a commutative diagram:

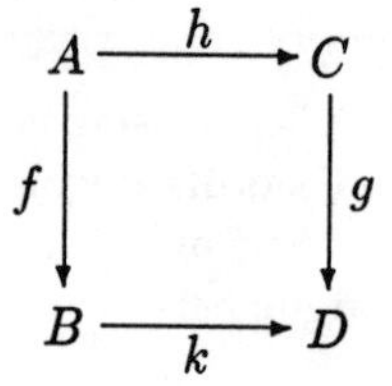

$$\begin{array}{ccc} A & \xrightarrow{h} & C \\ {\scriptstyle f}\big\downarrow & & \big\downarrow{\scriptstyle g} \\ B & \xrightarrow[k]{} & D \end{array} \tag{4.17}$$

The category of models is called the **arrow category** of $\mathcal{C}$; it is often denoted $\mathcal{C}^{\to}$.

4.2.18 Natural isomorphisms A natural transformation is called a **natural isomorphism** if every component of it is an isomorphism. To be precise, if $F, G : \mathcal{G} \to \mathcal{D}$ are models of a graph $\mathcal{G}$ and $\alpha : F \to G$ is a natural transformation, then α is a natural isomorphism if for each node (or object) a of $\mathcal{G}$, αa is an isomorphism of $\mathcal{D}$. Natural isomorphisms are often called **natural equivalences.**

4.2.19 Example The arrow $(h, k) : f \to g$ in the arrow category of a category $\mathcal{C}$, as shown in 4.17, is an isomorphism if and only if h and k are both isomorphisms in $\mathcal{C}$. This is a special case of the most important fact about natural isomorphisms, which we now state.

4.2.20 Theorem *Suppose $F : \mathcal{G} \to \mathcal{D}$ and $G : \mathcal{G} \to \mathcal{D}$ are models of $\mathcal{C}$ and $\alpha : F \to G$ is a natural isomorphism of models. Then there is a unique natural transformation $\beta : G \to F$ such that the composites $\alpha \circ \beta$ and $\beta \circ \alpha$ are the identity natural transformations of G and F respectively. Hence α is an isomorphism of the category* **Mod**$(\mathcal{G}, \mathcal{C})$.

Proof. The component of β at a node a is defined by letting $\beta a = (\alpha a)^{-1}$. This is the only possible definition, but it must be shown to be natural. Let $f : a \to b$ be an arrow of the domain of F and G. Then we have

$$\begin{aligned} Ff \circ (\alpha a)^{-1} &= (\alpha b)^{-1} \circ (\alpha b) \circ Ff \circ (\alpha a)^{-1} \\ &= (\alpha b)^{-1} \circ Gf \circ (\alpha a) \circ (\alpha a)^{-1} \\ &= (\alpha b)^{-1} \circ Gf \end{aligned}$$

which says that β is natural. The second equality uses the naturality of α. □

4.2.21 Exercises

1. Let $\mathcal{C}$ be a category with object B. Exhibit the slice category $\mathcal{C}/B$ as a subcategory of the arrow category of $\mathcal{C}$ defined in 4.2.17 (see 2.6.9). Is it full?

2. Let $\mathcal{G}$ be the graph with two nodes and no arrows, and $\mathcal{C}$ any category. Show that **Mod**$(\mathcal{G}, \mathcal{C})$ is isomorphic to $\mathcal{C} \times \mathcal{C}$.

3. Prove that in diagram (4.17), the arrow (h, k) is an isomorphism in the arrow category of $\mathcal{C}$ if and only if h and k are both isomorphisms in $\mathcal{C}$.

4. Let $\mathcal{C}$ and $\mathcal{D}$ be categories and $F : \mathcal{C} \to \mathcal{D}$ an equivalence. Show that every arrow of $\mathcal{D}$ is isomorphic in the arrow category of $\mathcal{D}$ to an arrow in the image of F. (In this sense, an equivalence of categories is 'surjective up to isomorphism' on both objects and arrows. See 3.4.2.)

5. Suppose $D, E : \mathcal{G} \to \mathcal{C}$ are two models of a graph in a category and $\alpha : D \to E$ is a natural equivalence. Suppose we let, for a a node of $\mathcal{G}$, $\beta a = (\alpha a)^{-1}$. Show that the collection of βa forms a natural transformation (a natural equivalence in fact) from E to D.

4.3 Natural transformations between functors

A functor is among other things a graph homomorphism, so a natural transformation between two functors is a natural transformation of the corresponding graph homomorphisms. The following proposition is an immediate consequence of 4.2.12.

4.3.1 Proposition *If $\mathcal{C}$ and $\mathcal{D}$ are categories, the functors from $\mathcal{C}$ to $\mathcal{D}$ form a category with natural transformations as arrows.*

We denote this category by $\mathbf{Fun}(\mathcal{C}, \mathcal{D})$. Other common notations for it are $\mathcal{D}^{\mathcal{C}}$ and $[\mathcal{C}, \mathcal{D}]$.

Of course, the *graph* homomorphisms from $\mathcal{C}$ to $\mathcal{D}$, which do not necessarily preserve the composition of arrows in $\mathcal{C}$, also form a category $\mathbf{Mod}(\mathcal{C}, \mathcal{D})$ (see 4.2.14), of which $\mathbf{Fun}(\mathcal{C}, \mathcal{D})$ is a full subcategory.

4.3.2 Natural isomorphisms A natural transformation from one functor to another is a special case of a natural transformation from one graph homomorphism to another, so Theorem 4.2.20 is true of natural transformations of functors. We restate it here because of its importance.

4.3.3 Theorem *Suppose $F : \mathcal{C} \to \mathcal{D}$ and $G : \mathcal{C} \to \mathcal{D}$ are functors and $\alpha : F \to G$ is a natural isomorphism. Then there is a unique natural transformation $\beta : G \to F$ such that the composites $\alpha \circ \beta$ and $\beta \circ \alpha$ are the identity natural transformations of G and F respectively. Hence α is an isomorphism of the category* $\mathbf{Fun}(\mathcal{C}, \mathcal{D})$.

If $\mathcal{C}$ is not a small category (see 2.1.5), then $\mathbf{Fun}(\mathcal{C}, \mathcal{D})$ may not be locally small (see 2.1.7). This is a rather esoteric question that will not concern us in this book since we will have no occasion to form functor categories of that sort.

We motivated the concept of natural transformation by considering models of graphs, and most of the discussion in the rest of this section concerns that point of view. Historically, the concept first arose for functors and not from the point of view of models.

4.3.4 Examples We have already described some examples of natural transformations, as summed up in the following propositions.

In 3.1.14, we defined the graph homomorphism $\eta\mathcal{G} : \mathcal{G} \to U(F(\mathcal{G}))$ which includes a graph $\mathcal{G}$ into $U(F(\mathcal{G}))$, the underlying graph of the free category $F(\mathcal{G})$.

4.3.5 Proposition *The family of arrows $\eta\mathcal{G}$ form a natural transformation from the identity functor on* **Grf** *to $U \circ F$, where U is the underlying graph functor from* **Cat** *to* **Grf**.

In 3.4.2, we defined the concept of equivalence of categories.

4.3.6 Proposition *A functor $F : \mathcal{C} \to \mathcal{D}$ is an equivalence of categories with pseudo-inverse $G : \mathcal{D} \to \mathcal{C}$ if and only if $G \circ F$ is naturally isomorphic to $\mathrm{id}_{\mathcal{C}}$ and $F \circ G$ is naturally isomorphic to $\mathrm{id}_{\mathcal{D}}$.*

We leave the proofs as exercises.

4.3.7 Example Let $\alpha : M \times S \to S$ and $\beta : M \times T \to T$ be two actions by a monoid M (see 3.2.1). Let $\phi : S \to T$ be an equivariant map. If F and G are the functors corresponding to α and β, as defined in 3.2.3, then ϕ is the (only) component of a natural transformation from F to G. Conversely, the only component of any natural transformation from F to G is an equivariant map between the corresponding actions.

4.3.8 Natural transformations of graphs We now consider some natural transformations involving the category **Grf** of graphs and homomorphisms of graphs.

4.3.9 Example In 3.1.9, we defined the functor $N : \mathbf{Grf} \to \mathbf{Set}$. It takes a graph $\mathcal{G}$ to its set G_0 of nodes and a homomorphism ϕ to ϕ_0. Now pick a graph with one node $*$ and no arrows and call it $\mathcal{E}$. Let $V = \mathrm{Hom}_{\mathbf{Grf}}(\mathcal{E}, -)$.

A graph homomorphism from the graph $\mathcal{E}$ to an arbitrary graph $\mathcal{G}$ is evidently determined by the image of $\mathcal{E}$ and that can be any node of $\mathcal{G}$. In other words, nodes of $\mathcal{G}$ are 'essentially the same thing' as graph homomorphisms from $\mathcal{E}$ to $\mathcal{G}$, that is, as the elements of the set $V(\mathcal{G})$.

We can define a natural transformation $\alpha : V \to N$ by defining

$$\alpha\mathcal{G}(f) = f_0(*)$$

where $\mathcal{G}$ is a graph and $f : \mathcal{E} \to \mathcal{G}$ is a graph homomorphism (arrow of **Grf**). There must be a naturality diagram (4.16) for each arrow of the source category, which in this case is **Grf**. Thus to see that α is natural, we require that for each graph homomorphism $g : \mathcal{G}_1 \to \mathcal{G}_2$, the diagram

$$\begin{array}{ccc} V\mathcal{G}_1 & \xrightarrow{Vg} & V\mathcal{G}_2 \\ {\scriptstyle \alpha\mathcal{G}_1}\downarrow & & \downarrow{\scriptstyle \alpha\mathcal{G}_2} \\ N\mathcal{G}_1 & \xrightarrow[Ng]{} & N\mathcal{G}_2 \end{array}$$

commutes. Now Ng is g_0 (the node map of g) by definition, and the value of V (which is a hom functor) at a homomorphism g composes g with a graph homomorphism from the graph $\mathcal{E}$. Then we have, for a homomorphism $f : \mathcal{E} \to V\mathcal{G}_1$ (i.e., an element of the upper left corner of the diagram),

$$(\alpha\mathcal{G}_2 \circ Vg)(f) = \alpha\mathcal{G}_2(g \circ f) = (g \circ f)_0(*)$$

while

$$(Ng \circ \alpha\mathcal{G}_1)(f) = Ng(f_0(*)) = g_0(f_0(*))$$

and these are equal from the definition of composition of graph homomorphisms.

The natural transformation α is in fact a natural isomorphism (Exercise 9). This shows that N is naturally isomorphic to a hom functor. Such functors are called 'representable', and are considered in greater detail in 4.5.1.

4.3.10 Connected components A node a can be **connected** to the node b of a graph $\mathcal{G}$ if it is possible to get from a to b following a sequence of arrows of $\mathcal{G}$ in either direction. Precisely, a is connected to b if there is a sequence $(c_0, c_1, \ldots, c_n)$ of arrows of $\mathcal{G}$ with the property that a is a node (either the source or the target) of c_0, b is a node of c_n, and for $i = 1, \ldots, n$, c_{i-1} and c_i have a node in common. We call such a sequence an **undirected path** between a and b.

It is a good exercise to see that 'being connected to' is an equivalence relation. (For reflexivity: a node is connected to itself by the empty sequence.) An equivalence class of nodes with respect to this relation is called a **connected component** of the graph $\mathcal{G}$, and the set of connected components is called $W\mathcal{G}$.

Connected components can be defined for categories in the same way as for graphs. In that case, each connected component is a full subcategory.

If $f : \mathcal{G} \to \mathcal{H}$ is a graph homomorphism and if two nodes a and b are in the same component of $\mathcal{G}$, then $f(a)$ and $f(b)$ are in the same component of $\mathcal{H}$; this is because f takes an undirected path between a and b to an undirected path between $f(a)$ and $f(b)$. Thus the arrow f

induces a function $Wf : W\mathcal{G} \to W\mathcal{H}$, namely the one which takes the component of a to the component of $f(a)$; and this makes W a functor from **Grf** to **Set**.

For a graph $\mathcal{G}$, let $\beta\mathcal{G} : N\mathcal{G} \to W\mathcal{G}$ be the set function which takes a node of $\mathcal{G}$ to the component of $\mathcal{G}$ that contains that node. (The component is the *value* of $\beta\mathcal{G}$ at the node, not the codomain.) Then $\beta : N \to W$ is a natural transformation. It is instructive to check the commutativity of the requisite diagram.

4.3.11 Example In 3.5.9, we described a functor Σ which provided a meaning in **Set** for each program in the programming language L of 2.2.5. A person more oriented to machine language might have preferred to give the meaning of all the data in terms of numbers, in particular the integers between 0 and 2^K for some fixed number $K \geq 7$ (the constraint is to accommodate the ASCII codes).

Thus one could define a functor Σ' for which

(i) $\Sigma'(\mathbf{NAT})$ is the set of integers between 0 and $2^K - 1$. Then the constant 0 would be the number 0, but $\Sigma'(\mathbf{succ})$ would have to calculate the successor modulo 2^K.
(ii) $\Sigma'(\mathbf{BOOLEAN})$ is the set $\{0, 1\}$, with $\mathbf{true} = 1$ and $\mathbf{false} = 0$.
(iii) For each character c, $\Sigma'(c)$ is the ASCII code for c. Then we would have to take $\Sigma'(\mathbf{CHAR})$ to be the set $A = \{n \in \mathbf{N} \mid 0 \leq n \leq 127\}$.

For each of the three types T in our language, we have a way of rewriting each datum in $\Sigma(T)$ to become the corresponding datum in $\Sigma'(T)$. This rewriting becomes a function $\beta_T : \Sigma(T) \to \Sigma'(T)$:

(i) $\beta_{\mathbf{NAT}}(n)$ is n modulo 2^K.
(ii) $\beta_{\mathbf{BOOLEAN}}(\mathbf{true}) = 1$ and $\beta_{\mathbf{BOOLEAN}}(\mathbf{false}) = 0$.
(iii) $\beta_{\mathbf{CHAR}}(c)$ is the ASCII code of c for each character c.

In order to preserve the intended meaning, $\Sigma'(\mathbf{succ})$ would have to be the successor function modulo 2^K, $\Sigma'(\mathbf{ord})$ would have to be the inclusion of A into $\Sigma'(\mathbf{NAT})$ and $\Sigma'(\mathbf{chr})$ would have to be the function from $\Sigma'(\mathbf{NAT})$ to A which takes the remainder modulo 128.

Preserving the meaning of **ord** (an informal idea) means formally that this diagram must commute, as it does with the definitions given of $\Sigma(\mathbf{ord})$ and $\Sigma'(\mathbf{ord})$:

$$\begin{array}{ccc}
\Sigma(\mathbf{CHAR}) & \xrightarrow{\beta_{\mathbf{CHAR}}} & \Sigma'(\mathbf{CHAR}) \\
{\scriptstyle \Sigma(\mathbf{ord})}\downarrow & & \downarrow{\scriptstyle \Sigma'(\mathbf{ord})} \\
\Sigma(\mathbf{N}) & \xrightarrow[\beta_{\mathbf{N}}]{} & \Sigma'(\mathbf{N})
\end{array}$$

Similar remarks apply to the preservation of the other operations. This is a special case of a general principle that, given two functors G and G' which are semantics in some sense, a natural transformation $\beta : G \rightarrow G'$ can be said to preserve the meaning.

The natural transformation β was constructed for the given data types. The only constructor in L, namely composition, does not destroy the natural transformation property: if the given β gives the naturality property for primitive operations, it does so for all their composites, as well. This is an instance of the following proposition.

4.3.12 Proposition *Let $F, G : \mathcal{C} \rightarrow \mathcal{D}$ be functors. Let C be a (possibly empty) set of arrows of $\mathcal{C}$ with the property that every arrow of $\mathcal{C}$ is a composite of arrows of C and let $\beta C : F(C) \rightarrow G(C)$ be an arrow of $\mathcal{D}$ for each object C of $\mathcal{C}$. Suppose for every arrow $f : A \rightarrow B$ of C this diagram commutes:*

$$\begin{array}{ccc}
F(A) & \xrightarrow{F(f)} & F(B) \\
{\scriptstyle \beta A}\downarrow & & \downarrow{\scriptstyle \beta B} \\
G(A) & \xrightarrow[G(f)]{} & G(B)
\end{array} \tag{4.18}$$

Then β is a natural transformation.

Proof: if $f : A \rightarrow B$ and $g : B \rightarrow C$ are arrows of C, then the outer rectangle below commutes because the two squares do:

$$\begin{array}{ccccc}
F(A) & \xrightarrow{F(f)} & F(B) & \xrightarrow{F(g)} & F(C) \\
{\scriptstyle \beta A}\downarrow & & \downarrow{\scriptstyle \beta B} & & \downarrow{\scriptstyle \beta C} \\
G(A) & \xrightarrow[G(f)]{} & G(B) & \xrightarrow[G(g)]{} & G(C)
\end{array}$$

The naturality diagram for the case $f = \mathrm{id}$ (that is, the empty composite) is automatic. The proof follows from these facts by induction. □

4.3.13 Exercises

1. Prove Proposition 4.3.5.

2. Prove Proposition 4.3.6.

3. Show that the family $\beta\mathcal{G}$ of arrows taking a node to its component defined in 4.3.10 is indeed a natural transformation.

4. Let $\mathcal{C}$ be a category. A **subfunctor** of a functor $F : \mathcal{C} \to \mathbf{Set}$ is a functor $G : \mathcal{C} \to \mathbf{Set}$ with the property that for each object C of $\mathcal{C}$, $G(C) \subseteq F(C)$ and such that for each arrow $f : C \to C'$ and each element $x \in GC$, we have that $Gf(x) = Ff(x)$. Show that the inclusion function $i_C : G(C) \to F(C)$ is a natural transformation.

5. Show that the map which takes an arrow of a graph to its source is a natural transformation from A to N. (See 3.1.9.) Do the same for targets. (Actually, every operation in any multisorted algebraic structure gives a natural transformation. See [Linton, 1969a, 1969b].)

6.† Show that if $\mathcal{C}$ is a discrete category with set of objects $\mathcal{C}_0$ (hence essentially a set), then $\mathbf{Fun}(\mathcal{C}, \mathbf{Set})$ is equivalent to the slice category $\mathbf{Set}/\mathcal{C}_0$. (See 2.6.9.)

7. For each set S, let $\{\}S : S \to \mathcal{P}S$ be the function which takes an element x of S to the singleton subset $\{x\}$.

a. Show that $\{\}$ is a natural transformation from the identity functor on **Set** to the direct image powerset functor $\mathcal{P}$. (See 3.1.16.)

b. Show that $\{\}$ is not a natural transformation from the identity functor on **Set** to the universal image powerset functor. (See 3.1.19.)

c. Explain why it does not even make sense to ask whether it is a natural transformation to the inverse image powerset functor.

8. Verify the claims in 4.3.7.

9. Show that the natural transformation of 4.3.9 is a natural isomorphism.

10. a. Show that for any integer k, the set of paths of length k of a graph is the object part of a functor $P_k : \mathbf{Grf} \to \mathbf{Set}$. (See 2.1.2.)

b. Show that P_0 is naturally isomorphic to the node functor N defined in 3.1.9.

c. Show that P_1 is naturally isomorphic to the arrow functor A of 3.1.9.

4.4 The Godement calculus of natural transformations

We collect here, mostly without proof, some of the basic combinatorial properties of functors and natural transformations. These rules were first codified by Godement [1958]. They are not used in the rest of the book, but verifying (some of) them is an excellent way to familiarize yourself with natural transformations.

4.4.1 Let $F : \mathcal{A} \to \mathcal{B}$ and $G : \mathcal{B} \to \mathcal{C}$ be functors. There is a composite functor $G \circ F : \mathcal{A} \to \mathcal{C}$ defined in the usual way by $G \circ F(A) = G(F(A))$. Similarly, let H, K and L be functors from $\mathcal{A} \to \mathcal{B}$ and $\alpha : H \to K$ and $\beta : K \to L$ be natural transformations. Recall that this means that for each object A of $\mathcal{A}$, $\alpha A : HA \to KA$ and $\beta A : KA \to LA$. Then as in 4.2.11, we define $\beta \circ \alpha : H \to L$ by

$$(\beta \circ \alpha)A = \beta A \circ \alpha A$$

Things get more interesting when we mix functors and natural transformations. For example, suppose we have three categories $\mathcal{A}$, $\mathcal{B}$ and $\mathcal{C}$, four functors, two of them, $F, G : \mathcal{A} \to \mathcal{B}$ and the other two $H, K : \mathcal{B} \to \mathcal{C}$, and two natural transformations $\alpha : F \to G$ and $\beta : H \to K$. We picture it as follows:

$$\mathcal{A} \;\overset{F}{\underset{G}{\rightrightarrows}}\; \Downarrow\alpha \;\; \mathcal{B} \;\overset{H}{\underset{K}{\rightrightarrows}}\; \Downarrow\beta \;\; \mathcal{C} \qquad (4.19)$$

4.4.2 Definition The natural transformation $\beta F : H \circ F \to K \circ F$ is defined by the formula $(\beta F)A = \beta(FA)$ for an object A of $\mathcal{A}$.

This means the component of the natural transformation β at the object FA. This is indeed an arrow from $H(F(A)) \to K(F(A))$ as required. Of course, it must also be verified to be natural, but we leave this to an exercise.

4.4.3 Definition The natural transformation $H\alpha : H \circ F \to H \circ G$ is defined by letting $(H\alpha)A = H(\alpha A)$ for an object A of $\mathcal{A}$, that is the value of H applied to the arrow αA.

Note that the definitions of βF and $H\alpha$ are quite different. The first is the natural transformation whose value at an object A of $\mathcal{A}$ is the component of β on the object FA while the value of the second is the result of applying the functor H to the component αA (which is an arrow of $\mathcal{B}$). Nevertheless, we use similar notations. The reason for this is that their formal properties are indistinguishable. In fact, even categorists quite commonly (though not universally) distinguish them by writing β_F but $H\alpha$. That notation emphasizes the fact that they are semantically different. The notation used here is chosen to emphasize the fact that they are syntactically indistinguishable. More precisely, the left/right mirror image of each of Godement's rules given below is again a Godement rule.

In a great deal of mathematical reasoning, one forgets the semantics of the situation except at the beginning and the end of the process, relying on the syntactic rules in the intermediate stages. This is especially true in the kind of 'diagram chasing' arguments so common in category theory. For that reason, the notation we have adopted emphasizes the syntactic similarity of the two constructions, rather than the semantic difference.

In Exercise 4, we give another, more sophisticated definition of βF and $H\alpha$ which shows that they can be thought of as *semantically* parallel, as well.

In the rest of the book, we will use the notation βF and $H\alpha$, but not the other material in this section.

4.4.4 There is a second way of composing natural transformations. The naturality of β in Diagram (4.19) implies that for any object A of $\mathcal{A}$, the diagram

$$\begin{array}{ccc} (H\circ F)A & \xrightarrow{(H\alpha)A} & (H\circ G)A \\ {\scriptstyle(\beta F)A}\downarrow & & \downarrow{\scriptstyle(\beta G)A} \\ (K\circ F)A & \xrightarrow[(K\alpha)A]{} & (K\circ G)A \end{array} \tag{4.20}$$

commutes. The two equal composites are defined to be $\beta * \alpha : H \circ F \to K \circ G$. It is left as an exercise to show that $\beta * \alpha$ is again a natural transformation and that this composition is associative.

We usually call $\beta \circ \alpha$ the **vertical composite** and $\beta * \alpha$ the **horizontal composite**. (Warning: some authors use $\circ$ for the horizontal composite.) One must keep careful track of the difference between them. Fortunately, the notations do not often clash, since usually only one makes sense.

There is one case in which they can clash. If $\mathcal{A} = \mathcal{B} = \mathcal{C}$ and $G = H$, then $\beta \circ \alpha : F \to K$, while $\beta * \alpha : G \circ F \to K \circ G$. This clash is exacerbated by the habit among many categorists of omitting the composition circle and $*$, except for emphasis. We will often omit the $*$, but not the circle. On the other hand, no confusion can possibly arise from the overloading of the circle notation to include composition of arrows, functors and natural transformations since their domains uniquely define what kind of composition is involved.

4.4.5 Godement's five rules There are thus several kinds of composites. There is a composite of functors, vertical and horizontal composite of natural transformations and the composite of a functor and a natural transformation in either order (although the latter is in fact the horizontal composite of a natural transformation and the identity natural transformation of a functor, a fact we leave to an exercise). The possibilities are sufficiently numerous that it is worth the effort to codify the rules.

Let $\mathcal{A}$, $\mathcal{B}$, $\mathcal{C}$, $\mathcal{D}$ and $\mathcal{E}$ be categories; $E : \mathcal{A} \to \mathcal{B}$, F_1, F_2 and $F_3 : \mathcal{B} \to \mathcal{C}$, G_1, G_2 and $G_3 : \mathcal{C} \to \mathcal{D}$, and $H : \mathcal{D} \to \mathcal{E}$ be functors; and $\alpha : F_1 \to F_2$, $\beta : F_2 \to F_3$, $\gamma : G_1 \to G_2$, and $\delta : G_2 \to G_3$ be natural transformations. This situation is summarized by the following diagram:

$$\mathcal{A} \xrightarrow{E} \mathcal{B} \begin{array}{c} \xrightarrow{F_1} \\ F_2 \ \Downarrow \alpha \\ \longrightarrow \\ F_3 \ \Downarrow \beta \\ \longrightarrow \end{array} \mathcal{C} \begin{array}{c} \xrightarrow{G_1} \\ G_2 \ \Downarrow \gamma \\ \longrightarrow \\ G_3 \ \Downarrow \delta \\ \longrightarrow \end{array} \mathcal{D} \xrightarrow{H} \mathcal{E}$$

Then

G–1 $(\delta \circ \gamma)(\beta \circ \alpha) = (\delta\beta) \circ (\gamma\alpha)$.

G–2 $(H \circ G_1)\alpha = H(G_1\alpha)$.

G–3 $\gamma(F_1 \circ E) = (\gamma F_1)E$.

G–4 $G_1(\beta \circ \alpha)E = (G_1\beta E) \circ (G_1\alpha E)$.

G–5 $\gamma\alpha = (\gamma F_2) \circ (G_1\alpha) = (G_2\alpha) \circ (\gamma F_1)$.

The expression $G_1(\beta \circ \alpha)E$ in G–4 is not ambiguous because of Exercise 3. G–1 is called the **Interchange Law**. It is the basis for the definition of 2-category, a concept which has some application to computer science. The most accessible introduction is that of [Kelly and Street, 1974]; much more is in [Gray, 1974] and [Kelly, 1982b]. The ultimate generalization to n-category and even further is in [Street, 1987]. Applications to computer science are discussed in [Seely, 1986], [Power, 1990] and [Power and Wells, 1990].

4.4.6 Exercises

1. Show that βF and $H\alpha$ of 4.4.2 and 4.4.3 are natural transformations.

2. a. Show that (4.20) commutes.

b. Show that the $\beta * \alpha$ defined there is a natural transformation.

c. Formulate and prove the associativity of the horizontal composition of natural transformations.

3. Show that, using the notation of the Godement rules, $(G_1\alpha)E = G_1(\alpha E)$.

4. Show that in the Diagram (4.19), the composites βF and $H\alpha$ are the horizontal composites $\beta * \mathrm{id}_F$ and $\mathrm{id}_H * \alpha$ respectively.

5. a. Show (using Exercise 4) that Godement's fifth rule is an instance of the first.

b. Show that Godement's fourth rule follows from the first and the associativity of horizontal composition.

4.5 The Yoneda Lemma and universal elements

For an arbitrary category $\mathcal{C}$, the functors from $\mathcal{C}$ to **Set** are special because the hom functors $\mathrm{Hom}(C, -)$ for each object C of $\mathcal{C}$ are set-valued functors. In this section, we introduce the concept of representable functor, the Yoneda Lemma, and universal elements, all of which are based on these hom functors. These ideas have turned out to be fundamental tools for categorists. They are also closely connected with the concept of adjunction, to be discussed later (note Theorem 12.3.2 and Proposition 12.3.5).

If you are familiar with group theory, it may be illuminating to realize that representable functors are a generalization of the regular representation, and the Yoneda embedding is a generalization of Cayley's Theorem.

We have already considered set-valued functors as actions in Section 3.2.

4.5.1 Representable functors A functor from a category $\mathcal{C}$ to the category of sets (a **set-valued functor**) is said to be **representable** if it is naturally isomorphic to a hom functor, see 3.1.20. A covariant functor is representable if it is naturally isomorphic to $\mathrm{Hom}(C, -)$ for some object C of $\mathcal{C}$; in this case one says that C **represents** the functor. A contravariant functor is representable if it is naturally isomorphic to $\mathrm{Hom}(-, C)$ for some object C (and then C represents the contravariant functor).

We have already looked at one example of representable functor in some detail in 4.3.9, where we showed that the set-of-nodes functor for graphs is represented by the graph with one node and no arrows. The set-of-arrows functor is represented by the graph with two nodes and one arrow between them (Exercise 2).

4.5.2 The Yoneda embedding Let $\mathcal{C}$ be a category. There is a functor $Y : \mathcal{C}^{\mathrm{op}} \to \mathbf{Fun}(\mathcal{C}, \mathbf{Set})$, the **Yoneda functor**, defined as follows.

Y–1 For an object C of $\mathcal{C}$, $Y(C) = \mathrm{Hom}(C, -)$.

Y–2 If $f : C \to D$ in $\mathcal{C}$ and A is an object of $\mathcal{C}$, then the component $Y(f)A : \mathrm{Hom}(D, A) \to \mathrm{Hom}(C, A)$ of the natural transformation $Y(f) : \mathrm{Hom}(D, -) \to \mathrm{Hom}(C, -)$ is defined by

$$[Y(f)A](h) = h \circ f$$

for $h : D \to A$ in $\mathcal{C}$.

To see that $Y(f)$ is a natural transformation requires checking that this diagram commutes for every arrow $k : A \to B$ of $\mathcal{C}$:

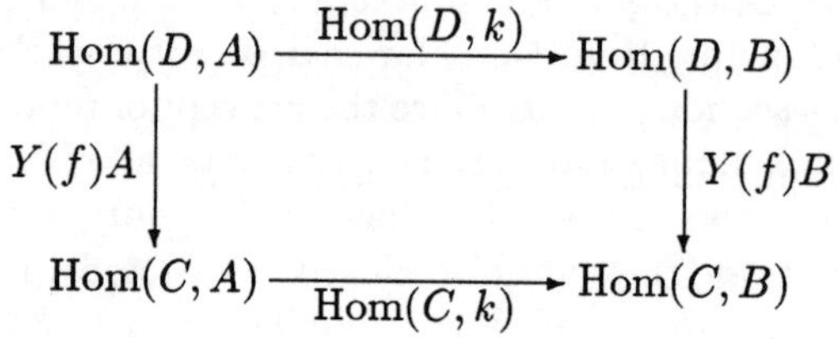

$$\begin{array}{ccc} \mathrm{Hom}(D, A) & \xrightarrow{\mathrm{Hom}(D, k)} & \mathrm{Hom}(D, B) \\ {\scriptstyle Y(f)A}\downarrow & & \downarrow{\scriptstyle Y(f)B} \\ \mathrm{Hom}(C, A) & \xrightarrow[\mathrm{Hom}(C, k)]{} & \mathrm{Hom}(C, B) \end{array}$$

To see that it commutes, start with $h : D \to A$, an arbitrary element of the northwest corner. The upper route takes this to $k \circ h$, then to $(k \circ h) \circ f$. The lower route takes it to $k \circ (h \circ f)$, so the fact that the diagram commutes is simply a statement of the associative law.

$Y(f) : \mathrm{Hom}(D, -) \to \mathrm{Hom}(C, -)$ is the **induced natural transformation** corresponding to f.

The main theorem concerning Y is the following.

4.5.3 Theorem $Y : \mathcal{C}^{\mathrm{op}} \to \mathbf{Fun}(\mathcal{C}, \mathbf{Set})$ *is a full and faithful functor.*

The fact that Y is full and faithful is encapsulated in the following remarkable corollary.

4.5.4 Corollary *Every natural transformation*

$$\mathrm{Hom}(C, -) \to \mathrm{Hom}(D, -)$$

is given by composition with a unique arrow $D \to C$. The natural transformation is an isomorphism if and only if the corresponding arrow

*$D \to C$ is an isomorphism. In particular, if $F : \mathcal{C} \to$ **Set** is represented by both C and D, then $C \cong D$.*

This means that you can construct an arrow in a category by constructing a natural transformation between hom functors. This is one of the most widely used techniques in category theory.

Proof. Theorem 4.5.3 is an immediate corollary of the Yoneda Lemma (4.5.7). We give a direct proof here. This proof is an excellent exercise in manipulating natural transformations and hom sets.

Let $f, g : C \to D$ in $\mathcal{C}$. The component

$$Y(f)D : \text{Hom}(D, D) \to \text{Hom}(C, D)$$

of the natural transformation $Y(f)$ at D takes id_D to f, and similarly $Y(g)D$ takes id_D to g. Thus if $f \neq g$, then $Y(f)D \neq Y(g)D$, so that $Y(f) \neq Y(g)$. Thus Y is faithful.

We must show that Y is full. Given $\phi : \text{Hom}(D, -) \to \text{Hom}(C, -)$, we get the required $f : C \to D$ by one of the basic tricks of category theory: We define $f = \phi D(\text{id}_D)$. The component of ϕ at D is a function $\phi D : \text{Hom}(D, D) \to \text{Hom}(C, D)$, so this definition makes sense.

To complete the proof, we must prove that if $k : D \to A$ is any arrow of $\mathcal{C}$, then $\phi A(k) = k \circ f : C \to A$. This follows from the fact that the following diagram commutes by naturality of ϕ:

$$\begin{array}{ccc}
\text{Hom}(D, D) & \xrightarrow{\text{Hom}(D, k)} & \text{Hom}(D, A) \\
{\scriptstyle \phi D}\downarrow & & \downarrow{\scriptstyle \phi A} \\
\text{Hom}(C, D) & \xrightarrow[\text{Hom}(C, k)]{} & \text{Hom}(C, A)
\end{array}$$

If you start in the northwest corner with id_D, the upper route takes you to $\phi A(k)$ in the southeast corner, whereas the lower route take you to $k \circ f$, as required. □

4.5.5 By replacing $\mathcal{C}$ by $\mathcal{C}^{\text{op}}$ in Theorem 4.5.3, we derive a second Yoneda functor $J : \mathcal{C} \to \textbf{Fun}(\mathcal{C}^{\text{op}}, \textbf{Set})$ which is also full and faithful. For an object C of $\mathcal{C}$, $J(C) = \text{Hom}(-, C)$, the contravariant hom functor. If $f : C \to D$ in $\mathcal{C}$ and A is an object of $\mathcal{C}$, then the component

$$J(f)A : \text{Hom}(A, C) \to \text{Hom}(A, D)$$

of the natural transformation $J(f) : \text{Hom}(-, C) \to \text{Hom}(-, D)$ is defined by

$$[J(f)A](h) = f \circ h$$

for $h : A \to C$ in $\mathcal{C}$.

The fullness of J means that an arrow $f : A \to B$ of $\mathcal{C}$ can be defined by giving a natural transformation from $\mathrm{Hom}(-, A)$ to $\mathrm{Hom}(-, B)$. This statement is the dual of Corollary 4.5.4. Such a natural transformation $\alpha : \mathrm{Hom}(-, A) \to \mathrm{Hom}(-, B)$ has a component $\alpha T : \mathrm{Hom}(T, A) \to \mathrm{Hom}(T, B)$ for each object T of $\mathcal{C}$. The effect of this is that you can define an arrow $f : A \to B$ by giving a function $\alpha T : \mathrm{Hom}(T, A) \to \mathrm{Hom}(T, B)$ for each object T which prescribes a variable element of B for each variable element of A (as described in 2.8.2), in such a way that for each $f : T' \to T$, the diagram

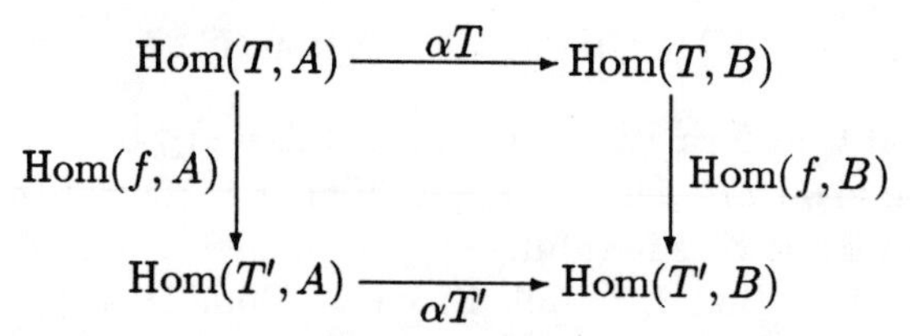

commutes. This can be summed up by saying, 'An arrow is induced by defining its value on each variable element of its domain, provided that the definition is natural with respect to change of parameters.'

4.5.6 Elements of a set-valued functor Corollary 4.5.4 says that any natural transformation from $\mathrm{Hom}(C, -)$ to $\mathrm{Hom}(D, -)$ is given by a unique arrow from D to C, that is, by an element of $\mathrm{Hom}(D, C)$, which is $\mathrm{Hom}(D, -)(C)$. Remarkably, the result remains true when $\mathrm{Hom}(D, -)$ is replaced by an arbitrary set-valued functor.

Suppose $F : \mathcal{C} \to \mathbf{Set}$ is a functor and C is an object of $\mathcal{C}$. An element $c \in FC$ induces a natural transformation from the representable functor $\mathrm{Hom}(C, -)$ to F by the formula

$$f \mapsto F(f)(c) \tag{4.21}$$

That is, if $f : C \to C'$ is an element of $\mathrm{Hom}(C, C')$, the definition of functor requires an induced function $F(f) : F(C) \to F(C')$ and this function can be evaluated at $c \in F(C)$. The naturality is left as Exercise 7. These natural transformations are the only ones.

4.5.7 Theorem (Yoneda Lemma) *Formula (4.21) defines a one to one correspondence between elements of $F(C)$ and natural transformations from* $\mathrm{Hom}(C, -)$ *to* F.

The proof is left as an exercise (Exercise 8).

4.5.8 Definition Let $F : \mathcal{C} \to \mathbf{Set}$ be a functor and let c be an element of $F(C)$ for some object C of $\mathcal{C}$. If the natural transformation from $\mathrm{Hom}(C, -)$ to F induced by c is an isomorphism, then c is a **universal element** of F.

The existence of a universal element means that F is representable (see 4.5.1). The converse is also true because a natural isomorphism $\alpha : \mathrm{Hom}(C, -) \to F$ is, from the Yoneda lemma, induced by a unique element c of $F(C)$ and by definition α is an isomorphism if and only if c is universal.

The unique element c can be calculated using the following fact.

4.5.9 Proposition *Let $\alpha : \mathrm{Hom}(C, -) \to F$ be a natural isomorphism. The unique element $c \in F(C)$ inducing α is $\alpha C(\mathrm{id}_C)$.*

Proof. For an arbitrary $f : C \to C'$, $\alpha C'(f) = F(f)(\alpha C(\mathrm{id}_C))$ because this diagram must commute (chase id_C around the square):

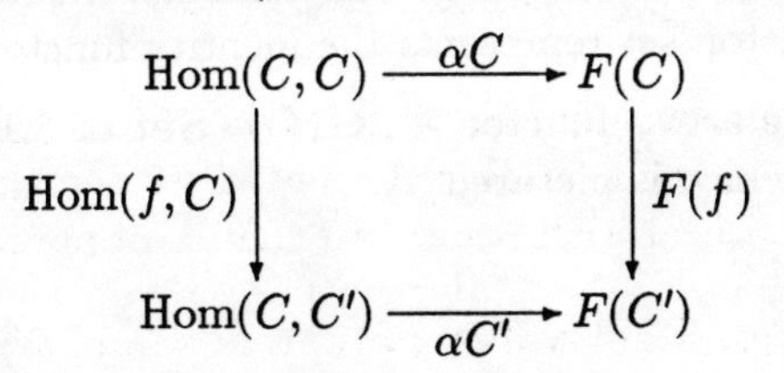

Then, by Formula (4.21), $\alpha C(\mathrm{id}_C)$ must be the required unique element c. □

A detailed example of the use of this construction is in the proof of Proposition 5.2.14.

4.5.10 The definition of universal element can be reworded in elementary terms using (4.21), as follows.

4.5.11 Proposition *Let $F : \mathcal{C} \to \mathbf{Set}$ be a functor, C an object of $\mathcal{C}$ and c an element of $F(C)$. Then c is a universal element of F if and only if for any object C' of $\mathcal{C}$ and any element $x \in F(C')$ there is a unique arrow $f : C \to C'$ of $\mathcal{C}$ for which $x = F(f)(c)$.*

Proof. If c is a universal element then the mapping (4.21) must be an isomorphism, hence every component must be bijective by Theorem 4.2.20. This immediately ensures the existence and uniqueness of the required arrow f. Conversely, the existence and uniqueness of f for each C' and $x \in F(C')$ means that there is a bijection $\alpha C' : \mathrm{Hom}(C, C') \to F(C')$ for every C' which takes $f : C \to C'$ to $F(f)(c)$. These are the components of a natural transformation (Exercise 7) which is therefore a natural isomorphism. □

In the case of a functor $F : \mathcal{C}^{\mathrm{op}} \to \mathbf{Set}$, c in $F(C)$ is a universal element if for any object C' of $\mathcal{C}$ and any element $x \in F(C')$ there is a unique arrow $f : C' \to C$ for which $x = F(f)(c)$.

4.5.12 Corollary *If $c \in F(C)$ and $c' \in F(C')$ are universal elements, then there is a unique isomorphism $f : C \to C'$ such that $F(f)(c) = c'$*

The proof is left as Exercise 10.

Universal elements are considered again in Proposition 12.3.5. the exposition in [Mac Lane, 1971] uses the concept of universal element (defined in the manner of the preceding proposition) as the central idea in discussing representable functors and adjunction.

4.5.13 Exercises

1. Show that $\mathrm{Hom}_{\mathbf{Set}}(1, -)$ is naturally isomorphic to the identity functor on **Set**. ('A set is its set of global elements.' In terms of 4.5.1 this says that a singleton set represents the identity functor on **Set**.)

2. Show that the arrow functor $A : \mathbf{Grf} \to \mathbf{Set}$ of 3.1.9 is represented by the graph **2** which is pictured as

$$0 \longrightarrow 1$$

3. Show that the set of objects of a small category is 'essentially the same thing' as the set of global elements of the category (as an object of **Cat**), and translate this into a natural isomorphism following the pattern of 4.3.9.

4. Is the set of arrows of a small category the object part of a functor? If it is, is it representable?

5. Prove that any set-valued functor $F : \mathcal{C} \to \mathbf{Set}$ is naturally isomorphic to a functor for which if C and D are distinct objects of $\mathcal{C}$, then $F(C)$ and $F(D)$ are disjoint sets.

6. Show that the second Yoneda embedding J defined in 4.5.5 is full and faithful.

7. Show that the function described by (4.21) is natural.

8. Prove the Yoneda Lemma.

9.† Formulate carefully and prove that the equivalence in the Yoneda Lemma is natural in both C and F.

10. Verify the claim made in 4.5.12 that if $c \in F(C)$ and $c' \in F(C')$ are both universal elements of the functor $F : \mathcal{C} \to \mathbf{Set}$, then there is a unique isomorphism $f : C \to C'$ such that $F(f)(c) = c'$.

4.6 Linear sketches (graphs with diagrams)

Specifications in mathematics and computer science are most commonly expressed using a formal language with rules spelling out the semantics. However, there are other objects in mathematics intended as specifications that are not based on a formal language. Many of these are tuple-based; for example the signature of an algebraic structure or the tuple specifying a finite state automaton.

A sketch is another kind of formal abstract specification of a mathematical structure; it is based on a graph rather than on a formal language or tuple. The semantics is often a functor; in other contexts it is a structure generalizing Goguen's initial algebra semantics.

Each sketch generates a (categorical) theory; this theory is a category that in a strong sense contains all the syntax implied by the sketch.

4.6.1 Linear sketches We construct a hierarchy of types of sketches here and in Chapters 7 and 9. Each new type uses additional categorical constructions to provide more expressive power than the preceding types.

What we can do now is describe a very simple type of structure using 'linear' sketches. It can describe multisorted algebraic structures with only unary operations. If you are not familiar with multisorted algebraic structures, it will not matter.

4.6.2 Definition A **linear sketch** $\mathcal{S}$ is a pair $(\mathcal{G}, \mathcal{D})$ where $\mathcal{G}$ is a graph and $\mathcal{D}$ is a set of diagrams in $\mathcal{G}$. Because of the motivating example of algebraic structures, the arrows of the graph of a sketch (not just a linear sketch) are often called **operations** of the sketch.

4.6.3 Definition A **model of a linear sketch** $\mathcal{S}$ in a category $\mathcal{C}$ is a model (graph homomorphism) $M : \mathcal{G} \to \mathcal{C}$ such that whenever $D : \mathcal{I} \to \mathcal{G}$ is a diagram in $\mathcal{D}$, then $M \circ D$ is a *commutative* diagram in $\mathcal{C}$. The diagrams represent the equations which have to be true in all models. The set of all models of $\mathcal{S}$ in $\mathcal{C}$ is denoted $\mathbf{Mod}(\mathcal{S}, \mathcal{C})$.

A model of a sketch $\mathcal{S}$ in **Set** is (among other things) a set indexed by the nodes of the graph of $\mathcal{S}$ as discussed in 2.6.9. If M is a model and c is a node of the graph, an element of $M(c)$ is an element indexed by c, or an element of type c.

4.6.4 Definition A **homomorphism of models** of a linear sketch $\mathcal{S}$, both models in the same category $\mathcal{C}$, is a natural transformation between the models. For given $\mathcal{S}$ and $\mathcal{C}$, the models therefore form a category with natural transformations as arrows; this category is a full subcategory of the category of all graph homomorphisms from $\mathcal{G}$ to $\mathcal{C}$ (which in general do not take the diagrams in $\mathcal{D}$ to commutative diagrams).

4.6.5 Example Any category $\mathcal{E}$ can be made into a linear sketch called the **underlying linear sketch** of $\mathcal{E}$ by taking for the diagrams the collection of all commutative diagrams in the category. A model of the underlying linear sketch in a category $\mathcal{C}$ is exactly a functor on the original category, since any functor must take commutative diagrams to commutative diagrams. Thus the category of models is the same as $\mathbf{Fun}(\mathcal{E}, \mathcal{C})$.

4.6.6 Example The linear sketch for u-structures has the graph with one node and one arrow as its graph, and no diagrams. Similarly, the linear sketch for graphs has (4.15) as its graph and no diagrams.

The construction in 4.2.7 gives the **sketch for graphs**. Its graph is 1.3.6 and it has no diagrams.

4.6.7 Example We now consider an example of a linear sketch which has diagrams. Suppose we wanted to consider sets with permutations as structures. This would be a u-structure (S, u) with u a bijection. We can force u to go to a bijection in **Set**-models by requiring that it have an inverse. Thus the sketch $\mathcal{P}$ of sets with permutations has as graph the graph $\mathcal{G}$ with one node e and two arrows u and v, together with this diagram D:

$$e \underset{v}{\overset{u}{\rightleftarrows}} e \tag{4.22}$$

based on the shape diagram

$$i \underset{y}{\overset{x}{\rightleftarrows}} j \tag{4.23}$$

A model M of this sketch in **Set** must have $M(e)$ a set, $M(u)$ and $M(v)$ functions from $M(e)$ to itself (since $D(i) = D(j) = e$), and because (4.22) must go to a commutative diagram, it must have

$$M(u) \circ M(v) = M(v) \circ M(u) = \mathrm{id}_{M(e)}$$

This says that $M(u)$ and M(v) are inverses to each other, so that they are permutations. Note that a model in any category is an object of that category together with an isomorphism of the object with itself and the inverse of that isomorphism.

If we had used as the only diagram the diagram with *one* node e and both arrows u and v, the result would have been a sketch in which any model M had the property that $M(u)$ and $M(v)$ are the identity.

4.6.8 Example Suppose we wanted to have a linear sketch for graphs which have at least one loop at every node. We could try the following construction, which contains a mild surprise. The sketch has (4.15), page 84 as its graph, with an arrow $s : N \to A$ added. The diagrams are

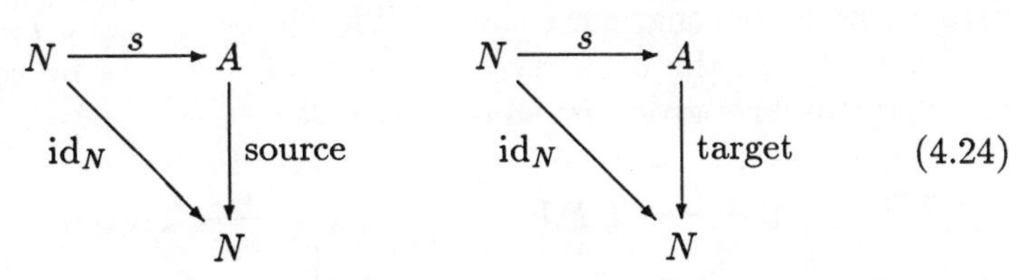

(4.24)

In a model M in sets, if n is a node, that is, $n \in M(N)$, then $M(s)(n)$ is an arrow. The commutativity of the diagrams (4.24) forces the source and target of $M(s)(n)$ to be n, so that $M(s)(n)$ is a loop on n.

The surprise is that a homomorphism $\alpha : M \to N$ of models of this sketch must take the particular loop $M(s)(n)$ to $N(s)(\alpha N(n))$. Of course, any homomorphism of graphs will take the loop $M(s)(n)$ to *some* loop on $\alpha N(n)$, but our homomorphisms are stricter than that. So what we really have are graphs with a *distinguished* loop at every node and homomorphisms which take distinguished loops to distinguished loops. These are called **reflexive** graphs. A node in a reflexive graph may have other loops but they are not part of the given structure.

If you want a sketch for graphs which have a loop on every node, but not a distinguished loop (so that a homomorphism takes the loop on n to some loop on $\alpha N(n)$), but it does not matter which one, you will have to wait until we can study regular sketches in 9.4.5.

4.6.9 The theory of a linear sketch You can reverse 4.6.5: given a linear sketch $\mathcal{S}$, there is a category $\mathbf{Th}(\mathcal{S})$, the **theory of** $\mathcal{S}$, which has a **universal model** M_0 of $\mathcal{S}$ in this sense: $M_0 : \mathcal{S} \to \mathbf{Th}(\mathcal{S})$ is a model, and for any other model $M : \mathcal{S} \to \mathcal{C}$, there is a unique functor $F : \mathbf{Th}(\mathcal{S}) \to \mathcal{C}$ such that $UF \circ M_0 = M$. Since UF is a model of the underlying sketch of the category $\mathbf{Th}(\mathcal{S})$ (see 4.6.5), M_0 induces a bijection between models of $\mathcal{S}$ and models of its theory.

$\mathbf{Th}(\mathcal{S})$ satisfies the following requirements:

LT–1 Every arrow of $\mathbf{Th}(\mathcal{S})$ is a composite of arrows of the form $M_0(f)$ for arrows a of $\mathcal{S}$.

LT–2 M_0 takes every diagram of $\mathcal{S}$ to a commutative diagram in $\mathbf{Th}(\mathcal{S})$.

4.6.10 Construction of the theory of a linear sketch Let $\mathcal{S}$ be a linear sketch. The idea behind the construction of $\mathbf{Th}(\mathcal{S})$ is: 'Freely compose the arrows of $\mathcal{G}$ and impose the diagrams as equations.' Formally, begin with the free category $F(\mathcal{G})$ generated by $\mathcal{G}$ and construct $\mathbf{Th}(\mathcal{S})$ and a functor $Q : F(\mathcal{G}) \to \mathbf{Th}(\mathcal{S})$ which make all the diagrams in $\mathcal{D}$ commute as described in 4.1.12. The universal model $M_0 : \mathcal{G} \to \mathbf{Th}(\mathcal{S})$ (really $M_0 : \mathcal{G} \to U(\mathbf{Th}(\mathcal{S})))$ is $UQ \circ \eta\mathcal{G}$; it takes each node to itself and each arrow to its congruence class in $\mathbf{Th}(\mathcal{S})$.

That M_0 has the property claimed in 4.6.9 follows by considering this diagram for a given model $M : \mathcal{G} \to U\mathcal{C}$:

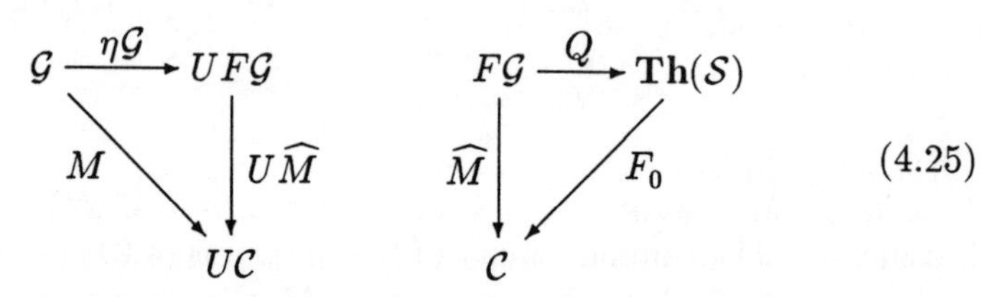

(4.25)

For a given M, it follows by Proposition 3.1.15 that there is a unique functor $\widehat{M}$ for which the left triangle commutes. Then by Proposition 3.5.4 (see also 4.1.12) there is a unique functor $F_0 : \mathbf{Th}(\mathcal{S}) \to \mathcal{C}$ making the right hand triangle commute. Apply U to the right hand triangle, put them together and let $M_0 = UQ \circ \eta\mathcal{G}$. Then we have a commutative triangle:

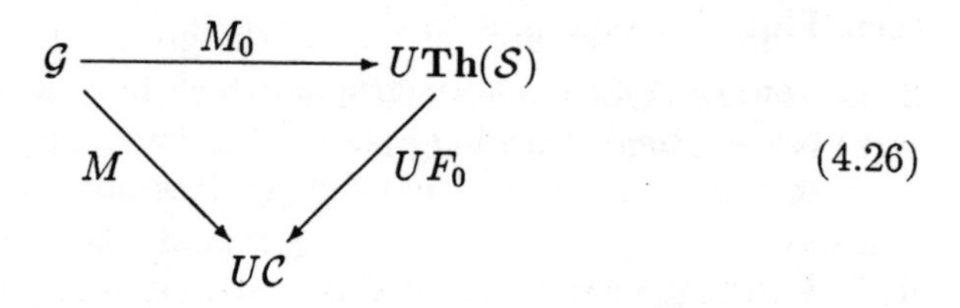

(4.26)

This shows the existence and we leave the uniqueness to the reader.

The construction of a semantics functor for a functional programming language illustrated in 3.5.9 is a special case of the construction just given.

4.6.11 Examples The theory which is generated by the underlying linear sketch of a category is isomorphic to the category itself (Exercise 2). The sketch for u-structures (see 4.2.5 and 4.6.6) generates a category with one node u_0, its identity arrow, and arrows e, $e \circ e$, $e \circ e \circ e$, and so on, all different because there are no equations to make them the same. In other words, it has one arrow e^n for each natural number n.

The sketch for permutations in 4.6.7 is more complicated. It has one node e and arrows u^n for all positive and *negative* integers n. This is essentially because v must be u^{-1} in the theory; see 4.7.12 for more details.

The reader may wonder why the term 'linear sketch' is appropriate. One way of thinking of linear sketches is that they are exactly the sketches for which the following are true:

(i) If M_1 and M_2 are models in the category of sets, then there is a model $M = M_1 + M_2$ defined by $M(a) = M_1(a) + M_2(a)$, $a \in G_0(\mathcal{S})$. ($S + T$ is the disjoint union of sets S and T.)

(ii) If M is a model and X is a set, there is a model $X \times M$ defined by $(X \times M)(a) = X \times M(a)$, $a \in G_0$.

The linear sketches with constants that we will consider in Section 4.7 below lack these 'linearity' properties. They should perhaps be called affine sketches. For the reader familiar with equational theories we observe that linear sketches have only unary operations while linear sketches with constants have, in addition, nullary operations.

4.6.12 Exercises

1. Describe a linear sketch whose models in **Set** are sets S with two arrows from S to S which commute with each other on composition.

2. Prove that if a category $\mathcal{C}$ is regarded as a linear sketch as in 4.6.5, then $\mathbf{Th}(\mathcal{C})$ is isomorphic to $\mathcal{C}$.

3. Let M and N be models of the sketch in 4.6.7. Show that $f : M(e) \to N(e)$ is a homomorphism of models if and only if $f \circ M(u) = N(u) \circ f$.

4.7 Linear sketches with constants: initial term models

In this section, we describe a semantics for linear sketches which is essentially a special case of Goguen's initial algebra semantics. We will treat a version equivalent to the general case in Section 7.6. We construct a specific model of the sketch in **Set** by using the ingredients of the sketch, in a similar spirit to mathematical logic wherein models of a theory ('Herbrand models') are constructed using the expressions in the language.

4.7.1 Definition A model of a sketch in a category $\mathcal{C}$ is called **initial** if it has exactly one arrow (natural transformation) to every model in $\mathcal{C}$. Any two initial objects of any category are isomorphic by a unique isomorphism, so initial models in a given category are unique up to isomorphism.

In the case of a linear sketch, to describe the initial model in the category of sets is easy; it is the model $M : \mathcal{S} \to \mathbf{Set}$ for which $M(a) = \emptyset$ for every node a of $\mathcal{S}$. When we discuss further sketches we will see that the initial models will be more interesting. At this point, we have to complicate things a bit to get interesting models.

4.7.2 Definition By a **linear sketch with constants** we mean a triple $\mathcal{S} = (\mathcal{G}, \mathcal{D}, C)$ where $(\mathcal{G}, \mathcal{D})$ is a linear sketch and C is a set of constants indexed (as in 2.6.9) by the set of nodes of $\mathcal{G}$. We let type : $C \to G_0$ be the indexing function, where G_0 denotes the set of nodes of $\mathcal{G}$. We will not formalize this further because we will have a more systematic way of doing this in Chapter 7. In particular, we could discuss models on categories other than that of sets, but there is no purpose in doing so.

The sketch $(\mathcal{G}, \mathcal{D})$ is called the **underlying linear sketch** of the linear sketch with constants.

4.7.3 Definition **A model** of this linear sketch with constants in the category of sets is a model $M : (\mathcal{G}, \mathcal{D}) \to \mathcal{C}$ together with a function $M : C \to \bigcup_{a \in G_0} M(a)$ such that for $x \in C$, $M(x) \in M(\text{type}(x))$.

(Note that we have overloaded M, and that M need not be injective.) Thus a model of a linear sketch with constants is a model of the underlying linear sketch together with elements chosen in the models of various types.

The existence of the constants means that the values need no longer be empty. In particular, the initial models may be interesting.

4.7.4 Example We give a somewhat arbitrary example which illustrates what happens. Let us add three constants to the sketch for graphs, a of type A and n_1, n_2 of type N. A set model of this sketch with constants will be a model M in **Set** of the sketch for graphs – in other words, a graph – with a distinguished arrow $M(a)$ and two distinguished nodes $M(n_1)$ and $M(n_2)$. It may happen that $M(n_1) = M(n_2)$ and there is no requirement that $M(n_1)$ or $M(n_2)$ be the source or target of $M(a)$.

4.7.5 Definition A homomorphism of models of a linear sketch with constants is a natural transformation between two models of the sketch that takes the values of the constants in the first model to the values in the second.

In other words, if M_1 and M_2 are models and $\phi : M_1 \to M_2$ is a homomorphism of models of the underlying linear sketch, then to be a homomorphism of models of the linear sketch with constants it must also satisfy the requirement that for any $x \in C$ of type a, $\phi a(M_1(x)) =$

$M_2(x)$. Thus in 4.7.4, a homomorphism $\alpha : M_1 \to M_2$ of models of that sketch with constants has to have $\alpha A(M_1(a)) = M_2(a)$, $\alpha N(M_1(n_1)) = M_2(n_1)$, and $\alpha N(M_1(n_2)) = M_2(n_2)$.

4.7.6 Term models and initial term models A model M of a sketch with constants is called a **term model** if for every node a of the underlying graph, every element of $M(a)$ is reachable by beginning with constants and applying various operations (arrows of the sketch). The constants you begin with do not have to be of type a, but the final operation will, of course, have to be one that produces an element of type a. The significance of this condition from the computational point of view is that elements that cannot be produced in this way might as well not be there.

4.7.7 Example Let us consider the linear sketch of u-structures with one constant which we will call zero. Let the graph have node n and call the only arrow succ. There are no diagrams. As this nomenclature suggests, one model of this sketch in **Set** is the natural numbers with the successor operation; the constant is 0. Other models are the integers and the integers modulo a fixed number k (in both cases, take the successor of x to be $x + 1$). However, the natural numbers are the unique (up to unique isomorphism) *initial* model.

To see this, suppose M is any other model. Let us use the same letter M to denote $M(n)$ since there are no other nodes (common practice when the sketch has only one node). Also, let $t : M \to M$ denote the value of M at succ and m_0 the value at zero. We let $\mathbf{N}$, succ and 0 denote the values of these things in the natural numbers. To show that $\mathbf{N}$ is the initial model we must define a natural transformation $f : \mathbf{N} \to M$ and show that it is the only one.

Define f as follows: let $f(0) = m_0$, as required if f is to be an arrow between linear sketches with constants. Then since f must commute with succ, we must have that $f(1) = t(m_0)$, $f(2) = t(t(m_0))$, and so on. This defines f inductively on the whole of $\mathbf{N}$. It is clearly unique and immediate to see that it is an arrow between models.

In Section 5.5, we base the definition of natural numbers object in an arbitrary category on this sketch.

4.7.8 Example The set of all integers is a model but not a term model of the linear sketch of u-structures with one constant. For imagine you have a computer that can store integers, but the only operation that can be carried out on them is that of increment (successor). Suppose, further, that the only natural number whose existence you are certain of is 0. Then you can certainly produce, in addition, 1, 2,..., but no negative numbers. Therefore, they may as well not be there. You can get them by, for example, adding a decrement operation, but as it stands they are inaccessible. They are what J. Goguen and J. Meseguer have called 'junk'.

4.7.9 Example The set $\mathbf{Z}_k$ of natural numbers $(\bmod k)$ is a term model of the sketch of 4.7.7, but not an initial model. For example, there is no arrow from the natural numbers $(\bmod k)$ to the natural numbers. In the first, the successor of $k-1$ is 0, while in the second it is nonzero. Thus no arrow could preserve successor at that point. What has happened here is that the model satisfies an additional equation $k = 0$ not required by the diagrams. This is an example of what Goguen and Meseguer call 'confusion'.

4.7.10 Construction of initial term models Linear sketches with constants always have initial models. When the sketch is finite, an initial model can always be constructed recursively as a term model. ('Finite' means finite number of nodes and arrows.) We now give this construction.

Let $\mathcal{S} = (\mathcal{G}, \mathcal{D}, C)$ be a linear sketch with constants. We define a model $I : \mathcal{S} \to \mathbf{Set}$ recursively as the model constructed by the following requirements I–1 through I–3. The elements are congruence classes of **terms** of $\mathcal{G}$ (composable strings of arrows, including constants, of $\mathcal{G}$); $[x]$ denotes the congruence class of a term x by the congruence relation generated by the relation $\sim$ constructed recursively in the model. By 'congruence relation', we mean congruence on the free category generated by C as described in 3.5.1. In particular, if (g, f) and (g', f') are both composable pairs and $[f] = [f']$ and $[g] = [g']$, then $[g \circ f] = [g' \circ f']$.

I–1 If a is a node of $\mathcal{G}$ and x is a constant of type a, then $[x] \in I(a)$.

I–2 If $f : a \to b$ is an arrow of $\mathcal{G}$ and $[x]$ is an element of $I(a)$, then $[fx] \in I(b)$ and $I(f)[x] = [fx]$. (Note that this constructs both an element and a value of the function $I(f)$ simultaneously).

I–3 If $(f_1, \ldots, f_m)$ and $(g_1, \ldots, g_k)$ are paths in a diagram in $\mathcal{D}$, both going from a node labeled a to a node labeled b, and $[x] \in I(a)$, then

$$(If_1 \circ If_2 \circ \ldots \circ If_m)[x] = (Ig_1 \circ Ig_2 \circ \ldots \circ Ig_k)[x]$$

in $I(b)$.

4.7.11 By 'the model constructed by' these requirements, we mean that

(i) no element is in $I(a)$ except congruence classes of the terms constructed in I–1 and I–2, and

(ii) two terms are equivalent if and only if they are forced to be equivalent by the congruence relation generated by I–3.

Requirement (i) means that the models have no elements not nameable in the theory ('no junk') and (ii) means that elements not provably the same are different ('no confusion'). Concerning (ii), see Exercise 1. It follows from requirement (ii) that if $[x] = [y]$ in $I(a)$ and $f : a \to b$ is an arrow of $\mathcal{G}$, then $[fx] = [fy]$ in $I(b)$.

Note that the models in 4.7.9 have terms giving the same element which are not forced to be equivalent by I–3.

4.7.12 Example Let us work out the initial term model of the sketch from 4.6.7 with one constant called x added. Since the sketch has only one node, the model has only one type. Thus in this case, there is only one set, call it S, and the arrows of the sketch lead to functions from S to S.

Then S has elements in accordance with the following rules:

Mod–1 There is an element $[x] \in S$.

Mod–2 If $[y] \in S$, then there are elements $[uy], [vy] \in S$.

Mod–3 If $[y] = [z]$, then $[uy] = [uz]$ and $[vy] = [vz]$.

Mod–4 For any $[y] \in S$, $[uvy] = [vuy] = [y]$.

It is clear that the set of all 'words' $[w_1 w_2 \ldots w_k x]$, where each w_i is either u or v, satisfies the first two rules above. In order to satisfy all four, we have to impose the equalities they force. In order to gain some insight into this, let us calculate some of the elements of S.

We observe that there must be elements

$$[x_0] = [x],\ [x_1] = [ux],\ [x_2] = [uux],\ \ldots\ ,\ [x_n] = [\underbrace{uu \cdots ux}_{n \text{ copies}}]$$

as well as elements we will denote

$$[x_{-1}] = [vx],\ [x_{-2}] = [vvx],\ \cdots\ ,\ [x_{-n}] = [\underbrace{vv \cdots vx}_{n \text{ copies}}]$$

We first explain why these elements exhaust S. We will not give a formal proof, but let us see which element is represented by an element chosen more or less at random, $[y] = [uvuvuuvux]$. Since $[vux] = [x]$, we have that $[y] = [uvuvuux] = [uvuvx_2]$. Since $[uvx_2] = [x_2]$, it follows that $[y] = [uvx_2]$ and then $[y] = [x_2]$, by another application of the same identity.

This kind of reasoning can be used to show that any application of u's and v's to $[x]$ gives the element $[x_k]$ where k is the number of u's less the number of v's.

In particular, $[ux_k] = [x_{k+1}]$ and $[vx_k] = [x_{k-1}]$ so that the set $\{x_k \mid -\infty < k < \infty\}$ is carried into itself by both u and v. It contains $[x] = [x_0]$ and so must be all of S.

There remains the question of all the $[x_k]$ being distinct; that is whether or not there are any identities among the $[x_k]$. There is a standard way of resolving this question: if there is an equation among two combinations of arrows from the sketch, that equation must hold in every model. Thus if the equation fails in any one model, it cannot be a consequence of the identities in the sketch. In this case, there is an easy model, namely the set $\mathbf{Z}$ of all integers. In the set $\mathbf{Z}$, we let u act by addition of the number 1 and v act by subtracting 1. Then any combination of actions by u and v is just addition of k, the difference between the number of u's and v's (which may be negative).

The discussion above suggests how to construct a bijection between S and $\mathbf{Z}$ which is an isomorphism of models. We must choose an element to correspond to $[x]$. A plausible, but by no means necessary, choice is to correspond $[x]$ to 0. If we do that then we must correspond $[x_1] = [ux]$ to $[u0] = 1$, $[x_2] = [uux]$ to $uu0 = 2$ and so on to correspond $[x_k]$ to k, for $k > 0$. For $k < 0$, the argument is similar, replacing u by v, to show that we correspond $[x_k]$ to k in that case as well.

The isomorphism just constructed takes each $[x_k]$ to the integer k, which implies that if $k \neq k'$, then $[x_k] \neq [x_{k'}]$. Thus S consists of precisely the distinct classes $[x_k]$, one for each integer $k \in \mathbf{Z}$.

4.7.13 Given the construction in 4.7.10 and any **Set** model M of the same sketch, the unique homomorphism $\alpha : I \to M$ is constructed inductively as follows:

M–1 If x is a constant of type a, then $\alpha a[x] = M(x)$.

M–2 If $f : a \to b$ in $\mathcal{G}$ and $[x] \in I(a)$, then $\alpha b([fx]) = M(f)([M(x)])$.

It is a straightforward exercise to show that this is well defined and is a homomorphism of models. It is clearly the only possible one.

The construction in 4.7.10 can be seen as the least fixed point of an operator on models of the sketch (without the constants) in the category of sets and *partial* functions. To any such model M, the operator adjoins an element $f(x)$ to $M(b)$ for any arrow $f : a \to b$ and any element $x \in M(a)$ for which $M(f)(x)$ is not defined. It forces $f(x)$ to be the same as some other element of $M(b)$ if the diagrams force that to happen (we leave the formal description of this to you). To get the model for a particular set of constants, you start with the model obtained by applying only I–1 (so that the sorts have only constants in them and all the arrows have empty functions as models). The least fixed point of this operator is the model in 4.7.10, up to isomorphism.

4.7.14 Free models Let $\mathcal{S}$ be a fixed linear sketch with just one node. With each set C we can associate a linear sketch with the set C of constants. Let us call it $\mathcal{S}(C)$. Let $F(C)$ denote an initial model of $\mathcal{S}(C)$. A model of $\mathcal{S}(C)$ is a model M of $\mathcal{S}$ together with a function $C \to M$. Here, as above, we will use the name of the model to denote the value at the single node of the sketch. To say that $F(C)$ is an initial model of $\mathcal{S}(C)$ is to say that given any model M of $\mathcal{S}$ together with a function $C \to M$, there is a unique arrow $F(C) \to M$ in the category of models for which

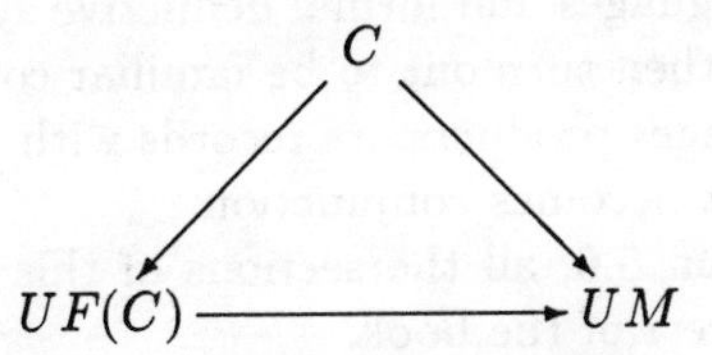

commutes. This property is summarized by saying that $F(C)$ is the **free model** of $\mathcal{S}$ generated by C. Note the similarity with Proposition 3.1.15. Freeness is given a unified treatment in Definition 12.2.1.

4.7.15 This notion of free models can be generalized to the case of many nodes. We indicate briefly how this can be done. Let $\mathcal{S}$ be a linear sketch and G_0 be the nodes of its graph. By a **G_0-indexed set**, we mean a set C together with a function $C \to G_0$ (see 2.6.9). Given any such set we can form a sketch $\mathcal{S}(C)$ which is the linear sketch with the set C of constants with the given function as type function. An initial model of this sketch is called the free model generated by the G_0-indexed set C.

Example 4.7.7 can now be seen as describing the free u-structure with one constant. Another example is the free graph on one node and one arrow. It has *three* nodes, and one arrow connecting two of them.

We describe sketches with more expressive power in Chapter 7.

4.7.16 Exercise

1. Show that requirement (ii) of 4.7.11 is equivalent to the following statement: two terms in the model I are equivalent if and only if every model of $\mathcal{S}$ takes them to the same function.

5

Products and sums

This chapter introduces products, which are constructions allowing the definition of operations of arbitrary arity, and sums, which allow the specification of alternatives. In **Set**, the product is essentially the cartesian product, and the sum is disjoint union.

Sections 5.1 through 5.3 introduce products, and Section 5.4 introduces sums. These ideas are used to define the important concept of natural numbers object in Section 5.5. The last section describes a way to regard formal languages and formal deductive systems as categories. Products and sums then turn out to be familiar constructions. Thus in programming languages products are records with fields, and in deductive systems product becomes conjunction.

Except for Section 5.6, all the sections of this chapter are used in many places in the rest of the book.

5.1 The product of two objects in a category

5.1.1 Definition If S and T are sets, the **cartesian product** $S \times T$ is the set of all ordered pairs with first coordinate in S and second coordinate in T; in other words, $S \times T = \{(s,t) \mid s \in S \text{ and } t \in T\}$. The coordinates are functions $\mathrm{proj}_1 : S \times T \to S$ and $\mathrm{proj}_2 : S \times T \to T$ called the **coordinate projections**, or simply **projections**.

We give a specification of product of two objects in an arbitrary category which will have the cartesian product in **Set** as a special case. This specification is given in terms of the coordinate projections, motivated by these two facts:

(i) you know an element of $S \times T$ by knowing what its two coordinates are, and

(ii) given any element of S and any element of T, there is an element of $S \times T$ with the given element of S as first coordinate and the given element of T as second coordinate.

5.1.2 The product of two objects Let A and B be two objects in a category $\mathcal{C}$. By a (*not* the) **product** of A and B, we mean an object

U together with arrows $\mathrm{proj}_1 : U \to A$ and $\mathrm{proj}_2 : U \to B$ that satisfy the following condition.

5.1.3 For any object V and arrows $q_1 : V \to A$ and $q_2 : V \to B$, there is a unique arrow $q : V \to U$:

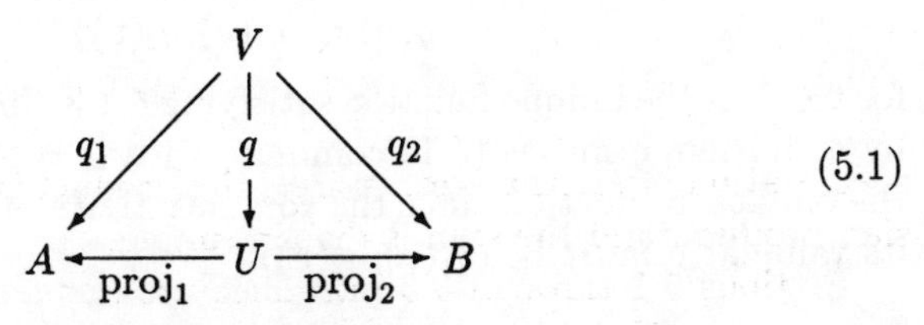

(5.1)

such that $\mathrm{proj}_1 \circ q = q_1$ and $\mathrm{proj}_2 \circ q = q_2$.

This specification gives the product as U together with proj_1 and proj_2. The corresponding diagram

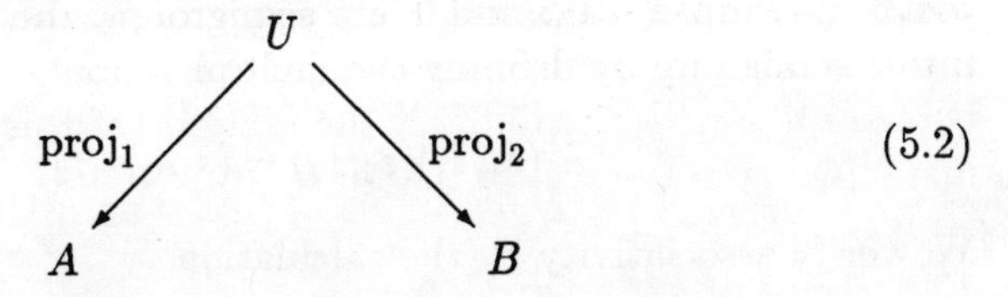

(5.2)

is called a **product diagram** or **product cone**, and the arrows proj_i are called the **projections**. The ordered pair (A, B) is called the **base** of the cone. It is not a set with two elements, but a diagram in $\mathcal{C}$ indexed by the discrete graph with two nodes 1 and 2 which is essentially the same thing as an ordered pair of objects. The diagram

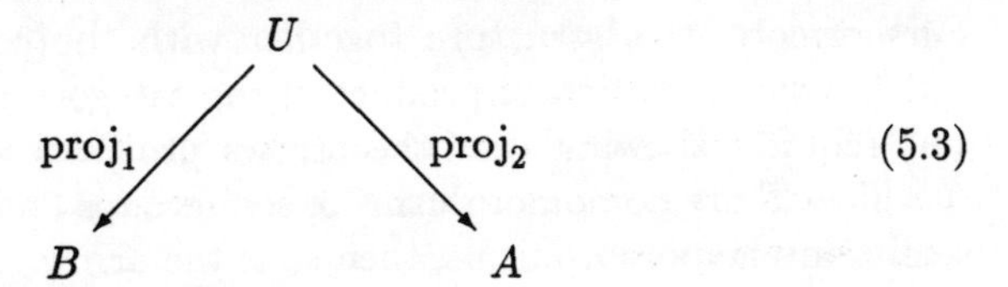

(5.3)

is regarded as a different product cone.

By a type of synecdoche, one often says that an object (such as U above) 'is' a product of two other objects (here A and B), leaving the projections implicit, but the projections are nevertheless part of the structure we call 'product'.

The existence of the unique arrow q with the property just given is called the **universal mapping property** of the product. Any object construction which is defined up to a unique isomorphism (see Theorem 5.2.2) in terms of arrows into or out of it is often said to be defined by a universal mapping property.

5.1.4 Products in Set If A and B are sets, then the cartesian product $A \times B$, together with the coordinate functions discussed in 5.1.1, is indeed a product of A and B in **Set**. For suppose we have a set V and two functions $q_1 : V \to A$ and $q_2 : V \to B$. The function $q : V \to A \times B$ defined by

$$q(v) = (q_1(v), q_2(v))$$

for $v \in V$ is the unique function satisfying 5.1.3. Since $\text{proj}_i(q(v)) = q_i(v)$ by definition, q makes (5.1) commute with $U = A \times B$, and it must be the only such function since the commutativity of (5.1) determines that its value at v *must* be $(q_1(v), q_2(v))$.

5.1.5 Products in categories of sets with structure In many, but not all, categories of sets with structure, the product can be constructed by endowing the product set with the structure in an obvious way.

5.1.6 Example If S and T are semigroups, then we can make $S \times T$ into a semigroup by defining the multiplication

$$(s_1, t_1)(s_2, t_2) = (s_1 s_2, t_1 t_2)$$

We verify associativity by the calculation

$$\begin{aligned} [(s_1,t_1)(s_2,t_2)](s_3,t_3) &= (s_1s_2, t_1t_2)(s_3,t_3) \\ &= ((s_1s_2)s_3, (t_1t_2)t_3) \\ &= (s_1(s_2s_3), t_1(t_2t_3)) \\ &= (s_1,t_1)(s_2s_3, t_2t_3) \\ &= (s_1,t_1)[(s_2,t_2)(s_3,t_3)] \end{aligned} \qquad (5.4)$$

Furthermore, this structure together with the coordinate projections satisfies the definition of product in the category of semigroups. To see this requires showing that the arrows $\text{proj}_1 : S \times T \to S$ and $\text{proj}_2 : S \times T \to T$ are homomorphisms of semigroups and that if q_1 and q_2 are semigroup homomorphisms, then so is the arrow q determined by 5.1.3. Both are necessary because the definition of product in a category $\mathcal{C}$ requires that the arrows occurring in Diagram (5.1) be arrows of the category, in this case, **Sem**. We leave both proofs as exercises (Exercise 5). A similar construction works for most other categories of sets with structure.

One example of a category of sets with structure which lacks products is the category of fields. We discuss this in 7.9.3.

5.1.7 Products in posets We have already seen in 2.3.1 that any poset (partially ordered set) has a corresponding category structure $C(P)$. Let P be a poset and x and y two objects of $C(P)$ (that is, elements of P). Let us see what, if anything, is their product. A product must be an element z together with a pair of arrows $z \to x$ and $z \to y$, which is just another way of saying that $z \leq x$ and $z \leq y$. The definition of product also requires that for any $w \in P$, given an arrow $w \to x$ and one $w \to y$, there is an arrow $w \to z$.

This translates to

$$w \leq x \text{ and } w \leq y \text{ implies } w \leq z$$

which, together with the fact that $z \leq x$ and $z \leq y$, characterizes z as the infimum of x and y. Thus the existence of products in such a category is equivalent to the existence of infimums. In particular, we see that products generalize a well-known construction in posets. Note that a poset that lacks infimums provides an easy example of a category without products.

5.1.8 Exercises

1. Show that the product of two categories, as in 2.6.6, is the product in the category of categories and functors.

2. Describe the product of two monoids in the category of monoids and monoid homomorphisms.

3. Describe the product of two posets in the category of posets and monotone functions.

4. Let $\mathcal{G}$ and $\mathcal{H}$ be two graphs. Show that the product $\mathcal{G} \times \mathcal{H}$ in the category of graphs and homomorphisms is defined as follows: $(\mathcal{G} \times \mathcal{H})_0 = G_0 \times H_0$. An arrow from (g, h) to (g', h') is a pair (a, b) with $a : g \to g'$ in $\mathcal{G}$ and $b : h \to h'$ in $\mathcal{H}$. The projections are the usual first and second projections.

5. Complete the proof that the product structure given in 5.1.6 for $S \times T$ makes it the product in the category of semigroups.

6. Show that if A is an object in a category with a terminal object 1, then

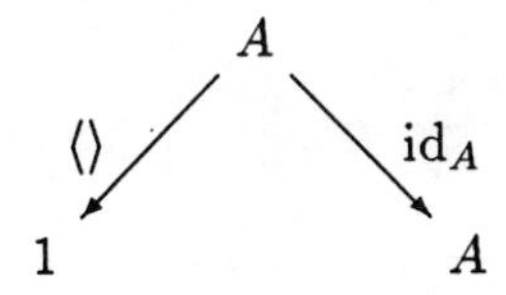

is a product diagram.

7. Give an example of a product diagram in a category in which at least one of the projections is not an epimorphism.

5.2 Notation for and properties of products

5.2.1 Consider sets $S = \{1,2,3\}$, $T = \{1,2\}$ and $U = \{1,2,3,4,5,6\}$. Define $\text{proj}_1 : U \to S$ and $\text{proj}_2 : U \to T$ by this table:

u	$\text{proj}_1(u)$	$\text{proj}_2(u)$
1	2	1
2	1	1
3	3	1
4	2	2
5	1	2
6	3	2

Since the middle and right columns give every possible combination of a number 1, 2 or 3 followed by a number 1 or 2, it follows that U, together with proj_1 and proj_2, is a product of S and T. For example, if $q_1 : V \to S$ and $q_2 : V \to T$ are given functions and $q_1(v) = 1$, $q_2(v) = 2$ for some v in V, then the unique function $q : V \to U$ satisfying 5.1.3 must take v to 5.

In effect, proj_1 and proj_2 code the ordered pairs in $S \times T$ into the set U. As you can see, any choice of a six-element set U and any choice of proj_1 and proj_2 which gives a different element of U for each ordered pair in $S \times T$ gives a product of S and T.

This example shows that the categorical concept of product gives a more general construction than the cartesian product construction for sets. One cannot talk about 'the' product of two objects, but only of 'a' product. However, the following theorem says that two products of the same two objects are isomorphic in a strong sense.

5.2.2 Theorem *Let $\mathcal{C}$ be a category and let A and B be two objects of $\mathcal{C}$. Suppose*

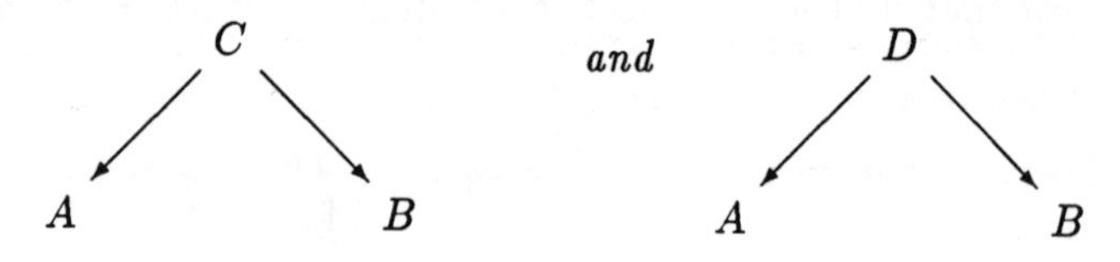

are both product diagrams. Then there is an isomorphism, and only one, between C and D such that

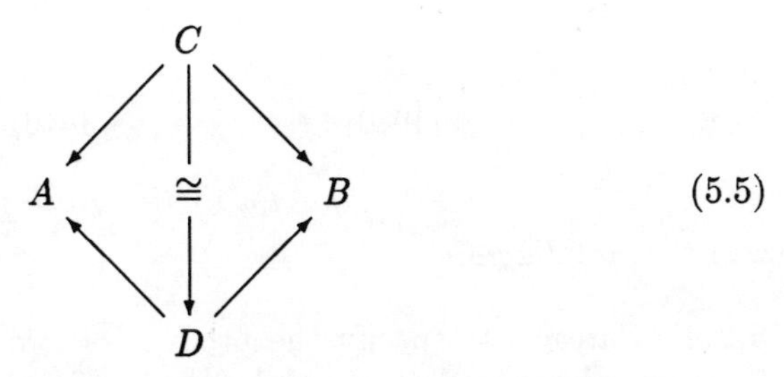

commutes.

The proof we give is quite typical of the kind of reasoning common in category theory and is worth studying, although not necessarily on first reading.

Proof. Let the projections be $p_1 : C \to A$, $p_2 : C \to B$, $q_1 : D \to A$ and $q_2 : D \to B$. In accordance with 5.1.3, there are unique arrows $p : C \to D$ and $q : D \to C$ for which

$$\begin{aligned} p_1 \circ q &= q_1 \\ p_2 \circ q &= q_2 \\ q_1 \circ p &= p_1 \\ q_2 \circ p &= p_2 \end{aligned} \tag{5.6}$$

Thus we already know there is exactly one arrow (namely p) making Diagram (5.5) commute; all that is left to prove is that p is an isomorphism (with inverse q).

The arrow $q \circ p : C \to C$ satisfies

$$p_1 \circ q \circ p = q_1 \circ p = p_1 = p_1 \circ \mathrm{id}_C$$

$$p_2 \circ q \circ p = q_2 \circ p = p_2 = p_2 \circ \mathrm{id}_C$$

and by the uniqueness part of 5.1.3, it follows that $q \circ p = \mathrm{id}_C$. If we exchange the p's and q's, we similarly conclude that $p \circ q = \mathrm{id}_D$ and hence that p and q are isomorphisms which are inverse to each other. □

There is a converse to Theorem 5.2.2 whose proof we leave as an exercise.

5.2.3 Proposition *Let*

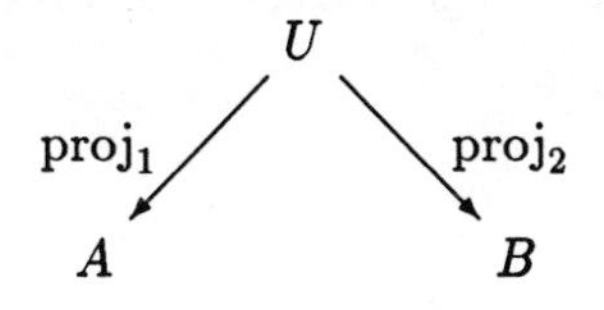

be a product diagram, and suppose that an object V *is isomorphic to* U *by an isomorphism* $i : V \to U$*. Then*

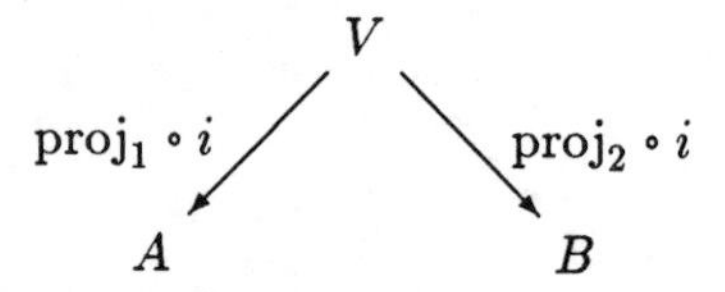

is a product diagram.

5.2.4 Categorists specify the product of two sets by saying that all they care about an element of the product is what its first coordinate is and what its second coordinate is. Theorem 5.2.2 says that two structures satisfying this specification are isomorphic in a unique way.

The name '(s,t)' represents the element of the product with first coordinate s and second coordinate t. In a different realization of the product, '(s,t)' represents the element of *that* product with first coordinate s and second coordinate t. The isomorphism of Theorem 5.2.2 maps the representation in the first realization of the product into a representation in the second. Moreover, the universal property of product says that any name '(s,t)' with $s \in S$ and $t \in T$ represents an element of the product: it is the unique element $x \in S \times T$ with $\text{proj}_1(x) = s$ and $\text{proj}_2(x) = t$.

In traditional approaches to foundations, the concept of ordered pair (hence the product of two sets) is defined by giving a specific model (or what a computer scientist might call an implementation) of the specification. Such a definition makes the product absolutely unique instead of unique up to an isomorphism. We recommend that the reader read the discussion of this point in [Halmos, 1960], Section 6, who gives a beautiful discussion of (what in present day language we call) the difference between a specification and an implementation.

In categories other than sets there may well be no standard implementation of products, so the specification given is necessary. In Chapter 14, we will discuss a category known as the category of modest sets in which any construction requires the choice of a bijection between $\mathbf{N}$ and $\mathbf{N} \times \mathbf{N}$. There are many such, and there is no particular reason to choose one over another.

5.2.5 Notation for products It is customary to denote a product of objects A and B of a category as $A \times B$. Precisely, the name $A \times B$

applied to an object means there is a product diagram

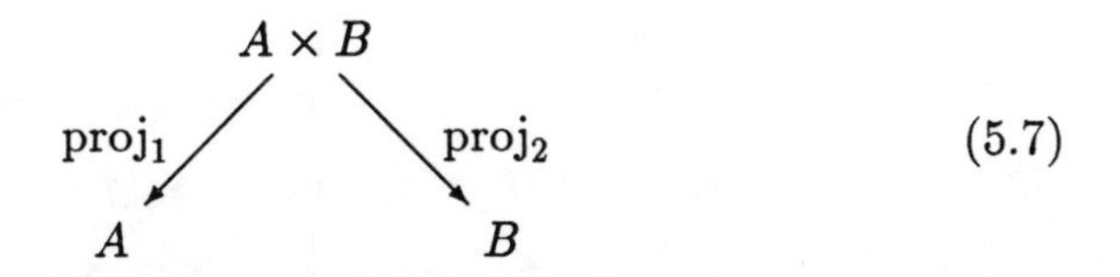

Using the name '$A \times B$' implies that there are specific, but unnamed, projections given for the product structure.

If $A = B$, one writes $A \times A = A^2$ and calls it the **cartesian square** of A.

5.2.6 The notations $A \times B$ and A^2 may be ambiguous, but because of Theorem 5.2.2, it does not matter *for categorical purposes* which product the symbol refers to. Even in the category of sets, you do not really know which set $A \times B$ is unless you pick a specific definition of ordered pair (and the average mathematician does not normally need to give any thought to the definition).

5.2.7 Binary operations A binary operation on a set S is a function from $S \times S$ to S. An example is addition on the natural numbers, which is a function $+ : \mathbf{N} \times \mathbf{N} \to \mathbf{N}$. This and other familiar binary operations are usually written in infix notation; one writes $3 + 5 = 8$, for example, instead of $+(3, 5) = 8$. In mathematics texts, the value of an arbitrary binary operation m at a pair (x, y) is commonly denoted xy, without any symbol at all.

Using the concept of categorical product, we can now define the concept of binary operation on any object S of any category provided only that there is a product $S \times S$: a **binary operation** on S is an arrow $S \times S \to S$.

The associative law $(xy)z = x(yz)$ can be described using a commutative diagram as illustrated in 4.1.10. In that section, the diagram is a diagram in **Set**, but now it has a meaning in any category with products. (The meaning of expressions such as mult $\times S$ in arbitrary categories is given in 5.2.17 below.)

5.2.8 Suppose we are given a product diagram (5.7). For each pair of arrows $f : V \to A$ and $g : V \to B$ requirement 5.1.3 produces a unique

$q : V \to A \times B$ making the following diagram commute.

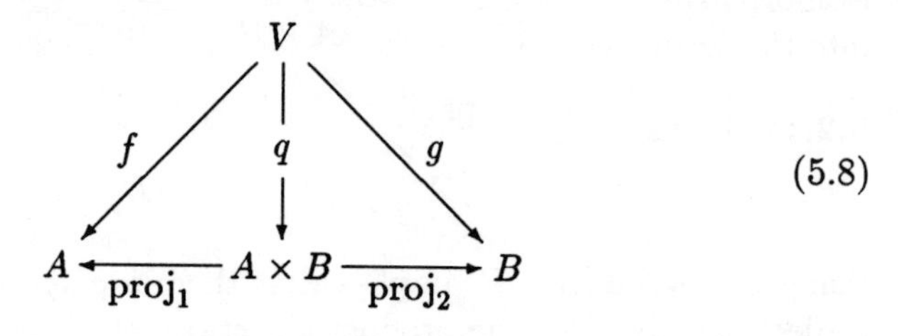

(5.8)

commute; in other words, it produces a function

$$\pi V : \mathrm{Hom}_{\mathcal{C}}(V, A) \times \mathrm{Hom}_{\mathcal{C}}(V, B) \to \mathrm{Hom}_{\mathcal{C}}(V, A \times B)$$

Thus $q = \pi V(f, g)$.

5.2.9 Proposition *The function πV is a bijection.*

Proof. πV is injective, since if (f, g) and (f', g') are elements of

$$\mathrm{Hom}_{\mathcal{C}}(V, A) \times \mathrm{Hom}_{\mathcal{C}}(V, B)$$

both of which produce the same arrow q making (5.9) commute, then $f = \mathrm{proj}_1 \circ q = f'$, and similarly $g = g'$.

It is also surjective, since if $r : V \to A \times B$ is any arrow of $\mathcal{C}$, then it makes

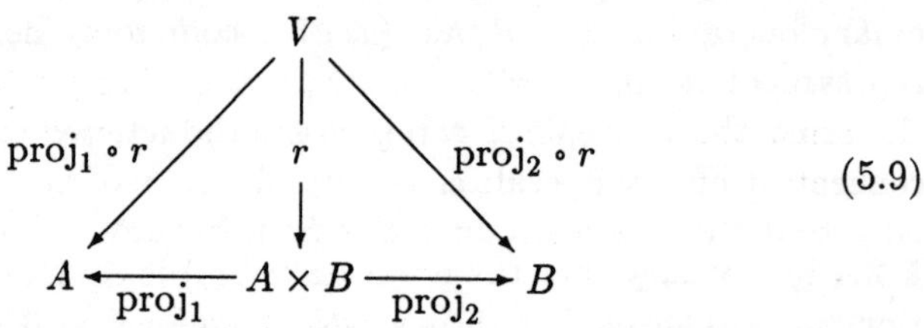

(5.9)

commute, and so is the image of the pair $(\mathrm{proj}_1 \circ r, \mathrm{proj}_2 \circ r)$ under πV. □

5.2.10 It is customary to write $\langle f, g \rangle$ for $\pi V(f, g)$. The *arrow* $\langle f, g \rangle$ internally represents the *pair* of arrows (f, g) of the category $\mathcal{C}$. Proposition 5.2.9 says that the representation is good in the sense that $\langle f, g \rangle$ and (f, g) each determine the other. Proposition 5.2.14 below says that the notation $\langle f, g \rangle$ is compatible with composition.

We have already used the notation '$\langle f, g \rangle$' in the category of sets in 1.2.8.

5.2.11 In the case of two products for the same pair of objects, the isomorphism of 5.2.2 translates the arrow named $\langle f, g\rangle$ for one product into the arrow named $\langle f, g\rangle$ for the other, in the following precise sense.

5.2.12 Proposition *Suppose*

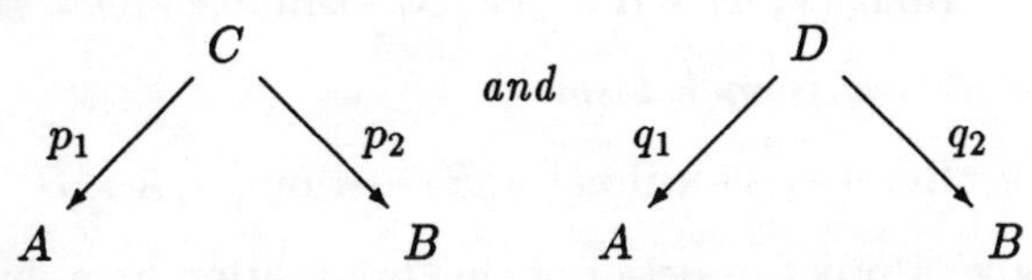

are two product diagrams and $\phi : C \to D$ *is the unique isomorphism given by Theorem 5.2.2. Let* $f : V \to A$ *and* $g : V \to B$ *be given and let* $u : V \to C$, $v : V \to D$ *be the unique arrows for which* $p_1 \circ u = q_1 \circ v = f$ *and* $p_2 \circ u = q_2 \circ v = g$. *Then* $\phi \circ u = v$.

Note that in the statement of the theorem, both u and v could be called '$\langle f, g\rangle$', as described in 5.2.10. The ambiguity occurs because the pair notation does not name which product of A and B is being used. It is rare in practice to have two different products of the same two objects under consideration at the same time.

Proof. By 5.2.2, $p_i = q_i \circ \phi$, $i = 1, 2$. Using this, we have $q_1 \circ \phi \circ u = p_1 \circ u = f$ and similarly $q_2 \circ \phi \circ u = g$. Since v is the unique arrow which makes $q_1 \circ v = u$ and $q_2 \circ v = g$, it follows that $\phi \circ u = v$. □

This theorem provides another point of view concerning elements (s, t) of a product $S \times T$. As described in 2.7.14, the element s may be represented by an arrow $s : 1 \to S$, and similarly t by $t : 1 \to T$. Then the arrow $\langle s, t\rangle : 1 \to S \times T$ represents the ordered pair (s, t) whichever realization of $S \times T$ is chosen.

5.2.13 To show that the notation $\langle q_1, q_2\rangle$ is compatible with composition, we will show that the arrows πV defined in 5.2.8 are the components of a natural isomorphism. To state this claim formally, we need to make $\mathrm{Hom}_{\mathcal{C}}(V, A) \times \mathrm{Hom}_{\mathcal{C}}(V, B)$ into a functor. This is analogous to the definition of the contravariant hom functor. A and B are fixed and the varying object is V, so we define the functor $\mathrm{Hom}_{\mathcal{C}}(-, A) \times \mathrm{Hom}_{\mathcal{C}}(-, B)$ as follows:

(i) $[\mathrm{Hom}_{\mathcal{C}}(-, A) \times \mathrm{Hom}_{\mathcal{C}}(-, B)](V) = \mathrm{Hom}_{\mathcal{C}}(V, A) \times \mathrm{Hom}_{\mathcal{C}}(V, B)$, the set of pairs (g, h) of arrows $g : V \to A$ and $h : V \to B$.

(ii) For $f : W \to V$, let $\mathrm{Hom}_{\mathcal{C}}(f, A) \times \mathrm{Hom}_{\mathcal{C}}(f, B)$ be the arrow

$$\mathrm{Hom}_{\mathcal{C}}(V, A) \times \mathrm{Hom}_{\mathcal{C}}(V, B) \to \mathrm{Hom}_{\mathcal{C}}(W, A) \times \mathrm{Hom}_{\mathcal{C}}(W, B)$$

that takes a pair (g, h) to $(g \circ f, h \circ f)$.

Now we can state the proposition.

5.2.14 Proposition *The family of arrows*

$$\pi V : \mathrm{Hom}_{\mathcal{C}}(V, A) \times \mathrm{Hom}_{\mathcal{C}}(V, B) \to \mathrm{Hom}_{\mathcal{C}}(V, A \times B)$$

constitutes a natural isomorphism

$$\pi : \mathrm{Hom}(-, A) \times \mathrm{Hom}(-, B) \to \mathrm{Hom}(-, A \times B)$$

Proof. We give a proof in detail of this proposition here, but you may want to skip it on first reading, or for that matter on fifteenth reading. We will not always give proofs of similar statements later (of which there are many).

Let $P_{A,B}$ denote the functor $\mathrm{Hom}(-, A) \times \mathrm{Hom}(-, B)$. The projections $p_1 : A \times B \to A$ and $p_2 : A \times B \to B$ from the product form a pair $(p_1, p_2) \in P_{A,B}(A \times B)$.

5.2.15 Lemma *The pair* (p_1, p_2) *is a universal element for* $P_{A,B}$.

Proof. The pair fits the requirements of Proposition 4.5.11 by definition of product: if $(q_1, q_2) \in P_{A,B}(V)$, in other words if $q_1 : V \to A$ and $q_2 : V \to B$, there is a unique arrow $q : V \to A \times B$ such that $P_{A,B}(q)(p_1, p_2) = (q_1, q_2)$, in other words, using 5.2.13(ii), $p_i \circ q = q_i$ for $i = 1, 2$. □

Note that this gives an immediate proof of Theorem 5.2.2. See 12.2.3 for another point of view concerning (p_1, p_2).

Continuing the proof of Proposition 5.2.14, it follows from Equation (4.21) that the natural isomorphism α from $\mathrm{Hom}_{\mathcal{C}}(V, A \times B)$ to $P_{A,B}$ induced by this universal element takes $q : V \to A \times B$ to $(p_1 \circ q, p_2 \circ q)$, which is $p_{A,B}(q)(p_1, p_2)$. Then by definition of π we have that $\alpha V = (\pi V)^{-1}$, so that π is the inverse of a natural isomorphism and so is a natural isomorphism. □

It is not hard to give a direct proof of Proposition 5.2.14 using Proposition 5.2.9 and the definition of natural transformation. That definition requires that the following diagram commute for each arrow $f : V \to W$:

$$\begin{array}{ccc}
\mathrm{Hom}(V, A) \times \mathrm{Hom}(V, B) & \xrightarrow{\pi V} & \mathrm{Hom}(V, A \times B) \\
\uparrow & & \uparrow \\
\mathrm{Hom}(W, A) \times \mathrm{Hom}(W, B) & \xrightarrow[\pi W]{} & \mathrm{Hom}(W, A \times B)
\end{array} \qquad (5.10)$$

We leave the details as Exercise 4.

5.2.16 If $f : V \to W$ and $q_1 : W \to A$ and $q_2 : W \to B$ determine $\langle q_1, q_2\rangle : W \to A \times B$, then the commutativity of (5.10) says exactly that $\langle q_1 \circ f, q_2 \circ f\rangle$ and $\langle q_1, q_2\rangle \circ f$ are the same arrow. In this sense, the $\langle f, g\rangle$ notation is compatible with composition.

Category theorists say that the single arrow $\langle q_1, q_2\rangle$ is the *internal* pair of arrows with first coordinate q_1 and second coordinate q_2. The idea behind the word 'internal' is that the category $\mathcal{C}$ is the workspace; inside that workspace the arrow $\langle q_1, q_2\rangle$ is the pair (q_1, q_2).

When you think of $\mathcal{C}$ as a structure and look at it from the outside, you would say that the arrow q *represents* the *external* pair of arrows (q_1, q_2).

5.2.17 The cartesian product of arrows The cartesian product construction 1.2.9 for functions in sets can also be given a categorical definition. Suppose that $f : S \to S'$ and $g : T \to T'$ are given. Then the composite arrows $f \circ \mathrm{proj}_1 : S \times T \to S'$ and $g \circ \mathrm{proj}_2 : S \times T \to T'$ induce, by the definition of product, an arrow denoted $f \times g : S \times T \to S' \times T'$ such that

$$\begin{array}{ccccc} S & \xleftarrow{\mathrm{proj}_1} & S \times T & \xrightarrow{\mathrm{proj}_2} & T \\ {\scriptstyle f}\downarrow & & \downarrow{\scriptstyle f \times g} & & \downarrow{\scriptstyle g} \\ S' & \xleftarrow[\mathrm{proj}_1]{} & S' \times T' & \xrightarrow[\mathrm{proj}_2]{} & T' \end{array} \tag{5.11}$$

commutes. Thus $f \times g = \langle f \circ \mathrm{proj}_1, g \circ \mathrm{proj}_2\rangle$. It is characterized by the properties

$$\mathrm{proj}_1 \circ (f \times g) = f \circ \mathrm{proj}_1; \quad \mathrm{proj}_2 \circ (f \times g) = g \circ \mathrm{proj}_2$$

Note that we use proj_1 and proj_2 for the product projections among different objects. This is standard and rarely causes confusion since the domains and codomains of the other arrows determine them. We will later call them p_1 and p_2, except for emphasis.

An invariance theorem similar to Proposition 5.2.12 is true of cartesian products of functions.

5.2.18 Proposition *Suppose the top and bottom lines of each diagram below are product cones, and that m and n are the unique arrows making the diagrams commute. Let $\psi : P \to Q$ and $\phi : C \to D$ be the unique isomorphisms given by Theorem 5.2.2. Then $\phi \circ m = n \circ \psi$.*

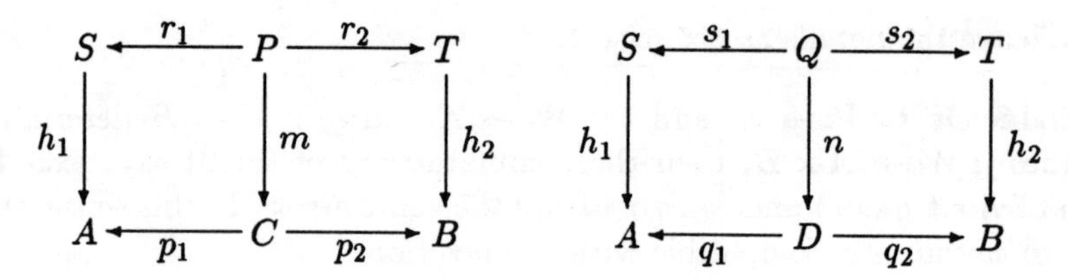

Proof. For $i = 1, 2$,

$$\begin{aligned} q_i \circ \phi \circ m \circ \psi^{-1} &= p_i \circ m \circ \psi^{-1} \\ &= h_i \circ r_i \circ \psi^{-1} \\ &= h_i \circ s_i \circ \psi \circ \psi^{-1} \\ &= h_i \circ s_i \end{aligned} \tag{5.12}$$

The first equality is the property of ϕ given by Theorem 5.2.2, the second by definition of product applied to C, and the third is defining property of ψ. It follows that $\phi \circ m \circ \psi^{-1}$ is the unique arrow determined by $h_1 \circ s_1 : P \to A$ and $h_2 \circ s_2 : P \to B$ and the fact D is a product. But n is that arrow, so that $\phi \circ m \circ \psi^{-1} = n$, whence the theorem. □

5.2.19 Exercises

1. Give explicitly the isomorphism claimed by Theorem 5.2.2 between $S \times T$ and the set $\{1, 2, 3, 4, 5, 6\}$ expressed as the product of $\{1, 2, 3\}$ and $\{1, 2\}$ using the projections in 5.2.1.

2. Given a two-element set A and a three-element set B, in how many ways can the set $\{1, 2, 3, 4, 5, 6\}$ be made into a product $A \times B$? (This refers to 5.2.1.)

3. Let $\mathcal{C}$ be a category with products, and f_1, f_2, g_1, g_2 arrows for which $g_1 \circ f_1$ and $g_2 \circ f_2$ are defined. Show that

$$(g_1 \circ f_1) \times (g_2 \circ f_2) = (g_1 \times g_2) \circ (f_1 \times f_2)$$

Deduce that there is a functor from $\mathcal{C} \times \mathcal{C}$ to $\mathcal{C}$ which takes a pair (C, D) of objects to a product $C \times D$.

4. Prove that Diagram (5.10) commutes.

5. Prove Proposition 5.2.3.

5.3 Finite products

Products of two objects, as discussed in the preceding sections, are called **binary products**. We can define products of more than two objects by an obvious modification of the definition.

For example, if A, B and C are three objects of a category, a product of them is an object $A \times B \times C$ *together with* three arrows:

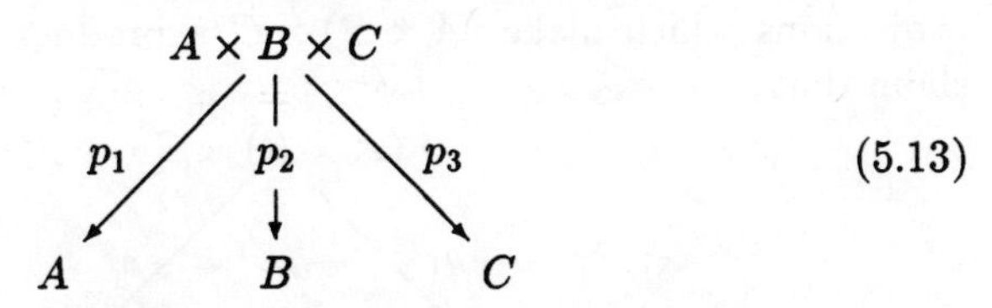

for which, given any other diagram

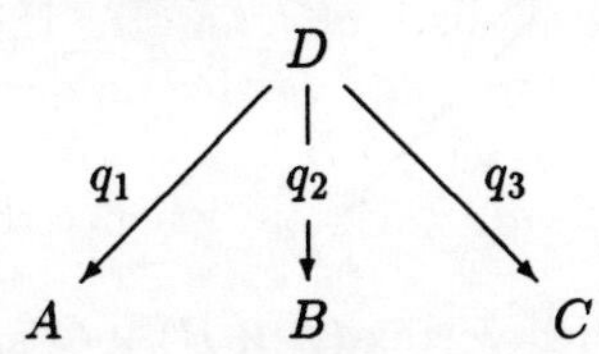

there exists a unique arrow $q = \langle q_1, q_2, q_3 \rangle : D \to A \times B \times C$ such that $p_i \circ q = q_i$, $i = 1$, 2, 3. A diagram of the form (5.13) is called a **ternary product diagram** (or ternary product cone). The general definition of product follows the same pattern.

5.3.1 Definition A **product** of a list $A_1, A_2, \ldots, A_n$ of objects (not necessarily distinct) of a category is an object V together with arrows $p_i : V \to A_i$, for $i = 1, \ldots, n$, with the property that given any object W and arrows $f_i : W \to A_i$, $i = 1, \ldots, n$, there is a unique arrow $\langle f_1, f_2, \ldots, f_n \rangle : W \to V$ for which $p_i \circ \langle f_1, f_2, \ldots, f_n \rangle = f_i$, $i = 1, \ldots, n$.

A product of such a list $A_1, A_2, \ldots, A_n$ is called an n**-ary product** when it is necessary to specify the number of factors. Such a product may be denoted $A_1 \times A_2 \times \cdots \times A_n$ or $\prod_{i=1}^{n} A_i$.

The following uniqueness theorem for general finite products can be proved in the same way as Theorem 5.2.2.

5.3.2 Theorem *Suppose $A_1, A_2, \ldots, A_n$ are objects of a category $\mathcal{C}$ and that C, with projections $p_i : C \to A_i$, and D, with projections $q_i : D \to A_i$, are products of these objects. Then there is a unique arrow $\phi : C \to D$ for which $q_i \circ \phi = p_i$ for $i = 1, \ldots, n$. Moreover, ϕ is an isomorphism.*

Propositions 5.2.3, 5.2.12 and 5.2.18 also generalize in the obvious way to n-ary products.

5.3.3 Binary products give ternary products An important consequence of the definition of ternary product is that in any category with binary products, and any objects A, B and C, either of $(A \times B) \times C$ and $A \times (B \times C)$ can be taken as ternary products $A \times B \times C$ with appropriate choice of projections.

We prove this for $(A \times B) \times C$. Writing p_i, $i = 1, 2$, for the projections which make $A \times B$ a product of A and B and q_i, $i = 1, 2$ for the projections which make $(A \times B) \times C$ a product of $A \times B$ and C, we claim that

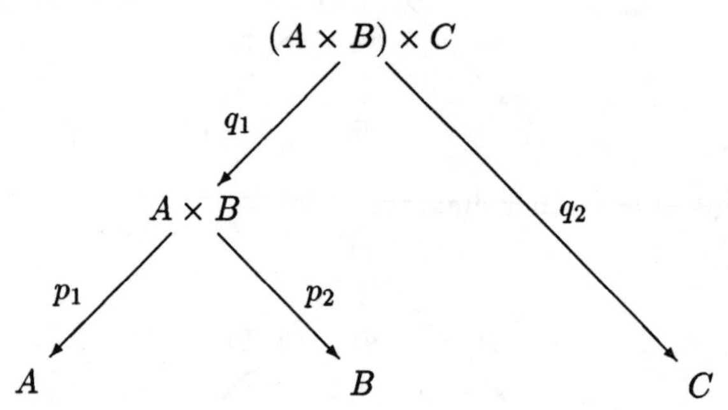

is a product diagram with vertex $(A \times B) \times C$ and projections $p_1 \circ q_1 : (A \times B) \times C \to A$, $p_2 \circ q_1 : (A \times B) \times C \to B$, and $q_2 : (A \times B) \times C \to C$.

Suppose that $f : V \to A$, $g : V \to B$, and $h : V \to C$ are given. We must construct an arrow $u : V \to (A \times B) \times C$ with the property that

(a) $p_1 \circ q_1 \circ u = f$,

(b) $p_2 \circ q_1 \circ u = g$, and

(c) $q_2 \circ u = h$.

Let v be the unique arrow making

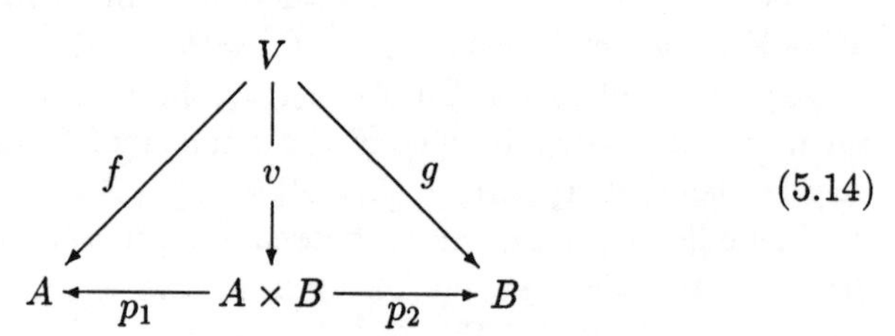

commute. This induces a unique arrow u making

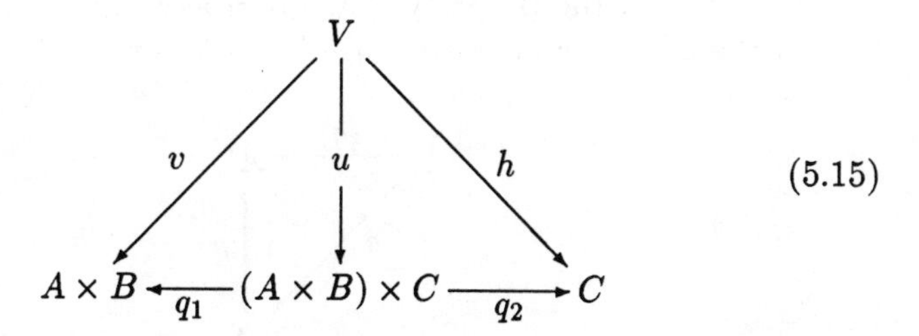

(5.15)

commute.

The fact that (a) through (c) hold can be read directly off these diagrams. For example, for (a), $p_1 \circ q_1 \circ u = p_1 \circ v = f$.

Finally, if u' were another arrow making (a) through (c) hold, then we would have $p_1 \circ q_1 \circ u' = f$ and $p_2 \circ q_1 \circ u' = g$, so by uniqueness of v as defined by (5.14), $v = q_1 \circ u'$. Since $q_2 \circ u'$ must be h, the uniqueness of u in (5.15) means that $u' = u$.

A generalization of this is stated in Proposition 5.3.10 below.

5.3.4 It follows from the discussion in 5.3.3 that the two objects $(A \times B) \times C$ and $A \times (B \times C)$ are pairwise canonically isomorphic (to each other and to any other realization of the ternary product) in a way that preserves the ternary product structure.

In elementary mathematics texts the point is often made that 'cartesian product is not associative'. When you saw this you may have thought in your heart of hearts that $(A \times B) \times C$ and $A \times (B \times C)$ are nevertheless really the same. Well, now you know that they are really the same in a very strong sense: they satisfy the *same specification* and so carry exactly the same information. The only difference is in *implementation*.

5.3.5 If all the factors in an n-ary product are the same object A, the n-ary product $A \times A \times \cdots \times A$ is denoted A^n. This suggests the possibility of defining the **nullary product** A^0 and the **unary product** A^1.

5.3.6 For nullary products, the definition is: given no objects of the category $\mathcal{C}$, there should be an object we will temporary call T and no arrows from it such that for any other object B and no arrows from B, there is a unique arrow $B \to T$ subject to no commutativity condition.

When the language is sorted out, we see that a nullary product in $\mathcal{C}$ is simply an object T with the property that every other object of the category has exactly one arrow to T. That is, T must be a terminal object of the category, normally denoted 1. Thus, for any object A of the category, we take $A^0 = 1$. (Compare 4.1.5.)

5.3.7 A unary product A^1 of a single object A should have an arrow $p : A^1 \to A$ with the property that given any object V and arrow $q : V \to A$ there is a unique arrow $\langle q \rangle : V \to A^1$ for which

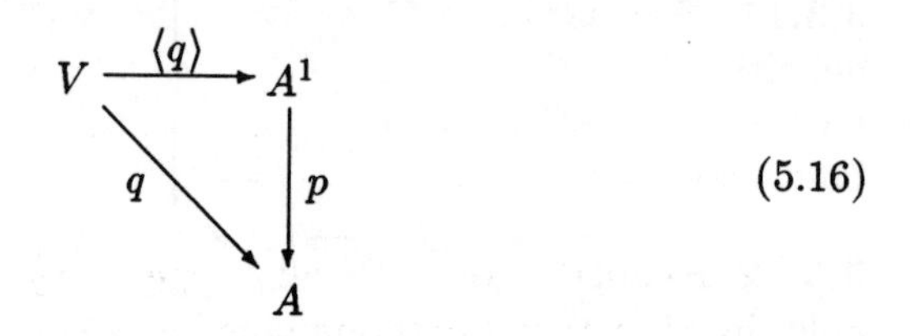

(5.16)

commutes. The identity $\mathrm{id} : A \to A$ satisfies this specification for p; given the arrow $q : V \to A$, we let $\langle q \rangle = q : V \to A$. The fact that $\mathrm{id} \circ \langle q \rangle = q$ is evident, as is the uniqueness of $\langle q \rangle$. It follows that A^1 can always be taken to be A itself, with the identity arrow as the coordinate arrow.

It is straightforward to show that in general an object B is a unary product of A with coordinate $p : B \to A$ if and only if p is an isomorphism (Exercise 1.b).

There is therefore a conceptual distinction between A^1 and A. In the category of sets, A^n is often taken to be the set of strings of elements of A of length exactly n. As you may know, in some computer languages, a string of characters of length one is not the same as a single character, mirroring the conceptual distinction made in category theory.

5.3.8 Definition A category has **binary products** if the product of any two objects exists. It has **canonical binary products** if a specific product diagram is given for each pair of objects. Thus a category with canonical binary products is a category with extra structure given for it. Precisely, a canonical binary products structure on a category $\mathcal{C}$ is a function from $\mathcal{C}_0 \times \mathcal{C}_0$ to the collection of product diagrams in $\mathcal{C}$ which takes a pair (A, B) to a diagram of the form (5.2).

The fact that A and id_A can be taken as a product diagram for any object A of any category means that every category can be given a canonical unary product structure. This is why the distinction between A and A^1 can be and often is ignored.

5.3.9 Definition A category has **finite products** or is a **cartesian category** if the product of any finite number of objects exists. This includes nullary products – in particular, a category with finite products has a terminal object. The category has **canonical finite products** if every finite list of objects has a specific given product.

The following proposition is proved using constructions generalizing those of 5.3.3. (See Exercises 1.c and 6.)

5.3.10 Proposition *If a category has a terminal object and binary products, then it has finite products.*

5.3.11 **Set**, **Grf** and **Cat** all have finite products. In fact, a choice of definition for ordered pairs in **Set** provides canonical products not only for **Set** but also for **Grf** and **Cat**, since products in those categories are built using cartesian products of sets.

5.3.12 Record types To allow operations depending on several variables in a functional programming language L (as discussed in 2.2.1), it is reasonable to assume that for any types A and B the language has a record type P and two field selectors $P.A : P \to A$ and $P.B : P \to B$. If we insist that the data in P be determined completely by those two fields, it follows that for any pair of operations $f : X \to A$ and $g : X \to B$ there ought to be a unique operation $\langle f, g \rangle : X \to P$ with the property that $P.A \circ \langle f, g \rangle$. In other words, that $\langle f, g \rangle; P.A$, is the same as f and that $P.B \circ \langle f, g \rangle$ is the same as g. This would make P the product of A and B with the selectors as product projections.

For example, a record type PERSON with fields NAME and AGE could be represented as a product cone whose base diagram is defined on the discrete graph with two nodes NAME and AGE. If HUMAN is a variable of type PERSON, then the field selector HUMAN.AGE implements the coordinate projection indexed by AGE. This example is closer to the spirit of category theory than the cone in Diagram (5.2); there, the index graph has nodes 1 and 2, which suggests an ordering of the nodes (and the projections) which is in fact spurious.

Thus to say that one can always construct record types in a functional programming language L is to say the corresponding category $C(L)$ has finite products. (See [Poigné, 1986a].)

5.3.13 Functors that preserve products Let $F : \mathcal{A} \to \mathcal{B}$ be a functor between categories. Suppose that

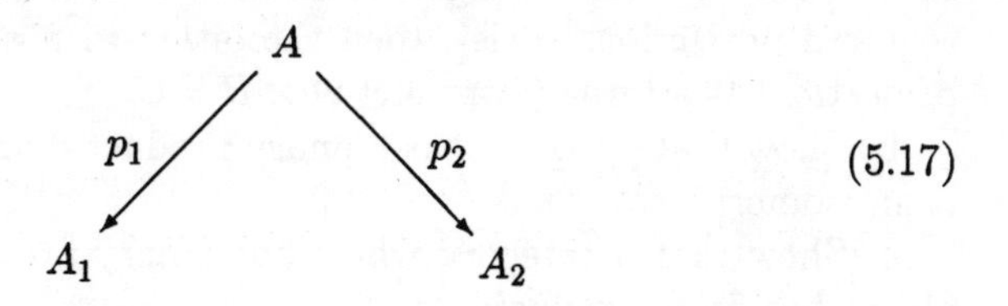

is a (binary) product diagram in $\mathcal{A}$. We say that F **preserves the product** if

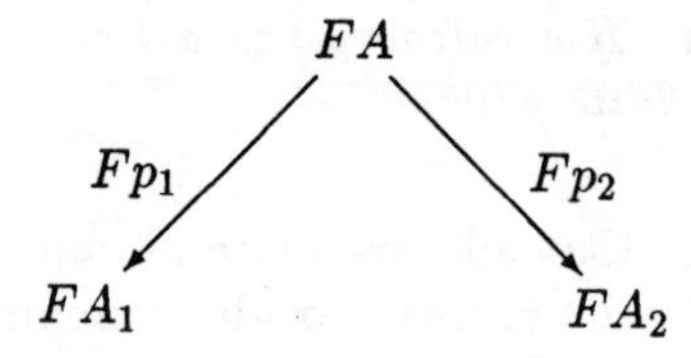

is a product diagram in $\mathcal{B}$. It is important to note that F must preserve the diagram, not merely the object A. It is possible for F to take A to an object which is isomorphic to a product, but do the wrong thing on projections (Exercise 4).

F **preserves canonical products** if $\mathcal{A}$ and $\mathcal{B}$ have canonical products and F preserves the canonical product diagrams.

Similar definitions can be made about a product diagram of any family of objects. A functor is said to **preserve finite products**, respectively **all products** if it preserves every finite product diagram, respectively all product diagrams. Preserving canonical product diagrams is defined analogously (but note Exercise 5.b).

We have a proposition related to Proposition 5.3.10.

5.3.14 Proposition *If a functor preserves terminal objects and binary products, it preserves all finite products.*

5.3.15 Infinite products There is no difficulty in extending the definition of products to allow infinitely many objects. Suppose I is an arbitrary set and $\{A_i\}$, $i \in I$ is an indexed set of objects of the category $\mathcal{C}$. (See 2.6.9.) A product of the indexed set, denoted $\prod_{i \in I} A_i$ or simply $\prod A_i$, is an object of $\mathcal{A}$ together with an indexed set of arrows $p_j : \prod A_i \to A_j$, $j \in I$, such that given any object A of $\mathcal{A}$ together with arrows $q_i : A \to A_i$, for $i \in I$, there is a unique arrow $q = \langle q_i \rangle : A \to \prod A_i$ such that $p_i \circ q = q_i$, for all $i \in I$.

5.3.16 Exercises

1. a. Show that assuming $A \times B$ and $A \times (B \times C)$ (along with the required projections) exist, then the latter with appropriately defined projections is a ternary product $A \times B \times C$.

b. Show that $p : B \to A$ is a unary product diagram *if and only if* p is an isomorphism.

c. Show that a category which has binary products and a terminal object has finite products.

2. Let $\mathcal{C}$ be a category with the following properties.

(i) For any two objects A and B there is an object $A \times B$ and arrows $p_1 : A \times B \to A$ and $p_2 : A \times B \to B$.

(ii) For any two arrows $q_1 : X \to A$ and $q_2 : X \to B$ there is an arrow $\langle q_1, q_2 \rangle : X \to A \times B$.

(iii) For any arrows $q_1 : X \to A$ and $q_2 : X \to B$, $p_1 \circ \langle q_1, q_2 \rangle = q_1$ and $p_2 \circ \langle q_1, q_2 \rangle = q_2$.

(iv) For any arrow $h : Y \to A \times B$, $\langle p_1 \circ h, p_2 \circ h \rangle = h$.

Prove that $\mathcal{C}$ has binary products. (This exercise shows that the property of having binary products can be expressed using rewrite rules.)

3. Show that the underlying functor $U : \mathbf{Cat} \to \mathbf{Grf}$ defined in 3.1.8 preserves products.

4.[†] Give an example of categories $\mathcal{C}$ and $\mathcal{D}$ with products and a functor $F : \mathcal{C} \to \mathcal{D}$ which does not preserve products, but for which nevertheless $F(A \times B) \cong F(A) \times F(B)$ for all objects A and B of $\mathcal{C}$. (Hint: consider the category whose objects are countably infinite sets and arrows are all functions between them.)

5. a. Show that a functor which preserves terminal objects and binary products preserves finite products.

b. Give an example of categories $\mathcal{C}$ and $\mathcal{D}$ with canonical finite products and a functor $F : \mathcal{C} \to \mathcal{D}$ which preserves canonical binary products which does not preserve canonical finite products.

6. Let $\mathbf{N}$ be the set of nonnegative integers with the usual ordering. Show that the category determined by $(\mathbf{N}, \leq)$ has all binary products but no terminal object.

5.4 Sums

A sum in a category is a product in the dual category. To be explicit, the **sum**, also called the **coproduct**, $A + B$ of two objects in a category consists of an object called $A + B$ together with arrows $i_1 : A \to A + B$ and $i_2 : B \to A + B$ such that given any arrows $f : A \to C$ and $g : B \to C$, there is a unique arrow $\langle f; g \rangle : A + B \to C$ for which

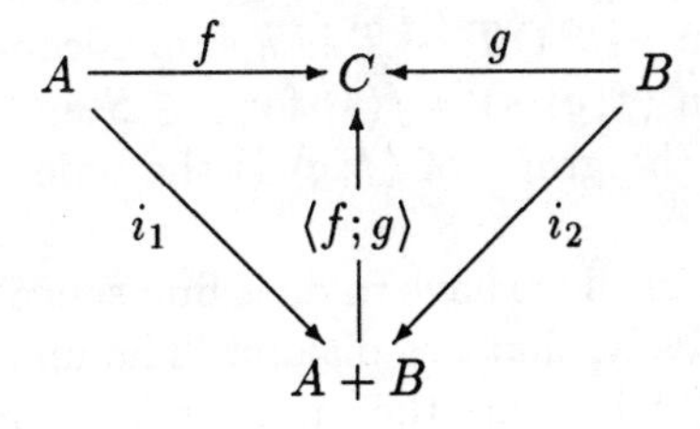

commutes. (The semicolon in the notation '$\langle f;g\rangle$' distinguishes it from the dual idea for products given in 5.2.10.)

The arrows i_1 and i_2 are called the **canonical injections** or the **inclusions** even in categories other than **Set**. These arrows need not be monomorphisms (Exercise 4).

5.4.1 More generally, one can define the sum of any finite or infinite indexed set of objects in a category. The sum of a family $A_1, \ldots, A_n$ is denoted $\sum_{i=1}^{n} A_n$ or $\coprod_{i=1}^{n} A_n$ (the latter symbol is the one typically used by those who call the sum the 'coproduct'). Theorems such as Theorem 5.2.2 and Propositions 5.2.3, 5.2.12 and 5.3.2 are stateable in the opposite category and give uniqueness theorems for sums (this depends on the fact that isomorphisms are isomorphisms in the opposite category). The functor represented by the sum of A and B is $\text{Hom}(A, -) \times \text{Hom}(B, -)$ and the universal element is the pair of canonical injections. (Compare Proposition 5.2.14 and the discussion which follows.)

5.4.2 Definition A **binary discrete cocone** in a category $\mathcal{C}$ is a diagram of the form

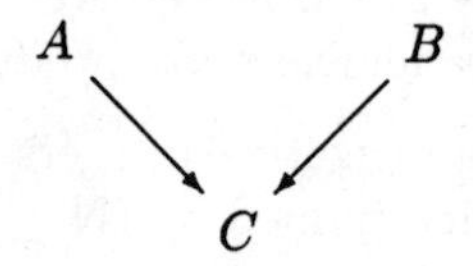

It is a **sum cocone** if the two arrows shown make C a sum of A and B.

5.4.3 The notions of a category having sums or of having canonical sums, and that of a functor preserving sums, are defined in the same way as for products. The notion corresponding to $f \times g$ for two arrows f and g is denoted $f + g$ (Exercise 1).

5.4.4 Sums in Set In the category of sets, one can find a sum of two sets in the following way. If S and T are sets, first consider the case that S and T are disjoint. In that case the set $S \cup T$, together with the inclusion functions $S \to S \cup T \leftarrow T$ is a sum cocone. Given $f : S \to C$ and $g : T \to C$, then $\langle f;g\rangle(s) = f(s)$ for $s \in S$ and $\langle f;g\rangle(t) = g(t)$ for $t \in T$. In this case, the graph of $\langle f;g\rangle$ is the union of the graphs of f and g.

In the general case, all we have to do is find sets S' and T' isomorphic to S and T, respectively, that are disjoint. The union of those two sets is a sum of S' and T' which are the same, as far as mapping properties, as S and T. The usual way this is done is as follows: let

$$S' = S_0 = \{(s,0) \mid s \in S\} \quad \text{and} \quad T' = T_1 = \{(t,1) \mid t \in T\}$$

These sets are disjoint since the first is a set of ordered pairs each of whose second entries is a 0, while the second is a set of ordered pairs each of whose second entries is a 1. The arrow $i_1 : S \to S' \cup T'$ takes s to $(s,0)$, and i_2 takes t to $(t,1)$. If $f : S \to C$ and $g : T \to C$, then $\langle f;g\rangle(s,0) = f(s)$ and $\langle f;g\rangle(t,1) = g(t)$.

Note that it will not do to write $S_0 = S \times \{0\}$ and $T_1 = T \times \{1\}$ since our specification of products does not force us to use ordered pairs and, in fact, S is itself a possible product of S with either $\{0\}$ or $\{1\}$.

5.4.5 Finite sums in programming languages In 5.3.12, we have described products in the category corresponding to a programming language as records. Sums play a somewhat subtler role.

If A and B are types, the sum $A + B$ can be thought of as the free variant or free union of the types A and B. If we are to take this seriously, we have to consider the canonical structure maps $i_1 : A \to A+B$ and $i_2 : B \to A+B$. These are type conversions; i_1 converts something of type A to something of the union type. Such type conversions are not explicit in languages such as Pascal or C, because in those languages, the free union is implemented in such a way that the type conversion $i_1 : A \to A + B$ can always be described as 'use the same internal representation that it has when it is type A'. Thus in those languages, the effect of an operation with domain '$A + B$' is implementation-dependent.

Thus sums and products provide some elementary constructions for functional programming languages. What is missing are some constructions such as **IF...THEN...ELSE** and **WHILE** loops or recursion which give the language the full power of a Turing machine. **IF...THEN...ELSE** can be handled using *disjoint sums*, discussed in 8.6.1. One approach to recursion is to provide a category equivalent to a typed λ-calculus, which is done in Chapter 6. Another is the approach of directly introducing recursion used by [Cockett, 1987, 1989], giving a **locos**. These are discussed in Section 13.2.

[Wagner, 1986b] discusses these constructions and others in the context of more traditional imperative programming languages.

5.4.6 Exercises

1. For $f : S \to S'$ and $g : T \to T'$ in a category with sums, the arrow $f + g : S + T \to S' + T'$ is the unique arrow making the opposite of Diagram (5.11) commute. Describe this construction in **Set**.

2. Show that the sum of two elements in a poset is the least upper bound of the elements.

3. Describe the sum of two posets in the category of posets and monotone maps.

4. Give an example of a category containing a sum with the property that one of the 'canonical injections' is not monic.

5.5 Natural numbers objects

Virtually all the categorical models for programming language semantics suppose there is an object of the category that allows a recursive definition of arrows. To define such an object, we must suppose that the category has a terminal object, which we denote by 1.

5.5.1 Definition An object $\mathbf{N}$, an arrow zero : $1 \to \mathbf{N}$ and an arrow succ : $\mathbf{N} \to \mathbf{N}$ constitute a **natural numbers object** in a category $\mathcal{C}$ if given any object A, any arrow $f_0 : 1 \to A$ and any arrow $t : A \to A$, there is a unique arrow $f : \mathbf{N} \to A$ such that the following diagram commutes.

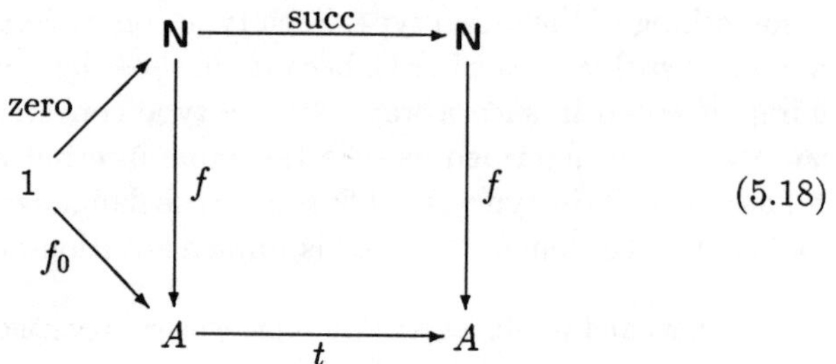

(5.18)

5.5.2 Example A natural numbers object in $\mathcal{C}$ is exactly an initial model in $\mathcal{C}$ of the sketch of Example 4.7.7 (Exercise 2). In particular, the set $\mathbf{N}$ of natural numbers with zero = 0 and succ the successor function is a natural numbers object in **Set**. Let the function zero choose the element $0 \in \mathbf{N}$ and let succ be the usual successor function, which we will denote by s. For any set A, a function $1 \to A$ is an element we may call $a_0 \in A$. The commutation of the triangle forces that $f(0) = a_0$ and then we have that $f(1) = f(s0) = t \circ f(0) = t(a_0)$, $f(2) = f(s1) = t \circ f(1) = t^2(a_0)$ and so on. Thus we can show by induction that any function f making the diagram commute must satisfy $f(n) = t^n(a_0)$. This proves both the existence and the uniqueness of f.

In practice, this definition is not strong enough to be really useful and must be strengthened as follows.

5.5.3 Definition An object $\mathbf{N}$ in a category $\mathcal{C}$ together with arrows zero : $1 \to \mathbf{N}$ and succ : $\mathbf{N} \to \mathbf{N}$ is called a **stable** natural numbers object if for all objects A and B and arrows $f_0 : B \to A$ and $t : A \to A$, there is a unique arrow $f : B \times \mathbf{N} \to A$ such that

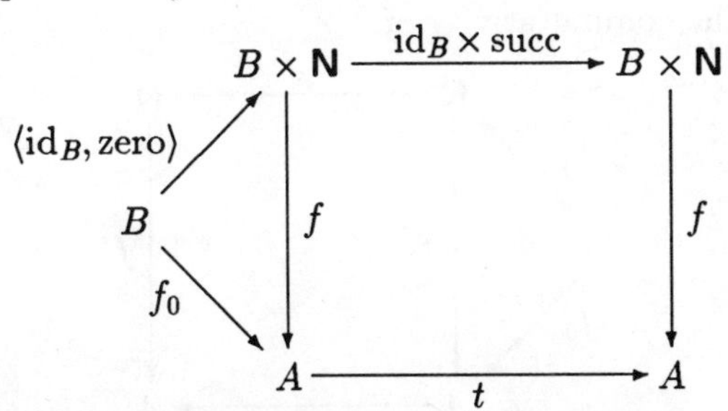

commutes. When $B = 1$, this reduces to the previous definition. We think of B as an object of parameters. For this reason, such an $\mathbf{N}$ is sometimes called a parametrized natural numbers object, even though it is not $\mathbf{N}$ that is parametrized.

This definition is equivalent to the statement that the product projection $B \times \mathbf{N} \to B$ is a natural numbers object in the category $\mathcal{C}/B$ for every object B. Thus we are using 'stable' in its usual sense of 'invariant under slicing'.

5.5.4 Remark In many sources, what we have called a stable natural numbers object is called, simply, a natural numbers object and the weaker concept is given no name. The reason for this is that the concept was originally defined in cartesian closed categories (the subject of Chapter 6) and in such categories the two notions coincide; see Exercise 8 of Section 6.1. In truth, natural numbers objects that are not stable are not very useful.

5.5.5 Theorem *Let $\mathcal{C}$ be a category with a terminal object 1 and $\mathbf{N}$ a sum of countably many copies of 1. Then $\mathbf{N}$ satisfies the specification of a natural numbers object.*

Proof. Let $u_i : 1 \to \mathbf{N}$ be the element of the cocone corresponding to the ith copy of the sum. Let $s : \mathbf{N} \to \mathbf{N}$ be the unique arrow defined by the formula $u_i \circ s = u_{i+1}$. That is the arrow s is the unique arrow such that for each i, the diagram

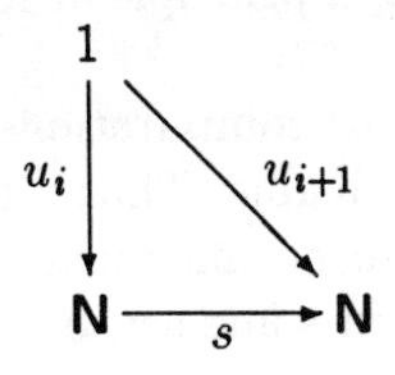

commutes. Now suppose $t : A \to A$ is an endomorphism and $f_0 : 1 \to A$ is a global element. We define a sequence of arrows $f_n : 1 \to A$ by ordinary induction by letting f_0 be the given map and $f_{n+1} = t \circ f_n$. Then we use the universal property of the sum to define $f : \mathbf{N} \to A$ by $f \circ u_n = f_n$. The commutativity of

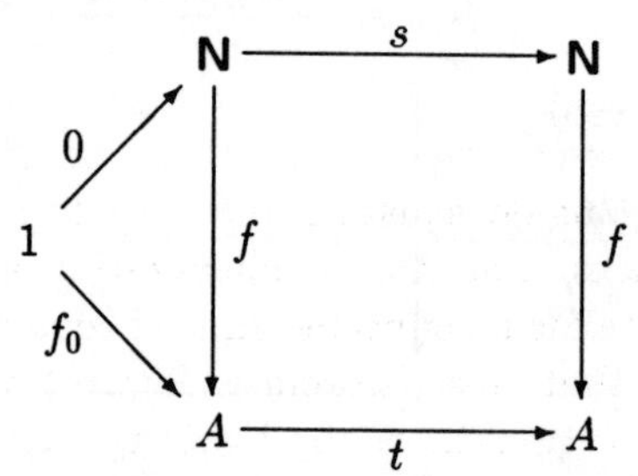

is evident. To show uniqueness, suppose $g : \mathbf{N} \to A$ is an arrow such that

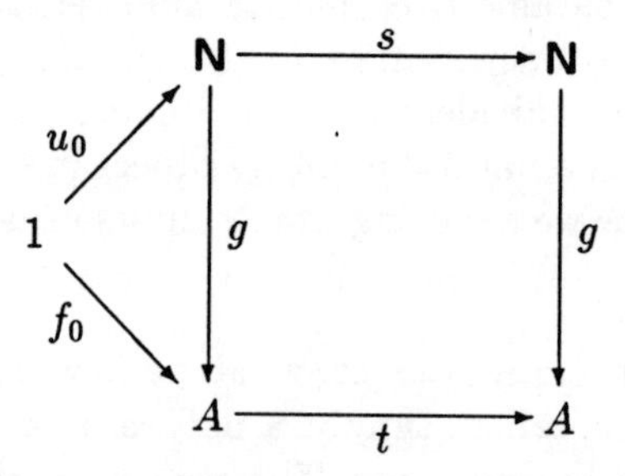

commutes, then the commutativity of the triangle implies that $g \circ u_0 = f_0$ and if we suppose inductively that $g \circ u_n = f_n = f \circ u_n$, then

$$g \circ u_{n+1} = g \circ s \circ u_n = t \circ g \circ u_n = t \circ f_n = f_{n+1} = f \circ u_{n+1}$$

But then it follows by the uniqueness of arrows from a sum that $f = g$. □

5.5.6 There are two comments we would like to make about Theorem 5.5.5.

First, there is a question of whether assuming countable sums is computationally plausible as a model of programming language semantics. We are not talking about the fact that computers are finite. Despite that, it is a reasonable idealization (and common practice) to allow computational models that, in principle at least, allow indefinitely large objects.

The problem is that these computations, although potentially infinite, should be finitely describable. Thus a function from natural numbers to natural numbers can be imagined computable if it can be described by a finite formula in some sense of the word formula. Since a formula can be described with a finite alphabet in a finite way, one can show that only countably many such formulas exist. On the other hand, if a countable sum of copies of 1 exists, that sum **N** can be shown to admit uncountably many functions to itself. It is evident that most of these cannot be given by a formula and make no sense in a model of computation.

Although one cannot actually prove it (owing to imprecision in the notion 'formula'), it seems likely that the class of functions that are computationally reasonable is just the recursive functions (Church's thesis). One should look on a natural numbers object as a computationally meaningful substitute for a countable sum of copies of 1.

The real reason that this theorem is interesting is that it tells us something about the strength of the hypothesis of existence of a natural numbers object; it is weaker than the assumption of existence of countable sums and that is usually considered by category theorists to be fairly weak. Of course, we have just argued above that it is fairly strong, but at any rate the hypothesis of a natural numbers object is weaker.

The second comment is that the construction in Theorem 5.5.5 does not generally give stable natural numbers. Many familiar categories, for example the category of semigroups, have countable (in fact arbitrary) sums and a terminal object and so a natural numbers object which, however, is not stable. In Section 6.1, Exercise 8, and Section 8.6, Exercise 7, we will see additional assumptions that force the existence of a stable natural numbers object.

5.5.7 Exercises

1. Show that the natural numbers, with zero and the successor function, form a stable natural numbers object in **Set**.

2. Show that a natural numbers object in a category is exactly a model of the sketch of Example 4.7.7 in that category. Show that it is in fact an initial model.

3. Suppose **N** is a stable natural numbers object in some category and A and B are arbitrary objects. Show that given any arrows $g : B \to A$ and $h : B \times \mathbf{N} \times A \to A$, there is a unique $f : B \times \mathbf{N} \to A$ such that $f(b, 0) = g(b)$ and $f(b, sn) = h(b, n, f(b, n))$. (This uses variable element notation, as discussed in 2.8.2. It is a great convenience here.) This is the usual formulation of induction, except that B is usually taken to be the cartesian product of a finite number of copies of **N**. (The way to do this is to find an appropriate arrow t from $B \times \mathbf{N} \times A$ to itself, to define $k_0 : B \to B \times \mathbf{N} \times A$ by $k_0(b) = (b, 0, g(0))$ and then let $k : B \times \mathbf{N} \to B \times \mathbf{N} \times A$ be the unique arrow so that

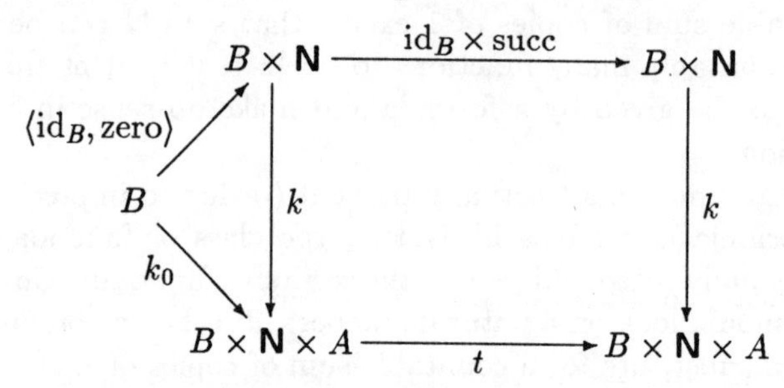

commutes. Then set $f = p_3 \circ k$.)

5.6 Deduction systems as categories

In this section, we describe briefly the connection between formal logical systems and categories.

5.6.1 **A deduction system** has **formulas** and **proofs**. The informal idea is that the formulas are statements, such as $x \leq 7$, and the proofs are valid lines of reasoning starting with one formula and ending with another; for example, it is valid in high school algebra (which is a deductive system) to prove that if $x \leq 7$ then $2x \leq 14$. We will write a proof p which assumes A and deduces B as $p : A \to B$.

Typically, *formal* deductive systems define the formulas by some sort of context free grammar, a typical rule of which might be: 'If A and B are formulas, then so is $A \wedge B$'. The valid proofs are composed of chains of applications of certain specified **rules of inference**, an example of which might be: 'From $A \wedge B$ it is valid to infer A'.

5.6.2 **Assumptions on a deductive system** As in the case of functional programming languages (see 2.2.4), certain simple assumptions on a deductive system, however it is defined, produce a category.

DS–1 For any formula A there is a proof $\mathrm{id}_A : A \to A$.

DS–2 Proofs can be composed: If you have proofs $p : A \to B$ and $q : B \to C$ then there is a valid proof $p; q : A \to C$.

DS–3 If $p : A \to B$ is a proof then $p; \mathrm{id}_B$ and $\mathrm{id}_A; p$ are both the *same* proof as p.

DS–4 If p, q and r are proofs, then $(p;q);r$ must be the same proof as $p;(q;r)$, which could be denoted $p;q;r$.

These requirements clearly make a deductive system a category.

One could take a deductive system and impose the minimum requirements just given to produce a category. One could on the other hand go all the way and make any two proofs A to B the same. In that case, each arrow in the category stands for the usual notion of deducible. The choice of an intermediate system of identification could conceivably involve delicate considerations.

5.6.3 Definition A **conjunction calculus** is a deductive system with a formula **true** and a formula $A \wedge B$ for any formulas A and B, which satisfies CC–1 through CC–5 below for all A and B.

CC–1 There is a proof $A \to$ **true**.

CC–2 If $u : A \to$ **true** and $v : A \to$ **true** are proofs, then $u = v$.

CC–3 There are proofs $p_1 : A \wedge B \to A$ and $p_2 : A \wedge B \to B$ with the property that given any proofs $q_1 : X \to A$ and $q_2 : X \to B$ there is a proof $\langle q_1, q_2 \rangle : X \to A \wedge B$.

CC–4 For any proofs $q_1 : X \to A$ and $q_2 : X \to B$, $p_1 \circ \langle q_1, q_2 \rangle = q_1$ and $p_2 \circ \langle q_1, q_2 \rangle = q_2$.

CC–5 For any proof $h : Y \to A \wedge B$, $\langle p_1 \circ h, p_2 \circ h \rangle = h$.

Similar constructions may be made using sums to get a disjunction calculus. Cartesian closed categories (the subject of Chapter 6) give an implication operator and quantifiers are supplied in a topos (Chapter 14). These topics are pursued in detail in [Lambek and Scott, 1986], [Makkai and Reyes, 1977] and [Bell, 1988].

5.6.4 Exercises

1. Prove that equations CC–1 through CC–5 make the given deductive system a category with finite products.

2. Prove that in any conjunction calculus, for any objects A, B and C, there are proofs

a. $A \wedge A \to A$.

b. $A \to A \wedge A$.

c. $A \wedge B \to B \wedge A$.

d. $(A \wedge B) \wedge C \to A \wedge (B \wedge C)$.

6

Cartesian closed categories

A cartesian closed category is a type of category that as a formal system has the same expressive power as a typed λ-calculus. In this chapter, we define cartesian closed categories and typed λ-calculus in the first two sections. Section 6.3 describes the constructions involved in translating from one formalism to the other, without proof.

Section 6.4 discusses the issues and technicalities involved in translating between a term in the λ-calculus and an arrow in the corresponding category.

Section 6.5 describes the construction of fixed points of endomorphisms of an ω-complete partially ordered object in a cartesian closed category. This is used to provide a semantics for If and While loops.

Most of the cartesian closed categories considered in computer science satisfy a stronger property, that of being 'locally cartesian closed', which is discussed in Chapter 12.4.

A basic reference for cartesian closed categories and their connection with logic is [Lambek and Scott, 1986]. See also [Huet, 1986] and [Mitchell and Scott, 1989]. [Cousineau, Curien and Mauny, 1985] and [Curien, 1986] describe a substantial application of the idea to computing.

In this chapter and later, we frequently use the name of an object to stand for the identity arrow on the object: thus 'A' means 'id_A'. This is common in the categorical literature because it saves typographical clutter.

6.1 Cartesian closed categories

6.1.1 Functions of two variables If S and T are sets, then an element of $S \times T$ can be viewed interchangeably as a *pair* of elements, one from S and one from T or as a *single* element of the product. If V is another set, then a function $f : S \times T \to V$ can interchangeably be viewed as a function of a single variable ranging over $S \times T$ or as a function of two variables, one from S and one from T. Conceptually, we must distinguish between these two points of view, but they are equivalent. (Compare the discussion in 5.2.7.)

In a more general category, the notion of function of two variables should be understood as meaning a function defined on a product. Under certain conditions, such a function can be converted to a function of one variable with values in a 'function object'. We now turn to the study of this phenomenon.

6.1.2 To **curry** a function of two variables is to change it into a function of one variable whose values are functions of one variable. Precisely, if $f : S \times T \rightarrow V$ is a function, and if $[S \rightarrow T]$ denotes the set of functions $S \rightarrow T$, then there is function $\lambda f : S \rightarrow [T \rightarrow V]$ defined by letting $\lambda f(s)$ be the function whose value at an element $t \in T$ is $f(s,t)$.

As an example, the definition on arrows of the functor F_α obtained from a monoid action as in 3.2.3 is obtained by currying α.

In the other direction, if $g : S \rightarrow [T \rightarrow V]$ is a function, it induces a function $f : S \times T \rightarrow V$ defined by $f(s,t) = [g(s)](t)$. This construction produces an inverse to λ that determines an isomorphism

$$\mathrm{Hom}_{\mathbf{Set}}(S \times T, V) \cong \mathrm{Hom}_{\mathbf{Set}}(S, [T \rightarrow V])$$

These constructions can readily be stated in categorical language. The result is a theory that is equivalent to the typed λ-calculus (in the sense of Section 6.2) and has several advantages over it.

6.1.3 Definition A category $\mathcal{C}$ is called a **cartesian closed category** if it satisfies the following:

CCC–1 There is a terminal object 1.

CCC–2 Each pair of objects A and B of $\mathcal{C}$ has a product $A \times B$ with projections $p_1 : A \times B \rightarrow A$ and $p_2 : A \times B \rightarrow B$.

CCC–3 For every pair of objects A and B, there is an object $[A \rightarrow B]$ and an arrow $\mathrm{eval} : [A \rightarrow B] \times A \rightarrow B$ with the property that for any arrow $f : C \times A \rightarrow B$, there is a unique arrow $\lambda f : C \rightarrow [A \rightarrow B]$ such that the composite

$$C \times A \xrightarrow{\lambda f \times A} [A \rightarrow B] \times A \xrightarrow{\mathrm{eval}} B$$

is f.

6.1.4 Terminology Traditionally, $[A \rightarrow B]$ has been denoted B^A and called the **exponential object**, and A is then called the exponent. The exponential notation is motivated by the following special case: if $\mathcal{C}$ is the category of sets and $n = \{0, 1, \ldots, n-1\}$ the standard set with n elements, then B^n is indeed the set of n-tuples of elements of B.

In CCC–3, there is a different arrow eval for each pair of objects A and B. They form the counit of an adjunction, to be described in 12.2.4. Because of that, λf is often called the **adjoint transpose** of f.

6.1.5 In view of Proposition 5.3.10, CC–1 and CC–2 could be replaced by the requirement that C have finite products. There is a slight notational problem in connection with this. We have used p_1 and p_2 for the two projections of a binary product. We will now use

$$p_j : \prod A_i \to A_j$$

for $j = 1, \ldots, n$ to denote the jth projection from the n-ary product. This usage should not conflict with the previous usage since it will always be clear what the domain is.

A related problem is this. Condition CC–3 appears to treat the two factors of $C \times A$ asymmetrically, which is misleading since of course $C \times A \cong A \times C$. Even that last isomorphism is misleading since $C \times A$ and $A \times C$ could be taken to be the same object. Products are of indexed sets of objects, not necessarily indexed by an ordered set, even though our notation appears to suggest otherwise. It gets even worse with n-ary products, so we spell out the notation we use more precisely.

6.1.6 Proposition *In any cartesian closed category, for any objects* $A_1, \ldots, A_n$ *and* A *and any* $i = 1, \ldots, n$, *there is an object* $[A_i \to A]$ *and an arrow*

$$\text{eval} : [A_i \to A] \times A_i \to A$$

such that for any $f : \prod A_j \to A$, *there is a unique arrow*

$$\lambda_i f : \prod_{j \neq i} A_j \to [A_i \to A]$$

such that the following commutes:

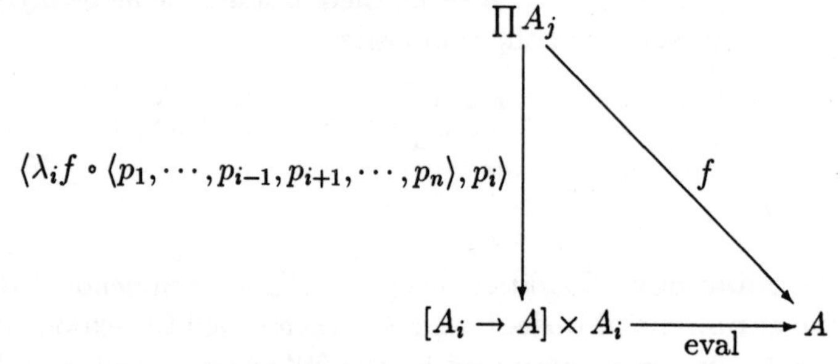

$$\begin{array}{ccc} \prod A_j & & \\ \Big\downarrow {\scriptstyle \langle \lambda_i f \circ \langle p_1, \cdots, p_{i-1}, p_{i+1}, \cdots, p_n \rangle, p_i \rangle} & \searrow^{f} & \\ [A_i \to A] \times A_i & \xrightarrow[\text{eval}]{} & A \end{array}$$

In **Set**, $\lambda_i f$ gives a function from A_i to A for each $n - 1$-tuple

$$(a_i, a_2, \ldots, a_{i-1}, a_{i+1}, \ldots, a_n)$$

6.1.7 Evaluation as universal element For fixed objects A and B of a cartesian closed category, let $F_{A,B}$ denote the functor $\mathrm{Hom}(- \times A, B)$, so that if $g : D \to C$, $F_{A,B}(g) : \mathrm{Hom}(C \times A, B) \to \mathrm{Hom}(D \times A, B)$ takes $f : C \times A \to B$ to $f \circ (g \times A)$. CCC–3 says that $\mathrm{eval} : [A \to B] \times A \to B$ is a universal element for $F_{A,B}$. This is true for every object A and B and implies by Proposition 4.5.11 that $\mathrm{Hom}(- \times A, B)$ is naturally isomorphic to $\mathrm{Hom}(-, [A \to B])$ for every A and B. It follows that a category with finite products can be a cartesian closed category in essentially only one way. The following proposition makes this precise.

6.1.8 Proposition *Let C be a category with finite products. Suppose that for every pair of objects A and B, there are objects $[A \to B]$ and $[A \to B]'$ and arrows $\mathrm{eval} : [A \to B] \times A \to B$ and $\mathrm{eval}' : [A \to B]' \times A \to B$. Suppose these have the property that for any arrow $f : C \times A \to B$, there are unique arrows $\lambda f : C \to [A \to B]$, $\lambda' f : C \to [A \to B]'$ for which $\mathrm{eval} \circ (\lambda f \times A) = \mathrm{eval}' \circ (\lambda' f \times A) = f$. Then for all objects A and B, there is a unique arrow $\phi(A, B) : [A \to B]' \to [A \to B]$ such that for every arrow $f : C \times A \to B$ the following diagrams commute:*

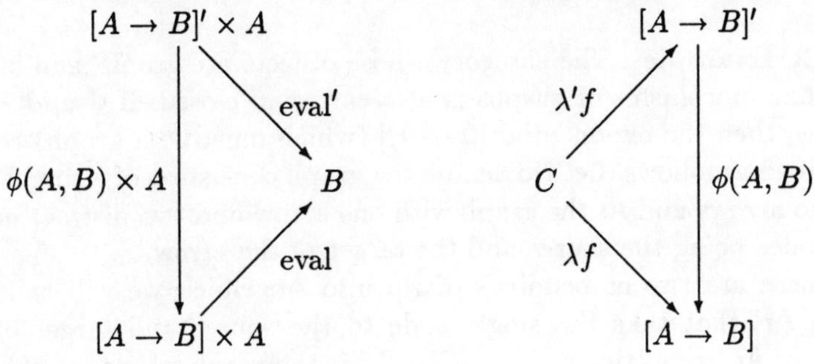

Moreover, ϕ is an isomorphism.

6.1.9 Example The first example of a cartesian closed category is the category of sets. For any sets A and B, $[A \to B]$ is the set of functions from A to B, and eval is the evaluation or apply function. The meaning of λf is discussed above in 6.1.2. Note that in the case of **Set**, $[A \to B]$ *is* $\mathrm{Hom}_{\mathbf{Set}}(A, B)$.

6.1.10 Example A poset B is a **Boolean algebra** if

BA–1 For all $x, y \in B$, there is a least upper bound (supremum) $x \vee y$.

BA–2 For all $x, y \in B$, there is a greatest lower bound (infimum) $x \wedge y$.

BA–3 $\wedge$ distributes over $\vee$: for all $x, y, z \in B$,

$$x \wedge (y \vee z) = (x \wedge y) \vee (x \wedge z)$$

BA–4 B contains two elements $\bot$ and $\top$ that are the least and greatest elements of B respectively.

BA–5 Each element x has a **complement** $\neg x$, meaning that $x \wedge \neg x = \bot$ and $x \vee \neg x = \top$.

It follows that a Boolean algebra has all finite (including empty) infimums and supremums, that both operations $\wedge$ and $\vee$ are commutative and associative and that $\vee$ distributes over $\wedge$. A Boolean algebra B is **trivial** if $\bot = \top$; in that case B contains only one element.

A Boolean algebra B is a poset, so corresponds to a category $C(B)$. This category is a cartesian closed category (Exercise 12).

6.1.11 Example A poset with all finite (including empty) infs and sups that is cartesian closed as a category is called a **Heyting algebra**. A Heyting algebra is thus a generalization of a Boolean algebra. Heyting algebras correspond to intuitionistic logic in the way the Boolean algebras correspond to classical logic. A Heyting algebra is **complete** if it has sups of all subsets. We use this concept in Chapter 14 (see Exercise 1 of Section 14.5).

6.1.12 Example The category whose objects are graphs and arrows are homomorphisms of graphs is also cartesian closed. If $\mathcal{G}$ and $\mathcal{H}$ are graphs, then the exponential $[\mathcal{G} \to \mathcal{H}]$ (which must be a graph) can be described as follows. Let No denote the graph consisting of a single node and no arrows and Ar the graph with one arrow and two distinct nodes, the nodes being the source and the target of the arrow.

There are two embeddings of No into Ar, which we will call $s, t : \mathsf{No} \to \mathsf{Ar}$, that take the single node to the source and target of the arrow of Ar, respectively. Then $[\mathcal{G} \to \mathcal{H}]$ is the graph whose set of nodes is the set $\mathrm{Hom}(\mathcal{G} \times \mathsf{No}, \mathcal{H})$ of graph homomorphisms from $\mathcal{G} \times \mathsf{No}$ to $\mathcal{H}$ and whose set of arrows is the set $\mathrm{Hom}(\mathcal{G} \times \mathsf{Ar}, \mathcal{H})$ with the source and target functions given by $\mathrm{Hom}(\mathcal{G} \times s, \mathcal{H})$ and $\mathrm{Hom}(\mathcal{G} \times t, \mathcal{H})$, respectively. In Exercise 10, we give a more elementary description of the cartesian closed structure. The description given here is a special case of a general result on categories of set-valued functors.

Note that in this case, the object $[\mathcal{G} \to \mathcal{H}]$ is *not* $\mathrm{Hom}(\mathcal{G}, \mathcal{H})$ with the structure of a graph on it. Since $[\mathcal{G} \to \mathcal{H}]$ is an object of the category, it must be a graph; but neither its set of objects nor its set of arrows is $\mathrm{Hom}(\mathcal{G}, \mathcal{H})$. In particular, to prove that a category of sets with structure is not cartesian closed, it is not enough to prove that an attempt to put a structure on the hom set must fail.

6.1.13 Example If $\mathcal{C}$ is a small category then $\mathbf{Fun}(\mathcal{C}, \mathbf{Set})$ is a category; the objects are functors and the arrows are natural transformations. (See 4.3.1.) $\mathbf{Fun}(\mathcal{C}, \mathbf{Set})$ is a cartesian closed category. If F, $G : \mathcal{C} \to \mathbf{Set}$ are two functors, the product is the functor $P : \mathcal{C} \to \mathbf{Set}$ defined by $P(C) = F(C) \times G(C)$ for an object C and for $f : C \to D$, $P(f) = F(f) \times G(f)$, the product of the functions $F(f)$ and $G(f)$. The exponential object $[F \to G]$ is the following functor. For an object C, $[F \to G](C)$ is the set of natural transformations from the functor $\mathrm{Hom}(C, -) \times F$ to G. For $f : C \to D$ an arrow of $\mathcal{C}$ and $\alpha : \mathrm{Hom}(C, -) \times F \to G$ a natural transformation, $[F \to G](f)(\alpha)$ has component at an object A that takes a pair (v, x) with $v : D \to A$ and $x \in F(A)$ to the element $\alpha A(v \circ f, x)$ of $G(A)$. See [Johnstone, 1977], p. 24 for a proof of a more general theorem of which this is a special case.

Example 6.1.12 is a special case of this example. This also implies, for example, that the category of u-structures described in 4.2.5 is cartesian closed, as is the arrow category of $\mathbf{Set}$ (see 4.2.17).

6.1.14 Example The category $\mathbf{Cat}$ of small categories and functors is cartesian closed. In this case, we have already given the construction in 4.3.1: for two categories $\mathcal{C}$ and $\mathcal{D}$, $[\mathcal{C} \to \mathcal{D}]$ is the category whose objects are functors from $\mathcal{C}$ to $\mathcal{D}$ and whose arrows are natural transformations between them.

If $F : \mathcal{B} \times \mathcal{C} \to \mathcal{D}$ is a functor, then $\lambda F : \mathcal{B} \to [\mathcal{C} \to \mathcal{D}]$ is the functor defined this way: if B is an object of $\mathcal{B}$, then for an object C of $\mathcal{C}$, $\lambda F(B)(C) = F(B, C)$, and for an arrow $g : C \to C'$, $\lambda F(B)(g)$ is the arrow $F(B, g) : (B, C) \to (B, C')$ of $\mathcal{D}$. For an arrow $f : B \to B'$ of $\mathcal{B}$, $\lambda F(f)$ is the natural transformation from $\lambda F(B)$ to $\lambda F(B')$ with component $\lambda F(f)(C) = F(f, C)$ at an object C of $\mathcal{C}$. It is instructive to check that this does give a natural transformation.

6.1.15 Example The category consisting of ω-CPOs (2.4.3) and continuous functions is a cartesian closed category, although the subcategory of strict ω-CPOs and strict continuous maps is not. The category of continuous lattices and continuous functions between them is cartesian closed; a readable proof is in [Scott, 1972], Section 3. Many other categories of domains have been proposed for programming languages semantics and many, but not all, are cartesian closed. More about this is in [Scott, 1982], [Smyth, 1983b] and [Dybjer, 1986].

6.1.16 Example In 2.2.2 we described how to regard a functional programming language as a category. If such a language is a cartesian closed category, then for any types A and B, there is a type $[A \to B]$ of functions from A to B. Since that is a type, one can apply programs to data of that type: that is, functions can be operated on by programs. This means that functions are on the same level as other data, often described by saying 'functions are first class objects'.

Proposition 6.1.8 puts strong constraints on making functions first class objects; if you make certain reasonable requirements on your types $[A \to B]$, there is essentially only one way to do it.

6.1.17 Example When a deductive system (Section 5.6) is a cartesian closed category, the constructions giving the exponential turn out to be familiar rules of logic.

Thus, if A and B are formulas, $[A \to B]$ is a formula; think of it as A implies B. Then eval is a proof allowing the deduction of B from A and $[A \to B]$; in other words, it is *modus ponens*. Given $f : C \times A \to B$, that is, given a proof that C and A together prove B, λf is a proof that deduces $[A \to B]$ from C. This is the rule of detachment.

6.1.18 Exercises

1. Let $\mathcal{C}$ be a cartesian closed category with objects A and B. Show that

$$\lambda(\text{eval}) = \text{id}_{[A \to B]} : [A \to B] \to [A \to B]$$

2. Let $g : C' \to C$ and $f : C \times A \to B$ in a cartesian closed category. Show that

$$\lambda(f \circ (g \times A)) = \lambda f \circ g : C' \to [A \to B]$$

3. Show that for any cartesian closed category $\mathcal{C}$, the global elements of $[A \to B]$ (see 2.7.14) are in one to one correspondence with the elements of $\text{Hom}_{\mathcal{C}}(A, B)$.

4.† Prove Proposition 6.1.8.

5. Check that the constructions in 6.1.14 make **Cat** a cartesian closed category.

6. For any objects A, B, C in a cartesian closed category, let ΛB be the map

$$f \mapsto \lambda f : \text{Hom}(C \times A, B) \to \text{Hom}(C, [A \to B])$$

a. Show that ΛB is a bijection.

b. For fixed A and C and $h : B \to B'$, define the function $\widehat{h}C : \text{Hom}(C, [A \to B]) \to \text{Hom}(C, [A \to B'])$ by

$$\widehat{h}C = \Lambda B' \circ \text{Hom}(C \times A, h) \circ (\Lambda B)^{-1}$$

Show that this makes $\widehat{h} : \text{Hom}(-, [A \to B]) \to \text{Hom}(-, [A \to B'])$ a natural transformation of contravariant functors to **Set**.

c. For fixed A and $f : B \to B'$, let $[A \to f] : [A \to B] \to [A \to B']$ be the unique arrow induced by Corollary 4.5.4 and the preceding part of this exercise. Show that Λ is a natural isomorphism from $\text{Hom}(C \times A, -)$ to $\text{Hom}(C, [A \to -])$. (This is an instance of Theorem 12.3.2.)

d. Prove that for any A and $f : B \to B'$,

$$[A \to f] = \lambda(f \circ \text{eval}) : [A \to B] \to [A \to B']$$

7. Show that for any objects A, B, C in a cartesian closed category,

a. $[A \to B] \times [A \to C] \cong [A \to B \times C]$.

b. $[A \times B \to C] \cong [A \to [B \to C]]$.

(Hint: Use Exercise 6 and the Yoneda embedding of 4.5.2.)

8.† Show that if $\mathcal{C}$ is a cartesian closed category with a natural numbers object, then $\mathcal{C}$ has a stable natural numbers object. (See Section 5.5 for the definitions. Note that in a cartesian closed category, to construct an arrow $B \times \mathbf{N} \to A$ is equivalent to constructing one $\mathbf{N} \to [B \to A]$.)

9. In the notation of 6.1.12, show that for any graph $\mathcal{G}$, $\mathcal{G} \times \mathsf{No}$ (the product in the category of graphs and graph homomorphisms) is essentially the set of nodes of $\mathcal{G}$, regarded as a graph with no arrows. (Precisely, find graph homomorphisms from the set of nodes of $\mathcal{G}$ to $\mathcal{G}$ and to No that make the set of nodes the product.)

10.† This exercise provides a description of the cartesian closed structure on the category of graphs and graph homomorphisms that is distinct from that given in 6.1.12. The preceding exercise gives part of the connection between the two descriptions. For two graphs $\mathcal{G}$ and $\mathcal{H}$, define the exponential $[\mathcal{G} \to \mathcal{H}]$ in the category of graphs and graph homomorphisms as follows. A node of $[\mathcal{G} \to \mathcal{H}]$ is a function from G_0 to H_0 (the nodes of $\mathcal{G}$ and $\mathcal{H}$, respectively). An arrow consists of three functions $f_1 : G_0 \to H_0$, $f_2 : G_0 \to H_0$ and $f_3 : G_1 \to H_1$ that satisfy the condition that for any arrow n of $\mathcal{G}$, $\text{source}(f_3(n)) = f_1(\text{source}(n))$ and $\text{target}(f_3(n)) = f_2(\text{target}(n))$. The source and target of this arrow (f_1, f_2, f_3) are f_1 and f_2, respectively. Define $\text{eval} : [\mathcal{G} \to \mathcal{H}] \times \mathcal{G} \to \mathcal{H}$ by

$$\begin{cases} \text{eval}_0(f : G_0 \to H_0, n) &= f(n) \\ \text{eval}_1((f_1, f_2, f_3), a) &= f_3(a) \end{cases}$$

If $f : \mathcal{C} \times \mathcal{G} \to \mathcal{H}$ is a graph homomorphism, define $\lambda f : \mathcal{C} \to [\mathcal{G} \to \mathcal{H}]$ to be the homomorphism that takes a node c of $\mathcal{C}$ to the function $\lambda f(c)$ for which $\lambda f(c)(g) = f(c, g)$ for any node g of $\mathcal{G}$, and that takes an arrow $a : c \to d$ to the arrow $(\lambda f(c), \lambda f(d), f_a)$, where for an arrow u of $\mathcal{G}$, $f_a(u) = f(a, u)$. Show that these constructions provide a cartesian closed structure on the category of graphs and graph homomorphisms.

11.† Describe the one to one correspondence of Exercise 3 for the category of graphs and graph homomorphisms.

12.† Let B be a Boolean algebra. Define $[a \to b]$ to be $\neg a \vee b$. Show that this definition of $[a \to b]$ makes $C(B)$ a cartesian closed category.

6.2 Typed λ-calculus

In order to describe the connection between cartesian closed categories and typed λ-calculus, we give a brief description of the latter. The material is essentially adapted from [Lambek and Scott, 1984, 1986]. It differs from the latter reference in that we do not suppose the existence of a (weak) natural numbers object. The brief discussion in [Lambek, 1986] may be helpful as an overview. The standard reference on the λ-calculus is [Barendregt, 1984], which emphasizes the untyped case.

There is a more general idea of typed λ-calculus in the literature that includes the idea presented here as one extreme. See [Hindley and Seldin, 1986], p. 168ff.

6.2.1 Definition A **typed λ-calculus** is a formal theory consisting of **types, terms, variables** and **equations**. To each term a, there corresponds a type A, called the type of a. We will write $a \in A$ to indicate that a is a term of type A. These are subject to the following rules:

TL–1 There is a type 1.

TL–2 If A and B are types, then there are types $A \times B$ and $[A \to B]$.

TL–3 There is a term $*$ of type 1.

TL–4 For each type A, there is a countable set of terms x_i^A of type A called **variables of type** A.

TL–5 If a and b are terms of type A and B, respectively, there is a term (a, b) of type $A \times B$.

TL–6 If c is a term of type $A \times B$, there are terms $\mathrm{proj}_1(c)$ and $\mathrm{proj}_2(c)$ of type A and B, respectively.

TL–7 If a is a term of type A and f is a term of type $[A \to B]$, then there is a term $f`a$ of type B.

TL–8 If x is a variable of type A and $\phi(x)$ is a term of type B, then $\lambda_{x\in A}\phi(x) \in^{\cdot} [A \to B]$.

The intended meaning of all the term-forming rules should be clear, with the possible exception of $f`a$ which is to be interpreted as f applied to a. The notation $\phi(x)$ for the term in TL–8 means that ϕ is a term that *might* contain the variable x; the '(x)' has no actual significance and is included only to make the expression look familiar.

6.2.2 Before stating the equations, we need some definitions. We omit, as usual, unnecessary subscripts.

If x is a variable, then x is **free** in the term x. If an occurrence of x is free in either of the terms a or b, then that occurrence of x is free in (a, b). If x occurs freely in either f or a, then that occurrence x is free in $f`a$. On the other hand, every occurrence of x is not free (and is called **bound**) in $\lambda_x\phi(x)$.

The term a is **substitutable** for x in $\phi(x)$ if no occurrence of a variable in a becomes bound when every occurrence of x in $\phi(x)$ (if any) is replaced by a. A term is called **closed** if no variable is free in it.

6.2.3 The equations take the form $a =_X a'$, where a and a' are terms of the same type and X is a finite set of variables among which are all variables occurring freely in either a or a'.

TL–9 The relation $=_X$ is reflexive, symmetric and transitive.

TL–10 If a is a term of type 1, then $a =_{\{\}} *$.

TL–11 If $X \subseteq Y$, then $a =_X a'$ implies $a =_Y a'$.

TL–12 $a =_X a'$ implies $f`a =_X f`a'$.

TL–13 $\phi(x) =_{X\cup\{x\}} \phi'(x)$ implies $\lambda_x\phi(x) =_X \lambda_x\phi'(x)$.

TL–14 $\mathrm{proj}_1(a, b) =_X a$; $\mathrm{proj}_2(a, b) =_X b$, for $a \in A$ and $b \in B$.

TL–15 $c =_X (\mathrm{proj}_1(c), \mathrm{proj}_2(c))$, for $c \in A \times B$.

TL–16 $\lambda_x\phi(x)`a =_X \phi(a)$, if a is substitutable for x in $\phi(x)$ and $\phi(a)$ is gotten by replacing every free occurrence of x by a in $\phi(x)$.

TL–17 $\lambda_{x\in A}(f`x) =_X f$ provided $x \notin X$ (and is thus not free in f).

TL–18 $\lambda_{x\in A}\phi(x) =_X \lambda_{x'\in A}\phi(x')$ if x' is substitutable for x in $\phi(x)$ and x' is not free in $\phi(x)$ and vice versa.

It is important to understand that the expression $a =_X a'$ does not imply that the terms a and a' are equal. Two terms are equal only if they are identical. The symbol '$=_X$' may be understood as meaning that in any formal interpretation of the calculus, the denotation of the two terms must be the same.

It should be emphasized that these type and term-forming rules and equations are not exhaustive. There may and generally will be additional types, terms and equations. It is simply that a typed λ-calculus must have at least the types, terms and equations described above.

We will usually abbreviate proj_1 and proj_2 by p_1 and p_2.

6.2.4 Exercises

1. Let c, c' be terms of type $A \times B$. Show that $c =_X c'$ implies that $\mathrm{proj}_1(c) =_X \mathrm{proj}_1(c')$ and $\mathrm{proj}_2(c) =_X \mathrm{proj}_2(c')$.

2. Let a, a' be terms of type A and b, b' terms of type B. Show that if $a =_X a'$ and $b =_X b'$ then $(a, b) =_X (a', b')$.

6.3 λ-calculus to category and back

Both of the concepts of typed λ-calculus and cartesian closed category are adapted to understanding the calculus of functions of several variables, so it is not surprising that they are equivalent. In this section and the next, we describe the constructions which give this equivalence.

6.3.1 Definition of the category Given a typed λ-calculus $\mathcal{L}$, the objects of the category $C(\mathcal{L})$ are the types of $\mathcal{L}$. An arrow from an object A to an object B is an equivalence class of terms of type B with one free variable of type A (which need not actually occur in the terms).

The equivalence relation is the least reflexive, symmetric, transitive relation induced by saying that two such terms $\phi(x)$ and $\psi(y)$ are equivalent if ϕ and ψ are both of the same type, x and y are both of the same type, x is substitutable for y in ψ, and $\phi(x) =_{\{x\}} \psi(x)$, where $\psi(x)$ is obtained from $\psi(y)$ by substituting x for every occurrence of y.

The reason we need equivalence classes is that any two variables of the same type must correspond to the same arrow, the identity, of that object to itself. If $\lambda_{x\in A} x : 1 \to [A \to A]$ is to name the identity arrow of A for any variable $x \in A$, as is intuitively evident, then the arrow corresponding to a variable x of type A must be the identity of A.

The equivalence relation also makes two terms containing a variable of type 1 equivalent (because of TL–10), thus ensuring that 1 will be a terminal object of the category.

6.3.2 Suppose ϕ is a term of type B with at most one free variable x of type A and ψ is a term of type C with at most one free variable y of type B. Note that by replacing, if necessary, x by a variable that is not bound in ψ, we can assume that ϕ is substitutable for y in ψ. We then define the composite of the corresponding arrows to be the arrow which is the equivalence class of $\psi(\phi)$.

6.3.3 Proposition *The category $C(\mathcal{L})$ is a cartesian closed category.*

We will not prove this here. The construction we have given follows Lambek and Scott [1986], who give a proof.

We do not need to say what the cartesian closed structure on $C(\mathcal{L})$ is by virtue of Proposition 6.1.8. Nevertheless, the construction is the obvious one. $A \times B$ with proj_1 and proj_2 is the product of A and B, and $[A \to B]$ is the exponential object. If $\phi(x)$ determines an arrow $f : C \times A \to B$, then x must be a variable of type $C \times A$. Using TL–15 we can substitute (z, y) for x in ϕ, getting $\phi(z, y)$ where z is of type C and y is of type A. Then λf is the equivalence class of $\lambda_z \phi(z, y)$.

Note that $\phi(z, y)$ is not in any equivalence class, since it has two free variables.

6.3.4 Cartesian closed category to λ-calculus Let $\mathcal{C}$ be a cartesian closed category. We will suppose that it has been equipped with given finite products (including, of course, their projections). This means that with each finite indexed set of objects $\{A_i\}$, $i \in I$, there is a given product cone with its projections $p_i : \prod_{i \in I} A_i \to A_i$.

The **internal language** of $\mathcal{C}$ is a typed λ-calculus $\mathsf{L}(\mathcal{C})$. We will describe this λ-calculus by following definition 6.2.1.

The types of $\mathsf{L}(\mathcal{C})$ are the objects of $\mathcal{C}$. The types required by TL–1 and TL–2 are the objects 1, $A \times B$ and $[A \to B]$. We will assume there is a countable set of variables x_i^A of type A for each object A as required by TL–4. The terms are defined by TL–3 through TL–8, and equality by TL–9 through TL–18.

6.3.5 Theorem *Let $\mathcal{C}$ be a cartesian closed category with internal language $\mathcal{L}$. Then $C(\mathcal{L})$ is a category equivalent to $\mathcal{C}$.*

[Lambek and Scott, 1986] prove much more than this. They define what it means for languages to be equivalent and show that if you start with a typed λ-calculus, construct the corresponding category, and then construct the internal language, the language and the original typed λ-calculus are equivalent. They state this in a more powerful way in the language of adjunctions.

6.3.6 Exercise

1. Let A and B be types in a λ-calculus. Show that in $C(\mathcal{L})$, eval : $[A \to B] \times A \to B$ can be taken to be the equivalence class of the term $(\text{proj}_1\, u)`(\text{proj}_2\, u)$, where u is of type $[A \to B] \times A$.

6.4 Arrows vs. terms

It may be instructive to compare what a simple function would look like when defined according to the two kinds of formalism, cartesian closed categories and the λ-calculus.

6.4.1 Example Consider the function $f : \mathbf{N} \times \mathbf{N} \to \mathbf{N}$ that we would traditionally define by the equation

$$f(x, y) = x^2 + 3xy$$

In the typed λ-calculus, this would appear as:

$$f = \lambda_{x \in \mathbf{N}} \lambda_{y \in \mathbf{N}}\ x^2 + 3xy$$

which is directly related to the traditional way of doing it. A rendition in a cartesian closed category is likely to look something like this (in which $*$ is multiplication):

$$\mathbf{N} \times \mathbf{N} \xrightarrow{\langle p_1, p_1, p_1, p_2, 3\rangle} \mathbf{N} \times \mathbf{N} \times \mathbf{N} \times \mathbf{N} \times \mathbf{N} \longrightarrow \cdots$$

$$\cdots \xrightarrow{* \times * \times \mathrm{id}} \mathbf{N} \times \mathbf{N} \times \mathbf{N} \xrightarrow{\mathrm{id} \times *} \mathbf{N} \times \mathbf{N} \xrightarrow{+} \mathbf{N}$$

which appears to be more complicated. However, the categorical formula could be reformulated as:

$$p_1^2 + 3p_1p_2$$

with an obvious semantics.

Thus we see that the only apparent difference between the two systems, at least for a simple example like this, is that we write p_1 and p_2 instead of x and y. The real difference is whether one chooses the formula or the computational process. It is worth noting that the categorical notation exposes clearly that two of the multiplications, at least, can be carried out in parallel. In fact, by emphasizing the process over the result, the categorical approach would appear to offer a natural way of exploring questions like that.

6.4.2 The time has come to explain one point that we have blurred up to now. If a is a term of type A in variables $x_1 \in A_1, \ldots, x_n \in A_n$, and b is a term of type B in variables $y_1 \in B_1, \ldots, y_m \in B_m$, then we described (a, b) as a term of type $A \times B$ in variables $x_1, \ldots, x_n, y_1, \ldots, y_m$. A problem arises with this if there is overlap among the variables. For example, if x is a variable of type A, then (x, x) is a term of type

$A \times A$. But there is only one free variable in the term. Thus the arrow $\langle x, x\rangle$ corresponds to an arrow $A \to A \times A$. The arrow in question is the diagonal arrow $\langle \mathrm{id}, \mathrm{id}\rangle$.

The method, using the notation of the preceding paragraph, is to rename and divide the variables into three classes according to whether they are free in both a and b, free only in a or free only in b.

Suppose that $x_1, \ldots, x_n$ of type $A_1, \ldots, A_n$ respectively are free in a alone, $y_1, \ldots, y_m$ of type $B_1, \ldots, B_m$ respectively are free in b alone and $z_1, \ldots, z_r$ of types $C_1, \ldots, C_r$, respectively are free in both. Then

$$\langle \tilde{a}, \tilde{b}\rangle : A_1 \times \cdots \times A_n \times B_1 \times \cdots \times B_m \times C_1 \cdots C_r \to A \times B$$

is defined as the composite given in (6.1).

$$\begin{array}{c} A_1 \times \cdots \times A_n \times B_1 \times \cdots \times B_m \times C_1 \cdots C_r \\ \Big\downarrow \langle p_1, \ldots, p_n, p_{n+m+1}, \ldots, p_{n+m+r}, p_{n+1}, \ldots, p_{n+m}, p_{n+m+1}, \ldots\rangle \\ A_1 \times \cdots \times A_n \times C_1 \cdots C_r \times B_1 \times \cdots \times B_m \times C_1 \cdots C_r \\ \Big\downarrow \langle a, b\rangle \\ A \times B \end{array} \tag{6.1}$$

In other words, a generalized diagonal is used on the repeated entries. It must be emphasized that this diagonal is used only when the variables are the same, not merely the same type. If x and y are distinct free variables of type A, then the term (x, y) corresponds to an arrow $A \times A \to A \times A$, namely the identity arrow.

A formal description of a conversion process similar to this in the case of signatures and equations is described in 7.7.6.

A similar identification must be made in the case of function application. The construction is essentially the same and need not be repeated.

6.4.3 The advantages of cartesian closed categories Of all the advantages of cartesian closed categories, the most important is that there being no variables, we never have to worry about any clash of variables. Consider the rule that says that under certain circumstances,

$$\lambda_{x \in A}\phi(x) =_X \lambda_{y \in A}\phi(y)$$

The categorical interpretation of this rule is simply that for

$$f : A_1 \times \cdots \times A_n \times A \to B$$

then

$$\lambda f = \lambda f : A_1 \times \cdots \times A_n \to [A \to B]$$

which is identically true. The rule in λ-calculus has to be circumscribed by conditions. If x and y are two variables of type A and $\phi(x) = y$, then $\lambda_x \phi(x) \neq \lambda_y \phi(y)$. The left hand side still has y free, while the right hand side has no free variables. The real problem comes from the use of variables as what are really only place holders.

Another advantage of the categorical approach, perhaps the most important in the long run, is that the composition is built in. This puts the entire structure of category theory, the machinery of commutative diagrams, etc. at the service of the enterprise.

6.4.4 Exercise

1. Translate the function f into an arrow to $\mathbf{N}$ with domain as shown:
   a. $f(x, y) = x^2 + y^2$, domain $\mathbf{N} \times \mathbf{N}$.
   b. $f(x, y, z) = x^2 + y^2$, domain $\mathbf{N} \times \mathbf{N} \times \mathbf{N}$.
   c. $f(x, y) = 5$, domain $\mathbf{N} \times \mathbf{N}$.

6.5 Fixed points in cartesian closed categories

The main result in this section is that the concept of ω-CPO can be defined in any cartesian closed category, and that the construction of fixed points for such objects can be carried out as in Section 2.4.8. This allows cartesian closed categories to be used as semantic models for functional programming languages.

6.5.1 One of the oft-cited reasons that theoretical computer scientists have studied untyped λ-calculus is the existence of a fixed point combinator. This is an element Y with the property that $x`(Y`x) = Y`x$; that is, $Y`x$ is a fixed point of x. This can hardly make sense in the typed λ-calculus. If, for example, there is a type of natural numbers, one could not expect the successor function to have a fixed point. On the other hand, without some kind of fixed point operator, there is no way of interpreting such a functional form as (in the notation of [Backus, 1981a])

$$f = p \Rightarrow q; H(f) \tag{6.2}$$

where $H(f)$ is some kind of function of f. This may be read, 'If p Then q Else $H(f)$'.

It should be noted that although we use Backus' notation for convenience, Backus is interested in actual convergence. Here we are studying the syntactic question; the fixed point combinator is certainly not guaranteed to give a terminating function. In the case of the usual fixed point combinator Y, the fixed point of the identity function is the function $(\lambda x.xx)`(\lambda x.xx)$, the typical example of a nonterminating loop.

In order to describe a program (in some pseudo-language) that actually implements this form, we have to know something about the nature of H. For example, if $H(f) = g \circ f \circ h$, then f in equation (6.2) stands for the infinite program:

```
If p(x) Then Output q(x)
Else If p∘h(x) Then Output g∘q∘h(x)
Else If p∘h∘h(x) Then Output g∘g∘q∘h∘h(x)
...
Else If p∘h^n(x) Then Output g^n∘q∘h^n(x)
...
```

Of course, if the condition is never satisfied, then the program runs forever. This can be programmed in closed form as

```
Proc f(x)
If p(x)
Then Output q(x)
Else Output H(f)(x)
EndIf
EndProc
```

One way around the dilemma mentioned at the beginning of this section is to recognize that one does not need all functions to have a fixed point, only certain ones. Then the question becomes to recognize the ones that do and show that there are enough of them to provide a description of all loops you will need. We turn to the first point first.

6.5.2 Partially ordered objects Suppose D is a partially ordered object in a cartesian closed category. This means that on every hom set $\text{Hom}(A, D)$, there is a partial order relation such that for any $f : B \to A$ and any $g, h : A \to D$, we have that $g \leq h$ in the hom set $\text{Hom}(A, D)$ implies $g \circ f \leq h \circ f$ in the hom set $\text{Hom}(B, D)$. We will similarly say that D is an **ω-complete partial ordered object** or ω-CPO object if it is a partial ordered object and each hom set is an ω-complete partial ordered set, meaning that every increasing countable chain has a least upper bound.

There is also a notion of an internally ω-complete partial order. Throughout this section, we understand all references as being to the external notion.

If D and D' are ω-complete partial orders, an arrow $f : D \to D'$ is ω-continuous if, for any object A and any sequence $g_0 \le g_1 \le \ldots$ of arrows $A \to D$ with supremum g, the arrow $f \circ g$ is the supremum of the sequence $f \circ g_0$, $f \circ g_1, \ldots$. Since $f \le g$ if and only if the supremum of the sequence $f, g, g, g, \ldots$ is g, it easily follows that an ω-continuous arrow is order-preserving.

A partially ordered object D is strict if there is an arrow $\bot : 1 \to D$ such that for any object A and any arrow $f : A \to D$, we have $\bot \circ \langle\rangle \le f$. In other words $\bot$ is the least element of D.

6.5.3 Proposition *Suppose D is a strict ω-CPO object and $f : D \to D$ is an ω-continuous arrow. Then there is an element* $\mathrm{fix}(f) : 1 \to D$ *such that* $f \circ \mathrm{fix}(f) = \mathrm{fix}(f)$. *The element* $\mathrm{fix}(f)$ *is the least element of D with this property.*

Proof. Since $\bot$ is the least element of D, we have that $\bot \le f \circ \bot$ (the terminal arrow on 1 is id_1). Since f is monotone, $f(\bot) \le f \circ f(\bot)$. An obvious induction allows us to prove that $f^n(\bot) \le f^{n+1}(\bot)$, where f^n denotes the n-fold composite of f with itself. Thus we get the sequence

$$\bot \le f(\bot) \le f^2(\bot) \le \cdots \le f^n(\bot) \le f^{n+1}(\bot) \le \cdots$$

We define $\mathrm{fix}(f)$ to be the least upper bound of this sequence. The fact that this is fixed is an immediate consequence of the fact that f preserves the least upper bound of ω-sequences. It is the least fixed point, for if $d : 1 \to D$ is another fixed point, one shows by induction beginning from $\bot \le d$ that $f^n(\bot) \le d$. □

6.5.4 Application of fixed point theory to programs It is not immediately evident what the above fixed point theory has to do with the interpretation of programs. Although most familiar data types have a partial order on them and may even be thought to be ω-complete, most functions do not preserve the order. For example, squaring does not preserve the natural order on the set of integers.

It is not the basic data types to which we want to apply the above constructions. It is rather the arrow types. We want to find a fixed point, not to the successor function, but rather to the function that assigns to each function f the function $p \Rightarrow q; H(f)$ as in Formula (6.2). Moreover, in thinking about this example, one sees that the appropriate order relation is the one in which $f \le g$ if the domain of definition of f

is included in that of g *and* if they agree on that domain. (Compare the discussion in 2.4.8.) This is how we get the fixed point by successively enlarging the domain. But once an element gets into the domain of definition, no further processing is applied to it. Here is one way of arranging this.

Consider some data type D, that is some object of our category of types. Forget about any order that may exist on D. Define a new data type we will denote $D_\perp$. It is just the object $D + \{\perp\}$ and the order relation is simply that $\perp \le d$ for all $d \in D$ and for $d \ne \perp$ and $d \ne d' \in D$, it is never the case that $d \le d'$. We will suppose that $D_\perp$ is ω-complete. If we make certain assumptions about our category that we will discuss in Section 8.6, it can be shown that this is so. These questions are addressed in [Barr, 1988], where the results of this section are 'internalized', that is to say, interpreted in the internal language of the category.

If we do make this assumption, then the data type $[A \to D_\perp]$ is also an ω-CPO object. In fact an increasing sequence of arrows $B \to [A \to D_\perp]$ is equivalent to an increasing sequence of arrows $B \times A \to D_\perp$ and that has a supremum by assumption. It follows that each continuous endoarrow on $[A \to D_\perp]$ has a fixed point.

To apply this to Backus' operator, we have only to show that the functional H is continuous. One interpretation of an arrow $A \to D_\perp$ is as a partial arrow from a complemented subobject (see 8.6.2 for the definition) of A to D. This is extended to an arrow that takes the value $\perp$ on the complementary subobject. An arrow $\phi : [A \to D] \to [A \to D]$ preserves order if and only if whenever g extends f, then $\phi(g)$ extends $\phi(f)$. As already mentioned, the order relation is that of extension of domain. So to apply Proposition 6.5.3, we have to show that if f is a restriction of g to a smaller domain, then $H(f)$ is a restriction of $H(g)$. It must also be shown that if f is the least upper bound of the increasing sequence

$$f_0 \le f_1 \le \cdots \le f_n \le \cdots$$

then $H(f)$ is the least upper bound of

$$H(f_0) \le H(f_1) \le \cdots \le H(f_n) \le \cdots$$

Now Backus considers the following possible functional forms for H (it is a little hard to apply this directly, since for Backus, there is just one data type, list):

FF–1 $H(f) = r$ (a constant; r is not constant, H is).

FF–2 $H(f) = f_i$ where $f = \langle f_1, \ldots, f_n \rangle$ (here he uses the fact that everything is a list).

FF–3 $H(f) = \Gamma \circ \langle E_1(f), \ldots, E_n(f) \rangle$, where $E_1, \ldots, E_n$ are (simpler) functional forms and Γ is a function.

All three of these rules generate ω-continuous functions on $[D \to D]$ and so are covered by Proposition 6.5.3. Of course, there are many other possibilities for H. It must be emphasized that Backus is mainly concerned about convergence and therefore puts other constraints on H, whereas we are dealing with programs.

7
Finite discrete sketches

A formal theory in mathematical logic is a specification method based on strings of symbols as the formal structure. A signature, with equations, is a specification based on tuples as the formal structure. A sketch is a specification based on graphs as the formal structure. As such, sketches are the intrinsically categorical way of providing a finite specification of a possibly infinite mathematical object or class of models.

In Section 4.6 we described a weak form of sketch, linear sketches, to illustrate the constructions involved. In this chapter we generalize the notion of linear sketch to allow operations with more than one argument and to allow choices. These examples show the power of the sketch concept in a way that linear sketches could not.

We discuss two major types of sketches. FP (finite product) sketches are developed in Section 7.1. The example in Section 7.2 shows the connection with signatures. A notation resembling the notation for signatures and equations is developed in Section 7.3, but this notation is intended to be informal. The formal object is the sketch, not the notation. The connection between FP sketches and the method of signatures and equations is made more explicit in Section 7.7 (which, by the way, is not needed in the rest of the book).

The semantics for a sketch are given by certain graph homomorphisms. The nature of FP sketches is such that models can be taken to be in an arbitrary category with finite products. We describe this in Section 7.4.

Each FP sketch gives rise to an FP theory (Section 7.5) which is a category with finite products. A model of an FP sketch in a category which has finite products is essentially the same as a functor that preserves finite products from the theory to the category. In this sense, the theory is analogous to the formal language of mathematical logic, and the sketch to the recursive rules which define the language.

Besides the functorial semantics described above, an FP sketch has an initial algebra semantics which is described in Section 7.6. We also discuss the construction of free algebras for an FP sketch; the free monoid construction is a special case of this.

Section 7.8 discusses FD (finite discrete) sketches, which allow the specification of objects which are sums as well as products. Section 7.9

gives a detailed description of the sketch for fields as an example of an FD sketch. The initial algebra construction for FD sketches works with some modification (Section 7.10). FD sketches have theories, too, but discussion of that is postponed until Chapter 9.

The concept of sketch is due to Charles Ehresmann and has been highly developed by his students in France. Their formalism is different from ours and is described in [Bastiani and Ehresmann, 1968] and (using more up-to-date notation) [Guitart and Lair, 1980]; the latter paper describes the connections with other formalisms in some detail.

7.1 Finite product sketches

We first define discrete cones, which are used to specify that an object of a sketch is a product.

7.1.1 A **finite discrete cone** in the graph $\mathcal{G}$ consists of a finite discrete graph $\mathcal{I}$ (a finite discrete graph is essentially the same thing as a set), a graph homomorphism $L : \mathcal{I} \to \mathcal{G}$, a node n of $\mathcal{G}$ and a collection of arrows $p_i : n \to Li$, one for each node $i \in \mathcal{I}$. The node n is called the **vertex** of the cone and the diagram L the **base** of the cone. Since $\mathcal{I}$ is discrete, the base of the cone is an indexed family of nodes. In particular it is possible to have $Li = Lj$ for $i \neq j$.

We will often use other labels for the nodes and arrows. Thus in the discrete cone

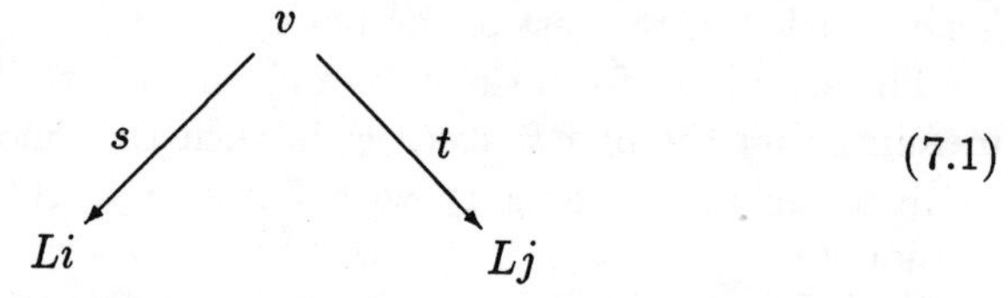

(7.1)

v is the vertex, s is the arrow p_i and t is the arrow p_j. Note that if Li and Lj are the same object but $i \neq j$, s and t will normally be distinct arrows.

When we study general limits we will introduce cones with arrows between the nodes of the base, so that the base is itself a diagram. This is why we call the cones here 'discrete' – the base is a discrete graph.

7.1.2 Definition A cone in a *category* is called a **product cone** if it is a product diagram in the category.

7.1.3 Definition A **finite product sketch** or **FP sketch** $\mathcal{S}$ is a triple $(\mathcal{G}, \mathcal{D}, \mathcal{L})$ where $\mathcal{G}$ is a finite graph, $\mathcal{D}$ a finite set of finite diagrams in $\mathcal{G}$ and $\mathcal{L}$ a finite set of finite discrete cones in $\mathcal{G}$.

7.1.4 Definition Let $\mathcal{S} = (\mathcal{G}, \mathcal{D}, \mathcal{L})$ be an FP sketch and $\mathcal{C}$ be a category. By a **model** of $\mathcal{S}$ in $\mathcal{C}$, we mean a model M of the linear sketch $(\mathcal{G}, \mathcal{D})$ in $\mathcal{C}$ that has the additional property that for any cone $L : \mathcal{I} \to \mathcal{G}$ in $\mathcal{L}$, the composite $M \circ L$ is a product cone. For example if the cone looks like

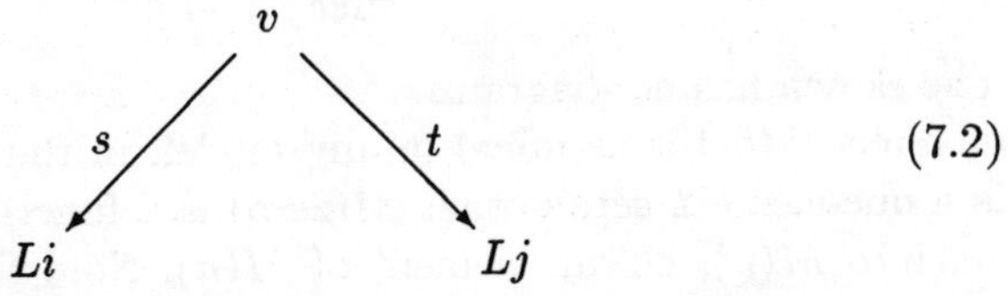

(7.2)

then our condition requires that

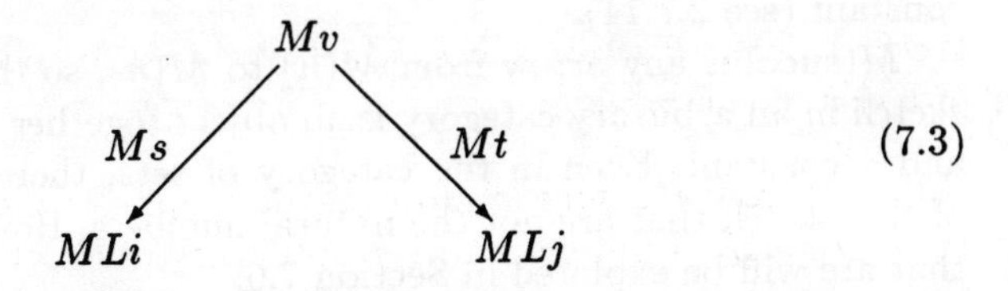

(7.3)

be a product cone in $\mathcal{C}$. If the cone is

$$v \tag{7.4}$$

meaning that the base is empty, then the cone

$$Mv \tag{7.5}$$

should also be a product cone, which means precisely that Mv should be a terminal object of $\mathcal{C}$.

When we refer to a model of a sketch without mentioning the category $\mathcal{C}$, we mean a model in **Set**.

7.1.5 A simple example of an FP sketch In 4.7.7, we described a linear sketch with constants for the natural numbers. Here we do the same thing with an FP sketch. The sketch has two nodes we will name 1 and n. There is just one cone, namely the cone with 1 as vertex and empty base. As we mentioned above, the result is that in any model M, $M(1)$ must be the terminal object of the value category. The graph has just two arrows,

$$\text{zero} : 1 \to n$$

$$\text{succ} : n \to n$$

The sketch has no diagrams.

Since $M(1)$ is terminal in any model, in the category of sets $M(1)$ is a one-element set, so that $M(\text{zero})$ is a function from a one-element set into $M(n)$, i.e. an element of $M(n)$. Note that this observation is independent of which one-element set $M(1)$ actually is. In an arbitrary category, $M(1)$ is terminal, and $M(\text{zero})$ is then a global element or constant (see 2.7.14).

$M(\text{succ})$ is any arrow from $M(n)$ to $M(n)$, so that a model of this sketch in an arbitrary category is an object together with an endoarrow and a constant. Even in the category of sets, there are many models of this sketch that are not the natural numbers. How to pick out those that are will be explored in Section 7.6.

7.1.6 Infinite lists Let S be the sketch with three nodes, 1, d and l, arrows $a, b : 1 \to d$, head : $l \to d$ and tail : $l \to l$, no diagrams, and two cones, one empty with 1 at the vertex and the other C given by

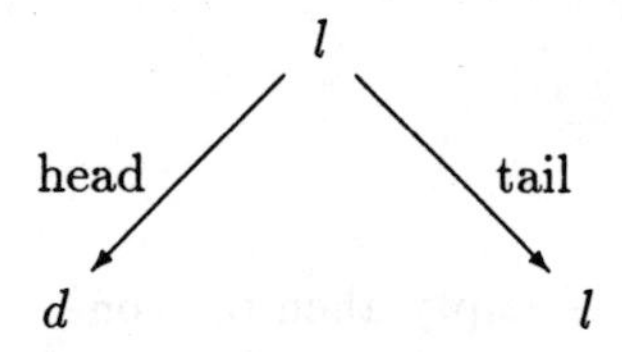

One model of this sketch is the model M with $M(l)$ the set of all infinite sequences of a's and b's. $M(\text{head})$ gives the first entry of a sequence and $M(\text{tail})$ gives the sequence obtained by deleting the first entry of the sequence.

If you try to define a model M of this sketch in which $M(l)$ is the set of finite lists of a's and b's and head and tail have their usual meaning, you will run into trouble eventually. The reason is that if a list is finite, then repeated applications of $M(\text{tail})$ to it will eventually give the empty list, but the empty list has no head, so there is no obvious way to allow it to be in $M(l)$. Various ways of handling this have been suggested in the literature. We will suggest one way using the concept of sum in Section 7.8.

7.1.7 We can introduce addition into the sketch in 7.1.5 by adding a binary operation, using our new ability to specify that an object be a cartesian product. We need an object $n \times n$ and a diagram

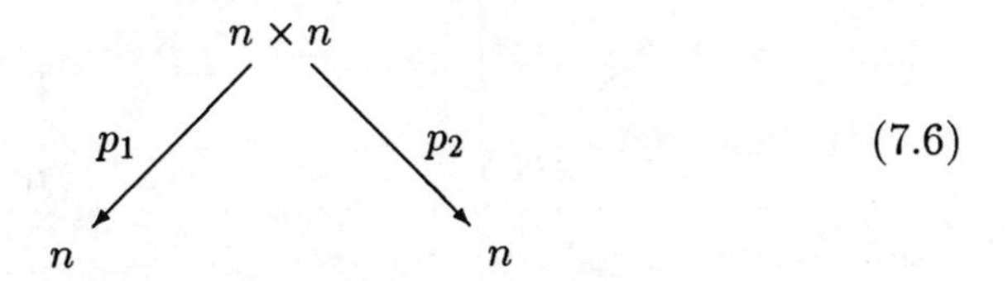

(7.6)

The notation $n \times n$ indicates our intention that the node become a product in a model. It cannot itself be a product, since it is a node in a graph, not an object of a category and products are defined only for categories.

We will also need new arrows:

$$\begin{aligned} z &: n \to n \\ \mathrm{id}_n &: n \to n \\ \langle z, \mathrm{id}_n \rangle &: n \to n \times n \\ + &: n \times n \to n \\ \mathrm{id}_n \times \mathrm{succ} &: n \times n \to n \times n \end{aligned}$$

Definition 7.1.4 will force $M(n \times n)$ to be isomorphic to $M(n) \times M(n)$ for a model M. To simplify the discussion, we assume $M(n \times n)$ is actually identical to $M(n) \times M(n)$. In 7.2.7 and 7.2.8 below we will show how this assumption is avoided.

The following diagrams ensure that this operation will satisfy the inductive definition of addition for integers:

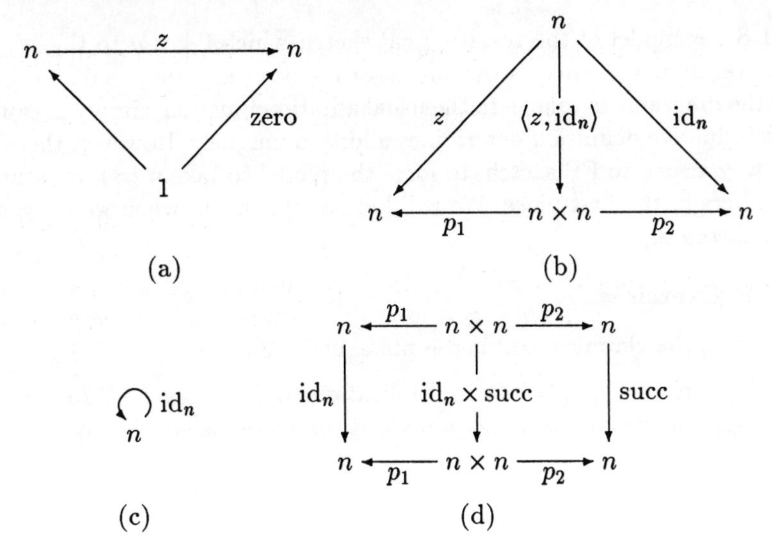

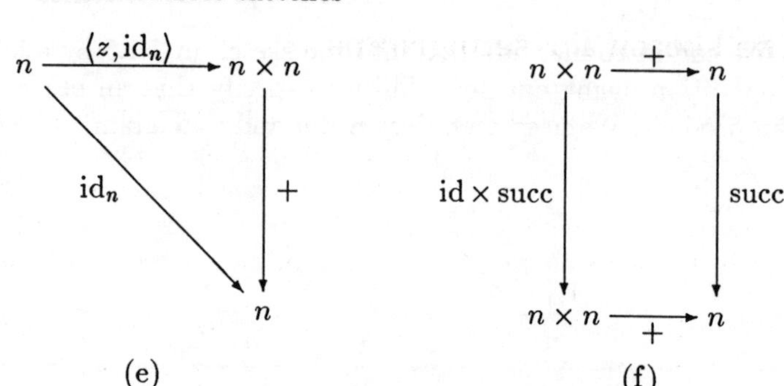

(e) (f)

In a model M, the commutativity of (a) forces the arrow $M(z)$ to be the constant map which takes any element of $M(n)$ to the element determined by zero. The commutativity of (b) forces the center arrow to go to the product suggested by its name: it forces

$$M\langle z, \mathrm{id}_n \rangle = \langle M(z), M(\mathrm{id}_n) \rangle = \langle M(z), \mathrm{id}_{M(n)} \rangle$$

The commutativity of (c) forces

$$M(\mathrm{id}_n) = \mathrm{id}_{M(n)}$$

and that of (d) forces

$$M(\mathrm{id}_n \times \mathrm{succ}) = \mathrm{id}_{M(n)} \times M(\mathrm{succ})$$

The last two force the inductive definition of addition to be true in a model: $\mathrm{zero} + k = k$ and $k + \mathrm{succ}(m) = \mathrm{succ}(k + m)$.

7.1.8 A model of the resulting FP sketch which takes n to the set of natural numbers, succ to the successor function and zero to 0 is forced by the diagrams to take $+$ to the usual addition function, simply because the inductive definition determines addition uniquely. However, there is no way, using an FP sketch, to force the model to take n to the natural numbers in the first place. We will look at this again when we consider FD sketches.

7.1.9 Exercises

1. Prove the claim concerning $+$ made in 7.1.8.

2. Explain how to adjoin a multiplication to Example 7.1.7 so that it becomes the usual multiplication when the model is actually **N**.

7.2 The sketch for semigroups

We now develop a substantial example of an FP sketch of a type of mathematical structure, namely the sketch for semigroups. It exhibits many of the issues and subtleties concerning presentation of structures by sketches. The presentation is quite lengthy; later we introduce a notation which makes such presentations shorter and easier to read.

7.2.1 We define an FP sketch $\mathcal{S} = (\mathcal{G}, \mathcal{D}, \mathcal{L})$. The graph $\mathcal{G}$ has three nodes we will give the names s, $s \times s$ and $s \times s \times s$. As in 7.1.7, these three nodes are not, indeed cannot be, products (which are defined only in categories). The names nevertheless suggest what they will become when a model is applied, as we will see.

The arrows in the graph are:

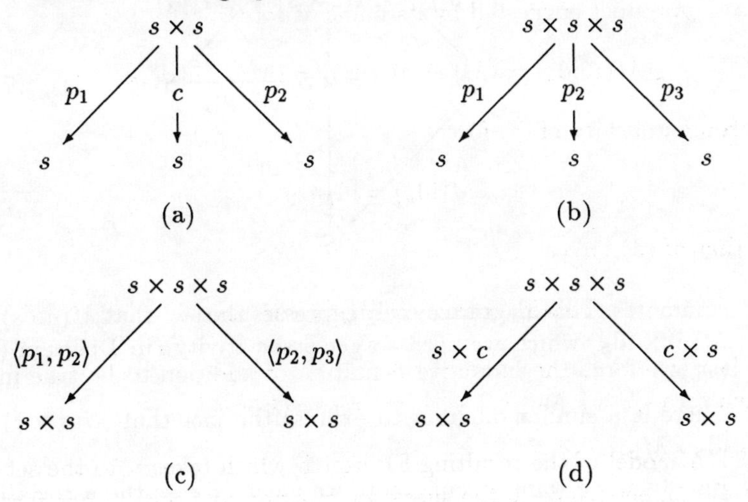

We have given the graph in four pictures, but it is just one graph with three nodes and ten arrows. Note that there are two arrows labeled p_1 and two labeled p_2.

7.2.2 Before giving the details of the diagrams and cones, we pause to explain the intent of this example. Up till now, we have described a graph. What is a model of this graph in the category **Set**? We need sets we call $S = M(s)$, $S^2 = M(s \times s)$ and $S^3 = M(s \times s \times s)$. For the moment, the exponents are simply superscripts. In addition we require functions $M(p_i) : S^2 \to S$, $i = 1, 2$, $M(p_i) : S^3 \to S$, $i = 1, 2, 3$, $M(c) : S^2 \to S$ and $M(s \times c)$ and $M(c \times s)$ from S^3 to S^2. So far, this is nothing familiar, but if we now suppose, as suggested by the notation, that S^2 and S^3 are actually the cartesian square and cube of S and if we make certain subsidiary assumptions given by diagrams to be described later, these data cause S to be a semigroup whose multiplication map is given by $M(c) : S \times S \to S$.

The subsidiary hypotheses are

(i) The various $M(p_i)$ are indeed the projections suggested by the notation.

(ii) $M(\langle p_1, p_2\rangle) = \langle M(p_1), M(p_2)\rangle$ and similarly for $\langle p_2, p_3\rangle$.

(iii) $M(c \times s) : S \times S \times S \to S \times S$ is the unique function (guaranteed by the specification for products) for which the diagram

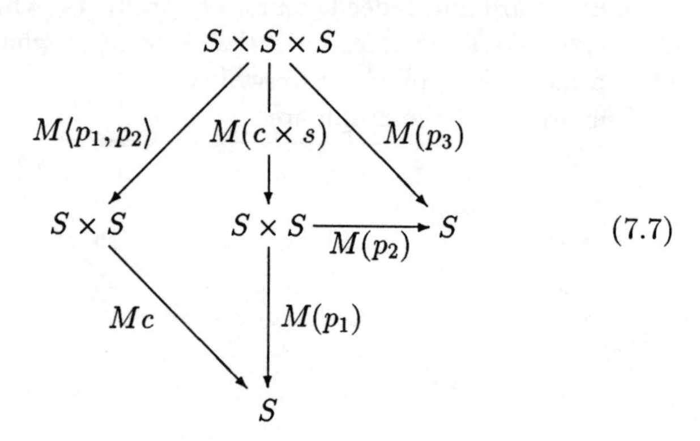

(7.7)

commutes. This diagram merely expresses the fact that $M(c \times s) = M(c) \times \mathrm{id}_S$, which we need to get associativity (in Diagram (d) below).

(iv) There is a similar diagram to express the fact that $M(s \times c) = \mathrm{id}_S \times M(c)$.

(v) Finally, if we want a semigroup, we must express the associative law of the multiplication. This is done by saying that the diagram

$$\begin{array}{ccc} S \times S \times S & \xrightarrow{M(s \times c)} & S \times S \\ {\scriptstyle M(c \times s)}\downarrow & & \downarrow{\scriptstyle M(c)} \\ S \times S & \xrightarrow[M(c)]{} & S \end{array} \qquad (7.8)$$

commutes.

7.2.3 Our task will be to express these requirements in our sketch. This is done as follows. We let $\mathcal{D}$ consist of the following diagrams:

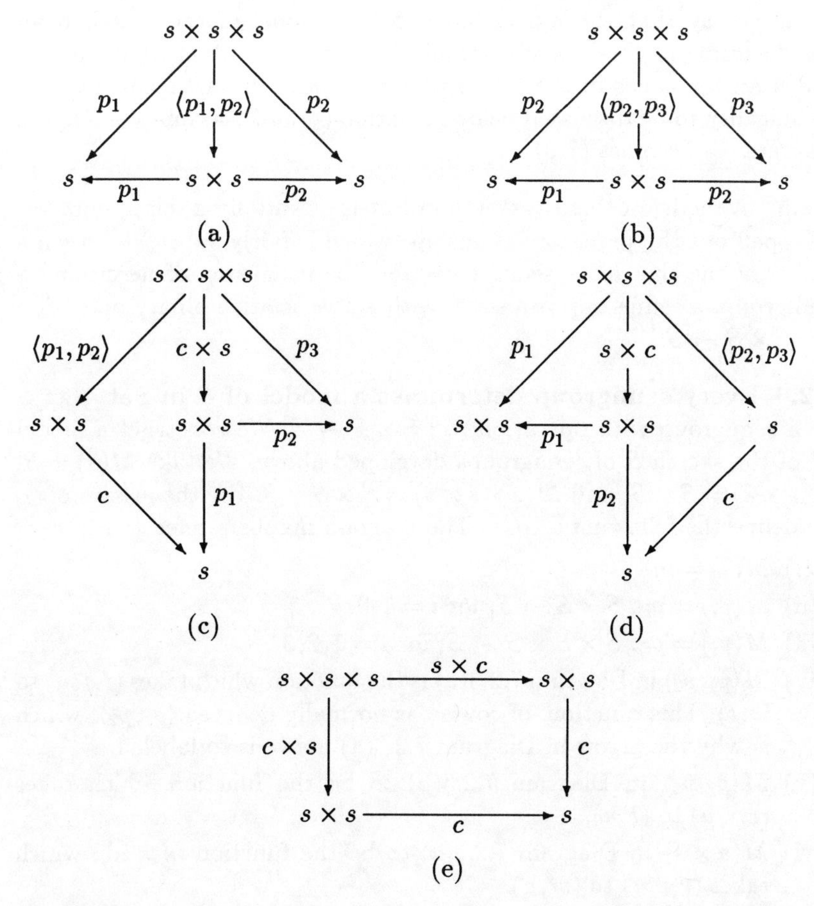

These five diagrams have (e) as their main statement; (c) and (d) are needed to define arrows which occur in (e), and (a) and (b) are needed to define arrows which occur in (c) and (d). This construction is reminiscent of the way you construct progressively higher level procedures in Pascal culminating in the procedure which actually does what you want.

7.2.4 The set $\mathcal{L}$ of cones consists of the following:

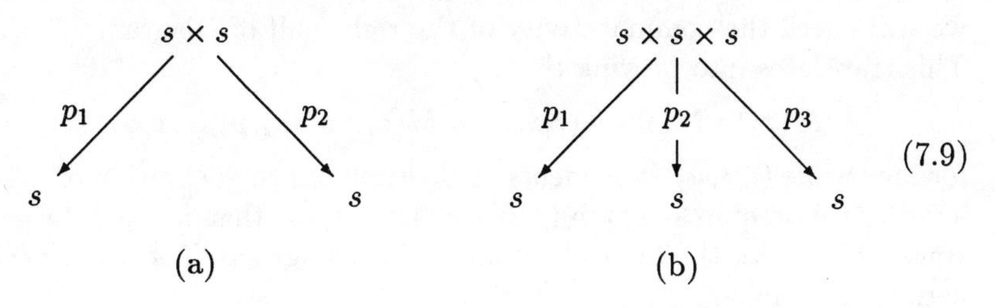

(7.9)

Then we say that the model M of $\mathcal{S}$ is a model of the sketch if all the diagrams in $\mathcal{D}$ become commutative diagrams when M is applied and if all the cones in $\mathcal{L}$ become product cones. In particular, up to a technicality to be discussed below, diagram (c) in $\mathcal{D}$ becomes (7.7), and diagram (e) becomes (7.8).

7.2.5 A model of this FP sketch in **Set** is essentially a semigroup. We will spell out the precise relationship (which is fairly subtle) between a model of the sketch for semigroups and the usual way of describing a semigroup: a semigroup is a set S with an associative binary operation $m : S \times S \to S$.

7.2.6 Every semigroup determines a model of $\mathcal{S}$ in Set Let S be a semigroup with operation $m : S \times S \to S$. We construct a model M of the sketch $\mathcal{S}$ of semigroups developed above. We take $M(s) = S$, $M(s \times s) = S \times S$, and $M(s \times s \times s) = S \times S \times S$. For the arrows of $\mathcal{G}$, we define the following arrows. The diagram numbers refer to 7.2.3.

(i) $M(c) = m$.

(ii) $M(p_i) = p_i : S \times S \to S$, for $i = 1, 2$.

(iii) $M(p_i) = p_i : S \times S \times S \to S$, for $i = 1, 2, 3$.

(iv) $M\langle p_1, p_2\rangle$ in Diagram 7.2.3(a) is the function which takes (r, t, u) to (r, t). This function, of course, is normally denoted $\langle p_1, p_2\rangle$, which is why the arrow in Diagram 7.2.3(a) above is so labeled.

(v) $M\langle p_2, p_3\rangle$ in Diagram 7.2.3(b) to be the function which takes (r, t, u) to (t, u).

(vi) $M(c \times s)$ in Diagram 7.2.3(c) to be the function $m \times \mathrm{id}_S$ which takes (r, t, u) to (rt, u).

(vii) $M(s \times c)$ in Diagram 7.2.3(d) to be the function $\mathrm{id}_S \times m$ which takes (r, t, u) to (r, tu), i.e., $(r, m(t, u))$.

Since we have required the arrows labeled p_i to be actual product projections, the two cones in the sketch both become limit cones in **Set**. Similarly, all five diagrams in 7.2.3 become commutative diagrams; in particular, Diagram 7.2.3(e) is just the associative law.

As an example of how one shows the diagrams become commutative, we will check the commutativity of the right half of Diagram 7.2.3(d). This translates into proving that

$$M(p_2) \circ M(s \times c)(r, t, u) = M(c) \circ M\langle p_2, p_3\rangle(r, t, u)$$

for any triple (r, t, u) of elements of S. Applying the definition of M on arrows, this requires that $p_2(r, tu) = M(c)(t, u)$, that is, that $tu = tu$ since $M(c) = m$, the multiplication of the semigroup. The other verifications are similar.

7.2.7 Every model of $\mathcal{S}$ in Set determines a semigroup If M is a model of $\mathcal{S}$ in **Set**, then $M(c): M(s^2) \to M(s)$ determines a binary operation on $M(s)$, since $M(s^2)$, with the functions $M(p_i): M(s^2) \to M(s)$ for $i = 1, 2$, is a product $M(s) \times M(s)$ in **Set**. (See the discussion in 5.2.7.)

Since M takes every diagram of $\mathcal{S}$ to a commutative diagram in **Set**, Diagram 7.8 commutes, and because diagrams (c) and (d) of 7.2.3 commute, we have that the following diagram commutes. Here we write $S \times S$ and $S \times S \times S$ for $M(s \times s)$ and $M(s \times s \times s)$, respectively. Because of the unique isomorphisms given by Theorem 5.2.2, it does not matter which particular product is used since if the diagram commutes with one, it will commute with any other one (see 5.2.6 and Proposition 5.2.18).

$$\begin{array}{ccc} S \times S \times S & \xrightarrow{M(s) \times M(c)} & S \times S \\ {\scriptstyle M(c) \times M(s)}\downarrow & & \downarrow{\scriptstyle M(c)} \\ S \times S & \xrightarrow[M(c)]{} & S \end{array} \tag{7.10}$$

But this is precisely the associativity of the multiplication.

7.2.8 Subtleties The preceding constructions exhibit the subtle relationship between the usual informal method of describing a mathematical structure and a model of a sketch of the structure.

In the first place, models do not have to take vertices of cones to canonical products; thus $M(s \times s)$ need not be the same as $M(s) \times M(s)$. This was just discussed in connection with Diagram 7.10. In any case, one can regard the map $M(c): M(s \times s) \to M(s)$ as a binary operation on $M(s)$ defined on a *coded* form of the cartesian product $M(s) \times M(s)$, as discussed in 5.2.1. The unique isomorphism between $M(s \times s)$ and the actual set of ordered pairs guarantees that the binary operation is uniquely defined on $M(s) \times M(s)$.

A deeper difference is that there are no distinguished nodes or operations in a sketch. The graph of the sketch for semigroups, for example, has three nodes, no one singled out, whereas in the usual definition of semigroup, the underlying set S (corresponding to $M(s)$ in a model of the sketch) is singled out and other things are defined in terms of it. Similarly, in the graph there are various arrows; c is just one of them.

The sketch approach requires you to construct everything you need explicitly from a few basic constants and operations. Consider the traditional definition of semigroup. You are given a set S and a binary operation on S. The set $S \times S$ is regarded as there for you to use in giving the binary operation. Similarly, when you state the associative law you use $S \times S \times S$ (or more traditionally, triples of variables from S) without comment: you *already have it available.*

By contrast, when you use a sketch to describe semigroups, you have to give explicitly everything you use and every property it has. Nothing exists until you construct it, as in lazy evaluation, where a datum is not constructed until you need it. Thus to define the binary operation, you need a node to become the cartesian product of $M(s)$ with itself, so you put it in the graph and you put in a cone making it a cartesian product. To state the associative law, you need $M(c) \times M(\mathrm{id}_{M(s)})$, so you have to put in the arrow $c \times s$ and Diagram 7.2.3(c); to construct *that*, you need $\langle p_1, p_2 \rangle$ and Diagram 7.2.3(a); and so on.

The result looks much more complicated than the usual way of defining a semigroup, even if you use the elaborate formal notation of signatures and equations. That is because everything you need is constructed from scratch. The sketch is the formal object corresponding to the informal specification for semigroups, and it is appropriate for the formal object to expose all the girders and braces, so to speak. *Sketches are not designed as notation, but as a mathematical structure embodying the formal syntax.* We introduce a usable informal notation in Section 7.3 below which allows the efficient description of sketches. The fact that the girders and braces are exposed is what allows you to define models in arbitrary categories.

The derived sorts and operations not needed in a sketch are still there potentially, just as they are for signatures. Where they exist is in the theory of the sketch, discussed in Section 7.5 below.

7.3 Notation for FP sketches

The reader will see that the cones and most of the diagrams used in the sketch for semigroups are required to specify that nodes are products of other nodes and that certain arrows into products are specified by their projections. Accordingly, we will adopt notation that will allow us in most cases to avoid writing down the cones and most of the diagrams. Although these are just notational conventions, they are extremely important for *they are what make the use of sketches feasible.*

In later sections, we will increase the expressive power of sketches in various ways, and extend the list of notations which begins below to cover those cases.

7.3.1 These conventions are as follows:

N–1 It is understood that in all the conventions below, a^n is a shorthand for $a \times a \times \cdots \times a$, with n copies of the node a.

N–2 If the sketch has nodes called a_1, $a_2, \ldots, a_n$ and $a_1 \times a_2 \times \cdots \times a_n$, then it is understood that there is a cone

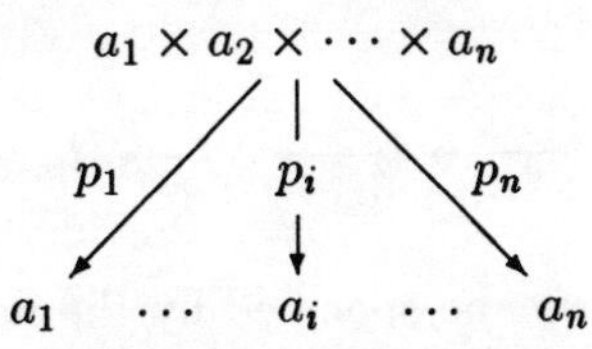

N–3 A node labeled 1 is assumed to imply the existence of a cone whose vertex is that node and whose base is the empty diagram.

N–4 If an arrow has the form

$$\langle f_1, f_2, \ldots, f_n \rangle : a \to b_1 \times b_2 \times \cdots \times b_n$$

where $f_i : a \to b_i$, $i = 1, \ldots, n$, then there are assumed to be diagrams

$$\begin{array}{ccc} & a & \\ {\scriptstyle \langle f_1, f_2, \ldots, f_n \rangle} \swarrow & & \searrow {\scriptstyle f_i} \\ b_1 \times b_2 \times \cdots \times b_n & \xrightarrow[p_i]{} & b_i \end{array}$$

for $i = 1, \ldots, n$.

N–5 If an arrow has the form

$$\langle f_1 \times f_2 \times \cdots \times f_n \rangle : a_1 \times a_2 \times \ldots \times a_n \to b_1 \times b_2 \times \cdots \times b_n$$

where $f_i : a_i \to b_i$, $i = 1, \ldots, n$ then there are assumed to be diagrams

$$\begin{array}{ccc} a_1 \times a_2 \times \cdots \times a_n & \xrightarrow{p_i} & a_i \\ {\scriptstyle f_1 \times f_2 \times \cdots \times f_n} \downarrow & & \downarrow {\scriptstyle f_i} \\ b_1 \times b_2 \times \cdots \times b_n & \xrightarrow[p_i]{} & b_i \end{array}$$

for $i = 1, \ldots, n$.

N–6 A diagram of the form

$$\begin{array}{ccccccccc} & & k_1 & \xrightarrow{s_2} & k_2 \to \cdots \to k_{n-2} & \xrightarrow{s_{n-1}} & k_{n-1} & & \\ & {\scriptstyle s_1}\nearrow & & & & & & \searrow{\scriptstyle s_n} & \\ i & & & & & & & & j \\ & {\scriptstyle t_1}\searrow & & & & & & \nearrow{\scriptstyle t_m} & \\ & & l_1 & \xrightarrow[t_2]{} & l_2 \to \cdots \to l_{m-2} & \xrightarrow[t_{m-1}]{} & l_{m-1} & & \end{array} \qquad (7.11)$$

may, if convenient, be specified by the notation

$$s_n \circ s_{n-1} \circ \cdots \circ s_1 = t_m \circ t_{m-1} \circ \cdots \circ t_1$$

The reader should note that this 'equation' has no meaning except as part of our convention since there is no composition of arrows in graphs.

The result of these notational conventions is that the sketch for semigroups may be specified by the graph as in 7.2.1, plus the single diagram

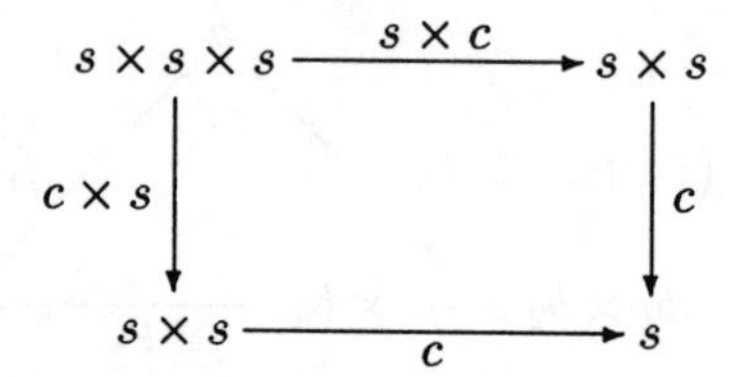

Equivalently, we may specify an equation

$$c \circ (s \times c) = c \circ (c \times s)$$

which is the usual form of the associative law.

7.3.2 There is at least one case in which the proper interpretation of some of these conventions may not be transparent; namely when $n = 0$. For example, the way to state the law for monoid identities, $ex = xe = x$, is that there are diagrams

$$s \xrightarrow{\langle s, e\langle\rangle\rangle} s \times s \xrightarrow{c} s$$

and

$$s \xrightarrow{\langle e\langle\rangle, s\rangle} s \times s \xrightarrow{c} s$$

each with shape graph

$$i \longrightarrow j \longrightarrow i$$

The 'same' diagram based on the shape graph

$$i \longrightarrow j \longrightarrow k$$

would have given no condition at all. We generally use one more notational convention to express this:

N–7 A commutative diagram of the form of Diagram (7.11) with $m = 0$ may, if convenient, be specified by the notation

$$s_n \circ s_{n-1} \circ \cdots \circ s_1 = \mathrm{id}$$

7.4 Arrows between models of FP sketches

If M and M' are two models of an FP sketch $\mathcal{S}$, a homomorphism $\alpha : M \to M'$ is defined to be a natural transformation between $M \to M'$, considered as models of the underlying graphs. Just as in 4.6.4, all the models of an FP sketch in a given category $\mathcal{C}$ form a category $\mathbf{Mod}(\mathcal{S},\mathcal{C})$ which is a full subcategory of the category of all graph homomorphisms from the graph of $\mathcal{S}$ to $\mathcal{C}$.

7.4.1 Let N be the model of the sketch in 7.1.5 for which $N(1) = \{*\}$, $N(n) = \mathbf{N}$ (the set of natural numbers), $N(\mathrm{succ})$ is the function taking n to $n+1$, and $N(\mathrm{zero})$ is the function picking out 0. Let M be the model which is the same for 1, for which $M(n) = \{0, 1, 2, 3\}$, $M(\mathrm{zero})$ is 0, and $M(\mathrm{succ})$ takes 0 to 1, 1 to 2, and 2 and 3 to 3. Then there is a homomorphism of models $\alpha : N \to M$ in which $\alpha n : N(n) \to M(n)$ takes 0, 1 and 2 to themselves and all other natural numbers to 3. What $\alpha 1$ does is forced.

The only nontrivial requirement forced by the definition of homomorphism is that for all natural numbers k,

$$M(\mathrm{succ})[\alpha n(k)] = \alpha n[N(\mathrm{succ})(k)]$$

which you should check.

7.4.2 In the case of sketches which specify mathematical structures, these homomorphisms are in most cases essentially the same as homomorphisms as usually defined. For example, we have the following proposition.

7.4.3 Proposition *Let M and N be two models of the sketch $\mathcal{S}$ for semigroups in* **Set** *(so that $M(s)$ and $N(s)$ are semigroups, see 7.2.7). Then if $\alpha : M \to N$ is a natural transformation, the component $\alpha s : M(s) \to N(s)$ is a semigroup homomorphism (see 2.5.1). Conversely, a semigroup homomorphism $h : M(s) \to N(s)$ induces a unique natural transformation $\alpha : M \to N$ for which $\alpha s = h$.*

Proof. For convenience, we will suppose that M and N take the cones in the diagram to canonical cones, but this can be avoided by using the device of 7.2.7. If M and N are models a natural transformation α between the sketches is given by three functions $\alpha(s) : M(s) \to N(s)$, $\alpha(s \times s) : M(s \times s) \to N(s \times s)$ and $\alpha(s \times s \times s) : M(s \times s \times s) \to N(s \times s \times s)$. There is a commutativity condition imposed for each arrow of the graph. The crucial one says that

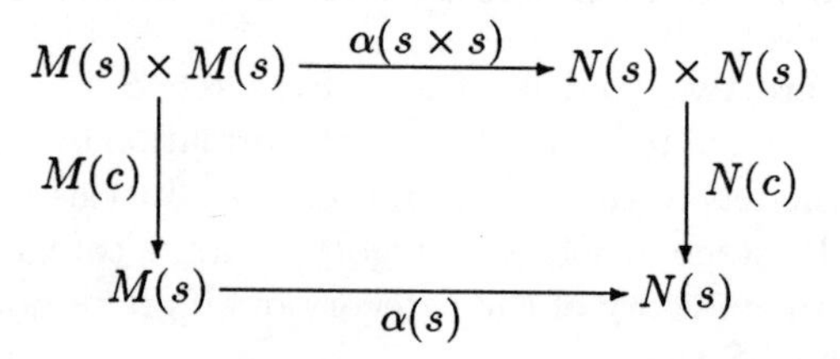

commutes. We can now see that the commutativity of the diagram above is exactly the definition of semigroup homomorphism: write h for $\alpha(s)$ and, for $(x, y) \in M(s) \times M(s)$, write xy for $M(c)(x, y)$. Then going south and east in the diagram gives $h(xy)$ and going east and south gives $N(c)[h \times h](x, y) = N(c)(h(x), h(y)) = h(x) \cdot h(y)$.

As for the converse, what must be shown is that a natural transformation $\alpha : M \to N$ with a homomorphism h as component at s must have $h \times h$ and $h \times h \times h$ as components at $s \times s$ and $s \times s \times s$ respectively. This is straightforward and is left as an exercise. □

7.4.4 Exercises

1. Prove the last sentence of Proposition 7.4.3.

2.† What modifications have to be made to the sketch for semigroups so that the category of models is that of monoids? You will have to add a new node, called 1, to the sketch and a cone over the empty diagram to force its image in a model to be a one-element set.

3.† What modifications would have to be made so that the category of models is that of real (or complex) vector spaces? (This sketch is not finite: you will need one unary operation of scalar multiplication for each scalar.)

7.5 The theory of an FP sketch

In 4.6.9, we constructed, for each linear sketch $\mathcal{S}$, a category $\mathbf{Th}(\mathcal{S})$ with the property that there is a model $M_0 : \mathcal{S} \to \mathbf{Th}(\mathcal{S})$ and such that for each category $\mathcal{C}$ and each model $M : \mathcal{S} \to \mathcal{C}$, there is a functor $F : \mathbf{Th}(\mathcal{S}) \to \mathcal{C}$ such that $F \circ M_0 = M$. The analog for FP sketches is given by the following.

7.5.1 Theorem *Given any FP sketch $\mathcal{S}$, there is a category $\mathbf{Th_{FP}}(\mathcal{S})$ with finite products and a model $M_0 : \mathcal{S} \to \mathbf{Th_{FP}}(\mathcal{S})$ such that for any model $M : \mathcal{S} \to \mathcal{C}$ into a category with finite products, there is a functor $F : \mathbf{Th_{FP}}(\mathcal{S}) \to \mathcal{C}$ that preserves finite products for which*

(i) *$F \circ M_0 = M$, and*

(ii) *if $F' : \mathbf{Th_{FP}}(\mathcal{S}) \to \mathcal{C}$ is another functor that preserves finite products for which $F' \circ M_0 = M$, then F and F' are naturally isomorphic.*

The proof of Theorem 7.5.1 is not given here. A proof that constructs $\mathbf{Th_{FP}}(\mathcal{S})$ as a full subcategory of a very special kind of category called a topos (which implies useful properties of $\mathbf{Th_{FP}}(\mathcal{S})$) may be found in [Barr and Wells, 1985], Section 4.3, Theorem 1 (p. 150), where it is also shown that the category of models of $\mathcal{S}$ is equivalent to the category of models of $\mathbf{Th_{FP}}(\mathcal{S})$. Toposes are discussed in Chapter 14. The construction of the theory is a special case of more general constructions in [Lair, 1971].

7.5.2 Terminology A category $\mathcal{T}$ with finite products is called an **FP theory**, and a **model** of an FP theory in a category $\mathcal{C}$ which has finite products is a functor $F : \mathcal{T} \to \mathcal{C}$ that preserves finite products. Theorem 7.5.1 can be described as saying that every sketch $\mathcal{S}$ has a **universal model** M_0 in an FP theory $\mathbf{Th_{FP}}(\mathcal{S})$ with the property that every model of $\mathcal{S}$ induces a model of $\mathbf{Th_{FP}}(\mathcal{S})$. This passage to induced models of the theory is actually an equivalence of categories of models.

7.5.3 The theory $\mathbf{Th_{FP}}(\mathcal{S})$ is defined up to equivalence by the following properties:

FPT–1 $\mathbf{Th_{FP}}\mathcal{S}$ has all finite products.

FPT–2 M_0 takes every diagram of $\mathcal{S}$ to a commutative diagram in $\mathbf{Th_{FP}}\mathcal{S}$.

FPT–3 M_0 takes every cone of $\mathcal{S}$ to a limit cone of $\mathbf{Th_{FP}}\mathcal{S}$.

FPT–4 No proper subcategory of $\mathbf{Th_{FP}}\mathcal{S}$ includes the image of M_0 and satisfies FPT–1, FPT–2 and FPT–3.

Two theories of an FP sketch $\mathcal{S}$ need not be isomorphic because, for example, in one of them $A \times B$ might be the same as $B \times A$, whereas in another the two might be different, although, of course, $A \times B$ and $B \times A$ are always isomorphic. As just stated, however, two theories of $\mathcal{S}$ are always *equivalent*. A category theorist normally regards the question as to whether two isomorphic objects are actually the same as irrelevant, and therefore will not usually care which of several equivalent categories she is working in. For this reason we customarily refer to the theory $\mathbf{Th_{FP}}(\mathcal{S})$ as 'the' theory of $\mathcal{S}$.

7.5.4 In general, giving an explicit description of the theory of a sketch is recursively unsolvable, since the general word problem for groups or semigroups can be expressed as the question of whether two arrows in the free category generated by the graph of a sketch generate the same arrow of a theory. However, we can give some explicit examples of theories.

For example, the theory of 7.1.5 has two objects, 1, which is a terminal object, and n. It has an arrow $0 : 1 \to n$, an arrow $\langle\rangle : n \to 1$, and a family $\mathrm{succ}^k : n \to n$ of arrows for $k = 1, 2, \ldots$, which are the n-fold composites of the arrow $\mathrm{succ} : n \to n$. And those, with the identity arrows, are all the arrows, and they are all different. With this description, M_0 is an inclusion.

In Section 4.7, we discussed linear sketches with constants. We can now see that these are special types of FP theories, representing each constant as an arrow from an object 1 which must be the vertex of a cone with empty base. We did not discuss the theory of a linear sketch with constants, but it is now clear that such sketches, since they are FP sketches, do have theories.

7.5.5 The theory of the sketch 7.2.1 for semigroups is more complicated. We will again describe it in such a way that M_0 is an inclusion.

It is easy to construct sketches for which M_0 is not injective on arrows, so cannot be treated as an inclusion, but this is not true of most interesting sketches – if you have two different arrows in a sketch, you probably already know a model where they are different, and then the universality of M_0 implies that M_0 has to take them to different arrows.

The theory consists of objects s^n for $n = 0, 1, \ldots$. It has as arrows all the projection arrows necessary to make s^n an n-fold product of s, the arrow $c : s^2 \to s$ for the binary operation, all the composites of these arrows with each other, and all induced arrows such as $\langle c, c\rangle : s^2 \to s^2$, $c \times p_1 : s^3 \to s^2$, and so on. Thus it has all the arrows you need to give all the diagrams in 7.2.3 and a great many more. Moreover, all the diagrams in 7.2.3 commute.

7.6 Initial term models for FP sketches

Just as in the case of a linear sketch with constants (see 4.7.10), a set-valued model of an FP sketch is called a **term model** if each element of the value at each node is forced to be there by the sketch. In the first instance, the node forced to be a terminal object is forced to have a unique element. But then by applying various operations, other elements are forced to exist. A term model has only those elements forced to exist in this way. This is spelled out precisely below in 7.6.5.

7.6.1 Definition A model of an FP sketch in an arbitrary category $\mathcal{C}$ is called an **initial model** if it has a unique homomorphism of models to each other model. It is thus an initial object in the category of models of the sketch in $\mathcal{C}$.

7.6.2 A particular example is a natural numbers object in a category $\mathcal{C}$ with terminal object. Using the characterization given in Section 5.5, it is not hard to see that a natural numbers object is an initial model of the sketch given in 7.1.5.

7.6.3 As with linear models with constants, an initial model of an FP sketch in the category of sets is necessarily a term model. For let M_0 be an initial model and suppose it is not a term model. Then the typed set of elements that are reachable beginning from the constants is certainly a model of the theory. It admits the constants and is closed, by definition, under the operations. Thus there is a term model $M_1 \subseteq M_0$. (The same argument, by the way, shows that every model includes a least submodel and that is a term model.)

Now we have an arrow (unique, actually) $f : M_0 \to M_1$ by the definition of initial model. That arrow, composed with the inclusion, gives an arrow $M_0 \to M_0$. But there is just one arrow from M_0 to itself, the identity (because M_0 is an initial model). Thus the composite must be the identity. But the image of f is included in M_1, so that $M_1 = M_0$, which means that M_0 is a term model, as claimed.

A colorful description of term models is that they have 'no junk', meaning no elements not describable by terms of the theory beginning with constants. Thus initial models have no junk.

7.6.4 Another property any initial model must have is: if t and u are two terms definable starting with constants, then in an initial model M_0, $M_0(t) = M_0(u)$ if and only if $M(t) = M(u)$ in every model M. The nontrivial direction of that statement follows from the observation that if $M(t) \neq M(u)$ in some model, then necessarily $N(t) \neq N(u)$ in any model N for which there is a homomorphism $N \to M$.

This property is described by saying that initial models have 'no confusion'. They are in fact characterized up to isomorphism by having no junk and no confusion. This terminology originated with J. Meseguer and J. Goguen.

7.6.5 Construction of initial models for finite FP sketches FP sketches always have initial models (see [Barr, 1986a] for a categorical proof). We now revise the construction of 4.7.10 to construct initial models for finite FP sketches.

Let $\mathcal{S} = (\mathcal{G}, \mathcal{D}, \mathcal{L})$ be an FP sketch. In the construction of the initial model of a linear sketch with constants, we constructed terms as strings of arrows of $\mathcal{G}$. We will now allow tuples of arrows in these strings. The set $A_{\mathcal{S}}$ consisting of all arrows of $\mathcal{G}$, all tuples (of finite length) of such arrows and the cones C of $\mathcal{C}$ is called the **alphabet** of the sketch $\mathcal{S}$. The rules construct an initial model I recursively using these data. The rules apply to each cone C in $\mathcal{L}$ of the form

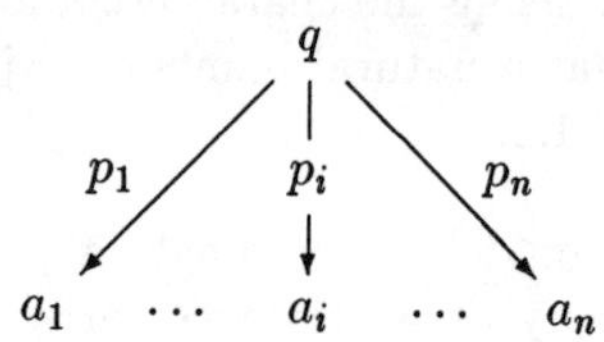

FP–1 If $f : a \to b$ is an arrow of $\mathcal{G}$ and $[x]$ is an element of $I(a)$, then $[fx] \in I(b)$ and $I(f)[x] = [fx]$.

FP–2 If $(f_1, \ldots, f_m)$ and $(g_1, \ldots, g_k)$ are paths in a diagram in $\mathcal{D}$, both going from a node labeled a to a node labeled b, and $[x] \in I(a)$, then

$$(If_1 \circ If_2 \circ \ldots \circ If_m)[x] = (Ig_1 \circ Ig_2 \circ \ldots \circ Ig_k)[x]$$

in $I(b)$.

FP–3 If for $i = 1, \ldots, n$, $[x_i]$ is an element in $I(a_i)$, then $[C(x_1, \ldots, x_n)]$ is an element of $I(q)$. (Note that $C(x_1, \ldots, x_n)$ is a string consisting of the cone C followed by a tuple of arrows.) In particular, if $n = 0$, there is a single element $[C\langle\rangle]$ in the empty product.

FP–4 for $i = 1, \ldots, n$, $[x_i]$ and $[y_i]$ are elements in $I(a_i)$ for which $[x_i] = [y_i]$, $i = 1, \ldots, n$, then

$$[C(x_1, \ldots, x_n)] = [C(y_1, \ldots, y_n)]$$

FP–5 For $i = 1, \ldots, n$, we require

$$[p_i C(x_1, \ldots, x_n)] = [x_i]$$

Thus FP–3 forces the vertex of the cone to contain an element representing each tuple of elements in the factors $a_1, \ldots, a_n$, and FP–5 forces the p_i to be the coordinate projections. Because it applies to empty cones, FP–3 subsumes I–1 of 4.7.10, which has no direct counterpart here. Observe that it follows from FP–1 and FP–5 that

$$I(p_i)[C(x_1, \ldots, x_n)] = [x_i]$$

for each i. FP–4 may be regarded as an extension of the definition of 'congruence relation' to cover the case of tuples.

7.6.6 Examples If M is the initial term model for an FP sketch $\mathcal{S}$, then for each sort g of the graph $\mathcal{G}$ of the sketch, $M(g)$ contains just those elements forced to be there by applying operations to constants. Indeed, up to isomorphism of models the elements *are* the formal applications of operations to constants, identifying those which the diagrams force to be the same.

Thus for the sketch in 7.1.5, $I(1)$ must be a singleton set, and $I(\text{zero})$ applied to that single element is an element of $I(n)$ which we may call 0. Then the elements of the sort $I(n)$ are just 0, succ(0), succ(succ(0)), succ(succ(succ(0))), and so on. These can be identified as the natural numbers, starting at zero.

From this point of view, 7.1.5 is a simple data type description, and the initial algebra is then the set of possible values of that type.

[Wells and Barr, 1987] describe a class of FP sketches whose initial models are all context free grammars.

Many other data types have been described using initial models using signatures and equations rather than FP sketches. (See the references at the end of this section.) We will present some of these in Section 7.8 after we have introduced finite sums, which allow us to distinguish exceptional cases such as an empty stack or a null node on a tree.

7.6.7 Free algebras Let $\mathcal{S}$ be a sketch with set S of nodes. Recall from 2.6.9 that an S-indexed set is a set X together with a typing function $\tau : X \to S$. Our point of view is that the nodes of the sketch represent types and that X is a set of typed constants. If $\tau : X \rightarrow S$ and $\tau' : X' \rightarrow S$ are sets typed by the same sketch $\mathcal{S}$, then a function $f : X \rightarrow X'$ is a **typed function** if $\tau = \tau' \circ f$. Sets typed by S and typed functions form the slice category $\mathbf{Set}/S$.

A particular example of a typed set is any model M of $\mathcal{S}$ in **Set** for which $M(c)$ and $M(d)$ have no elements in common for distinct nodes c and d. Any model is isomorphic to such a model, obtained by taking the disjoint union of the values of M at the different nodes of $\mathcal{S}$.

A model is a family of sets, indexed by S, but we can as well think of it as a single set (the union) typed by S. In any case, the underlying (family of) set(s) of a model is an object of the slice category $\mathbf{Set}/S$.

Now given an S-indexed set X, let $\mathcal{S}_X$ be the sketch constructed by adding to the graph of $\mathcal{S}$ a set of arrows $x : 1 \rightarrow s$ for each element $x \in X$ of type s. These are *in addition* to any constants of type S already given in the sketch.

An initial model of $\mathcal{S}_X$, if one exists, is called the **free algebra** generated by the typed set X. We use the definite article because, although not unique, it is unique up to a unique isomorphism that preserves the set X for the same reason that initial algebras are always unique.

The following theorem gives the main existence result.

7.6.8 Theorem *Let $\mathcal{S}$ be an FP sketch. Then for any typed set X, there is a free algebra generated by X.*

The free algebra on X is denoted $F(X)$. If the typed set X is finite, $F(X)$ exists because $\mathcal{S}_X$ is itself an FP sketch. In fact, an initial algebra exists for any (not necessarily finite) sketch with only diagrams and cones (no cocones), from which the preceding theorem follows immediately. An accessible proof is in [Barr, 1986a].

7.6.9 The map lifting property Free algebras have the map lifting property described in 3.1.14. This topic is continued in Section 9.2 and Section 12.2.

7.6.10 Theorem *Let $\mathcal{S}$ be an FP sketch, X a typed set and M a model of $\mathcal{S}$ in sets. Then any typed function $f : X \rightarrow M$ has a unique extension to an arrow between models $F(X) \rightarrow M$.*

7.6.11 Initial algebra semantics in the literature The initial algebra semantics of multisorted signatures and equations is equivalent to the initial algebra semantics of FP sketches. This approach has been developed extensively by Goguen and others. The idea is to give an algebraic specification of a data type (using a signature or an FP sketch) which explicitly specifies all the behavior of the type accessible to the program writer. The corresponding initial algebra is then an implementation of this specification: the terms in the algebra (arrows) are in fact the data inhabiting that type. The point is that the terms in the initial algebra, as we have seen in 7.6.5, can be constructed entirely from the given description. See [Gray, 1988a].

The approach via signatures is described in [Goguen, Thatcher and Wagner, 1978], [Goguen, 1978], [Zilles, Lucas and Thatcher, 1982] and [Meseguer and Goguen, 1985]. In many cases, extensions to the idea have been proposed, for example in [Goguen, Thatcher, Wagner and Wright, 1977], [Goguen, Jouannaud and Meseguer, 1985] and [Lellahi, 1987]. More references are given at the end of the next section.

7.6.12 Exercise

1. Let X be a set with two elements. Explain how the free model of Example 7.1.5 on X can be thought of as the disjoint union of *three* copies of $\mathbf{N}$.

7.7 Signatures and FP sketches

We will describe a formal version of the method of signatures and equations similar to that used by many authors in the literature and show how to produce an FP sketch with essentially the same models in the sense of 7.2.8.

7.7.1 Definition A **signature** $S = (\Sigma, \Omega)$ is a pair of finite sets Σ and Ω. The elements of Σ are **sorts**, which will usually be denoted by lower case Greek letters. The elements of Σ^* (the set of words of finite length in Σ – including the empty word) are called **arities**. The elements of Ω are **operations**, denoted by lower case English letters. Each operation $f \in \Omega$ has an arity in Σ^* and a sort in Σ. The arity gives the input type and the sort is the output type.

A **model** M of the signature S is a collection of sets M_σ indexed by Σ and a collection M_f of functions indexed by Ω with the property that if $\sigma_1\sigma_2\ldots\sigma_n$ is the arity of f and τ is the sort of f, then $M_f : M_{\sigma_1} \times M_{\sigma_2} \times \cdots \times M_{\sigma_n} \to M_\tau$. When $n = 0$, this M_f interpreted to mean a function from the one-element set to M_τ, that is an element of M_τ.

7.7.2 Terms In general, one is interested in structures defined by a signature which satisfy certain equations. For example, to describe a semigroup, we need a binary operation that sends (x, y) to $x * y$ and we also have to express the associative law that says $x * (y * z) = (x * y) * z$, that is that two terms are equal in any model. To do this in a formal way requires that we define terms for a signature.

Given a set V of variables $x, y, \ldots$ sorted by Σ (meaning that each variable is assigned a specific sort in Σ) we can define a set of **terms** of S recursively by T–1 and T–2 below.

Each term t has an **arity** $\mathsf{A}(t) = w \in \Sigma^*$, a **sort** $\mathsf{S}(t) = \sigma \in \Sigma$, and a **variable list** $\mathsf{VL}(t) = v \in V^*$. T–1 and T–2 define the arity, sort and variable list. Maintaining the information about the arity is redundant (but convenient) since the arity of the term is the list of sorts (in order) of the variables in the variable list and so is recoverable from the variable list. The reverse is not true: the arity does not indicate which variables are repeated.

T–1 If x is a variable of sort σ, then x is a term of arity $\mathsf{A}(x) =$ 'σ', sort $\mathsf{S}(x) = \sigma$ and variable list $\mathsf{VL}(x) =$ 'x'.

T–2 If for $i = 1, \ldots, n$, e_i is a term of arity $w_i \in \Sigma^*$, sort $\sigma_i \in \Sigma$ and variable list $l_i \in V^*$, and f is an operation of arity $\sigma_1\sigma_2\ldots\sigma_n$ and sort τ in Ω, then $t = f(e_1, e_2, \ldots, e_n)$ is a term of arity $\mathsf{A}(t) = w_1 w_2 \cdots w_n$ (this is the concatenate of the *strings* w_i), sort $\mathsf{S}(t) = \tau$ and variable list $\mathsf{VL}(t) = l_1 l_2 \ldots l_n$.

Terms stand in much the same relation to signatures as arrows of the theory of the sketch do to the graph of the sketch.

7.7.3 Example To give you a clue as to what this definition does, here is an example: if we have a signature S with

(i) sorts σ and τ in Σ,

(ii) operations m of arity '$\sigma\sigma$' and sort τ and f of arity '$\sigma\tau\tau$' and sort σ, and

(iii) variables x, y of sort σ and t of sort τ in V,

then $f(y, m(x, y), t)$ is a term of arity '$\sigma\sigma\sigma\tau$', sort τ and variable list '$yxyt$'.

7.7.4 Equations An **equation** for the signature S is a formula of the form $t = u$, where t and u are terms of the same sort. They do not have to have the same arity or variable list.

An equation has an obvious interpretation in any model of the signature. Such a model M **satisfies** a set E of equations if for every equation $t = u$ in E, t and u have the same interpretations in M. The models for a signature satisfying a set of equations form an **equational variety**.

7.7.5 The sketch corresponding to a signature with equations Given a signature S with equations E, we can construct a sketch $\mathcal{S}(S,E)$ whose models in **Set** which take products to canonical products are exactly the models of (S,E).

To begin with, the graph of $\mathcal{S}(S,E)$ should have a node labeled σ for each sort σ of S, plus a node labeled $\sigma_1 \times \cdots \times \sigma_n$ for each string $\sigma_1 \cdots \sigma_n$ which is the arity of an operation. For each such arity, $\mathcal{S}(S,E)$ must have a cone with vertex $\sigma_1 \times \cdots \times \sigma_n$ and base $\sigma_1, \ldots, \sigma_n$. There must be an arrow with the proper source and target for each operation of S.

The equations in E must be specified by diagrams of $\mathcal{S}(S,E)$, a complicated process which in general also requires adding arrows and nodes to its graph. There are two problems with this construction. The first is that in any equation $t = u$, the variables that appear in t and u are not necessarily the same. The second is that a variable may appear more than once in a term (compare 6.4.2). It is interesting to note that these two possibilities are what distinguish ordinary logic – based on cartesian product – from Girard's linear logic [1987] – based on another kind of product.

7.7.6 In order to describe how to express the equations in a sketch, we will associate with each term t a path in a certain graph. Since there are infinitely many terms, there will be infinitely many such paths. On the other hand, it is convenient to keep the graph finite. Thus in the sketch associated to a signature with equations, we will add only such nodes and arrows as are necessary to state the given equations. If there are just a finite set of operations and equations, then the resultant sketch will be finite.

In the descriptions below are a number of nodes and arrows which involve formal products, identities and projections. We will suppose without mention that the appropriate cones and diagrams to express the fact that these become actual products, identities and projections are included in the sketch.

Our approach will be to associate with each term t a 'pseudo-path' $\mathbf{Q}(t)$ that is the same as the path of the term with the same operations but no repeated variables. Then we show how to turn these pseudo-paths into diagrams in a way which takes into account the repeated variables. Note that the actual path of a term t can be recovered by applying this construction to the equation $t = t$, which results in a diagram with two identical paths, each the path of t.

Now let t be a term of sort σ, arity '$\sigma_1 \cdots \sigma_n$' and variable list '$v_1 \cdots v_n$'. We will describe a path $\mathbf{Q}(t)$ that starts at $\sigma_1 \times \cdots \times \sigma_n$ and ends at σ. This is defined inductively as follows. If x is a variable of type τ, then $\mathbf{Q}(x) = \mathrm{id} : \tau \to \tau$. If $t_1, \ldots, t_m$ are terms of sort $\tau_1, \ldots, \tau_m$, respectively, and if the concatenate of their arities is '$\sigma_1 \cdots \sigma_n$', and if f is an operation of arity '$\tau_1 \cdots \tau_m$' and sort σ, then $\mathbf{Q}(f(t_1, \ldots, t_m))$ is the composite path

$$\sigma_1 \times \cdots \times \sigma_n \xrightarrow{\mathbf{Q}(t_1) \times \cdots \times \mathbf{Q}(t_m)} \tau_1 \times \cdots \times \tau_m \xrightarrow{f} \sigma$$

Now suppose that we have an equation $t = u$. Suppose that the arities of t and u are '$\sigma_1 \cdots \sigma_n$' and '$\tau_1 \cdots \tau_m$' respectively. Let $X = \{x_1, \ldots, x_k\}$ be the set of variables that appear anywhere in t or u and suppose that x_i has sort ρ_i. The sketch will have an arrow

$$v : \rho_1 \times \cdots \times \rho_k \to \sigma_1 \times \cdots \times \sigma_n$$

and, for each $i = 1, \ldots k$, a diagram $p_i \circ v = p_j$, where j is chosen so that the jth variable in the variable list of t is x_i. We similarly have an arrow

$$w : \rho_1 \times \ldots \times \rho_k \to \tau_1 \times \ldots \times \tau_m$$

and a similar set of diagrams. Finally, we add the diagram $\mathbf{Q}(t) \circ v = \mathbf{Q}(u) \circ w$ to the set of diagrams. It is this diagram that expresses the equation that $t = u$.

7.7.7 An example The above may appear rather complicated; here is how the construction looks when applied to a specific case. Let the signature S, the operations f and m and the term t be as in 7.7.3. Let u be the term x of arity σ and sort 'σ' and variable list 'x'. We will work out the diagram corresponding to the equation $t = u$, that is, $f(y, m(x, y), t) = x$. The path $\mathbf{Q}(t)$ is

$$\sigma \times \sigma \times \sigma \times \tau \xrightarrow{\mathrm{id} \times m \times \mathrm{id}} \sigma \times \tau \times \tau \xrightarrow{f} \sigma$$

The path of u is $\mathrm{id} : \tau \to \tau$. The variable list of t is the string '$yxyt$' and that of u is the string 'x'. Thus the set of all variables that appear in either list is $\{x, y, t\}$. The sorts of the variables are σ, σ and τ, respectively. Accordingly, we form the diagram

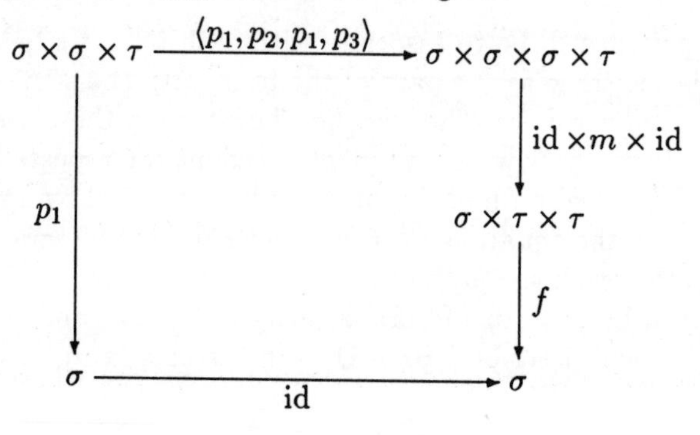

The top arrow $\langle p_1, p_2, p_1, p_3\rangle$ is exactly what is needed to transform the variable set $\{x, y, t\}$ into the variable list $xyxt$. Of course a set is not ordered, but our notation forces us to choose an ordering of it. If we had chosen to write the set as $\{t, x, y\}$, then the top arrow would have been

$$\langle p_2, p_3, p_2, p_1\rangle : \tau \times \sigma \times \sigma \to \tau \times \sigma \times \sigma \times \sigma$$

The bottom arrow labeled identity could have been omitted – or rather replaced by the null arrow. However, this would have made the inductive step harder.

7.7.8 FP sketches and multisorted algebraic theories The discussion in this section shows informally that FP sketches have the same expressive power as multisorted algebraic theories, at least as far as models in **Set** are concerned, and the considerable literature on applications of multisorted algebra to computer science provide examples of applications of FP sketches. See in particular [Ehrig and Mahr, 1985], [Wagner, Bloom and Thatcher, 1985], [Bloom and Wagner, 1985] and [Ehrich, 1986]. Other references, concentrating on initial algebra semantics, were mentioned in 7.6.11. The approach via sketches makes it natural to consider models in an arbitrary category, for example a category such as modest sets designed specifically for programming language semantics. This is an advantage over the usual formulation of multisorted universal algebra, which considers only models in sets. One would expect that a reformulation of universal algebra which puts models in other categories on the same footing as models in sets would wind up looking rather like FP sketches.

7.8 FD sketches

Using sum cocones as well as product cones in a sketch allows one to express alternatives as well as n-ary operations.

7.8.1 A **finite discrete sketch**, or **FD sketch**, $\mathcal{S} = (\mathcal{G}, \mathcal{D}, \mathcal{L}, \mathcal{K})$, consists of a finite graph $\mathcal{G}$, a finite set of finite diagrams $\mathcal{D}$, a finite set $\mathcal{L}$ of discrete cones with finite bases, and a finite set $\mathcal{K}$ of discrete cocones with finite bases. A **model** of the FD sketch $\mathcal{S}$ in **Set** is a graph homomorphism $M : \mathcal{G} \to$ **Set** which takes the diagrams in $\mathcal{D}$ to commutative diagrams, the cones in $\mathcal{L}$ to product cones, and the cocones in $\mathcal{K}$ to sum cocones.

7.8.2 The sketch for lists The FD sketch for finite lists illustrates how FD sketches can be used to specify a data type which has an exceptional case for which an operation is not defined. (Compare 7.1.6.) The graph has nodes 1, D (the data), L (the lists) and L^+ (the nonempty lists). There are no diagrams. There is a cone with empty base expressing the fact that 1 must become a terminal object in a model, another cone

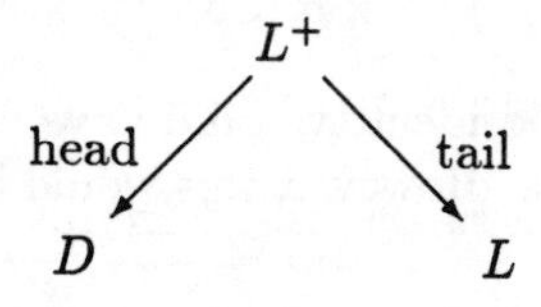

and one cocone

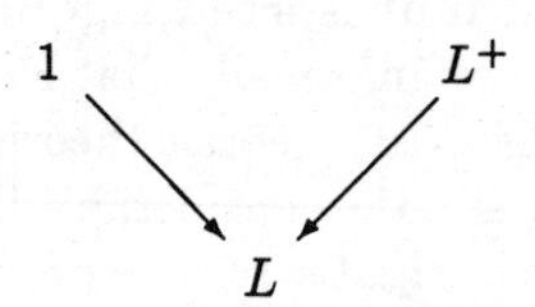

These can be summed up in the expressions $L^+ = D \times L$ (a nonempty list has a datum as head and a list as tail) and $L = 1 + L^+$: (a list is either the empty list – the single member of the terminal set – or a nonempty list). The fact that a sum becomes a disjoint union in a model in **Set** enables us to express alternatives in the way exhibited by this description. We thereby avoid the problem mentioned in 7.1.6: we can define head for only nonempty lists because we have a separate sort for nonempty lists.

If S is any finite set, there is a model M of this sketch for which $M(D) = S$, $M(1)$ is the singleton set containing only the empty list $\langle\rangle$, $M(L)$ is the set of finite lists of elements of S, and $M(L^+)$ is the set of finite nonempty lists of elements of S. $M(\text{head})$ and $M(\text{tail})$ pick out the head and tail of a nonempty list. This is an initial term algebra (to be defined in Section 7.10) for the sketch extended by one constant for each element of S.

7.8.3 Given a finite set S, the model M just mentioned is certainly not the only model of the sketch for which $M(D)$ is S. Another model has $M(L)$ all finite and infinite lists of elements of S, with $M(L^+)$ the nonempty lists and $M(\text{head})$ and $M(\text{tail})$ as before. This model contains elements inaccessible by applying operations to constants: you cannot get infinite lists by iterating operations starting with constants. They are junk as defined in 7.6.3.

7.8.4 The sketch for natural numbers We will now modify the sketch 7.1.5 in a way which shows how FD sketches can model overflow.

The graph contains nodes we will call 1, n, n_{over} and $n+n_{\mathrm{over}}$, which we interpret, respectively, as a one-element set, a set of natural numbers (of which there may be only finitely many), an overflow element and the set of all natural numbers. The only cone is the one implicit in the name 1: the cone has 1 as the vertex and the empty base. There is similarly a (discrete) cocone implicit in the name $n + n_{\mathrm{over}}$. There is a constant operation $\mathrm{zero} : 1 \to n$ and a unary operation $\mathrm{succ} : n \to n+n_{\mathrm{over}}$. There are no diagrams.

There are many models of this sketch in **Set**. Here are three of them. They are all term models.

7.8.5 The first model we consider of this sketch is the set of natural numbers. The value of zero is 0 and $\mathrm{succ}(i) = i + 1$. The model takes n_{over} to the empty set. This model illustrates the fact that models are permitted to take the empty set as value on one or more nodes. Classical model theorists have not usually allowed sorts in a model to be empty.

7.8.6 The second model takes n to the set of integers up to and including an upper bound N. The value at n_{over} is a one- element set we will call ∞, although any other convenient label (including $N+1$) could be used instead. As above, the value of zero is 0. As for succ, it is defined by $\mathrm{succ}(i) = i+1$ for $i < N$, while $\mathrm{succ}(N) = \infty$. Note that the domain of succ does not include ∞.

7.8.7 The third model takes for the value of n also the set of integers up to a fixed bound N. The value of zero is again 0 and of n_{over} is empty. The operation succ is defined by $\mathrm{succ}(i) = i+1$, for $i < N$ while $\mathrm{succ}(N) = 0$. This is arithmetic modulo n.

This third model could be varied by letting $\mathrm{succ}(N) = N$ or, for that matter, any intermediate value. This shows that these models really have at least two parameters: the value of N and what happens to the successor of N.

7.8.8 Exercise

1. Construct an FD sketch whose only model is the two-element Boolean algebra.

7.9 The sketch for fields

Another example of an FD sketch is the sketch for the mathematical structure known as a **field**. The concept of field abstracts the properties of the arithmetic of numbers. We will describe it in some detail here.

This section is used later only as an example in Section 7.10. Rather than define 'field' we will describe the sketch and say that a field is a model of the sketch in sets.

The nodes are 1, u, f, $f \times f$ and $f \times f \times f$. There are operations

FO–1 $0 : 1 \to f$.

FO–2 $1 : 1 \to u$.

FO–3 $+ : f \times f \to f$.

FO–4 $* : f \times f \to f$.

FO–5 $- : f \to f$.

FO–6 $(\ )^{-1} : u \to u$.

FO–7 $j : u \to f$.

The reader should note that the two 1's in FO–2 above are, of course, different. One of them is the name of a node and the other of an operation. This would normally be considered inexcusable. The reasons we do it anyway are (a) each usage is hallowed by long tradition and (b) because they are of different type, there is never any possibility of actual clash. The reader may think of it as an early example of overloading.

The diagrams are:

FE–1 (associativity of +): $+ \circ (\mathrm{id} \times +) = + \circ (+ \times \mathrm{id}) : f \times f \times f \to f$.

FE–2 (associativity of *): $* \circ (\mathrm{id} \times *) = * \circ (* \times \mathrm{id}) : f \times f \times f \to f$.

FE–3 (commutativity of +): $+ \circ \langle p_2, p_1 \rangle = + : f \times f \to f$.

FE–4 (commutativity of *): $* \circ \langle p_2, p_1 \rangle = * : f \times f \to f$.

FE–5 (additive unit): $+ \circ \langle p_1, 0 \circ \langle\rangle \rangle = + \circ \langle 0 \circ \langle\rangle, p_2 \rangle = \mathrm{id} : f \to f$.

FE–6 (multiplicative unit): $* \circ \langle j, 1 \circ \langle\rangle \rangle = * \circ \langle 1 \circ \langle\rangle, j \rangle = j : u \to f$.

FE–7 (additive inverse): $+ \circ (\mathrm{id} \times -) = + \circ (- \times \mathrm{id}) = 0 \circ \langle\rangle : f \to f$.

FE–8 (multiplicative inverse): $* \circ (j \times j) \circ (\mathrm{id} \times \langle\rangle^{-1}) =$
$* \circ (j \times j) \circ (\langle\rangle^{-1} \times \mathrm{id}) = 1 \circ \langle\rangle : u \to u$.

FE–9 (distributive): $+ \circ (* \times *) \circ \langle p_1, p_2, p_1, p_3 \rangle = * \circ (\mathrm{id} \times +)$;
$+ \circ (* \times *) \circ \langle p_1, p_3, p_2, p_3 \rangle = * \circ (+ \times \mathrm{id}) : f \times f \times f \to f$.

There are cones defined implicitly and one cocone:

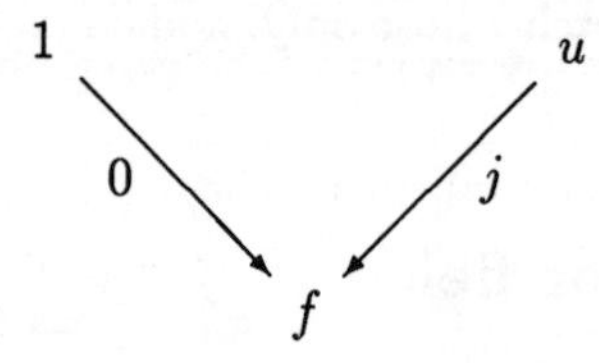

Intuitively, this says that each element of a field is either zero or else is an element of u. u is interpreted as the set of multiplicatively invertible elements in a model.

7.9.1 An example of a field is the set of ordinary rational numbers. Expressed as a model M of the sketch, the numbers $M(0)$ and $M(1)$ are the usual ones, $M(f)$ and $M(u)$ are the sets of all rationals and nonzero rationals, respectively, and $M(j)$ is the inclusion. The arithmetic operations are the usual ones. Other familiar examples are the real and complex numbers.

7.9.2 In building a model of this sketch, we would have to start with the elements 0 and 1 and then begin adding and multiplying them to get new elements. For example, we could let $2 = 1 + 1$. Now since $M(f) = M(u) + \{0\}$, we have to decide whether $2 \in M(u)$ or $2 \in \{0\}$, that is, whether or not $2 = 0$. Quite possibly, the only fields the reader has seen have the property that $2 \neq 0$. Even so, one cannot exclude out of hand the possibility that $2 = 0$ and it is in fact possible: there is a model of this sketch whose value at f is $\{0, 1\}$ with the values of the operations being given by the tables:

+	0	1
0	0	1
1	1	0

*	0	1
0	0	0
1	0	1

(These tables are addition and multiplication $(\bmod 2)$.) If, on the other hand, $2 \neq 0$, we can form $3 = 1 + 2$. The same question arises: is $3 = 0$ or not? It is not hard to write down a field in which $3 = 0$.

Suppose that neither 2 nor 3 is 0? We would define $4 = 1 + 3$ (and prove, by the way, that $2 + 2 = 2 * 2 = 4$) and ask whether $4 = 0$. It turns out that if $2 \neq 0$ then $4 \neq 0$. In fact, it is not hard to show that any product of elements of u also lies in u. Thus the first instance of an integer being 0 must happen at a prime. But, of course, it need never happen, since no number gotten by adding 1 to itself a number of times is zero in the rational, real or complex field.

7.9.3 A homomorphism between fields must preserve all operations. Since it preserves the operation $(\ )^{-1}$, no nonzero element can be taken to zero. Consequently, two distinct elements cannot go to the same element since their difference would be sent to zero. Thus field homomorphisms are injective.

From this, it follows that there can be no field that has a homomorphism both to the field of two elements and to the rationals, since in the first $1 + 1 = 0$, which is not true in the second. This implies that the category of fields and field homomorphisms does not have products.

7.9.4 Exercises

1. Prove that in a field, if two elements each have multiplicative inverses, then so does their product. Deduce that if $4 = 0$, then $2 = 0$.

2. Prove that the next to last sentence of 7.9.3 implies that the category of fields and field homomorphisms does not have products.

7.10 Term algebras for FD sketches

A complication arises in trying to extend the construction of initial term algebras to FD sketches. As we see from the examples of natural numbers and fields, an operation taking values in the vertex of a discrete cocone forces us to choose in which summand the result of any operation shall be. The choice, in general, leads to nonisomorphic term models which are nevertheless initial in a more general sense which we will make precise.

7.10.1 Dæmons How to make the choice? Clearly there is no systematic way. One way of dealing with the problem is to take *all* choices, or at least to explore all choices. In the example in Section 7.9 of fields, we mentioned that not all choices are possible; once $2 \neq 0$, it followed that also $4 \neq 0$. (Recall that in that example, saying that something is zero is saying that it is in one of two summands.)

More generally, suppose we have an FD sketch and there is an operation $s : a \to b$ and a cocone expressing $b = b_1 + b_2 + \ldots + b_n$. If we are building a model M of this sketch and we have an element $x \in M(a)$, then $M(s)(x) \in M(b)$, which means that we must have a unique i between 1 and n for which $M(s)(x) \in M(b_i)$. (For simplicity, this notation assumes that $M(b)$ is the actual union of the $M(b_i)$.)

Now it may happen that there is some equation that forces it to be in one rather than another summand, but in general there is no such indication. For example, the result of a push operation on a stack may or may not be an overflow, depending on the capacity of the machine or other considerations. Which one it is determines the particular term model we construct and it is these choices that determine which term model we will get.

Our solution is basically to try all possible sequences of choices; some such sequences will result in a model and others will abort. Thus as we explore all choices, some will eventually lead to a model; some will not. The theoretical tool we use to carry out this choice we call a **dæmon**. Just as a Maxwell Dæmon chooses, for each molecule of a gas, whether it goes into one chamber or another, our dæmon chooses, for each term of a model, which summand it goes into. The following description spells this out precisely.

7.10.2 Definition Let $\mathcal{S}$ be an FD sketch and suppose the maximum number of nodes in the base of any cocone is κ. A **dæmon** for $\mathcal{S}$ is a function D from the set of all strings in the alphabet $A_{\mathcal{S}}$ (see 7.6.5) of the underlying FP sketch (in other words, forget the cocones) to the initial segment $\{1..\kappa\}$ of the positive integers.

7.10.3 We will use a dæmon this way. We assume that the nodes in the base of each cocone of $\mathcal{S}$ are indexed by $1, 2, \ldots, k$ where $k \leq \kappa$ is the number of nodes in that cocone. In constructing an initial algebra, if a string w must be in a sort which is the vertex of a cocone (hence in the model it must be the disjoint union of no more than κ sorts), we will choose to put it in the $D(w)$th summand. If $D(w) > k$, the construction aborts. We will make this formal.

7.10.4 Construction of initial term models for FD sketches This construction includes the processes in 4.7.10 and 7.6.5; we repeat them here modified to include the effects of a dæmon D. The alphabet is the same as in 7.6.5.

If a node b is the vertex of a cocone with $k = k(b)$ summands, the summands will be systematically denoted $b^1, \ldots, b^k$ and the inclusion arrows $u^i : b^i \to b$ for $i = 1, \ldots, k$. If b is not the vertex of a cocone, then we take $k(b) = 1$, $b^1 = b$ and $u^1 = \mathrm{id}_b$. We denote the congruence relation by $\sim$ and the congruence class containing the element x by $[x]$. Rules FD–3, FD–4 and FD–5 refer to a cone C in $\mathcal{L}$ of the form:

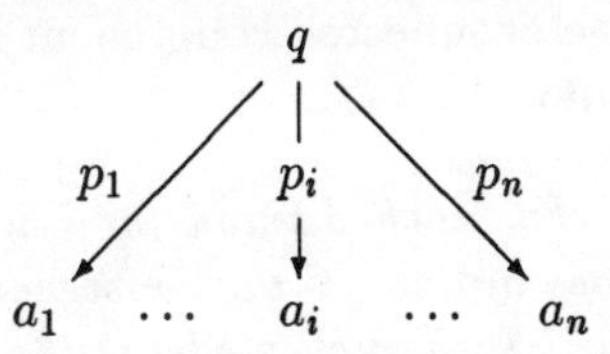

FD–1 If $u^i : a^i \to a$ is an inclusion in a cocone and $[x] \in I(a^i)$, then $[u^i x] \in a$ and $I(u^i)[x] = [u^i x]$. (Thus we ignore the wishes of the dæmon in this case.)

FD–2 Suppose $f : a \to b$ in $\mathcal{G}$, f is not an arrow of the form u^i, and $[x] \in I(a)$. Let $j = D(fx)$ (we ask the dæmon what to do). If $j > k(b)$, the construction aborts. Otherwise, we let $[fx] \in I(b^j)$ and $I(f)[x] = [u^j fx]$.

FD–3 For $i = 1, \ldots, n$, let $[x_i]$ be a term in $I(a_i)$. Let

$$j = D(C(x_1, \ldots, x_n))$$

If $j > k(q)$, the construction aborts. If not, put $[C(x_1, \ldots, x_n)]$ in $I(q^j)$.

FD–4 If for $i = 1, \ldots, n$, $[x_i]$ and $[y_i]$ are elements in $I(a_i)$ for which $[x_i] = [y_i]$, $i = 1, \ldots, n$, then

$$[C(x_1, \ldots, x_n)] = [C(y_1, \ldots, y_n)]$$

FD–5 For $i = 1, \ldots, n$, $I(p_i)([C(x_1, \ldots, x_n)]) = [x_i]$.

FD–6 If $\langle f_1, \ldots, f_m \rangle$ and $\langle g_1, \ldots, g_k \rangle$ are paths in a diagram in $\mathcal{D}$, both going from a node labeled a to a node labeled b, and $[x] \in I(a)$, then

$$(If_1 \circ If_2 \circ \ldots \circ If_m)[x] = (Ig_1 \circ Ig_2 \circ \ldots \circ Ig_k)[x]$$

in $I(b)$. If

$$D(f_1 f_2 \ldots f_m x) \neq D(g_1 \ldots g_k x)$$

(causing $(If_1 \circ If_2 \circ \ldots \circ If_m)[x]$ and $(Ig_1 \circ Ig_2 \circ \ldots \circ Ig_k)[x]$ to be in two different summands of $I(b)$), then the construction aborts.

This construction gives a term model if it does not abort. It is an initial model for only part of the category of models, however. To make this precise, we recall the definition of connected component from 4.3.10. It is easy to see that each connected component is a full subcategory of the whole category of models.

7.10.5 Proposition *For each dæmon for which the construction in FD–1 through FD–5 does not abort, the construction is a recursive definition of a model I of $\mathcal{S}$. Each such model is the initial model of a connected component of the category of models of $\mathcal{S}$, and there is a dæmon giving the initial model for each connected component.*

We will not prove this theorem here. However, we will indicate how each model determines a dæmon which produces the initial model for its component. Let M be a model of an FD sketch $\mathcal{S}$. Every string w which determines an element of a sort $I(a)$ in a term model as constructed above corresponds to an element of $M(a)$. That element must be in a unique summand of a; if it is the ith summand, then define $D(w) = i$. On strings not used in the construction of the term models, define $D(w) = 1$, not that it matters.

Our definition of dæmon shows that one can attempt a construction of an initial model without already knowing models. In concrete cases, of course, it will often be possible to characterize which choices give initial models and which do not.

7.10.6 Confusion maybe, junk no The slogan, 'No junk, no confusion' is only half true of the initial models for FD theories. The 'No junk' half of the slogan expresses exactly what we mean when we say that every element is reachable. There are no extraneous elements. 'No confusion' means no relations except those forced by the equations in the theory. As we will show by example it may happen that some initial models have confusion and others not. Later we give an example of a sketch that has more than one unconfused initial model and one that has no unconfused initial (or noninitial) model.

If there is just one unconfused initial model, that one may be thought of as a 'generic' model. The others remain nonetheless interesting. In fact, it is likely that the generic model is the one that cannot be accurately modeled on a real machine.

7.10.7 Example A typical example of a sketch with many initial models is the sketch for natural numbers with overflow. The generic model is easily seen to be the one in 7.8.5 in which the overflow state is empty. The models with overflow in 7.8.6 are all initial algebras for some component of the category of models, but they have confusion, since nothing in the sketch implies that the successor of any element can be the same element. None of these models have junk.

The modular arithmetic models of 7.8.7 are not initial models; in fact they are all in the same component as the natural numbers since the remainder map (mod N) is a morphism of models. They also have no junk.

7.10.8 Example Here is a simple sketch with no generic model. It has two initial models, each satisfying an equation the other one does not. There are five nodes $a = b + c$, d and 1. There is one constant x of type d, and a single operation $s : d \to a$. The initial models have one element – the constant – of type d. One of the initial models has an element of type b and the other an element of type c.

By modifying this example, we can get forced confusion. Add constants y and z of type b and c, respectively, and cones forcing b and c to be terminal. Now there are two initial models, one in which $s(x) = y$ and another in which $s(x) = z$. Since there is a model in which $s(x) \neq y$, there can be no equation that forces $s(x) = y$ and there is similarly no equation that forces $s(x) = z$. But one or the other equation must hold in any model.

7.10.9 Example It is well known and proved in abstract algebra texts that the initial fields are (a) the rational numbers and (b) the integers mod p for each prime p. (The word for initial model in these texts is 'prime field'.) A field is in the component of the integers mod p if and only if $1+1+\ldots+1$ (sum of p 1's) is zero. These fields have confusion. Otherwise the field is in the component of the rational numbers, which have no confusion (nor junk).

The real numbers and the complex numbers form fields with the usual operations. The irrational real numbers constitute junk.

7.10.10 Example The example in 7.8.2 has only one component and hence a single initial model in which all sorts are empty except the singleton 1. If you add constants to D, the initial model is just the set of lists of finite length of elements of D. In the model discussed there which also has all infinite sequences, the infinite sequences are junk.

7.10.11 Binary trees We now describe a sketch for ordered rooted binary trees (called trees in this discussion). 'Binary' means that each node has either no children or two children, and 'ordered' means that the children are designated left and right. This is an example which uses cocones to treat exceptional cases, in this case the empty tree.

Trees are parametrized by the type of data that are stored in them. We will say nothing about this type of data, supposing only that it is a type for which there is an initial model. The way in which the parametrized data type is filled in with a real one is described, for example, in Section 10.2. We would like to thank Adam D. Barr for helpful discussions on how binary trees operate (especially their error states) on real machines.

The sketch will have sorts 1, t, s, d. Informally, t stands for tree, s for nonempty tree and d for datum. We have the following operations:

$$\text{empty} : 1 \to t$$
$$\text{incl} : s \to t$$
$$\text{val} : s \to d$$
$$\text{left} : s \to t$$
$$\text{right} : s \to t$$

The intended meaning of these operations is as follows.

Empty$\langle\rangle$ is the empty tree; incl is the inclusion of the set of nonempty trees in the set of trees; val(S) is the datum stored at the root of S;

left(S) and right(S) are the right and left branches (possibly empty) of the nonempty tree S, respectively.

We require that

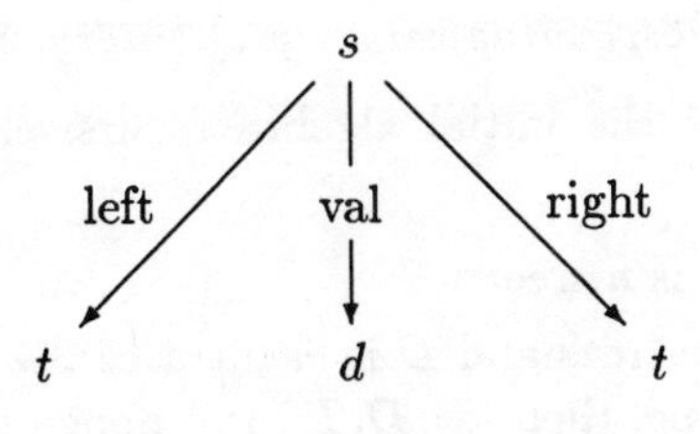

be a cone and that

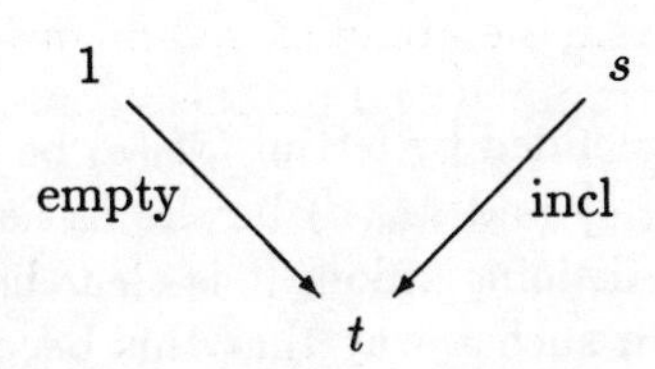

be a cocone.

There are no diagrams.

The cocone says that every tree is either empty or nonempty. This cocone could be alternatively expressed $t = s + \{\text{empty}\}$. The cone says that every nonempty tree can be represented uniquely as a triplet $(\text{left}(S), \text{val}(S), \text{right}(S))$ and that every such triplet corresponds to a tree. Note that this implies that left, val and right become coordinate projections in a model.

Using this, we can define subsidiary operations on trees. For example, we can define an operation of left attachment, $\text{lat} : t \times s \to s$ by letting

$$\text{lat}(T, (\text{left}(S), \text{val}(S), \text{right}(S))) = (T, \text{val}(S), \text{right}(S))$$

This can be done without elements: lat is defined in any model as the unique arrow making the following diagram commute (note that the horizontal arrows are isomorphisms):

$$\begin{array}{ccc}
M(t) \times M(s) & \xrightarrow{M(t) \times \langle \text{left,val,right} \rangle} & M(t) \times M(t) \times M(d) \times M(t) \\
\Big\downarrow \text{lat} & & \Big\downarrow \langle p_1, p_3, p_4 \rangle \\
M(s) & \xrightarrow[\langle \text{left,val,right} \rangle]{} & M(t) \times M(d) \times M(t)
\end{array}$$

In a similar way, we can define right attachment as well as the insertion of a datum at the root node as operations definable in any tree. These operations are implicit in the sketch in the sense that they occur as arrows in the theory generated by the sketch (see 4.6.9), and therefore are present in every model.

7.10.12 Proposition *Supposing there is an initial algebra for the data type, then the category of binary trees of that type has an initial algebra. If the data type has (up to isomorphism) a unique initial algebra, then so does the corresponding category of binary trees.*

Proof. We construct the initial algebra recursively according to the rules:

(i) The empty set is a tree;

(ii) If T_l and T_r are trees and D is element of the initial term algebra for the data type, then (T_l, D, T_r) is a nonempty tree;

(iii) Nothing else is a tree.

This is a model M_0 defined by letting $M_0(s)$ be the set of nonempty trees, $M_0(t) = M_0(s) + \{\}$ and $M_0(d)$ be the initial model of the data type. Here '+' denotes disjoint union. It is clear how to define the operations of the sketch in such a way that this becomes a model of the sketch.

Now let M be any model with the property that $M(d)$ is a model for the data type. Then there is a unique morphism $f(d) : M_0(d) \to M(d)$ that preserves all the operations in the data type. We also define $f(t)\{\}$ to be the value of $M(\text{empty}) : 1 \to M(t)$. Finally, we define

$$f(s)((T_l, D, T_r)) = (f(t)(T_l), f(d)(D), f(t)(T_r))$$

where $f(t)$ is defined recursively to agree with $f(s)$ on nonempty trees. It is immediately clear that this is a morphism of models and is unique. In particular, if the data type has, up to isomorphism, only one initial model then M_0 is also unique up to isomorphism. □

7.10.13 In Pascal textbooks a definition for a tree type typically looks like this:

```
type TreePtr = ^Tree;
     Tree = record LeftTree, RightTree : TreePtr;
          Datum : integer
     end;
```

Note that from the point of view of the preceding sketch, this actually defines nonempty trees. The empty tree is referred to by a null pointer. This takes advantage of the fact that in such languages defining a pointer to a type D actually defines a pointer to what is in effect a variant record (union structure) which is either of type D or of 'type' null.

7.10.14 Exercise

1.[†] **a.** Show that if $\mathcal{S}$ is an FD sketch and $f : M \to N$ is a homomorphism between models in the category of sets, then the image of f is a submodel of N.

b. Show that every model in the category of sets of an FD sketch has a smallest submodel.

8
Limits and colimits

A limit is the categorical version of the concept of an equationally defined subset of a product. For example, a circle of radius 1 is the subset of $\mathbf{R}\times\mathbf{R}$ ($\mathbf{R}$ is the set of real numbers) satisfying the equation $x^2+y^2=1$. Another example is the set of pairs of arrows in a category for which the target of the first is the source of the second: this is the set of pairs for which the composition operation is defined.

A different kind of example is division of one integer by another, which requires that the second argument be nonzero. This can be made an equational condition by building in a Boolean type and a test; the equation then becomes $[y=0]=\textbf{false}$. (This can also be handled using finite sums: see Section 7.9.)

A colimit is similarly the categorical version of a quotient of a sum by an equivalence relation. The quotient category constructed in Section 3.5 is an example of a colimit in the category of categories.

This chapter discusses finite limits and colimits in detail, concentrating on certain useful special cases. Infinite limits and colimits are widely used in mathematics, but they are conceptually similar to the finite case, and are not described here.

Sections 8.1, 8.2 and 8.3 discuss limits. Sections 8.4 and 8.5 describe colimits. Section 8.6 describes certain properties of sums which are desirable in programming language semantics. The chapter concludes with an application expressing unification (as used in Prolog) as a kind of colimit. This last section is not needed to read the rest of the book.

Limits and colimits are used in all the remaining chapters except Chapter 11. In particular, they are used in Chapter 9 to describe more expressive types of sketches.

This chapter may be read right after Chapter 5, except that Section 8.7 requires familiarity with FP sketches (Section 7.1).

8.1 Equalizers

If S and T are sets and $f,g: S\to T$ are functions, it is a familiar fact that we can form the subset $Eq(f,g)\subseteq S$ consisting of all elements $s\in S$ for which $f(s)=g(s)$. This concept can be made into a categorical concept by changing it to a specification which turns out to determine a subobject. We will look more closely at the construction in **Set** to see how to make the general categorical construction.

Suppose that $j : Eq(f,g) \to S$ is the inclusion function: $j(u) = u$ for $u \in Eq(f,g)$. Let $h : V \to S$ be a function. Then h factors through $j : Eq(f,g) \to S$ if and only if the image of the function h lies in the subset $Eq(f,g)$ (see 2.8.6). This is the key to the categorical specification of $Eq(f,g)$, Definition 8.1.2 below.

8.1.1 Definition Two arrows $f : A \to B$ and $g : A \to B$ of a category (having the same domain and the same codomain) are a **parallel pair** of arrows. An arrow h with the property that $f \circ h = g \circ h$ is said to **equalize** f and g.

8.1.2 Definition Let $\mathcal{C}$ be a category and $f, g : A \to B$ be a parallel pair of arrows. An **equalizer** of f and g is an object E *together with* an arrow $j : E \to A$ with the property that for each arrow $h : C \to A$ such that $f \circ h = g \circ h$, there is an arrow $k : C \to E$, and only one, such that $j \circ k = h$.

Frequently, E is referred to as the equalizer of f and g without referring to j. Nevertheless, j is a crucial part of the data.

8.1.3 Examples of equalizers In **Set**, the equalizer of the function from $\mathbf{R} \times \mathbf{R}$ to $\mathbf{R}$ which takes (x, y) to $x^2 + y^2$ and the function which takes (x, y) to the constant 1 is the circle $x^2 + y^2 = 1$. The arrow j is the inclusion.

Given a graph $\mathcal{G}$, the inclusion of the set of loops in the graph is the equalizer of the source and target functions. This equalizer will be empty if the graph has no loops.

In a monoid M regarded as a category, any two elements of the monoid form a parallel pair of arrows which may or may not have an equalizer. (See Exercise 4.)

A theorem like Theorem 5.2.2 is true of equalizers as well.

8.1.4 Proposition *If $j : E \to A$ and $j' : E' \to A$ are both equalizers of $f, g : A \to B$, then there is a unique isomorphism $i : E \to E'$ for which $j' \circ i = j$.*

Proof. We give two proofs. The first uses the concept of universal element. Let $\mathcal{C}$ be the category containing the equalizers given. Let $F : \mathcal{C}^{\text{op}} \to \mathbf{Set}$ be the functor for which $F(C) = \{u : C \to A \mid f \circ u = g \circ u\}$, the set of arrows from C which equalize f and g. If $h : D \to C$ let $F(h)(u) = u \circ h$. This makes sense, because if $u \in F(C)$, then $f \circ u \circ$

$h = g \circ u \circ h$, so $u \circ h \in F(D)$. Note that this makes F a subfunctor of the contravariant hom functor $\mathrm{Hom}(-, A)$ as in Exercise 4 of Section 4.3. The definition of the equalizer of f and g can be restated this way: an equalizer of f and g is an element $j \in F(E)$ for some object E such that for any $u \in F(C)$ there is a unique arrow $k : C \to E$ such that $F(k)(j) = u$. This means j is a universal element of F, so by Corollary 4.5.12, any two equalizers $j \in F(E)$ and $j' \in F(E')$ are isomorphic by a unique arrow $i : E \to E'$ such that $F(i)(j') = j$. That is, $j = j' \circ i$, as required.

This method of constructing a functor of which the given limit is a universal element is a standard method in category theory. We have already seen it used in the proof of Proposition 5.2.14 and in the discussion of the uniqueness of eval in 6.1.7.

Here is a direct proof not using universal elements: the fact that $f \circ j' = g \circ j'$ implies the existence of a unique arrow $h : E' \to E$ such that $j' = j \circ h$. The fact that $f \circ j = g \circ j$ implies the existence of a unique arrow $h' : E \to E'$ such that $j = j' \circ h'$. Then $j \circ h \circ h' = j' \circ h' = j = j \circ \mathrm{id}_E$ and the uniqueness part of the definition of equalizer implies that $h \circ h' = \mathrm{id}_E$. By symmetry, $h' \circ h = \mathrm{id}_{E'}$. □

Any equalizer is a monomorphism (Exercise 1), and in fact the following holds.

8.1.5 Corollary *Any two equalizers of the same pair of arrows belong to the same subobject.*

Proof. In the notation of Proposition 8.1.4, i and i^{-1} are the arrows required by the definition of subobject in 2.8.8, since $j' \circ i = j$ and $j \circ i^{-1} = j'$. □

8.1.6 More generally, if $f_1, \ldots, f_n$ are all arrows from A to B, then an object E together with an arrow $j : E \to A$ is the equalizer of $f_1, \ldots, f_n$ if it has the property that an arbitrary arrow $h : C \to A$ factors uniquely through j if and only if $f_1 \circ h = \cdots = f_n \circ h$. Having equalizers of parallel pairs implies having equalizers of all finite lists (Exercise 2).

8.1.7 Exercises

1. Show that if $j : E \to A$ is the equalizer of the pair of arrows $f, g : A \to B$ then j is a monomorphism. (Hint: use the uniqueness property of the definition.)

2. Show that a category with equalizers of all parallel pairs of arrows has equalizers of every finite list $f_1, \ldots, f_n : A \to B$ of arrows.

3.† Prove that in the category of monoids every pair of parallel arrows has an equalizer.

4. a. Prove that if M is a free monoid, then in $C(M)$ no two different elements have an equalizer.

b. Prove the same statement for finite monoids. (Hint: see Exercise 3 of Section 2.9.)

5. A monomorphism $e : S \to T$ in a category is **regular** if e is the equalizer of a pair of arrows.

a. Show that every monomorphism in **Set** is regular.

b. Show that an arrow in a category that is both an epimorphism and a regular monomorphism is an isomorphism.

c. Suppose that $\mathcal{A}$ and $\mathcal{B}$ are categories with two objects and whose only nonidentity arrows are as shown:

$$C \xrightarrow{\ u\ } D \qquad\qquad C \underset{v}{\overset{u}{\rightleftarrows}} D$$

$$\mathcal{A} \qquad\qquad\qquad \mathcal{B}$$

Show that the inclusion functor of $\mathcal{A}$ into $\mathcal{B}$ is an epimorphism in **Cat**. Conclude it cannot be a regular monomorphism. (Hint: the arrows u and v are inverse to each other.)

8.2 The general concept of limit

Products and equalizers are both examples of the general concept of limit.

8.2.1 Definition Let $\mathcal{G}$ be a graph and $\mathcal{C}$ be a category. Let $D : \mathcal{G} \to \mathcal{C}$ be a diagram in $\mathcal{C}$ with shape $\mathcal{G}$. A **cone** with base D is an object C of $\mathcal{C}$ together with a family $\{p_a\}$ of arrows of $\mathcal{C}$ indexed by the nodes of $\mathcal{G}$, such that $p_a : C \to Da$ for each node a of $\mathcal{G}$. The arrow p_a is the **component** of the cone at a.

The cone is **commutative** if for any arrow $s : a \to b$ of $\mathcal{G}$, the diagram

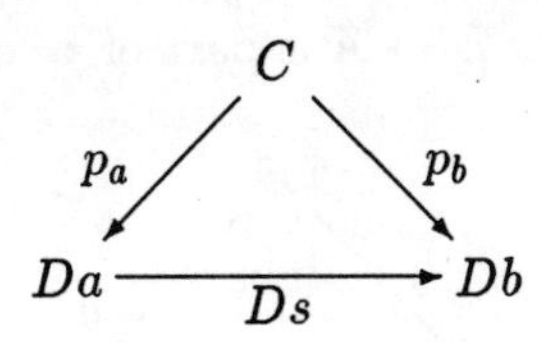

commutes. Note: the diagram D is *not* assumed to commute.

Such a diagram will also be termed a commutative cone **over** D (or **to** D or with **base** D) with vertex C. We will write it as $\{p_a\} : C \to D$ or simply $p : C \to D$. It is clear that if $p : C \to D$ is a cone over D and $f : C' \to C$ is an arrow in $\mathcal{C}$, then there is a cone $p \circ f : C' \to D$ whose component at a is $p_a \circ f$. Note that a cone over a discrete diagram (see 5.1.3) is vacuously commutative.

8.2.2 Definition If $p' : C' \to D$ and $p : C \to D$ are cones, an **arrow** from the first to the second is an arrow $f : C' \to C$ such that for each node a of $\mathcal{G}$, the diagram

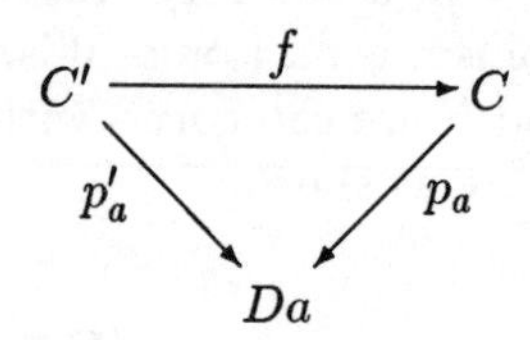

commutes.

8.2.3 Definition A commutative cone over the diagram D is called **universal** if every other commutative cone over the same diagram has a unique arrow to it. A universal cone, if such exists, is called a **limit** of the diagram D.

8.2.4 It is worth spelling out in some detail the meaning of a limit. To say that $p : C \to D$ is a limit means that there is given a family $p_a : C \to Da$, indexed by the nodes of $\mathcal{G}$, for which

L–1 Whenever $s : a \to b$ is an arrow of $\mathcal{G}$, then $Ds \circ p_a = p_b$.

L–2 If $p' : C' \to D$ is any other such family with the property that $Ds \circ p'_a = p'_b$ for every $s : a \to b$ in $\mathcal{G}$, then there is one and only one arrow $f : C' \to C$ such that for each node a of $\mathcal{G}$, $p_a \circ f = p'_a$.

8.2.5 Examples A limit cone over a finite discrete diagram is a product cone: here, L–1 and the commutativity condition $Ds \circ p'_a = p'_b$ in L–2 are vacuous.

Equalizers are also limits. Let $f, g : A \to B$ be parallel arrows in a category. An equalizer $e : E \to A$ is part of a cone

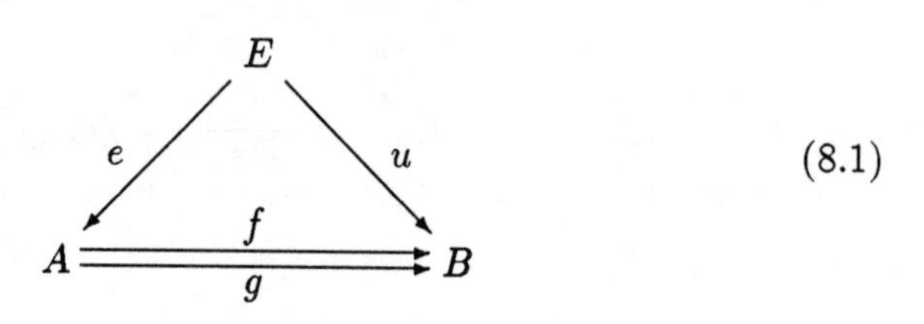

where $u = f \circ e = g \circ e$. This cone is commutative, and it is a limit cone if and only if e is an equalizer of f and g. Since u is determined uniquely by f and e (or by g and e), it was not necessary to mention it in Definition 8.1.2.

8.2.6 Equivalent definitions of limit There are two more equivalent ways to define limits. Let $D : \mathcal{G} \to \mathcal{C}$ be a diagram. There is a category we call cone(D) defined as follows. An object of this category is a cone $\{p_i : C \to Di\}$ and an arrow to $\{p'_i : C'_i \to Di\}$ is a family $\{f_i : C \to C'\}$ such that $p'_i \circ f_i = p_i$ for all i. It is evident that the identity is such an arrow and that the composite of two such arrows is another such. A terminal object in cone(D), if one exists, is a cone over D to which every other cone over D has a unique arrow. Thus we have shown that the existence of a limit is equivalent to the existence of a terminal object of cone(D).

The second equivalent construction associates to D a functor we call cone$(-, D) : \mathcal{C}^{\text{op}} \to \mathbf{Set}$. For an object C of $\mathcal{C}$, cone(C, D) is the set of commutative cones $p : C \to D$. If $f : C' \to C$, cone$(f, D) :$ cone$(C, D) \to$ cone(C', D) is defined by cone$(f, D)(p) = p \circ f$. It is straightforward to verify that this is a functor. A universal element of this functor is an object C and an element $p \in$ cone(C, D) such that for any C' and any element $p' \in$ cone(C', D) there is a unique arrow $f : C' \to C$ such that $p \circ f = p'$. But this is just the definition of a limit cone. (Compare the proof of Proposition 8.1.4.)

We have now sketched the proof of the following.

8.2.7 Theorem *The following three are equivalent for a diagram D in a category $\mathcal{C}$:*

(i) *a limit of D;*

(ii) *a terminal object of* cone(D)*;*

(iii) *a universal element of the functor* cone$(-, D)$.

In particular, one can see products and equalizers as terminal objects (which we have not mentioned) or as universal elements (which we have mentioned). One immediate consequence of this, in light of 4.5.12, is that a limit of a diagram is characterized uniquely up to a unique isomorphism in the same way that a product is.

8.2.8 Theorem *Let D be a diagram and $p : C \to D$ and $p' : C' \to D$ limits. Then there is a unique arrow $i : C \to C'$ such that for every node a of the shape graph of D, $p'_a \circ i = p_a$ and i is an isomorphism.*

A category is said to have all limits if every diagram has a limit. It has all finite limits if every diagram whose domain is a finite graph has a limit. The following theorem gives a useful criterion for the existence of all limits or all finite limits. Another way of getting all finite limits is given by Proposition 8.3.7.

8.2.9 Theorem *Suppose every set (respectively, every finite set) of objects of $\mathcal{C}$ has a product and every parallel pair of arrows has an equalizer. Then every diagram (respectively, every finite diagram) in $\mathcal{C}$ has a limit.*

The proof of the main claim is given in Exercise 3. The case for finite limits is exactly the same, specialized to that case. As an immediate corollary, using Exercise 1.c of Section 5.3, we have the following.

8.2.10 Corollary *A category $\mathcal{C}$ with the following properties*

(i) *$\mathcal{C}$ has a terminal object;*

(ii) *every pair of objects has a product; and*

(iii) *every parallel pair of arrows has an equalizer*

has all finite limits.

8.2.11 Exercises

1. Prove that Diagram (8.1) is a limit cone if and only if e is an equalizer of f and g.

2. Let $f : S \to T$ be a set function. Construct limits of each of these diagrams in **Set**. In (c) the arrows are the unique functions to the terminal object.

$$S \xrightarrow{f} T \qquad S \xrightarrow{f} T \xleftarrow{f} S \qquad S \longrightarrow 1 \longleftarrow S$$

(a) (b) (c)

3. In this exercise, $\mathcal{C}$ is a category in which every set of objects has a product and every pair of arrows with the same source and target has an equalizer. The purpose of this exercise is to prove Theorem 8.2.9.

a. Suppose that $\mathcal{I}$ is a graph that may have an arbitrary set of nodes, but just one nonidentity arrow $a : j \to k$. Show that for any $D : \mathcal{I} \to \mathcal{C}$, the equalizer

$$E \longrightarrow \prod_{i \in \mathcal{I}} Di \;\underset{\text{proj}_k}{\overset{Da \circ \text{proj}_j}{\rightrightarrows}}\; Dk$$

is a limit of D.

b. Suppose that $\mathcal{I}$ has just two nonidentity arrows $a : j \to k$ and $b : l \to m$ (no assumption of distinctness is made). Let $r_1 = Da \circ \mathrm{proj}_j : \prod Di \to Dk$, $s_1 = \mathrm{proj}_k : \prod Di \to Dk$, $r_2 = Db \circ \mathrm{proj}_l : \prod Di \to Dm$ and $s_2 = \mathrm{proj}_m : \prod Di \to Dm$. Let $r = \langle r_1, r_2 \rangle : \prod Di \to Dk \times Dm$ and $s = \langle s_1, s_2 \rangle : \prod Di \to Dk \times Dm$. Show that if

$$E \longrightarrow \prod Di \underset{s}{\overset{r}{\rightrightarrows}} Dk \times Dm$$

is an equalizer, then E is a limit of D.

c. Let $\mathcal{I}$ be an arbitrary graph and $D : \mathcal{I} \to \mathcal{C}$ a diagram in $\mathcal{C}$. Let $A = \prod Di$, taken over the objects of $\mathcal{I}$ and $B = \prod(D(\mathrm{target}\, a))$, the product taken over the set of all arrows of $\mathcal{I}$. Let $r : A \to B$ be such that $\mathrm{proj}_a \circ r = Da \circ \mathrm{proj}_{\mathrm{source}(a)}$ and $s : A \to B$ be such that $\mathrm{proj}_a \circ s = \mathrm{proj}_{\mathrm{target}(a)}$. Let

$$E \longrightarrow A \underset{s}{\overset{r}{\rightrightarrows}} B$$

be an equalizer. Then E is a limit of D.

8.3 Pullbacks

Here is another example of a finite limit that will be important to us. Consider an object P and arrows $A \xleftarrow{p_1} P \xrightarrow{p_2} B$ such that the diagram

$$\begin{array}{ccc} P & \xrightarrow{p_1} & A \\ {\scriptstyle p_2}\downarrow & & \downarrow{\scriptstyle f} \\ B & \xrightarrow[g]{} & C \end{array} \qquad (8.2)$$

commutes. This does not appear directly to be a cone, because there is no arrow from the vertex P to C. It is understood to be a commutative cone with the arrow from P to C being the composite $f \circ p_1$, which by definition is the same as the composite $g \circ p_2$. It is common to omit such forced arrows in a cone. (Compare the discussion after Diagram (8.1).) It is more cone-like if we redraw the diagram as

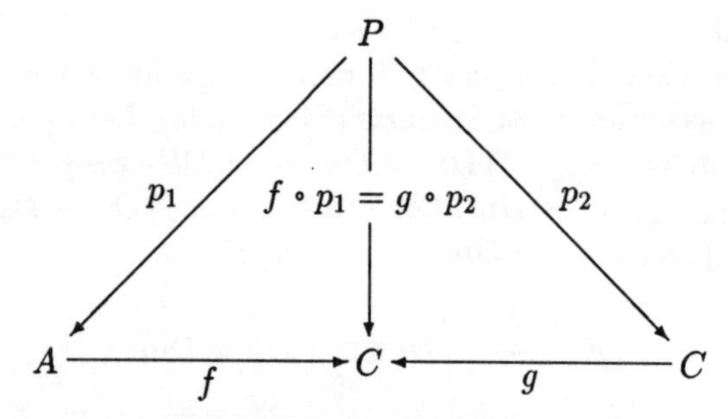

However, the square shape of (8.2) is standard.

If this commutative cone is universal, then we say that P together with the arrows p_1 and p_2 is a **pullback** or **fiber product** of the pair. We also say that p_1 is the **pullback of** f **along** g, and that (8.2) is a **pullback diagram**. We often write $P = A \times_C B$, although this notation omits the arrows which are as important as the object C.

8.3.1 Example In **Set**, if $f : S \to T$ and $g : U \to T$ are functions, then the pullback

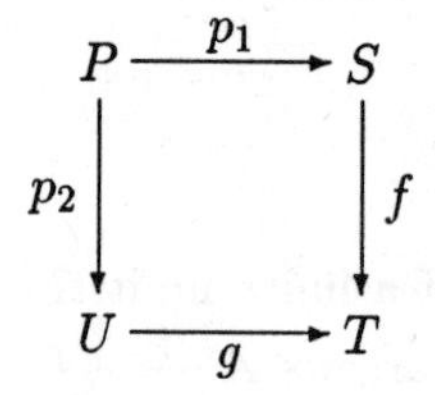

is constructed by setting

$$P = \{(s, u) \mid f(s) = g(u)\}$$

with $p_1(s, u) = s$ and $p_2(s, u) = u$.

8.3.2 Example The inverse image of a function is a special case of a pullback. Suppose $g : S \to T$ is a set function and $A \subseteq T$. Let $i : A \to T$ be the inclusion. Let $g^{-1}(A) = \{x \in S \mid g(x) \in A\}$ be the inverse image of A, j be its inclusion in S, and h be the restriction of g to $g^{-1}(A)$. Then the following is a pullback diagram.

$$\begin{array}{ccc} g^{-1}(A) & \xrightarrow{h} & A \\ {\scriptstyle j}\downarrow & & \downarrow{\scriptstyle i} \\ S & \xrightarrow[g]{} & T \end{array} \qquad (8.3)$$

Observe that Example 8.3.1 gives a general construction for pullbacks in sets, and the present example gives a construction for a special type of pullback, but this construction *is not a special case* of the construction in 8.3.1. By Theorem 8.2.8, there must be a unique function $u : \{(s,a) \mid g(s) = a\} \to g^{-1}(A)$ for which $g(j(u(s,a))) = i(h(u(s,a)))$. By the definitions of the functions involved, $u(s,a)$ must be s.

The reader may wish to know the origin of the term 'fiber product'. Consider a function $f : S \to T$. Of the many ways to think of a function, one is as determining a partition of S indexed by the elements of T. (See 2.6.9.) For $t \in T$, let

$$S_t = f^{-1}(t) = \{s \in S \mid f(s) = t\}$$

This is family of disjoint subsets of S (some of which may be empty) whose union is all of S. This is sometimes described as a **fibration** of S and S_t is called the **fiber** over t. Now if $S \to T$ and $U \to T$ are two arrows and $S \times_T U$ is the pullback, then it is not hard to see that

$$S \times_T U = \bigcup\{S_t \times U_t \mid t \in T\}$$

In other words, $S \times_T U$ is the fibered set whose fiber over any $t \in T$ is the product of the fibers. This is the origin of the term and of the notation. We will not use the term in this book, but the notation has become standard and is too useful to abandon. Fibrations can be constructed for categories as well as sets; they are considered in Section 11.1.

Pullbacks can be used to characterize monomorphisms in a category.

8.3.3 Theorem *These three conditions are equivalent for any arrow $f : A \to B$ in a category $\mathcal{C}$:*

(a) *f is monic.*

(b) *The diagram*

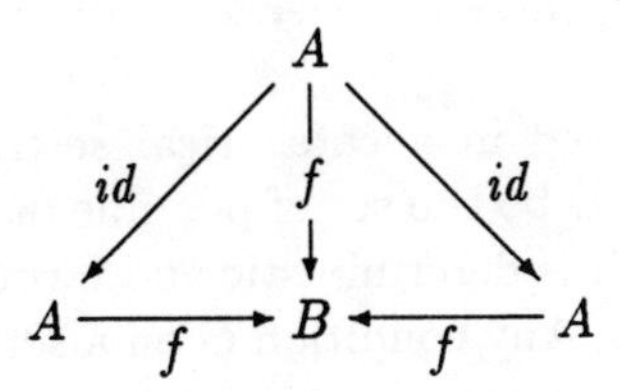

is a limit cone.

(c) *There is an object P and an arrow $g : P \to A$ for which*

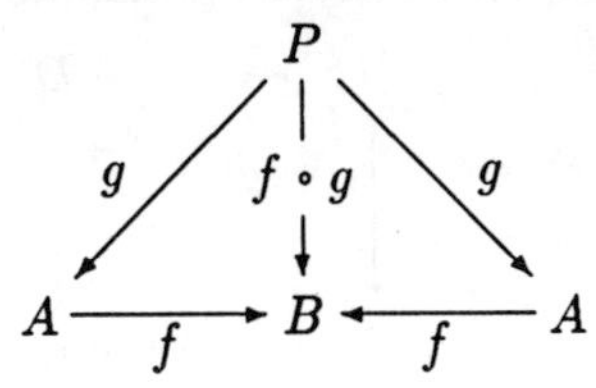

is a limit cone.

(The operative condition in (c) is that the same arrow g appears on both slant lines.)

Another connection between pullbacks and monomorphisms is given by the following.

8.3.4 Proposition *In Diagram (8.2), if the arrow f is monic, then so is p_2.*

This proposition is summed up by saying, 'A pullback of a monic is a monic.' Contrast it with the situation for epis given in Exercises 6 and 7.

The proofs of 8.3.3 and 8.3.4 are left as exercises.

8.3.5 Weakest precondition as pullback A widely used approach to program verification is to attach preconditions and postconditions to program fragments. An example is

```
1. {X < 3}
2. X := X + 1;                (8.4)
3. {X < 24}
```

where X is an integer variable.

Statement 1 is called a **precondition** and statement 3 is a **postcondition**. The whole expression (8.4) is an **assertion** about the program: if the precondition is true before the program fragment is executed, the postcondition must be true afterward.

Clearly (8.4) is correct but a stronger assertion can be made. For the given postcondition, the weakest possible precondition is {X < 23}. In a very general setting, there is a weakest precondition for every postcondition.

This can be placed in a categorical setting. Let D be the set of possible inputs and E be the set of possible outputs. Then the program fragment, provided it is deterministic and terminating, can be viewed as an arrow $f : D \to E$. Any condition C on a set X can be identified with the subset $X_0 = \{x \mid x \text{ satisfies } C\}$. In particular, the postcondition is a subset $S \subseteq E$ and the weakest precondition for that postcondition is the inverse image $f^{-1}(S)$, which is the unique subobject of D for which

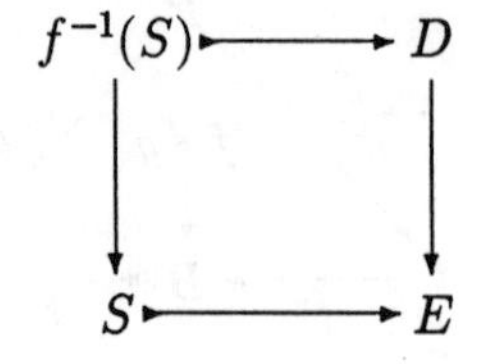

is a pullback. In the example, $f(x) = x + 1$, $S = \{x \mid x < 24\}$ and $f^{-1}(S) = \{x \mid x + 1 < 24\} = \{x \mid x < 23\}$.

This is easily checked in the case of **Set** and is a plausible point of view for other categories representing data and program fragments.

In the general case, one would expect that an assertion would be some special kind of subobject. For example, Manes [1986] requires them to be 'complemented', a concept to be discussed in 8.6.2 below, and Wagner [1987] requires them to be regular (see Exercise 5 of 8.1). Naturally, if one requires that assertions be subobjects with a certain property, one would want to work in a category in which pullbacks of such subobjects also had the property. In a topos (toposes will be discussed in Chapter 14), for example, pullbacks of complemented subobjects are complemented, and all monomorphisms are regular.

8.3.6 Constructing all finite limits revisited Corollary 8.2.10 described a class of limits whose existence guarantees the existence of all finite limits. A similar construction is possible for pullbacks.

8.3.7 Proposition *A category that has a terminal object and all pullbacks has all finite limits.*

The proof is left as an exercise.

8.3.8 Exercises

1. Prove that (8.3) is a pullback diagram.

2. Prove Theorem 8.3.3.

3. Prove Proposition 8.3.4.

4. Show that equalizers can be constructed out of products and pullbacks. (Hint: try to work it out in the category of sets and then do it in general categories.)

5. Prove Proposition 8.3.7.

6. Prove that in **Set** if f in Diagram (8.2) is an epimorphism, then so is p_1. (One says that a pullback of an epimorphism is an epimorphism in **Set**.)

7.† Give an example of a category with an epimorphism whose pullback along at least one arrow is not an epimorphism.

8. Prove that the pullback of a split epimorphism is a split epimorphism in any category.

9. Show that if the left and right squares below are pullbacks then so is the outer rectangle. What does this say about weakest preconditions?

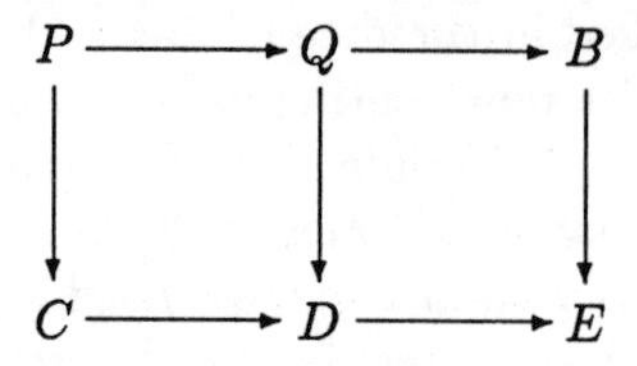

8.4 Coequalizers

In 8.1.2 we introduced the notion of equalizer. The dual notion is called coequalizer. Explicitly, consider $f, g : A \rightrightarrows B$. An arrow $h : B \to C$ is called a **coequalizer** of f and g provided $h \circ f = h \circ g$ and for any arrow $k : B \to D$ for which $k \circ f = k \circ g$, there is a unique arrow $l : C \to D$ such that $l \circ h = k$.

The way to think about the coequalizer of f and g is as the quotient object made by forcing f and g to be equal. In the category of sets, this observation becomes the construction that follows.

8.4.1 Coequalizers in Set Let $f, g : S \rightrightarrows T$ be a pair of arrows in the category of sets. The conditions a coequalizer $h : T \to U$ must satisfy are that $h \circ f = h \circ g$ and that given any arrow $k : T \to V$, the equation $k \circ f = k \circ g$ implies the existence of a unique arrow $l : U \to V$ such that $l \circ h = k$. (If such an arrow l exists, then necessarily $k \circ f = k \circ g$.)

The pair f and g determine a relation $R \subseteq T \times T$ which consists of

$$\{(f(s), g(s)) \mid s \in S\}$$

The equation $k \circ f = k \circ g$ is equivalent to the statement that $k(t_1) = k(t_2)$ for all $(t_1, t_2) \in R$.

In general R is not an equivalence relation, but it can be completed to one by forming the reflexive, symmetric and transitive closures of R in that order. The reflexive, symmetric closure is the union $R_1 = R \cup \Delta \cup R^{\mathrm{op}}$ where $\Delta = \{(t,t) \mid t \in T\}$ and $R^{\mathrm{op}} = \{(t_2, t_1) \mid (t_1, t_2) \in R\}$. If we inductively define R_{i+1} to be the set

$$\{(t_1, t_2) \mid \exists t \in T((t_1, t) \in R_1 \text{ and } (t, t_2) \in R_i)\}$$

for each $i \in \mathbf{N}$ and define $\widetilde{R} = \bigcup R_n$, then $\widetilde{R}$ is the transitive closure of R_1 and is the least equivalence relation containing R.

When S and T are finite, the least equivalence relation of the previous paragraph can be computed efficiently by Warshall's Algorithm ([Sedgewick, 1983], p. 425).

Let U be the set of equivalence classes modulo the equivalence relation $\widetilde{R}$ and let $h : T \to U$ be the function that sends an element of T to the class that contains that element.

8.4.2 Proposition *The arrow h is a coequalizer of f and g.*

Proof. Since $R \subseteq \widetilde{R}$, the fact that $h(t_1) = h(t_2)$ for all $(t_1, t_2) \in \widetilde{R}$ implies, in particular, that the same equation is satisfied for all $(t_1, t_2) \in R$; this means that for all $s \in S$, $h(f(s)) = h(g(s))$, so that $h \circ f = h \circ g$.

Now let $k : T \to V$ satisfy $k \circ f = k \circ g$. We claim that for $(t_1, t_2) \in \widetilde{R}$, $k(t_1) = k(t_2)$. In fact, this is true for $(t_1, t_2) \in R$. It is certainly true for $t_1 = t_2$, i.e. $(t_1, t_2) \in \Delta$ implies that $k(t_1) = k(t_2)$. Since $k(t_1) = k(t_2)$ implies that $k(t_2) = k(t_1)$, we now know that the assertion is true for $(t_1, t_2) \in R_1$. Since $k(t_1) = k(t_2)$ and $k(t_2) = k(t_3)$ imply $k(t_1) = k(t_3)$, one may show by induction that the assertion is true for all the R_n and hence for $\widetilde{R}$. This shows that k is constant on equivalence classes and hence induces a function $l : U \to V$ such that $l \circ h = k$. Since h is surjective, l is unique. □

8.4.3 Regular epimorphisms In Exercise 1 of Section 8.1 it is asserted that all equalizers are monomorphisms. The dual is of course also true; all coequalizers are epimorphisms. The converse is not true; not every epimorphism is a coequalizer of some pair of arrows. For example, in the category of monoids, the inclusion of the nonnegative integers into the set of all integers is an epimorphism but is not the coequalizer of any pair of arrows. Those epimorphisms that are the coequalizer of some pair of arrows into their domain are special and this is marked by giving them a special name: they are called **regular epimorphisms.**

In the category of sets, every epimorphism is regular (Exercise 2). Because of Exercise 6 of Section 8.3, it follows that in the category of sets, the pullback of a regular epi is regular.

A category is called a **regular category** if it has finite limits, if every parallel pair of arrows has a coequalizer, and if the arrow opposite a regular epimorphism in a pullback diagram is a regular epimorphism. A functor between regular categories is called a **regular functor** if it preserves finite limits and regular epis.

It is an old theorem (stated in different language) that all categories of models of FP sketches, in other words, all varieties of multisorted algebraic structures, are regular ([Barr and Wells, 1985], Theorem 1 of Section 8.4.)

8.4.4 Exercises

1. Show that any coequalizer is an epimorphism.

2. Prove that an epimorphism in the category of sets is regular. (Hint: its image is isomorphic to its quotient.)

3. Let **N** denote the monoid of natural numbers on addition and **Z** the monoid of integers on addition. Show that the inclusion function is an epimorphism (in the category of monoids) which is not regular.

4.† Prove that in the category of monoids an epimorphism f is regular if and only $U(f)$ is surjective, where U is the underlying functor to **Set**.

5. If $f : A \to B$ is an arrow in a category, a kernel pair

$$K \underset{d^1}{\overset{d^0}{\rightrightarrows}} A$$

is a limit

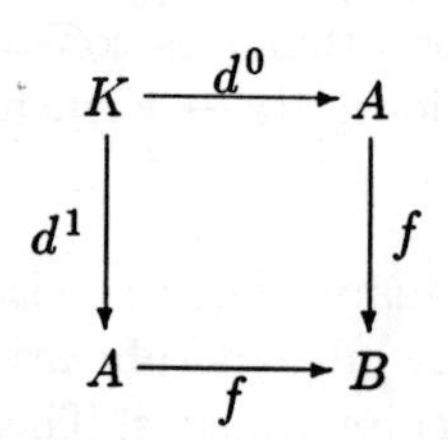

Show that if f has a kernel pair d^0, d^1, then f is regular epimorphism if and only if f is the coequalizer of its kernel pair.

8.5 Cocones

Just as there is a general notion of limit of which products and equalizers are special cases, so there is a general notion of colimit.

8.5.1 Definition A **cocone** is a cone in the dual graph. Spelling the definition out, it is a diagram $D : \mathcal{G}_0 \to \mathcal{G}$, a node g of $\mathcal{G}$ together with a family $u_a : Da \to g$ of arrows indexed by the objects of $\mathcal{G}_0$. The diagram is called the **base** of the cocone. If the diagram is discrete, it is called a **discrete cocone.**

A cocone in a category is called **commutative** if it satisfies the dual condition to that of commutative cone. Explicitly, if the diagram is $D : \mathcal{G} \to \mathcal{C}$ and the cone is $\{u_a : Da \to C \mid a \in \mathrm{Ob}(\mathcal{G})\}$, then what is required is that for each arrow $s : a \to b$ of $\mathcal{G}$, $u_b \circ Ds = u_a$. The commutative cocone is called a **colimit** cocone if it has a unique arrow to every other commutative cocone with the same base.

8.5.2 Pushouts For example, dual to the notion of pullback is that of pushout. In detail, a commutative square

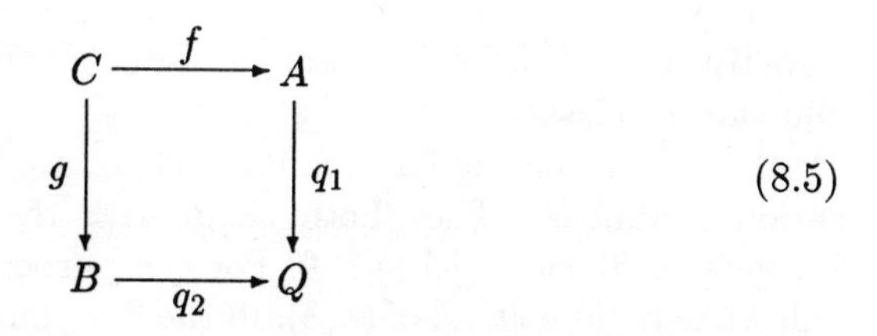

(8.5)

is called a **pushout** if for any object R and any pair of arrows $r_1 : A \to R$ and $r_2 : B \to R$ for which $r_1 \circ f = r_2 \circ g$ there is a unique arrow $r : Q \to R$ such that $r \circ q_i = r_i$, $i = 1, 2$.

8.5.3 Pushouts as amalgamations If the effect of coequalizers is to force identifications, that of pushouts is to form what are called amalgamated sums. We illustrate this with examples.

First, consider the case of a set S and three subsets S_0, S_1 and S_2 with $S_0 = S_1 \cap S_2$ and $S = S_1 \cup S_2$. Then the diagram

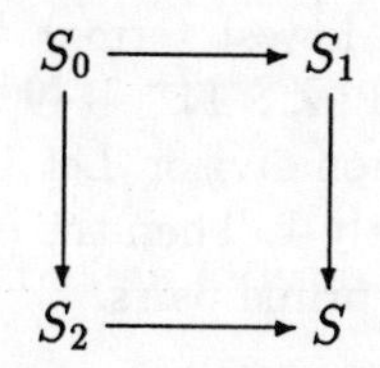

with all arrows inclusion, is a pushout. (It is a pullback as well; such diagrams are often referred to as **Doolittle diagrams.**) The definition of pushout translates to the obvious fact that if one is given functions $f_1 : S_1 \to T$ and $f_2 : S_2 \to T$ and $f_1|S_0 = f_2|S_0$, then there is a unique function $f : S \to T$ with $f|S_1 = f_1$ and $f|S_2 = f_2$.

Now consider a slightly more general situation. We begin with sets S_0, S_1 and S_2 and functions $g_1 : S_0 \to S_1$ and $g_2 : S_0 \to S_2$. If g_1 and g_2 are injections, then, up to isomorphism, this may be viewed as the same as the previous example. Of course, one cannot then form the union of S_1 and S_2, but rather first the disjoint sum and then, for $s \in S_0$, identify the element $g_1(s) \in S_1$ with $g_2(s) \in S_2$. In this case, the pushout is called an **amalgamated sum** of S_1 and S_2.

More generally, the g_i might not be injective and then the amalgamation might identify two elements of S_1 or of S_2, but the basic idea is the same; *pushouts are the way you identify part of one object with a part of another.*

8.5.4 Equations, equalizers and coequalizers One way of thinking about an equalizer is as the largest subobject on which an equation or set of equations is true. A coequalizer, by contrast, is the least destructive identification necessary to force an equation to be true on the equivalence classes.

Here is an instructive example. There are two ways of defining the rational numbers. They both begin with the set $\mathbf{Z} \times \mathbf{N}^+$ of pairs of integers (a, b) for which $b > 0$. For the purpose of this illustration, we will write (a/b) instead of (a, b). In the first, more familiar, construction, we identify (a/b) with (c/d) when $a * d = b * c$. One way of describing this is as the coequalizer of two arrows

$$T \rightrightarrows \mathbf{Z} \times \mathbf{N}^+$$

where T is the set of all $(a, b, c, d) \in \mathbf{Z} \times \mathbf{N}^+ \times \mathbf{Z} \times \mathbf{N}^+$ such that $a*d = b*c$ and the first arrow sends (a, b, c, d) to (a/b) while the second sends it to (c/d). The effect of the coequalizer is to identify (a/b) with (c/d) when $a * d = b * c$.

The second way to define the rationals is the set of pairs that are relatively prime (that is, in lowest terms). This can be realized as an equalizer as follows. Let $\text{gcd} : \mathbf{Z} \times \mathbf{N}^+ \to \mathbf{N}^+$ take a pair of numbers to their positive greatest common divisor. Let $1 \circ \langle\rangle : \mathbf{Z} \times \mathbf{N}^+ \to \mathbf{N}^+$ denote the function that is constantly 1. Then the equalizer of gcd and $1 \circ \langle\rangle$ is exactly the set of relatively prime pairs.

8.5.5 Exercises

1. Show that in any category, if both squares in

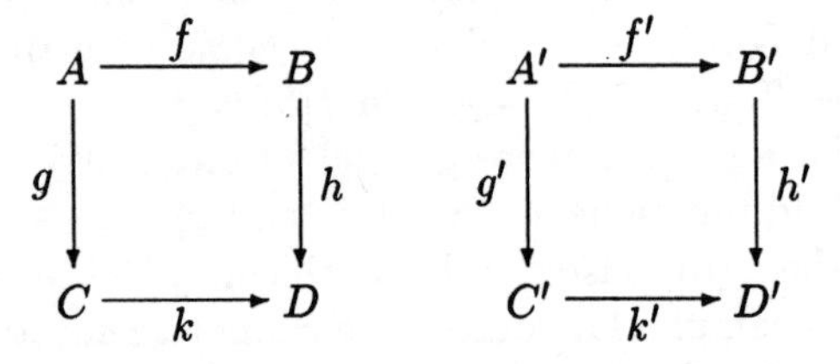

are pushouts, then so is

$$\begin{array}{ccc} A + A' & \xrightarrow{\;f + f'\;} & B + B' \\ {\scriptstyle g + g'}\downarrow & & \downarrow{\scriptstyle h + h'} \\ C + C' & \xrightarrow[\;k + k'\;]{} & D + D' \end{array}$$

2. Let A, B, C be sets and $f : C \to A$, $e : C \to B$ functions with e injective. Let $e(C)$ denote the image of e and $X = B - e(C)$ the complement of the image in B. Let $i : A \to A + X$ and $j : X \to A + X$ be the canonical injections.

a. Show that the diagram below is a pushout.

$$\begin{array}{ccc} C & \xrightarrow{\;e\;} & B = e(C) + X \\ {\scriptstyle f}\big\downarrow & & \big\downarrow{\scriptstyle i \circ f \circ e^{-1} + j} \\ A & \xrightarrow[\;i\;]{} & A + X \end{array}$$

b. Conclude that in **Set** *the pushout of a monomorphism is a monomorphism.*

3. a. Show that $\mathbf{Z}$ can be constructed from $\mathbf{N}$ as the coequalizer of two arrows $g, h : \mathbf{N} \times \mathbf{N} \times \mathbf{N} \to \mathbf{N} \times \mathbf{N}$, defined by $g(a, b, c) = (b, c)$ and $h(a, b, c) = (a + b, a + c)$ (Hint: addition is coordinatewise; 1 and -1 are the classes containing $(1, 0)$ and $(0, 1)$, respectively.)

b. Show that the integers can also be constructed as a subset of $\mathbf{N} \times \mathbf{N}$ consisting of all (n, m) such that $m = 0$ or $n = 0$. The sum is the usual sum followed by the application of the function $r(n, m) = (n \dot{-} m, m \dot{-} n)$, where $\dot{-}$ is subtraction at 0. This function is the additive analog of reduction to lowest terms by dividing by the greatest common divisor.

4. Let $f, g : \mathbf{N}^+ \times \mathbf{Z} \times \mathbf{N}^+ \to \mathbf{Z} \times \mathbf{N}^+$ be defined by $f(a, b, c) = (b, c)$ and $g(a, b, c) = (ab, ac)$. Show that the coequalizer of f and g can also be thought of as the rational numbers (see 8.5.4).

5. Show that in the category of sets, if both squares in

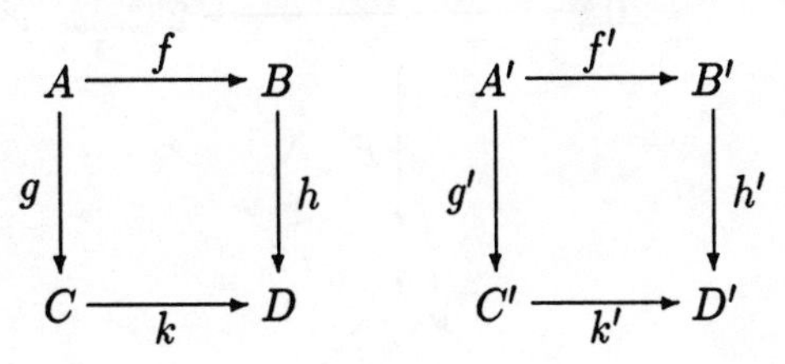

are pullbacks, then so is

$$\begin{array}{ccc} A + A' & \xrightarrow{\;f + f'\;} & B + B' \\ {\scriptstyle g + g'}\big\downarrow & & \big\downarrow{\scriptstyle h + h'} \\ C + C' & \xrightarrow[\;k + k'\;]{} & D + D' \end{array}$$

6.† **a.** Let **2** be the category with two objects and one nonidentity arrow between them and **N** be the category with one object and the natural numbers as its set of arrows, with 0 the identity arrow and $n \circ m = n + m$. Show that the functor $q : \mathbf{2} \to \mathbf{N}$ that takes the nonidentity arrow of **2** to 1 is a regular epi, even though it is not surjective on arrows.

b. Let $t : \mathbf{N} \to \mathbf{N}$ be the functor that assigns to the integer n the number $2n$. Let $\mathbf{1} + \mathbf{1}$ be the category with two objects and two identity arrows and none other. Show that there is a pullback diagram

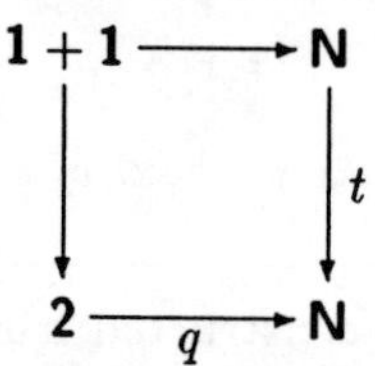

and that the top arrow is not a regular epi.

c. Let **3** be the category with three objects, say the numbers 0, 1 and 2 and three nonidentity arrows, one from 0 to 1, one from 1 to 2 and the composite of them from 0 to 2. Let $\mathcal{C}$ be the category with four objects we will call 0, 1, 1′ and 2 and three nonidentity arrows, one from 0 to 1, one from 1′ to 2 and one from 0 to 2. Show that the functor $F : \mathcal{C} \to \mathbf{3}$ pictured in Diagram (8.6) is not a regular epi, even though it is surjective on both objects and arrows.

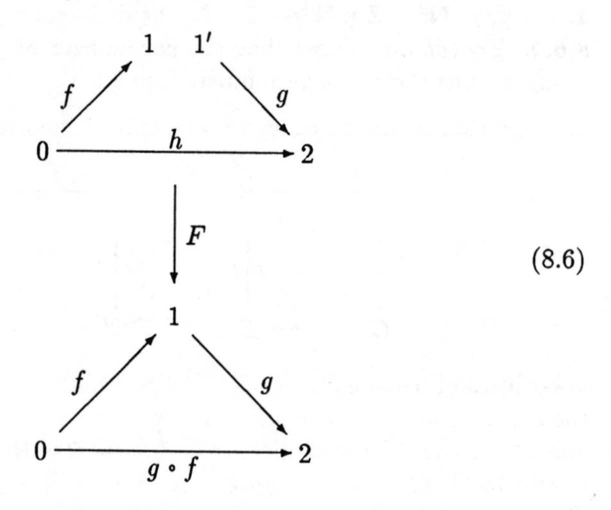

(8.6)

d. Show that a regular epi in **Cat** is **stable**, that is every pullback of it is a regular epi, if and only if it induces a surjection on the set of composable pairs of arrows.

8.6 More about sums

Sums in the category of sets have special properties they do not have in most other categories, including most categories of mathematical structures. Two of these special properties, that sums are disjoint and that they are universal, are widely assumed in the study of categories suitable for programming language semantics. They are defined in this section.

8.6.1 Disjoint sums Suppose that A and B are two objects in a category with an initial object 0 and a sum $A + B$. Then we have a commutative diagram

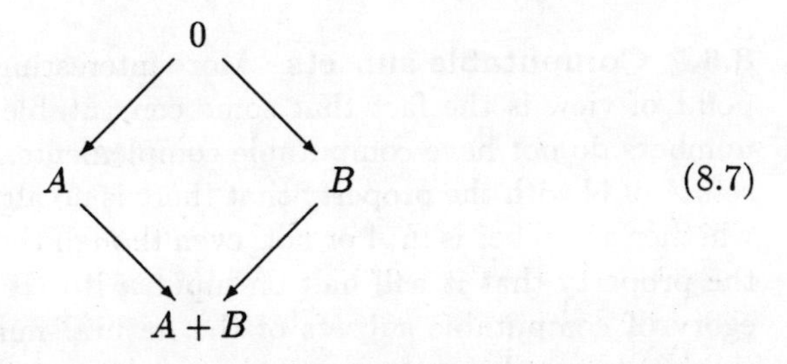

If this diagram is a pullback, and the canonical injections $A \to A + B$ and $B \to A + B$ are monic, then the sum is said to be **disjoint**. It is easy to see that sums in the category of sets are disjoint. In fact, that is essentially how they are defined.

8.6.2 Diagram (8.7) is a pushout in any case. If a diagram

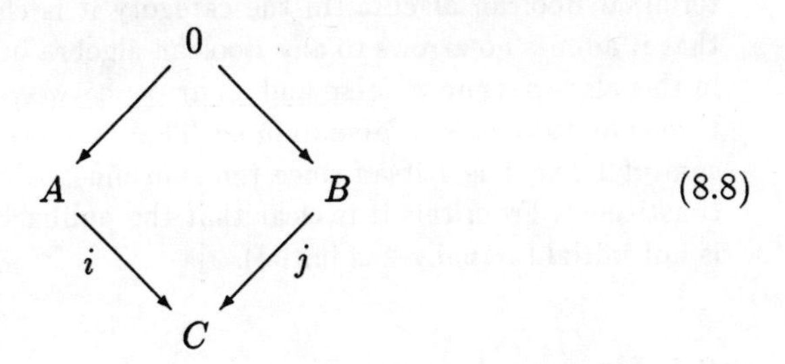

is both a pullback and a pushout, then C is the sum of A and B with the arrows i and j as canonical injections. If i and j are monic, then A and B are subobjects of C and the sum is disjoint. In this situation B is said to be the **complement** of A in C (and conversely), and A and B are said to be a **complemented subobjects**.

8.6.3 Example In **Set**, every subobject is complemented: the complement is what is usually meant by complement.

8.6.4 Example In **Mon** there are many subobjects which are noncomplemented; for example, the set of even integers is a submonoid of the monoid of all integers on addition, but it has no complement. Its set-theoretic complement, the set of odd integers, is not a submonoid: it does not contain the identity element (zero) and the sum of two odd integers is *never* odd. Even if its set-theoretic complement were a submonoid (impossible since only one of the two could contain the unit element), it still would not necessarily be the complement in the category.

A semigroup *can* have two subsemigroups whose underlying sets are complements but which is not the sum in the category of semigroups of those two subsemigroups. See Exercise 5.

8.6.5 Computable subsets More interesting from a computational point of view is the fact that some computable subsets of the natural numbers do not have computable complements. That is, there are subsets A of $\mathbf{N}$ with the property that there is no algorithm for determining whether a number is in A or not, even though there is an algorithm with the property that it will halt on input n if n *is* in A. Thus in the category of computable subsets of the natural numbers and computable maps between them, there are noncomplemented subobjects. Naturally, one would not want to give a *computable* meaning to an expression such as If n In A Then Do P unless A were complemented.

Here is an example of a category in which sums are not disjoint. The category is that of Boolean algebras. (They are defined in 6.1.10.) Although many books do not allow the one-element Boolean algebra, we do. It is the terminal Boolean algebra. In the category it is characterized by the fact that it admits no arrows to any Boolean algebra but itself. This is because in this algebra **true** = **false** and an arrow between Boolean algebras must take **true** to **true** and **false** to **false**. That being the case, the only possible sum of 1 and 1 is 1 itself since the sum must admit an arrow from each constituent. From this it is clear that the pullback in question is 1 which is not initial (actually 2 is initial).

8.6.6 Universal sums Finite sums in a category are **universal** if whenever

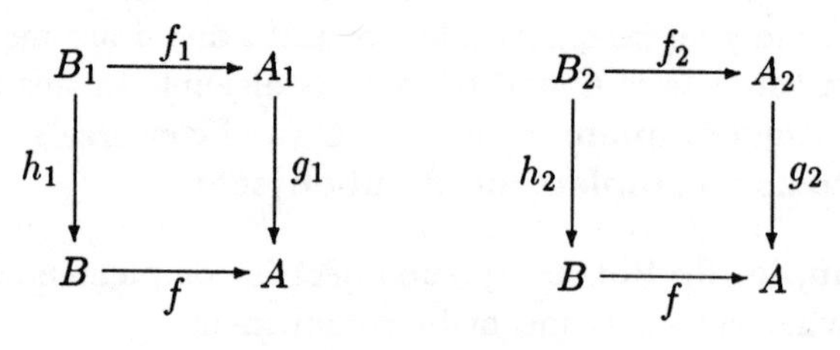

are pullbacks, then so is

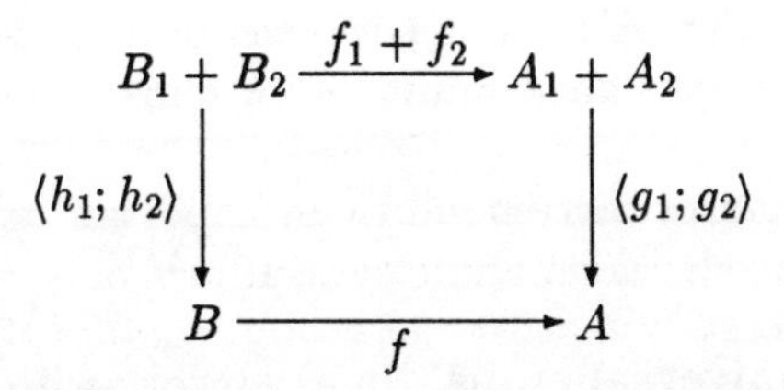

In addition it is required that if 0 denotes the initial object of the category, then for any arrow $B \to A$, the diagram

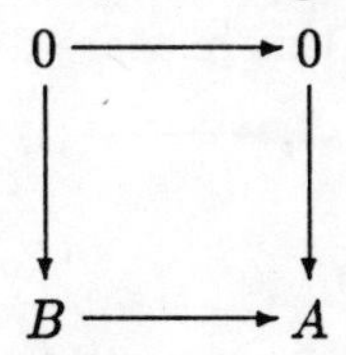

must be a pullback. The meaning of these two conditions together can be summarized by the statement that pullbacks preserve finite sums. A finite sum is either of no objects, one object, or two or more. The sum of no objects is the initial object; of one it is the object in question and the condition is vacuous since the identity functor preserves everything and the condition for more than two follows immediately from the one for two.

Although finite (and infinite) sums are universal in the category of sets, they are not in many familiar categories. In the category of monoids, for example, the initial monoid 0 is the monoid with one element. The sum $\mathbf{N} + \mathbf{N}$ is the free monoid on two generators, call them x and y. If we define the arrow $\mathbf{N} + \mathbf{N} \to \mathbf{N}$ by letting the generator x go to the number 0 and the generator y go to the number 1, then the pullback of

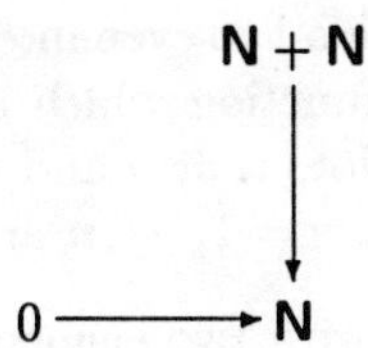

consists of all the elements of $\mathbf{N} + \mathbf{N}$ that go to 0 and is actually the free monoid generated by x, which is isomorphic to $\mathbf{N}$. Since $\mathbf{N}$ is not isomorphic to $0 + 0$ (which is 0), this pullback does not preserve finite sums. (See Exercise 4.)

8.6.7 Definition A category is called an **FLS category** (for finite limits and sums) if it has finite limits and finite sums and the sums are disjoint and universal.

8.6.8 Coherent categories A category is called **coherent** if it is both FLS and regular. A functor between coherent categories is called **coherent** if it preserves finite limits, finite sums and regular epimorphisms.

Coherent categories correspond to an important type of logic called **positive logic**. See [Makkai and Reyes, 1977].

8.6.9 Infinite universal sums In a category with pullbacks, an infinite sum $A = \sum A_i$ with transition arrows $u_i : A_i \to A$ is universal if for any arrow $B \to A$, we have that the natural arrow $B \to B \times_A A_i$ is an isomorphism. To explain this in more detail, for each i, let

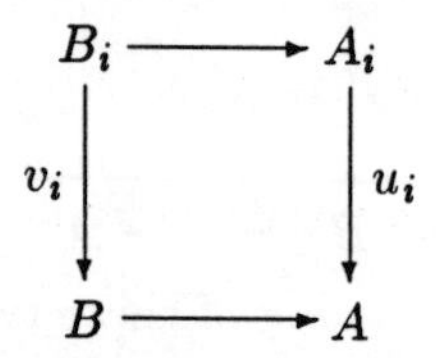

be a pullback. Then the assertion that the sum is universal is equivalent to the assertion that the cocone with elements $v_i : B_i \to B$ is a sum.

8.6.10 Exercises

1. Prove that Diagram (8.7) is a pushout in any category.

2. Prove that if Diagram (8.8) is a pushout then C is the sum of A and B with canonical injections i and j.

3. Prove that **Set** has universal sums.

4.† Let E denote the one-element semigroup. Note that it is a monoid. Show that in the category of semigroups, $E + E$ is infinite, but in the category of monoids, $E + E = E$. (Hint for the first part: let $u : \mathbf{N} \to \mathbf{N}$ denote the function which adds 1 to even integers but leaves odd integers along, and $v : \mathbf{N} \to \mathbf{N}$ the function which adds 1 to odd integers but leaves even integers alone. Note that u and v are both idempotent, but that the composites $(u \circ v)^n$, $n = 1, \cdots, n$ are all different.)

5.† Let S be the semigroup with two elements 0 and 1 under the usual multiplication of numbers. Show that even though both $\{0\}$ and $\{1\}$ are subsemigroups and each is the set-theoretic complement of the other, they are not complementary subobjects of S and that, in fact, their sum in the category of semigroups is infinite.

6.[†] (For those who know something about computability.)

a. Prove that there is a category whose only object is the set of natural numbers and whose arrows are the computable (recursive) functions with the usual composition of functions as composition.

b. Show that in this category there are subobjects without complements.

7.[†] Show that a category with 1 and universal countable sums has a stable natural numbers object. (See Section 5.5 and especially Theorem 5.5.5.)

8.7 Unification as coequalizer

A **unification** of two expressions is a common substitution which results in the two expressions becoming the same. This may not be possible, but when it is, there is a least such expression. We shall formulate this as the existence of a coequalizer (Theorem 8.7.4 below).

8.7.1 As an example, suppose we have a structure with three sorts, A, C and S, and two operations $m : A \times A \to S$ and $t : C \times C \to A$. If $a, b \in A$ and $u, v, w \in C$, then we can construct terms

$$\begin{aligned} x_1 &= m(t(u,v),a) \\ y_1 &= m(a,a) \\ x_2 &= m(b,t(u,w)) \\ y_2 &= m(b,a) \end{aligned} \tag{8.9}$$

all of type S. As it happens, the substitutions

$$\begin{aligned} w &\leftarrow v \\ a &\leftarrow t(u,v) \end{aligned} \tag{8.10}$$

in (8.9) produce

$$\begin{aligned} x'_1 &= m(t(u,v),t(u,v)) \\ y'_1 &= m(t(u,v),t(u,v)) \\ x'_2 &= m(b,t(u,v)) \\ y'_2 &= m(b,t(u,v)) \end{aligned} \tag{8.11}$$

so that $x'_i = y'_i$ for $i = 1,2$. Moreover, *any other substitution in (8.9) which causes those equations to be true can be obtained by a further substitution in (8.11).*

8.7.2 Free sketches Let us say that an FP sketch with no diagrams or cocones is a **free FP sketch**. The nodes look like $s_1 \times s_2 \times \cdots \times s_n$ and although the operations generally have the form

$$f : s_1 \times \cdots \times s_n \to t_1 \times \cdots \times t_m$$

we can without loss of generality suppose that the target of any operation is actually not a product node. The reason is that a function to a product is, by Definition 5.3.1, completely determined by the functions obtained by composing it with the projections, so you can always replace a function to a product of n objects by n functions, one to each object in the product.

The lack of equations in a free FP sketch implies that there are no nonobvious identifications among composites of these operations. This fact can be formalized in the following.

8.7.3 Proposition *Every element of any sort in a free algebra for a free FP sketch can be expressed in exactly one way as a term not involving projection arrows.*

The proof is an induction based on the fact that in 7.6.5, FP–2 never applies when there are no diagrams, so that the only identification of two terms is by FP–5 in 7.6.5, which eliminates an occurrence of a projection arrow.

This proposition gives us a formalism to deal with terms. In particular, the two sets of terms $\{x_1, x_2\}$ and $\{y_1, y_2\}$ in 8.7.1 can be seen as two arrows between free algebras:

$$x, y : F(\{1,2\}) \rightrightarrows F(\{a, b, u, v, w\})$$

Here 1 and 2 both represent variables of type S, x is the unique arrow that takes 1 to $m(t(u,v), a)$ and 2 to $m(b, t(u,v))$ and y the unique arrow taking 1 to $m(a,a)$ and 2 to $m(b,a)$. A unification is then an arrow which coequalizes these two arrows. In particular, in this example the arrow d in the theorem below is the substitution (8.10).

8.7.4 Theorem *Let $\mathcal{S}$ be a free FP sketch and $\mathcal{C}$ be the category of models of $\mathcal{S}$. Suppose X, Y and Z are sets typed by the objects of $\mathcal{S}$. Suppose there is a diagram in $\mathcal{C}$,*

$$F(X) \underset{d^1}{\overset{d^0}{\rightrightarrows}} F(Y) \xrightarrow{\ d\ } F(Z) \tag{8.12}$$

with $d \circ d^0 = d \circ d^1$. Then the coequalizer of the arrows d^0 and d^1 is also a free algebra.

Proof. This proof is adapted from that of [Rydeheard and Burstall, 1986]. The proof makes use of the fact that F preserves sums, which follows from Theorem 12.3.6.

The proof will be by a double induction. The first induction is on the complexity of the arrows d^0 and d^1 (complexity is defined below). The second induction is on the number of elements of X. Although this proves it only when X is finite, the general case can be carried out by transfinite induction.

For each $x \in X$, $d^0(x)$ is $f(a_1, \ldots, a_n)$ for some operation f and terms $a_1, \ldots, a_n$, and similarly for d^1. By the complexity of d^i, for $i = 1, 2$, we mean the maximum of the lengths of the induction needed to represent $d^i(x)$ in the form $f(a_1, \ldots, a_n)$ using FP–1 through FP–5 in 7.6.5. The primary induction in this proof is on the maximum of the complexities of d^0 and d^1.

We consider first the case that X consists of a single element x of some type. Let $d^0(x) = f^0(a_1, \ldots, a_n)$ and $d^1(x) = f^1(b_1, \ldots, b_m)$ where f^0 and f^1 are operations of the sketch and $a_1, \ldots, a_n, b_1, \ldots, b_m$ are terms. Note that it is possible that $n = m = 0$.

Now we have

$$d \circ d^0(x) = df^0(a_1, \ldots, a_n) = f^0(da_1 \ldots, da_n)$$

and similarly,

$$d \circ d^1(x) = f^1(db_1, \ldots, db_m)$$

Setting them equal and using Proposition 8.7.3, we conclude that $m = n$, $f^0 = f^1$ and $da_i = db_i$ for $i = 1, \ldots, n$.

Now d preserves type so the type of a_i is the same as that of b_i. Let W be the set $\{w_1, \ldots, w_n\}$ where w_i is a variable of the same type as a_i. (If a_i and a_j are of the same type for some distinct i and j, w_i and w_j are nevertheless distinct variables.) We have arrows $e^0, e^1 : F(W) \rightrightarrows F(Y)$ given by $e^0(w_i) = a_i$ and $e^1(w_i) = b_i$. This gives us a commutative diagram

$$F(W) \underset{e^1}{\overset{e^0}{\rightrightarrows}} F(Y) \xrightarrow{d} F(V) \tag{8.13}$$

The complexities of e^0 and e^1 are evidently less than those of d^0 and d^1 and so we can suppose that the coequalizer of e^0 and e^1 is free on some set V. Thus there is a coequalizer diagram

$$F(W) \underset{e^1}{\overset{e^0}{\rightrightarrows}} F(Y) \xrightarrow{e} F(V) \tag{8.14}$$

Let $h : F(\{x\}) \to F(W)$ take x to $f^0(w_1, \ldots, w_n)$, which we have seen is the same as $f^1(w_1, \ldots, w_n)$. Then h is epimorphic by Exercise 3, so by Exercise 1,

$$F(\{x\}) \underset{e^1 \circ h}{\overset{e^0 \circ h}{\rightrightarrows}} F(Y) \xrightarrow{\;e\;} F(V) \tag{8.15}$$

is a coequalizer diagram.

Now we suppose that X has more than one element and that the conclusion is true for any set with fewer elements. Let $x \in X$ be arbitrary. Since we are supposing that $d \circ d^0 = d \circ d^1$, we can compose both arrows with the inclusion $g : F(\{x\}) \to F(X)$ induced by the inclusion of $\{x\}$ into X and use the single element case to get a coequalizer

$$F(\{x\}) \underset{d^1 \circ h}{\overset{d^0 \circ h}{\rightrightarrows}} F(Y) \xrightarrow{\;e\;} F(V) \tag{8.16}$$

Now we let $h : X - \{x\} \to X$ be the inclusion and invoke the second induction on the number of elements of X to conclude that there is also a coequalizer diagram

$$F(X - \{x\}) \underset{e \circ d^1 \circ h}{\overset{e \circ d^0 \circ h}{\rightrightarrows}} F(V) \xrightarrow{\;c\;} F(U) \tag{8.17}$$

It now follows from this, the fact that F preserves sums, and Exercise 2 that $c \circ e$ is the coequalizer of the original pair of arrows. □

This proof is made into a formal algorithm in [Burstall and Rydeheard, 1986]. See also [Goguen, 1988] and [Rydeheard and Stell, 1987].

8.7.5 Exercises

1. Show that in any category if

$$A \underset{g}{\overset{f}{\rightrightarrows}} B \xrightarrow{\;c\;} C$$

is a coequalizer diagram and $e : D \to A$ is an epimorphism, then so is

$$D \underset{g \circ e}{\overset{f \circ e}{\rightrightarrows}} B \xrightarrow{\;c\;} C$$

2. Show that in any category in which the sums exist, if for $i = 1, 2$

$$A_i \overset{f_i}{\underset{g_i}{\rightrightarrows}} B \xrightarrow{c_i} C$$

are both coequalizer diagrams, then so is

$$A_1 + A_2 \overset{f_1 + f_2}{\underset{g_1 + g_2}{\rightrightarrows}} B_1 + B_2 \xrightarrow{c_1 + c_2} C_1 + C_2$$

3.† Show that an arrow $h : F(X) \to F(W)$ between free algebras for a free theory is either epimorphic or factors through a subalgebra of the form $F(W')$ for some proper subset $W' \subseteq W$. Conclude that h, as defined after Diagram (8.14), is an epimorphism.

9

More about sketches

This chapter develops the concept of sketch in several ways. The first three sections describe a generalization of FP theories called FL theories which allow the use of equalizers, pullbacks and other limits in the description of a structure. These theories have expressive power which includes that of universal Horn theories.

The last section gives further generalizations of the concept of sketch. These generalizations are described without much detail since they do not appear to have many applications (yet!) in computer science.

9.1 Finite limit sketches

A cone is called finite if its shape graph is finite, meaning that it has only finitely many nodes and arrows.

9.1.1 Definition A **finite limit** or **FL sketch** $\mathcal{S} = (\mathcal{G}, \mathcal{D}, \mathcal{L})$ is a graph together with a set $\mathcal{D}$ of diagrams and a set $\mathcal{L}$ of finite cones. A **model** of $\mathcal{S}$ is a model of $\mathcal{G}$ that takes all the diagrams in $\mathcal{D}$ to commutative diagrams and all the cones in $\mathcal{L}$ to limit cones.

For historical reasons, FL sketches are also known as **left exact sketches** or **LE sketches**.

FL sketches allow the specification of structures or data types with sorts that include equationally specified subsorts, by using equalizers. More generally, you can specify an operation whose domain, in a model, will be an equationally defined subobject of another sort. FL theories can express anything expressible by universal Horn theories, but in general FL theories are more powerful (see [Barr, 1989]). For example, small categories and functors can be described by an FL theory (which we give in 9.1.4 below) but not by a universal Horn theory.

In practice it is generally sufficient to restrict the types of cones to a few simple types, products, pullbacks and equalizers. In principle, an FL sketch can always be replaced by one which has an equivalent category of models and which has only these three types of cones, but there might some case in which this is not the most efficient approach.

9.1.2 Other approaches In the introduction to Chapter 7, we mentioned three approaches to formalization: logical theories, signatures and equations, and sketches. Systems equivalent to FL sketches have been developed for both the other approaches. [Coste, 1979], [Cartmell, 1986] and [McLarty, 1986] give logical systems and [Reichel, 1987] generalizes signatures. The book by Reichel has many examples of applications to computer science.

9.1.3 Notation for FL sketches We extend the notational conventions in Section 7.3 to cover products, pullbacks and equalizers.

First, the notation of N–2 in 7.3.1 using product projections is extended to cover the arrows from the limit to any of the nodes in the diagram: that is they are all denoted by an appropriate p; moreover, the notation of N–4 in 7.3.1 is extended to arrows into the limit in terms of the composite with the projections. See Example 9.1.4 below of the sketch of categories to see how this is used. Of course, it is sometimes necessary to be more explicit than this.

We add the following to the notational conventions of Section 7.3.

N–8 If we have a node labeled $a \times_c b$ this implies the existence of a cone

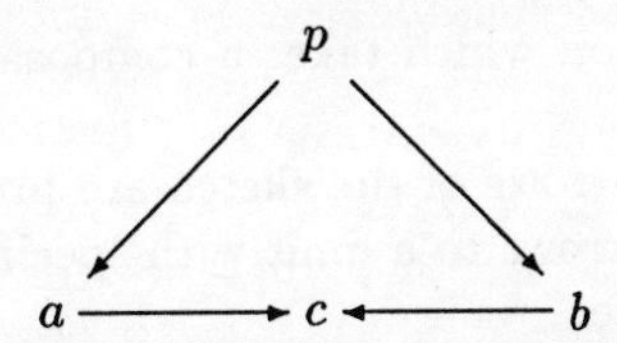

Of course, this notation is not self-contained since it is necessary to specify the arrows $a \rightarrow c \leftarrow b$. In many cases, these arrows are clear, but it may be necessary to specify them explicitly.

N–9 If we have an arrow labeled $s : a \rightarrowtail b$ (recall from 2.8.2 that in a category this notation means an arrow that is a monomorphism), then we are implicitly including a cone

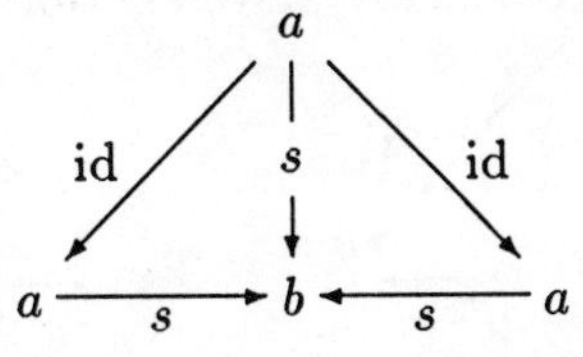

This notation makes sense because of Theorem 8.3.3.

We could go on and turn our notational conventions into a formal language, but have chosen not to do so because we do not know what it would add to the theory. For us, they remain a set of notational conventions; the real object of study is the whole sketch or else the theory it generates.

9.1.4 The sketch for categories We give here an FL sketch whose models in the category of sets are small categories and arrows between the models are functors. Models in an arbitrary category $\mathcal{C}$ with finite limits are called **category objects** in $\mathcal{C}$. These are used in Section 14.7.

The sketch has nodes c_0, c_1, c_2 and c_3 which stand for the objects, the arrows, the composable pairs and the composable triples of arrows respectively. The arrows of the sketch are:

$u : c_0 \to c_1$

$s, t : c_1 \to c_0$

$p_1, p_2, c : c_2 \to c_1$

$p_1, p_2, p_3 : c_3 \to c_1$

$\langle p_1, p_2\rangle, \langle p_2, p_3\rangle, \langle p_1, c\rangle, \langle c, p_3\rangle : c_3 \to c_2$

$\langle u \circ t, \mathrm{id}\rangle, \langle \mathrm{id}, u \circ s\rangle : c_1 \to c_2$

The intention is that in a **Set**-model, the arrows of the sketch will be interpreted as follows:

u is the unit function which assigns to each object its identity,

s and t are the source and target functions from the set of arrows to the set of objects, and

c is the function which takes a composable pair of arrows to its composite.

The remaining arrows of the sketch are projections from a limit or are interpreted as arrows to a limit with specified projections.

The diagrams are

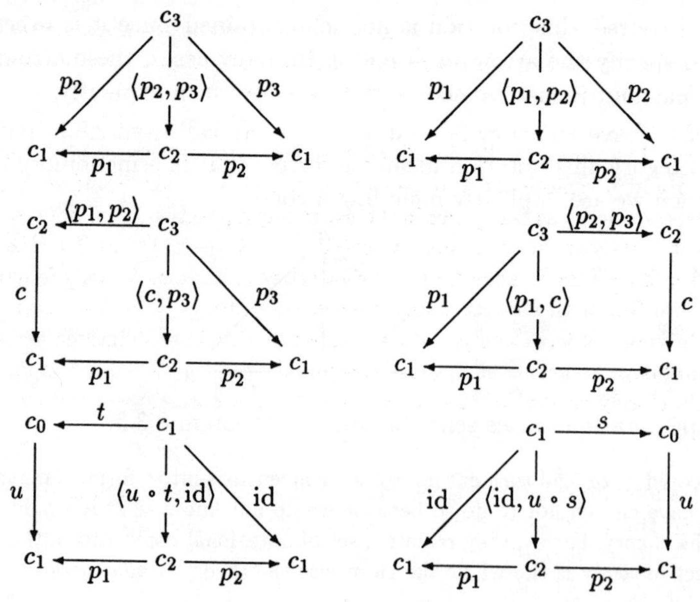

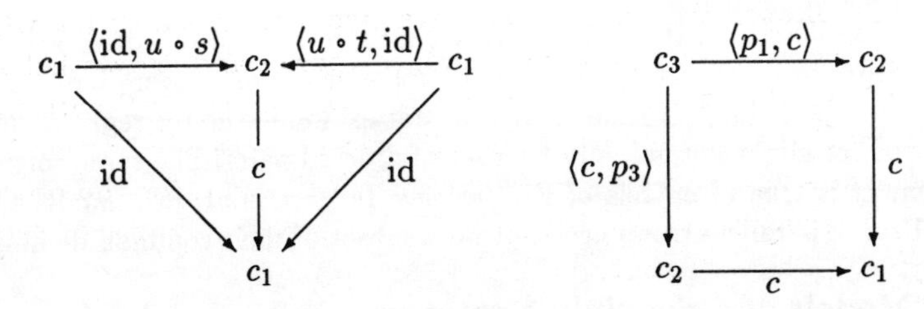

Finally, there are two cones that say that c_2 and c_3 are interpreted as the objects of composable pairs and triples or arrows, respectively.

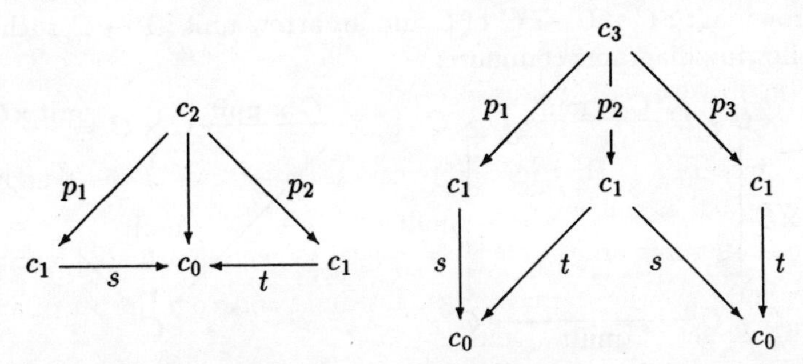

This sketch is the **sketch of categories** and is one of the simpler FL sketches around. It was originally described by Lawvere in somewhat different terms (the language of sketches not being available) in [1966].

The category of models (in **Set**) of an FP sketch must be regular (see [Barr and Wells, 1985], Theorem 1 of Section 8.4) and it can be proved that the category of categories and functors is not regular (see Exercise 6 of Section 8.6). It follows that categories and functors cannot be the models of an FP sketch.

The way Lawvere actually described the sketch for categories is this. Let **1**, **2**, **3** and **4** denote the total orders with one, two, three and four elements, considered as categories in the usual way a poset is. Let $\mathcal{L}$ denote the opposite category. In that opposite category, it turns out that $\mathbf{3} = \mathbf{2} \times_{\mathbf{1}} \mathbf{2}$ and $\mathbf{4} = \mathbf{2} \times_{\mathbf{1}} \mathbf{2} \times_{\mathbf{1}} \mathbf{2}$. Then **Cat** is the category of limit-preserving functors of $\mathcal{L}$ into **Set** with natural transformations as arrows. It follows that the sketch whose objects and arrows are those of $\mathcal{L}$, whose diagrams are the commutative diagrams of $\mathcal{L}$ and cones are the limit cones of $\mathcal{L}$ is another sketch, closely related in fact to the one above, whose category of models is **Cat**.

We show how binary trees can be described as models of an FL sketch in the next section.

In 7.2.8 we described the way the choice of products and terminal objects in a model to represent the vertices of cones in a sketch were irrelevant but could result in the technicality that, for example, the set representing the product might not actually be a set of ordered pairs. The same sort of statement is true of models of FL theories. In particular, in a model the equalizer of parallel arrows need not be a subset of their common domain.

9.1.5 Models of a sketch in a category If $\mathcal{S}$ is any sketch, we can consider models of $\mathcal{S}$ in any category. For example, a model of the sketch of monoids in a category $\mathcal{C}$ consists of an object C of $\mathcal{C}$ together with an arrow $\text{mult} : C \times C \to C$ of $\mathcal{C}$ and an arrow $\text{unit} : 1 \to C$ such that the following diagrams commute:

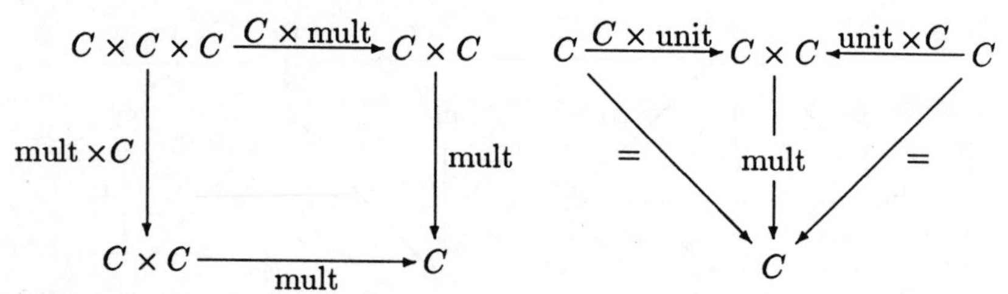

The category $\mathcal{C}$ can be quite arbitrary. Of course, if the category fails to have certain structures, there may be very few models. For example, there can be no models of the monoid sketch if the category lacks a terminal object. Similarly, in order for there to be a model with underlying object C, both $C \times C$ and $C \times C \times C$ must exist.

Suppose that $(C, \text{mult}_C, \text{unit}_C)$ and $(D, \text{mult}_D, unit_D)$ are two models of the sketch for monoids in the category $\mathcal{C}$. An arrow $f : C \to D$ will be said to preserve the monoid structure if the diagrams

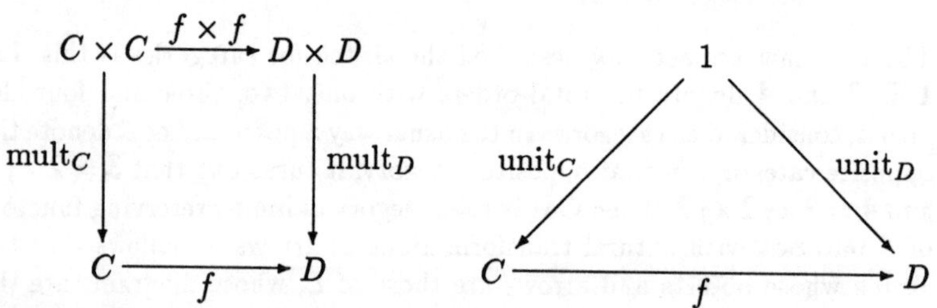

commute. The result is a category $\textbf{Mon}(\mathcal{C})$ of monoids of $\mathcal{C}$. This category comes equipped with an underlying functor $\textbf{Mon}(\mathcal{C}) \to \mathcal{C}$. In many categories $\mathcal{C}$, there is a functor which behaves like the free monoid functor in **Set** discussed in 12.1.2.

More generally, if $\mathcal{S}$ is any sketch and $\mathcal{C}$ is any category, we define the **category of models** to be the category whose objects are models and whose arrows are the natural transformations between models, considered as models of the underlying graph. We denote this category by $\mathbf{Mod}(\mathcal{S},\mathcal{C})$.

We will look at models of the sketch for categories in an arbitrary category in Section 14.7.

9.1.6 Exercise

1. Show that in the category of sets, the definition of homomorphism between two models of the sketch for categories gives the usual definition of functor.

9.2 Initial term models of FL sketches

Like FP sketches, FL sketches always have initial models. The construction 7.6.5 produced an initial term algebra for each finite FP sketch. A modification of the last three rules in 7.6.5 is sufficient to construct an initial term algebra for each finite FL sketch. This will be described below.

A different construction is in Barr [1986a], where it is proved that in fact FL sketches have free algebras on any typed set X (see Section 12.2.) Volger [1987] gives a logic-based proof for the special case of Horn theories (see [Volger, 1988] for applications).

The modification of 7.6.5 is required by the fact that the base diagram D of a cone in $\mathcal{L}$ need not be discrete in the case of general limits; that is, the shape graph $\mathcal{I}$ of the diagram D in Section 8.2 may have nontrivial arrows $u : i \to j$. A limit of such a cone in the category of sets is not just any tuple of elements of the sets corresponding to the nodes of $\mathcal{I}$, but only tuples which are compatible with the arrows of $\mathcal{I}$.

9.2.1 Precisely, suppose $E : \mathcal{I} \to \mathbf{Set}$ is a finite diagram (we use the letter E instead of D to avoid confusion below). A **compatible family** of elements of E is a sequence $(x_1, \ldots, x_n)$ indexed by the nodes of $\mathcal{I}$ for which

C–1 $x_i \in E(i)$ for each node i.

C–2 If $u : i \to j$ in $\mathcal{I}$, then $E(u)(x_i) = x_j$.

An initial term algebra of an FL sketch $\mathcal{S} = (\mathcal{G}, \mathcal{D}, \mathcal{L})$ is then the least model satisfying the following requirements.

FL–1 If $f : a \to b$ is an arrow of $\mathcal{G}$ and $[x]$ is an element of $I(a)$, then $[fx] \in I(b)$ and $I(f)[x] = [fx]$.

FL–2 If $(f_1, \ldots, f_m)$ and $(g_1, \ldots, g_k)$ are paths in a diagram in $\mathcal{D}$, both going from a node labeled a to a node labeled b, and $[x] \in I(a)$, then

$$(If_1 \circ If_2 \circ \ldots \circ If_m)[x] = (Ig_1 \circ Ig_2 \circ \ldots \circ Ig_k)[x]$$

in $I(b)$.

FL–3 If $D : \mathcal{I} \to \mathcal{G}$ is a diagram, $p : q \to D$ is a cone over D in $\mathcal{L}$ and $([x_1], \ldots, [x_n])$ is a compatible family of elements of $I \circ D$, then $[C(x_1, \ldots, x_n)]$ is an element of $I(q)$.

FL–4 If $D : \mathcal{I} \to \mathcal{G}$, $p : q \to D$ and $([x_1], \ldots, [x_n])$ are as in FL–3, and $x'_1, \ldots, x'_n$ is a compatible family of elements of $I \circ D$ for which $[x_i] = [x'_i]$ for $i = i, \ldots, n$, then

$$[C(x_1, \ldots, x_n)] = [C(x'_1, \ldots, x'_n)]$$

FL–5 For $i = 1, \ldots, n$, we require

$$[p_i C(x_1, \ldots, x_n)] = [x_i]$$

for any compatible family of elements of $I \circ D$.

Compatibility implies that for any arrow $u : i \to j$ of $\mathcal{I}$,

$$I(D(u))([x_i]) = [x_j]$$

As in 7.6.5, it follows that

$$I(p_i)([C(x_1, \ldots, x_n)]) = [x_i]$$

for each i.

9.2.2 Binary trees We describe an FL sketch whose initial term algebra is the set of binary trees of integers. We gave an FD sketch for binary trees in 7.10.11; the sketch given here illustrates a different approach to the problem that operations such as taking the datum at the root or producing the left or right subtrees are not defined on the empty tree.

We have the following basic nodes in the sketch: t, t^+, b, n. These should be thought of as representing the types of binary trees, nonempty binary trees, the Boolean algebra 2 and the natural numbers, respectively. We have the following operations:

empty : $1 \to t$	empty? : $t \to b$	incl : $t^+ \to t$
val : $t^+ \to n$	left : $t^+ \to t$	right : $t^+ \to t$
zero : $1 \to n$	succ : $n \to n$	true : $1 \to b$
and : $b \times b \to b$	not : $b \to b$	

The intended meaning of these operations is as follows: the constant empty⟨⟩ is the empty tree; empty? is the test for whether a tree is the empty tree; incl is the inclusion of the set of nonempty trees in the set of trees; val(T) is the datum stored at the root of the nonempty tree T; left(T) and right(T) are the right and left branches (possibly empty) of the nonempty tree T, respectively. The remaining operations are the standard operations appropriate to the natural numbers and the Boolean algebra 2.

We require that

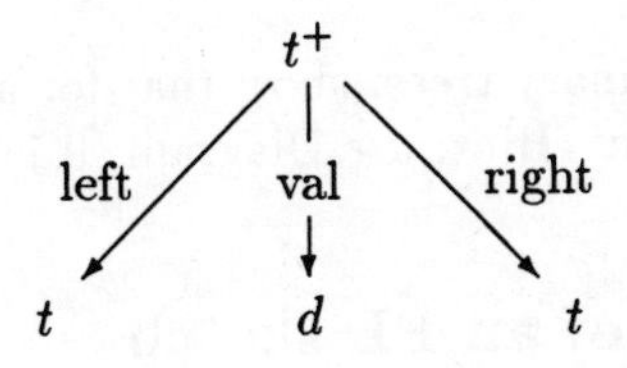

and

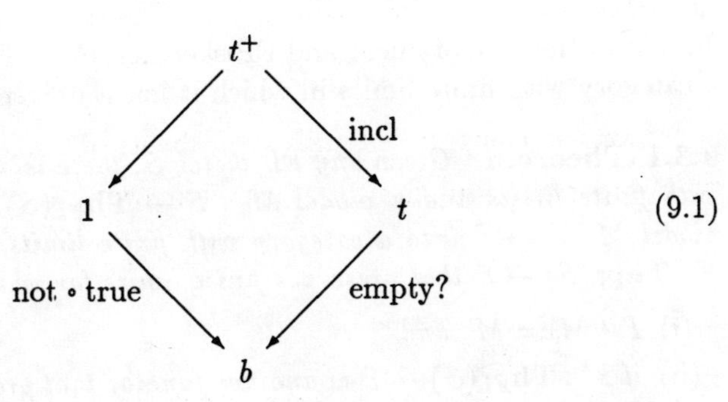

be cones and that

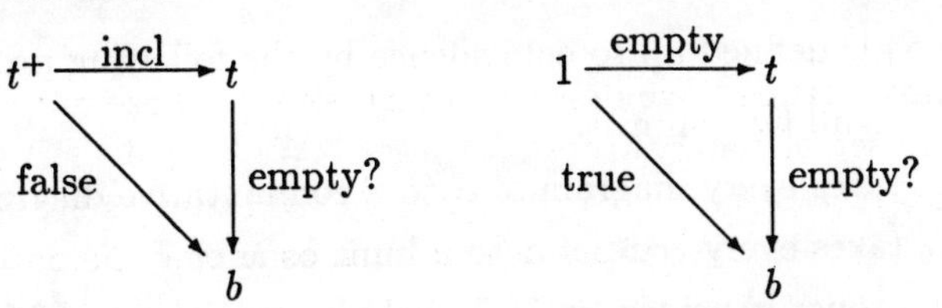

be diagrams.

In the cone (9.1), there should be an arrow from the vertex to the node b. It will appear in a model as either of the two (necessarily equal) composites. Since its value is forced, it is customary to omit it from the cone; however, there actually does have to be such an arrow there to complete the cone (and the sketch). In omitting it, we have conformed to the standard convention of showing explicitly only what it is necessary to show.

As in 7.10.11, the first cone says that every nonempty tree can be represented uniquely as a triplet

$$(\text{left}(T), \text{val}(T), \text{right}(T))$$

The fact that (9.1) is a cone requires that in a model M, $M(t^+)$ be exactly the subset of $M(t)$ of those elements which evaluate to false under $M(\text{empty?})$.

9.2.3 Exercise

1. In the sketch for binary trees, show that for any model M, $M(\text{incl})$ is an injective function. (Hint: use Diagram (9.1).)

9.3 The theory of an FL sketch

Just as in the case of linear and FP sketches, every FL sketch generates a category with finite limits in which it has a universal model.

9.3.1 Theorem *Given any FL sketch $\mathcal{S}$, there is a category $\mathbf{Th_{FL}}(\mathcal{S})$ with finite limits and a model $M_0 : \mathcal{S} \to \mathbf{Th_{FL}}(\mathcal{S})$ such that for any model $M : \mathcal{S} \to \mathcal{E}$ into a category with finite limits, there is a functor $F : \mathbf{Th_{FL}}(\mathcal{S}) \to \mathcal{E}$ that preserves finite limits for which*

(i) $F \circ M_0 = M$, *and*

(ii) *if $F' : \mathbf{Th_{FL}}(\mathcal{S}) \to \mathcal{E}$ is another functor that preserves finite limits for which $F' \circ M_0 = M$, then F and F' are naturally isomorphic.*

$\mathbf{Th_{FL}}(\mathcal{S})$ is defined up to equivalence by the following properties:

FLT–1 $\mathcal{T}$ has all finite limits.

FLT–2 M_0 takes every diagram of $\mathcal{S}$ to a commutative diagram in $\mathcal{T}$.

FLT–3 M_0 takes every cone of $\mathcal{S}$ to a limit cone of $\mathcal{T}$.

FLT–4 No proper subcategory of $\mathcal{T}$ includes the image of M_0 and satisfies FLT-1, FLT-2 and FLT-3.

The comments concerning Theorem 7.5.1 apply here too. See [Barr and Wells, 1985], Section 4.4, Theorem 2 (p. 156). It would be incorrect to assume in general (although it is true in familiar examples) that every node of $\mathcal{T}$ is a limit of a diagram in the graph of $\mathcal{S}$.

9.3.2 Exercise

1. Show that any model M of a sketch $\mathcal{S}$ is isomorphic to a model M' for which $M'(c)$ and $M'(d)$ have no elements in common if $c \neq d$.

9.4 General definition of sketch

Although we will not be using the most general notion of sketch, we give the definition here.

9.4.1 Definition A **sketch** $\mathcal{S} = (\mathcal{G}, \mathcal{D}, \mathcal{L}, \mathcal{K})$ consists of a graph $\mathcal{G}$, a set $\mathcal{D}$ of diagrams in $\mathcal{G}$, a set $\mathcal{L}$ of cones in $\mathcal{G}$ and a set $\mathcal{K}$ of cocones in $\mathcal{G}$.

9.4.2 Definition A model M of a sketch $\mathcal{S} = (\mathcal{G}, \mathcal{D}, \mathcal{L}, \mathcal{K})$ in a category $\mathcal{A}$ is a homomorphism from $\mathcal{G}$ to the underlying graph of $\mathcal{A}$ that takes every diagram in $\mathcal{D}$ to a commutative diagram, every cone in $\mathcal{L}$ to a limit cone and every cocone in $\mathcal{K}$ to a colimit cocone.

9.4.3 Definition Let M, N be models of a sketch $\mathcal{S}$ in a category $\mathcal{A}$. A **homomorphism of models** $\alpha : M \to N$ is a natural transformation from M to N.

Sketches in general do not have initial algebras, or even families of initial algebras. What they do have is 'locally free diagrams' as described in [Guitart and Lair, 1981, 1982]. Generalizations of the concept of sketch are in [Lair, 1987], [Wells, 1990] and [Power and Wells, 1990].

9.4.4 Regular sketches Chapter 8 of [Barr and Wells, 1985] describes special classes of sketches with cones and cocones that have in common the fact that their theories are embedded in a topos, and so inherit the nice properties of a topos. (Toposes are discussed in Chapter 14.) These are the regular sketches, coherent sketches and geometric sketches. We describe regular sketches here to illustrate the general pattern.

9.4.5 An arrow $f : A \to B$ is a regular epimorphism if and only if there is a cocone diagram

$$C \underset{h}{\overset{g}{\rightrightarrows}} A \xrightarrow{f} B$$

It follows that the predicate of being a regular epimorphism can be stated within the semantics of a sketch.

Let us say that a **regular cocone** is one of the form

$$c \rightrightarrows a \to b$$

9.4.6 Definition A **regular sketch** $\mathcal{S} = (\mathcal{G}, \mathcal{D}, \mathcal{L}, \mathcal{K})$ consists of a graph, a set of diagrams, a set of finite cones and a set of regular cocones.

There is one undesirable feature to the definition above. The introduction of a regular sketch requires the introduction of a sort c and two arrows $c \rightrightarrows a$. A model will have to provide a value for c as well as the two arrows. This is generally irrelevant information that one would not normally want to have to provide. Worse, the definition of natural transformation is such that arrows between models will have to preserve this additional information and that is definitely undesirable. The way in which this is usually dealt with is by adding, with each cocone

$$c \underset{d^1}{\overset{d^0}{\rightrightarrows}} a \xrightarrow{d} b$$

the cone

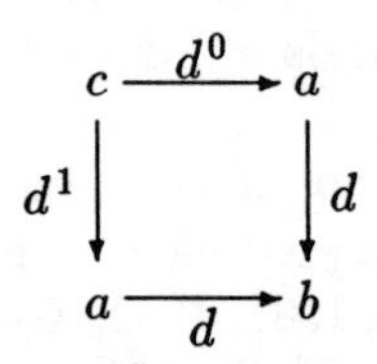

It follows from Exercise 5 of Section 8.4 that if d has a kernel pair, then d is a regular epimorphism if and only if it is the coequalizer of that kernel pair. The addition of this cone thus adds no new data, supposing the kernel pair exists. Moreover, the preservation of the kernel pair and the two arrows by homomorphisms of models is automatic, given the universal mapping properties of limits.

A regular sketch allows you to state that something is one of several types of things where overlap is possible between the types of things. (If the possibilities did not overlap one might be able to use finite sums.) For example, a poset $(P, \leq)$ is said to be **totally ordered** if for any elements $x, y \in P$ either $x \leq y$ or $y \leq x$. Thus $P \times P$ is the union (but not disjoint union) of two subsorts, the set of pairs (x, y) for which $x \leq y$ and the pairs (x, y) for which $y \leq x$.

It is not appropriate to define a total order in terms of a trichotomy, which could easily be done with an FD sketch, writing the square $P \times P$ as a sum of three parts. The trouble come in such a simple example as the monoid $\mathbf{N}$ of natural numbers on addition. The square $\mathbf{N} \times \mathbf{N}$ is, as a set, but not as a monoid, the sum of the three subsets of pairs (x, y) for which $x < y$, for which $x = y$ and for which $y < x$. However, the first and third of them are not submonoids (they lack unit elements) and so this will not give a model of the FD sketch. The construction in the preceding paragraph gives two overlapping submonoids in this case, and so is more appropriate.

9.4.7 Theories of sketches with colimits. The reader may have noticed that in Chapter 7, we discussed FP sketches, their initial algebras and their theories, and we discussed FD sketches and their initial algebras. We did not discuss theories for FD sketches. In fact, theories exist for FD sketches, and indeed for all sketches. This is because sketches can themselves be modeled by an FL theory, and the theory of a sketch can then be realized as an initial algebra (see Section 9.2). See [Wells, 1990].

Here, we state the universal properties for FD sketches for models in categories with finite products and finite disjoint universal sums. The resulting theory has these properties, too, and is embedded in a topos. The proof by initial algebras just mentioned does not give such an embedding.

9.4.8 Theorem *Let $\mathcal{S}$ be an FD sketch. Then there is a category denoted $\mathbf{Th_{FD}}(\mathcal{S})$ with finite disjoint universal sums and a model $M_0 : \mathcal{S} \to \mathbf{Th_{FD}}(\mathcal{S})$ such that for any model $M : \mathcal{S} \to \mathcal{E}$ into a category with finite products and finite disjoint universal sums, there is a functor $F : \mathbf{Th_{FD}}(\mathcal{S}) \to \mathcal{E}$ that preserves finite products and finite sums for which*

(i) *$F \circ M_0 = M$, and*

(ii) *If $F' : \mathbf{Th_{FD}}(\mathcal{S}) \to \mathcal{E}$ is another functor that preserves finite products and finite sums for which $F' \circ M_0 = M$, then F and F' are naturally isomorphic.*

A proof may be found in [Barr and Wells, 1985], Proposition 1 of Section 8.2. (FD sketches are called FS sketches there.)

The situation is similar for regular sketches.

9.4.9 Theorem *Given any regular sketch $\mathcal{S}$, there is a regular category $\mathbf{Th_{Reg}}(\mathcal{S})$ and a model $M_0 : \mathcal{S} \to \mathbf{Th_{Reg}}(\mathcal{S})$ such that for any model $M : \mathcal{S} \to \mathcal{E}$ into a regular category with finite sums, there is a regular functor $F : \mathbf{Th_{Reg}}(\mathcal{S}) \to \mathcal{E}$ such that*

(i) *$F \circ M_0 = M$, and*

(ii) *if $F' : \mathbf{Th_{Reg}}(\mathcal{S}) \to \mathcal{E}$ is another regular functor for which $F' \circ M_0 = M$, then F and F' are naturally isomorphic.*

9.4.10 Exercises

1. Give a regular sketch for totally ordered sets.

2. Give a regular sketch for reflexive graphs, as discussed in 4.6.8. (Hint: the composite of source and the equalizer of source and target should be a regular epimorphism.)

10
The category of sketches

The first section of this chapter defines the concept of homomorphism of sketches, yielding a category of sketches. In Section 10.2 we describe a formalism for defining parametrized data types using sketch homomorphisms. In Section 10.3 we develop the theory of sketches further, showing that a homomorphism of sketches induces a contravariant functor between the model categories and making contact with Goguen and Burstall's concept of institution.

Section 10.3 requires only Section 10.1 to read. Nothing in this chapter is needed later in the book.

Much more is known about the category of sketches than is mentioned here. It is cartesian closed, for example. A basic study in English of the category of sketches which is oriented toward computer science is given by Gray [1989].

10.1 Homomorphisms of sketches

10.1.1 Let $\mathcal{S} = (\mathcal{G}, \mathcal{D}, \mathcal{L}, \mathcal{K})$ and $\mathcal{S}' = (\mathcal{G}', \mathcal{D}', \mathcal{L}', \mathcal{K}')$ be sketches. A graph homomorphism $F : \mathcal{G} \to \mathcal{G}'$ takes a diagram $D : \mathcal{I} \to \mathcal{G}$ to a diagram $F \circ D : \mathcal{I} \to \mathcal{G}'$ which is called the **image** of D in $\mathcal{G}'$. It takes a cone $p : v \to D$ to a cone $F(p) : F(v) \to F \circ D$ where, for each node a of the shape graph of D, $F(p)_a : F(v) \to F(D(a))$ is defined to be $F(p_a)$. It is defined on cocones similarly.

An arrow or **homomorphism of sketches** $F : \mathcal{S} \to \mathcal{S}'$ is a graph homomorphism from the graph $\mathcal{G}$ to the graph $\mathcal{G}'$ for which, if D is a diagram in $\mathcal{D}$ then $F \circ D$ lies in $\mathcal{D}'$; if $p : v \to D$ is a cone in $\mathcal{L}$ then $F(p)$ is a cone in $\mathcal{L}'$; and if $c : D \to v$ is a cocone in $\mathcal{K}$ then $F(c)$ is a cocone in $\mathcal{K}'$.

Note that if in the sketch $\mathcal{S}$ the sets of diagrams, cones and cocones are all empty, then an arrow from $\mathcal{S}$ to $\mathcal{S}'$ is precisely a graph homomorphism from $\mathcal{G}$ to $\mathcal{G}'$.

10.1.2 The sketch underlying a category Let $\mathcal{C}$ be a category. There is an underlying sketch of $\mathcal{C}$, call it $\text{Sk}(\mathcal{C})$, whose graph consists of the objects of $\mathcal{C}$ as nodes and the arrows of $\mathcal{C}$ as arrows. This underlying sketch is not in general finite or even small. The commutative diagrams of this sketch are all diagrams that are commutative in $\mathcal{C}$. Similarly we take for cones all those that are limit cones in $\mathcal{C}$ and for cocones all those that are colimit cocones. An arrow from $\mathcal{S}$ to $\mathrm{Sk}(\mathcal{C})$ is then exactly what we have called a model of $\mathcal{S}$ in $\mathcal{C}$ in 9.4.2.

We note that although the composition in $\mathcal{C}$ has been forgotten, it can be completely recovered from the knowledge of which diagrams commute. For example, the information $f \circ g = h$ is equivalent to the information that

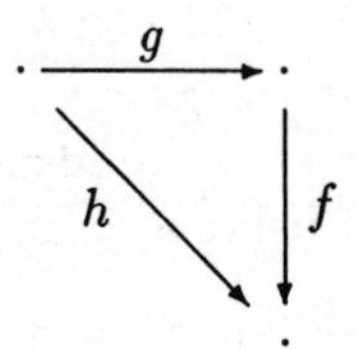

commutes so that we could recover the category $\mathcal{C}$ entirely from the underlying sketch (in fact, just from the graph and the commutative triangles).

10.1.3 The category of sketches With the definition of arrow between sketches given above, sketches themselves form a category which we will call **Sketch**. By restricting the shape graphs for the diagrams, cones and cocones one obtains many full subcategories of **Sketch**, for example the category of FP sketches, the category of FD sketches, and so on.

One consequence of this is that all constructions that we can carry out in any category make sense in the category of sketches. The particular construction that interests us here is the formation of colimits, particularly pushouts.

10.2 Parametrized data types as pushouts

One thing a theory of data types should do is give a way of saying how the data types of stacks of integers, say, is like the type of stacks of reals or for that matter of stacks of arrays of trees of characters. In other words, we need a way of talking about an abstract stack in a way that leaves it open to fill in the blank corresponding to the thing that it is stacks of. The data type is the parameter.

The way we do this is to describe a sketch for stacks of d where d stands for an abstract data type and then use a pushout construction to identify d with a concrete data type in any particular application. The point is that pushouts are the general way we use to identify things. We give several illustrations.

10.2.1 Abstract stacks Consider the sketch whose graph consists of nodes s, t, d, $d \times s$ and 1. The idea is that s stands for the set of stack configurations, t the set of nonempty stack configurations, and d the data. There are operations push : $d \times s \to t$, pop : $t \to d \times s$, empty : $1 \to s$ and incl : $t \to s$. In order to state the equations we also have two arrows $\mathrm{id}_{d\times s}$ and id_t which will be forced to be identity arrows.

There are four equations (diagrams). Two of them, $\mathrm{id}_{d\times s} = \langle\rangle_{d\times s}$ and $\mathrm{id}_t = \langle\rangle_t$, force those arrows to be identities, and the following two, which express the essence of being a stack:

$$\mathrm{pop} \circ \mathrm{push} = \mathrm{id}_{d\times s}$$

and

$$\mathrm{push} \circ \mathrm{pop} = \mathrm{id}_t$$

There are the cones to express 1 as terminal and $d \times s$ as a product, and one cocone

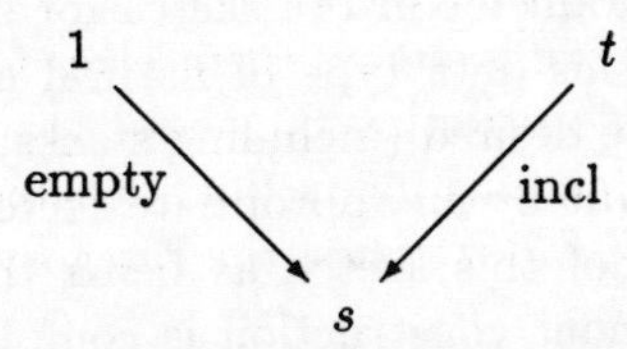

So far, this sketch is not very interesting, since the type of d is still undetermined. In fact its initial model I has $I(d) = I(t) = \emptyset$ and $I(s) = \{\mathrm{empty}\}$. Note that $\{\mathrm{empty}\}$ is not empty. It contains one element which we interpret as representing the empty stack. Since the type d of data is empty, the empty stack is the only kind of stack there is in the model I.

If there were constants in the data type d, then $I(s)$ would be the set of all possible configurations of a stack of data of type $I(d)$, and $I(t)$ would be the set of all possible configurations other than the empty stack.

10.2.2 Stacks of natural numbers In 4.7.7, we described the sketch with two nodes, 1 and n, two operations zero : $1 \to n$ and succ : $n \to n$. There are no equations (diagrams), just the cone describing 1 as terminal and no cocones. This sketch, which we called **Nat**, has models that can be reasonably viewed as describing natural numbers. We described a more elaborate sketch in 7.8.4 which could also be used in what follows.

Now let $\mathcal{E}$ denote the impoverished sketch which has one node we will call e and no arrows, diagrams, cones or cocones. Any assignment of e to a node of a sketch is an arrow of sketches. We are interested in the following diagram in the category of sketches, where the arrows F and G are defined by letting $Fe = n$ and $Ge = d$.

$$\begin{array}{ccc} \mathcal{E} & \xrightarrow{F} & \mathbf{Nat} \\ {\scriptstyle G}\downarrow & & \\ \mathbf{Stack} & & \end{array} \qquad (10.1)$$

A pushout of this diagram is a sketch **Stack(Nat)** made by forming the union of **Nat** and **Stack** and then identifying the node n of **Nat** with the node d of **Stack**.

From the definition of pushout, it follows that a model of this sketch is a model of the sketch for natural numbers that is simultaneously a model of the sketch for stacks whose value at the node d of **Stack** is the same as its value at the node n of **Nat**. Models of this sketch can be identified as stacks of natural numbers. The effect of this construction is to fill in the data parameter in the sketch for stacks with an actual data type. Of course, this data type of natural numbers could be replaced by any data type desired (including stacks of natural numbers!) by replacing $F : \mathcal{E} \to \mathbf{Nat}$ by an appropriate sketch homomorphism.

Although we think of this sketch as being that of stacks of natural numbers, the pushout construction is completely symmetric and any asymmetry is imposed by our way of looking at it. It is caused by our (quite reasonable) perception that the node d of **Stack** is an input parameter and the node s represents the actual data type.

10.2.3 Binary trees, revisited again Here is a second example. In 7.10.11, we described a sketch called **BinTree** and then another sketch of the same name in 9.2.2. Here we describe yet another sketch we call **BinTree**. This third and final approach is fully parametrized and represents exactly what we mean by trees, no more and no less. We use nodes t^+, t and d. Now the only operations we put in are empty, incl, val, left and right.

Then if $F : \mathcal{E} \to \mathbf{Nat}$ is as above and $H : \mathcal{E} \to \mathbf{BinTree}$ is defined by $H(e) = d$, then the pushout of

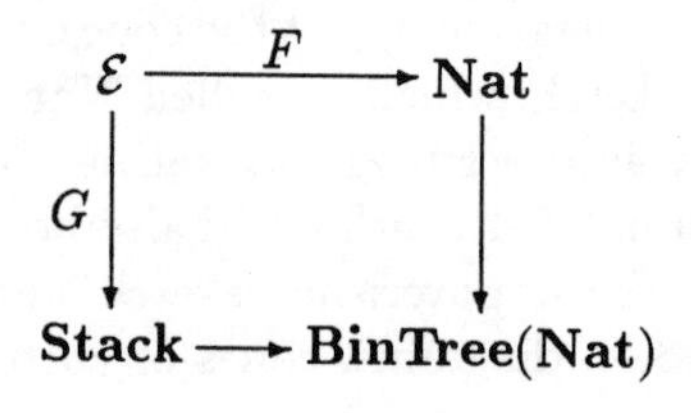

is a sketch whose models can be interpreted as binary trees of natural numbers.

10.2.4 Further combinators This operation can be iterated. For example, we can form the pushout

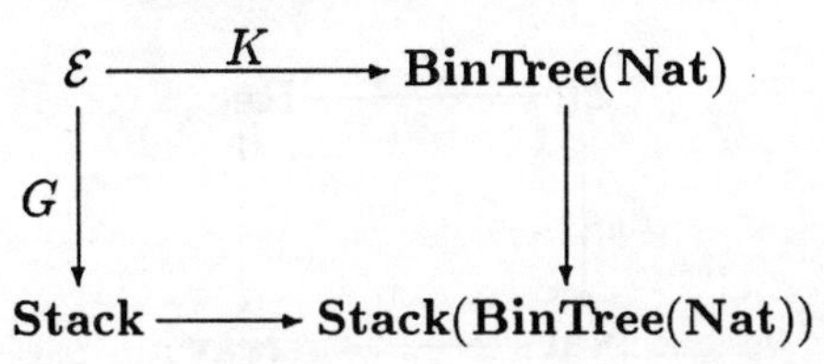

with $Ke = t$ or even

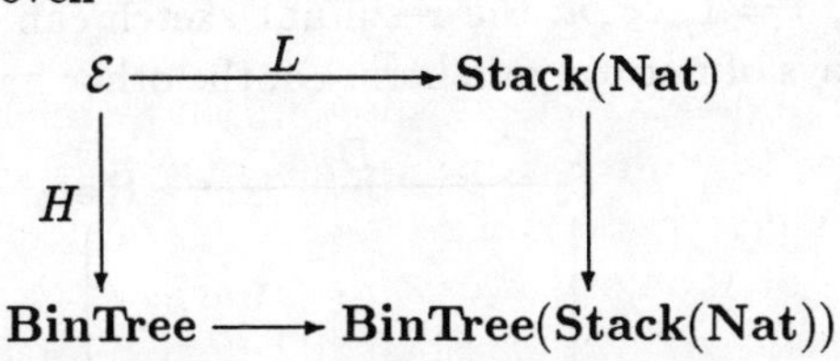

with $Le = s$. Models of these types will be interpreted as stacks of binary trees of natural numbers, respectively binary trees of stacks of natural numbers. Clearly what is at issue here is a notion of a type having an input node and an output node. However, things are not quite so simple, as the following example shows.

10.2.5 The basic idea of this type is that of n place records (Pascal) or structures (C). We leave aside here the use of variant records (which can be adequately handled by judicious use of finite sums as C's union type shows). Another question we do not tackle is that of parametrizing n. At this point, we make no attempt to relate the sketches for two-place and three-place records, say, although it seems clear that in a mature theory this will be done. Parametrizing n is discussed in [Wagner, 1986b] and [Gray, 1989].

So we describe a sketch we will call $\mathbf{Rec}_n$. It has nodes r and $d_1, d_2, \ldots, d_n$ and one cone

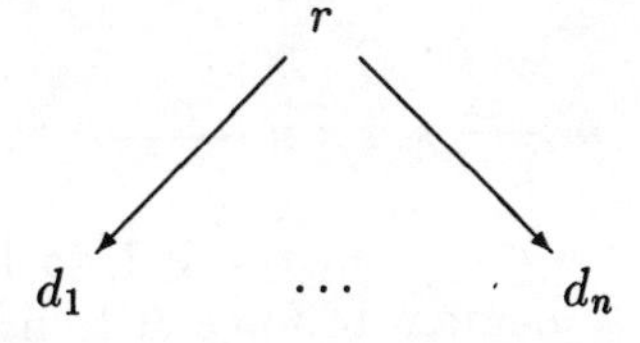

As it stands, the initial model of this sketch has all values empty. It is really a shell of a sketch. To give it content, we must fill in values for the data. To do this we must first recognize that the sketch has n input

types and these must all be specified. This could be done in n steps, but it is more coherent to do it all at once.

Let $\mathcal{E}_n$ denote the discrete graph with nodes $e_1, \ldots, e_n$. There is an obvious arrow $D_n : \mathcal{E}_n \to \mathbf{Rec}_n$ given by $De_i = d_i$. If, for example, we formed the pushout

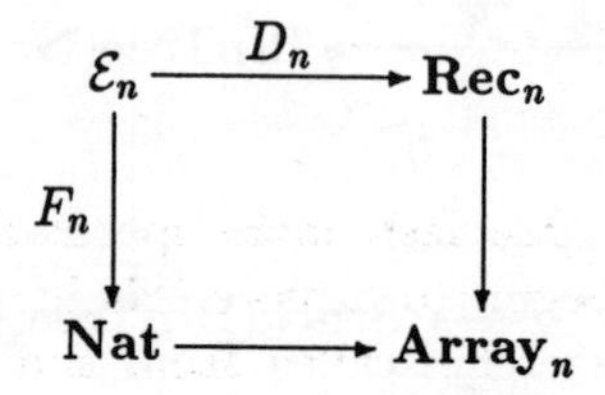

where $F_n e_i = n,\ i = 1, \ldots, n$, the resultant sketch can be evidently interpreted as arrays of natural numbers. On the other hand, the pushout

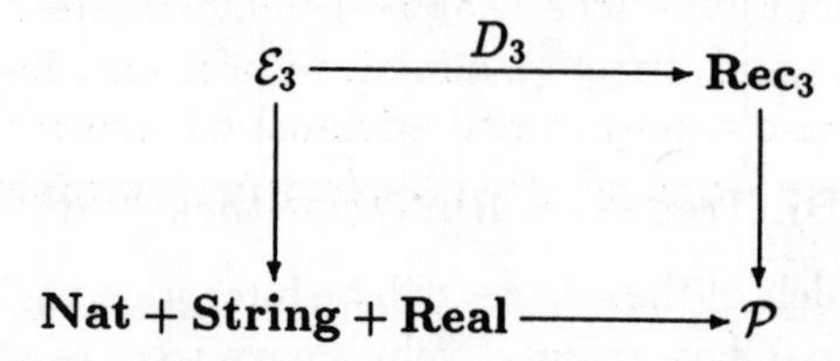

(where we suppose that sketches for **String** and **Real** have already been defined) is a sketch for three-place records, of which the first field is natural numbers, the second is strings and the third is floating point reals.

The ideas in this section are discussed without the explicit use of sketches in [Thatcher, Wagner and Wright, 1982], [Ehrig, Kreowski, Thatcher, Wagner and Wright, 1984] and [Wagner, 1986b].

10.2.6 Parametrizing operations The category $\mathcal{E}$ in the upper left corner of the pushout diagrams can contain operations. For example, let $\mathbf{Nat}^+$ denote the sketch for natural numbers with an added operation $+ : n \times n \to n$ with diagrams forcing it to be addition in the initial model, and $\mathbf{Nat}^*$ a similar sketch with an added operation $* : n \times n \to n$ forced to be multiplication there. Let **Diag** be the sketch whose graph contains operations

$$n \xrightarrow{\Delta} n \times n \xrightarrow{m} n$$

a cone with arrows $p_i : n \times n \to n,\ i = 1, 2$, to force $n \times n$ to be the indicated product and a diagram to force Δ to be the diagonal map in a model. Let **BinOp** be the sketch containing $n \times n \to n$ and similar data making $n \times n$ the indicated product. **BinOp** is included in both **Diag** and in each of $\mathbf{Nat}^+$ and $\mathbf{Nat}^*$. A pushout

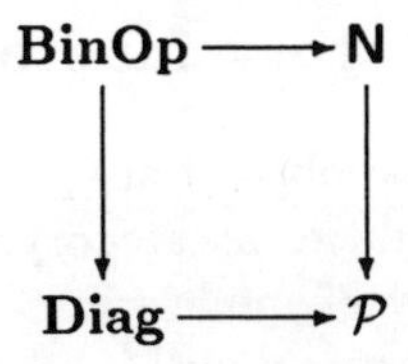

with $\mathbf{N} = \mathbf{Nat}^+$ would add a doubling operator to $\mathbf{Nat}^+$ and with $\mathbf{N} = \mathbf{Nat}^*$ would add a squaring operator to $\mathbf{Nat}^*$.

10.2.7 Arithmetic One last, more speculative example will show that we have barely begun to explore this idea. Let us suppose that we have defined two sketches called **Real** and **Complex** that implement the arithmetic operations (addition, subtraction, multiplication and division) of real and complex numbers, respectively, as well as absolute value. While it is true that real numbers are a special case of complex numbers and therefore do not, in principle, have to be treated differently, it is also true that real arithmetic is much easier and faster and virtually every language (that implements a complex number type) treats them separately.

Suppose, further, there is a sketch **Trig** whose operations implement the trigonometric functions by means of power series or other approximations which have the same algorithm for real and for complex numbers. This sketch will use four formal arithmetic operations on an abstract data type d which admits the arithmetic operations and absolute value, but is not otherwise specified. These five operations are determined by a sketch **Arith** which has an arrow to all of **Real, Complex, Trig.**

If we now form the pushout

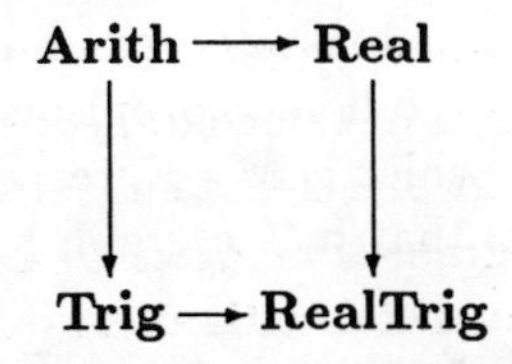

we get a sketch for the real type with trigonometric functions and if we replace **Real** by **Complex**, we get a sketch for complex trigonometry.

10.2.8 Exercises

1. Let $\mathcal{E}$ be the sketch with one node and no arrows, diagrams, cones or cocones. Show that the category of models of $\mathcal{E}$ in a category $\mathcal{C}$ is isomorphic to $\mathcal{C}$.

2. If $\mathcal{S}$ is the pushout of Diagram (10.1), describe precisely the nodes in $\mathcal{S}$ which must become a singleton in a model in **Set**.

10.3 The model category functor

Sketch is a category whose objects are sketches. If $\mathcal{C}$ is a fixed category, then each sketch produces a category of models of the sketch in $\mathcal{C}$ (see 9.1.5). The category of models in $\mathcal{C}$ of a sketch $\mathcal{S}$ we will call $\mathbf{Mod}_{\mathcal{C}}(\mathcal{S})$. ($\mathbf{Mod}_{\mathcal{C}}(\mathcal{S})$ might very well be the empty category.)

Most of the examples in this book have had $\mathcal{C} = \mathbf{Set}$, which regrettably obscures one of the major advantages sketches have over standard logical theories: the fact that their models can be in any suitable category.

10.3.1 For each sketch homomorphism $F : \mathcal{S} \to \mathcal{T}$ we will now describe a functor $\mathbf{Mod}_{\mathcal{C}}(F) : \mathbf{Mod}_{\mathcal{C}}(\mathcal{T}) \to \mathbf{Mod}_{\mathcal{C}}(\mathcal{S})$ (note the reversal). The functor $\mathbf{Mod}_{\mathcal{C}}(F)$ is defined by MF–1 and MF–2 below, in which we write $F^*(M)$ for $[\mathbf{Mod}_{\mathcal{C}}(F)](M)$, where M is a model of $\mathcal{T}$. $F^*(M)$ is to be a model of $\mathcal{S}$.

MF–1 The object function of F^* is defined by $F^*(M) = M \circ F$. Thus if g is a node or arrow of the graph of $\mathcal{S}$ and M is a model of $\mathcal{T}$, then $F^*(M)(g) = M(F(g))$.

MF–2 If $\alpha : M \to N$ is a homomorphism of models of $\mathcal{T}$, define $F^*(\alpha)$ to be the natural transformation αF as defined in Section 4.4. Thus at a node g of the graph of $\mathcal{S}$,

$$F^*(\alpha)(g) = (\alpha F)g = \alpha F(g) : M(F(g)) \to N(F(g))$$

10.3.2 Proposition *For each model M of $\mathcal{T}$ and homomorphism $F : \mathcal{T} \to \mathcal{S}$, $F^*(M)$ is a model of $\mathcal{S}$.*

Proof. Since M is among other things a graph homomorphism and so is F, and the composite of graph homomorphisms is a graph homomorphism, MF–1 makes $F^*(M)$, which is $M \circ F$, respect the source and target of arrows of the graph, so that it is a graph homomorphism to $\mathcal{C}$ as a model should be.

If D is a diagram of $\mathcal{S}$, then because F is a homomorphism of sketches, $F \circ D$ is a diagram of $\mathcal{T}$. Since M is a model of $\mathcal{T}$, $F^*(M) \circ D = M \circ F \circ D$ commutes.

Suppose $v \to D$ is a cone of $\mathcal{S}$. Then $F(v) \to F \circ D$ is a cone of $\mathcal{T}$ which must be taken to a limit cone by M. Since $F^*(M)(v \to D) = M(F(v)) \to M \circ F \circ D$, $F^*(M)$ takes $v \to D$ to a limit cone. A similar argument deals with cocones. □

10.3.3 Proposition *For each homomorphism $\alpha : M \to N$ of models of $\mathcal{T}$ and homomorphism $F : \mathcal{T} \to \mathcal{S}$ of sketches,* $\mathbf{Mod}_{\mathcal{C}}(F)(\alpha)$ *as defined by MF–2 is a homomorphism of models of $\mathcal{S}$.*

Proof. By MF–2, $F^*(\alpha)$ is the natural transformation αF that was defined in 4.4.1. Since its domain and codomain are models, it is a homomorphism of models by definition. □

For any sketch $\mathcal{S}$ we have defined $\mathbf{Mod}_{\mathcal{C}}(\mathcal{S})$ to be the category of models of $\mathcal{S}$ in $\mathcal{C}$. If $F : \mathcal{S} \to \mathcal{T}$ is a homomorphism of sketches, we have defined a functor $\mathbf{Mod}_{\mathcal{C}}(F) : \mathbf{Mod}_{\mathcal{C}}(\mathcal{T}) \to \mathbf{Mod}_{\mathcal{C}}(\mathcal{S})$.

10.3.4 Proposition $\mathbf{Mod}_{\mathcal{C}} : \mathbf{Sketch}^{\mathrm{op}} \to \mathbf{Cat}$ *is a functor.*

The proof involves some simple checking and is left as an exercise.

It is almost immediate that all three propositions, 10.3.2, 10.3.3 and 10.3.4, remain true if the category **Sketch** is replaced by the category of sketches with diagrams based on a specific class of shape graphs, cones to specific class of shape graphs not necessarily the same as that of the diagrams, and cocones from a specific class of shape graphs not necessarily the same as either of the other two. For example, taking any shapes for the diagrams, finite discrete shapes for the cones, and no shapes for the cocones gives the category of FP sketches and homomorphisms between them, so all these propositions are true of FP sketches.

10.3.5 Sentences For this subsection only, we will say a **sentence** in a sketch $\mathcal{S}$ is a diagram, cone or cocone in the graph of $\mathcal{S}$, *not* necessarily one in the specified set of diagrams, cones or cocones. The sentence is **satisfied** in a model M of $\mathcal{S}$ if it commutes when M is applied (if it is a diagram) or if it is a limit (co)cone when M is applied (if it is a (co)cone).

If $F : \mathcal{S} \to \mathcal{T}$ is a sketch homomorphism, and σ is a sentence of $\mathcal{S}$, then $F(\sigma)$ is a sentence of $\mathcal{T}$ where $F(\sigma)$ is defined by composition: this follows from the definition of sketch homomorphism.

10.3.6 Proposition *If $F : \mathcal{S} \to \mathcal{T}$ is a sketch homomorphism and σ is a sentence of $\mathcal{S}$, then $F(\sigma)$ is satisfied in a model M of $\mathcal{T}$ in a category $\mathcal{C}$ if and only if σ is satisfied in* $\mathbf{Mod}_{\mathcal{C}}(F)(M)$.

This follows directly from the definitions and is left as an exercise.

It follows from Propositions 10.3.4 and 10.3.6 that the category of sketches with designated shape graphs for each of the diagrams, cones and cocones, together with sentences, form a 'simple institution' in the sense of [Goguen and Burstall, 1986].

10.3.7 This result shows the way in which sketches are a very different method for describing mathematical structures from the theories of traditional mathematical logic. Of course, sketches use graphs, diagrams, cones and cocones instead of variables, symbols, expressions and formulas, but that difference, although significant, is not the biggest difference. When you use sketches, you factor the entire descriptive process differently. We have already discussed the differences to some extent for FP sketches in 7.2.8.

An example of the subtlety of the difference is exemplified by our word 'sentence' above. Diagrams correspond to universally quantified sentences, and the correspondence, which is not entirely trivial (see Section 7.7), is nevertheless exact. But to use a cone as a sentence corresponds to a statement ***about a type*** in multisorted logic, rather than statements about terms. ([McLarty, 1986] makes this explicit.) Such statements have not played a big role in classical logic. In classical logic, you do not say (with an axiom) that a type is (for example) a product type; the product type is implicit in the existence of some given term which has a sequence of variables of specific types.

On the other hand, our sentences do not provide a way of giving universal Horn clauses in (for example) an FL sketch. When using sketches, universal Horn clauses are usually implicit. Such a clause which must be satisfied in all models is given by an arrow built into the graph of the sketch which causes an operation to factor through some limit type. Thus if you wanted to require

$$a = b \Rightarrow g(f(a,b)) = h(f(a,b))$$

where a and b are of type A, f is an operation of type C and g and h are operations of type D, you build the graph of your sketch with arrows $f : A \times A \to C$, $g, h : C \to D$, $e : E \to C$, $u : A \to E$ and $d : A \to A \times A$, with cones requiring $A \times A$ to be the required product, e to be the equalizer of g and h, d to be the diagonal map, and this diagram

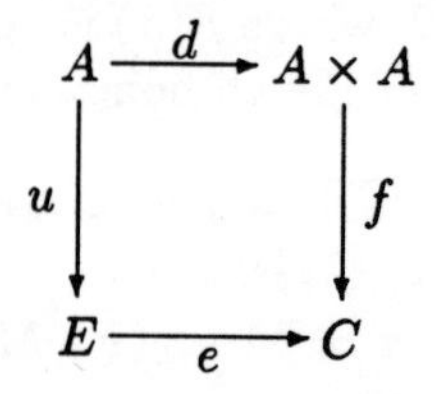

which requires $f \circ d$ to factor through the equalizer of g and h. It is built into the graph in a way analogous to the way in classical logic you build the product types implicitly by incorporating specific terms.

There is presumably a theory of sentences built in this way which makes FL sketches an institution, but it will require work to produce it: it does not fit well with the ingredients of a sketch. In the same way, building an institution out of classical logic using a definition of sentence which allows you to state that a type is a certain equalizer or pullback requires work because it does not fit the traditional way of doing things. (It is *not* impossible, it merely does not fit well.) *The classical approach and the sketch approach make different things easy.*

10.3.8 Exercises

1. Give an example of a sketch which has no models in **Set**. (Hint: it has to have cocones.)

2. Prove that MF–2 makes F^* a natural transformation.

3. Prove Proposition 10.3.4.

4. Prove Proposition 10.3.6.

11
Fibrations

A category is both a generalized poset and a generalized monoid. Many constructions in category theory can be understood in terms of the constructions in posets that they generalize, so that it is generally good advice when learning about a new categorical idea to see what it says about posets. Seeing what a construction says about monoids has not usually been so instructive.

However, certain concepts used to study the algebraic structure of monoids generalize to categories in a natural way, and often the theorems about them remain true. In addition, applications of monoids to the theory of automata have natural generalizations to categories, and some work has been done on these generalized ideas.

In this chapter we describe some aspects of categories as generalized monoids. We begin in Section 11.1 with the concept of fibration, which has been used in recent research on polymorphism. One way of constructing fibrations is by the Grothendieck construction, described in Section 11.2, which is a generalization of the semidirect product construction for monoids. Section 11.3 gives an equivalence between certain types of fibrations and category-valued functors. Section 11.4 describes the wreath product of categories, a generalization of the concept of the same name for monoids; some applications of the construction are mentioned.

The Grothendieck construction is used in Section 14.7. The rest of the material is not used elsewhere in the book.

11.1 Fibrations

In this section, we describe fibrations, which are special types of functors important in category theory and which have been proposed as useful in certain aspects of computer science. See in particular the connection with polymorphism in [Coquand, 1988], [Ehrhard, 1988] and [Hyland and Pitts, 1989].

The next section gives a way of constructing fibrations from set or category-valued functors.

11.1.1 Fibrations and opfibrations Let $P : \mathcal{E} \to \mathcal{C}$ be a functor between small categories, let $f : C \to D$ be an arrow of $\mathcal{C}$, and let $P(Y) = D$. An arrow $u : X \to Y$ of $\mathcal{E}$ is **cartesian** for f and Y if

CA–1 $P(u) = f$.

CA–2 For any arrow $v : Z \to Y$ of $\mathcal{E}$ and any arrow $h : P(Z) \to C$ of $\mathcal{C}$ for which $f \circ h = P(v)$, there is a unique $w : Z \to X$ in $\mathcal{E}$ such that $u \circ w = v$ and $P(w) = h$.

Similarly, if $f : C \to D$ and $P(X) = C$, then an arrow $u : X \to Y$ is **opcartesian** for f and X if

OA–1 $P(u) = f$.

OA–2 For any arrow $v : X \to Z$ of $\mathcal{E}$ and any arrow $k : D \to P(Z)$ for which $k \circ f = P(v)$, there is a unique $w : Y \to Z$ in $\mathcal{E}$ for which $w \circ u = v$ and $P(w) = k$.

If $P : \mathcal{E} \to \mathcal{C}$ is a functor, categorists often think of $\mathcal{E}$ as being above $\mathcal{C}$. (This is also common for functions between spaces, which is what originally suggested the ideas in this section.) For example, if $P(Y) = D$, one says that Y **lies over** D. Similar terminology is used for arrows. Thus a cartesian arrow for f must lie over f (CA–1) and one refers to CA–2 as a 'unique lifting' property.

11.1.2 Definition A functor $P : \mathcal{E} \to \mathcal{C}$ is a **fibration** if there is a cartesian arrow for every $f : C \to D$ in $\mathcal{C}$ and every object Y of $\mathcal{E}$ for which $P(Y) = D$. P is an **opfibration** if there is an opcartesian arrow for every $f : C \to D$ in $\mathcal{C}$ and every object X of $\mathcal{E}$ for which $P(X) = C$. It follows that $P : \mathcal{E} \to \mathcal{C}$ is a fibration if and only if $P^{\mathrm{op}} : \mathcal{E}^{\mathrm{op}} \to \mathcal{C}^{\mathrm{op}}$ is an opfibration.

If $P : \mathcal{E} \to \mathcal{C}$ is a fibration, one also says that $\mathcal{E}$ is **fibered over** $\mathcal{C}$. In that case, $\mathcal{C}$ is the **base category** and $\mathcal{E}$ is the **total category** of the fibration.

11.1.3 Definition A **cleavage** for a fibration $P : \mathcal{E} \to \mathcal{C}$ is a function γ that takes an arrow $f : C \to D$ and object Y such that $P(Y) = D$ to an arrow $\gamma(f, Y)$ of $\mathcal{E}$ that is cartesian for f and Y. Similarly an opcleavage κ takes $f : C \to D$ and X such that $P(X) = C$ to an arrow $\kappa(f, X)$ that is opcartesian for f and X.

The cleavage γ is a **splitting** of the fibration if it satisfies the following two requirements.

SC–1 Let D be an object of $\mathcal{C}$ and let Y be an object of $\mathcal{E}$ for which $P(Y) = D$. Then $\gamma(\mathrm{id}_D, Y) = \mathrm{id}_Y$.

SC–2 Suppose $f : C \to D$ and $g : D \to E$ in $\mathcal{C}$ and suppose Y and Z are objects of $\mathcal{E}$ for which $P(Z) = E$ and Y is the domain of $\gamma(g, Z)$. Then

$$\gamma(g, Z) \circ \gamma(f, Y) = \gamma(g \circ f, Z)$$

Note that under the assumptions in SC–2, $P(Y) = D$, so that $\gamma(f, Y)$ and $\gamma(g, Z) \circ \gamma(f, Y)$ are defined.

A fibration is **split** if it has a splitting.

Similarly, an opcleavage κ is a splitting of an opfibration $P : \mathcal{E} \to \mathcal{C}$ if $\kappa(\mathrm{id}_C, X) = \mathrm{id}_X$ whenever $P(X) = C$ and

$$\kappa(g, Y) \circ \kappa(f, X) = \kappa(g \circ f, X)$$

whenever $f : C \to D$ and $g : D \to E$ in $\mathcal{C}$, $P(X) = C$ and Y is the codomain of $\kappa(f, X)$. A split opfibration is again one which has a splitting.

11.1.4 Example Let $\mathcal{A}$ and $\mathcal{C}$ be any categories. Then the second projection $p_2 : \mathcal{A} \times \mathcal{C} \to \mathcal{C}$ is both a split fibration and a split opfibration. To see that it is a fibration, suppose that $f : C \to C'$ in $\mathcal{C}$ and let $Y = (A, C')$ be an object of $\mathcal{A} \times \mathcal{C}$. Then we can take $\gamma(f, Y)$ to be the arrow $(\mathrm{id}_A, f) : (A, C) \to (A, C')$. If $(g, h) : (A', C'') \to (A, C')$, and $u : C'' \to C'$ satisfies $f \circ u = h$ (note that $p_2(g, h) = h$), then the unique arrow from (A', C'') to (A, C') required by CA–2 is (g, u).

11.1.5 Example If $\mathcal{C}$ is a category, the **arrow category** of $\mathcal{C}$ (which we have already mentioned in 4.2.17) has as objects the *arrows* of $\mathcal{C}$. An arrow from $f : A \to B$ to $g : C \to D$ is a pair (h, k) of arrows with $h : A \to C$, $k : B \to D$ for which

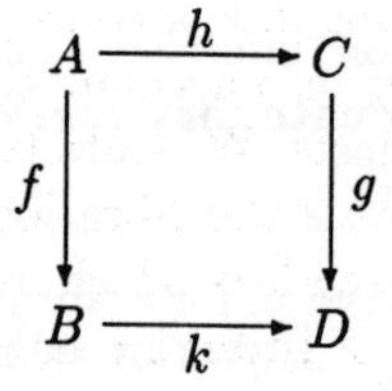

$$\begin{array}{ccc} A & \xrightarrow{h} & C \\ {\scriptstyle f}\downarrow & & \downarrow{\scriptstyle g} \\ B & \xrightarrow[k]{} & D \end{array} \qquad (11.1)$$

commutes.

If $\mathcal{A}$ is the arrow category of $\mathcal{C}$, there is a functor $P : \mathcal{A} \to \mathcal{C}$ which takes $f : A \to B$ to B and $(h, k) : f \to g$ to k. If $\mathcal{C}$ has pullbacks, this functor is a fibration. For a given $f : C \to D$ in $\mathcal{C}$ and object $k : B \to D$

of $\mathcal{A}$, a cartesian arrow for f and k is any (u, f) given by a pullback

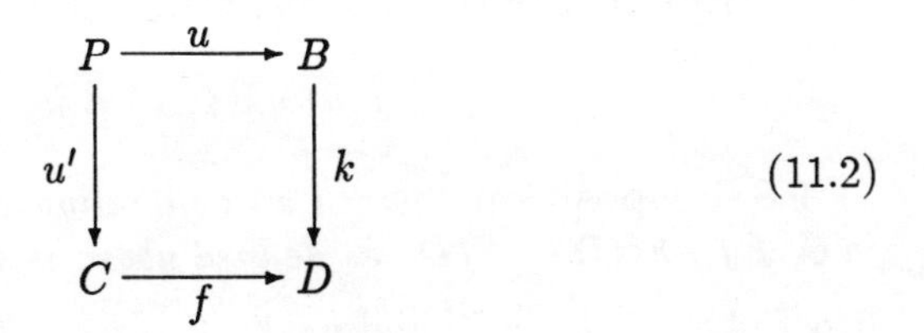

$$\begin{array}{ccc} P & \xrightarrow{u} & B \\ {\scriptstyle u'}\downarrow & & \downarrow{\scriptstyle k} \\ C & \xrightarrow[f]{} & D \end{array} \tag{11.2}$$

The verification is left as an exercise (Exercise 5).

11.1.6 Fibers For any functor $P : \mathcal{E} \to \mathcal{C}$, the **fiber over** an object C of $\mathcal{C}$ is the set of objects X for which $P(X) = C$ and arrows f for which $P(f) = \mathrm{id}_C$. This fiber is in fact a subcategory of $\mathcal{E}$ (Exercise 1).

In the case of Example 11.1.4, the fibers are all the same: each one is isomorphic to the category $\mathcal{A}$. This suggests thinking of an arbitrary fibration as a type of generalized product, in which the first coordinates come in general from varying sets depending on the second coordinate. This observation can also be made concerning the relationship between a set product $S \times T$ and a general T-indexed set.

On the other hand, the fiber of the fibration in Example 11.1.5 over an object A of $\mathcal{C}$ is the slice category $\mathcal{C}/A$. Since an object of $\mathcal{C}/A$ can be thought of as an indexed family of objects of $\mathcal{C}$, indexed by A, this example has been referred to as $\mathcal{C}$ 'fibered over itself'.

11.1.7 Cleavages induce functors If $\mathcal{E}$ is fibered over $\mathcal{C}$, then the fibers form an indexed set of categories (indexed by the objects). Given a cleavage, the arrows of $\mathcal{C}$ induce functors between the fibers. In this way fibrations or opfibrations give a concept like that of indexed sets, in which the indexing takes into account the arrows of the underlying categories as well as the objects. We state two propositions spelling this out for opfibrations and leave the fibrations to you (Exercise 3). An alternative approach to these ideas which explicitly follows the indexed set analogy is the concept of **indexed category** (see [Johnstone and Paré, 1978]).

In the rest of this chapter, when $F : \mathcal{C} \to \mathbf{Cat}$ or $F : \mathcal{C} \to \mathbf{Set}$ is a functor, we will normally write Ff for $F(f)$.

Let $P : \mathcal{E} \to \mathcal{C}$ be an opfibration with cleavage κ. Define $F : \mathcal{C} \to \mathbf{Cat}$ by

FF–1 $F(C)$ is the fiber over C for each object C of $\mathcal{C}$.

FF–2 For $f : C \to D$ in $\mathcal{C}$ and X an object of $F(C)$, $Ff(X)$ is defined to be the codomain of the arrow $\kappa(f, X)$.

FF–3 For $f : C \to D$ in $\mathcal{C}$ and $u : X \to X'$ in $F(C)$, $Ff(u)$ is the unique arrow from $Ff(X)$ to $Ff(X')$ given by OA–2 for which

$$Ff(u) \circ \kappa(f, X) = \kappa(f, X') \circ u$$

11.1.8 Proposition *Given an opfibration $P : \mathcal{E} \to \mathcal{C}$ and $f : C \to D$ in $\mathcal{C}$, $Ff : F(C) \to F(D)$ as defined above is a functor.*

Proof. Let $u : X \to X'$ and $v : X' \to X''$ in $F(C)$. Then

$$\begin{aligned} Ff(v) \circ Ff(u) \circ \kappa(f, X) &= Ff(v) \circ \kappa(f, X') \circ u \\ &= \kappa(f, X'') \circ v \circ u \end{aligned} \tag{11.3}$$

by two applications of FF–3. But then by the uniqueness part of FF–3, $Ff(v) \circ Ff(u)$ must be $Ff(v \circ u)$. This proves Ff preserves composition. We leave the preservation of identities to you. □

11.1.9 Proposition *Let $P : \mathcal{E} \to \mathcal{C}$ be a split opfibration with splitting κ. Then F as defined by FF–1 through FF–3 is a functor from $\mathcal{C}$ to* **Cat**.

Proof. Let $f : C \to D$ and $g : D \to E$ in $\mathcal{C}$. Let $u : X \to X'$ in $F(C)$. Then $F(g \circ f)(u)$ is the unique arrow from $F(g \circ f)(X)$ (the codomain of $\kappa(g \circ f, X)$) to $F(g \circ f)(X')$ (the codomain of $\kappa(g \circ f, X')$) for which

$$F(g \circ f)(u) \circ \kappa(g \circ f, X) = \kappa(g \circ f, X') \circ u$$

Since κ is a splitting, this says

$$F(g \circ f)(u) \circ \kappa(g, Ff(X)) \circ \kappa(f, X) = \kappa(g, Ff(X')) \circ \kappa(f, X') \circ u$$

By FF–3, the right side is

$$\kappa(g, Ff(X')) \circ Ff(u) \circ \kappa(f, X)$$

Applying FF–3 with g and $Ff(u)$ instead of f and u, this is the same as

$$Fg[Ff(u)] \circ \kappa(g, Ff(X) \circ \kappa(f, X))$$

which is

$$Fg[Ff(u)] \circ \kappa(g \circ f, X)$$

because κ is a splitting. Using the uniqueness requirement in FF–3, this means $F(g \circ f)(u) = Fg[Ff(u)]$, so that F preserves composition. Again, we leave preservation of the identity to you. □

In a similar way, split fibrations give functors $F^{\mathrm{op}} \to$ **Cat** (Exercise 2).

11.1.10 Exercises

1. Verify that for any functor $P : \mathcal{E} \to \mathcal{C}$ and object C of $\mathcal{C}$, the fiber over an object C is a subcategory of $\mathcal{E}$.

2. State and prove a proposition analogous to Proposition 11.1.9 which associates a contravariant **Cat**-valued functor to each split fibration.

3. State and prove a proposition like Proposition 11.1.8 for fibrations.

4. Let $\phi : \mathbf{Z}_4 \to \mathbf{Z}_2$ be the homomorphism defined in Exercise 2 of Section 2.9.

a. Show that the functor from $C(\mathbf{Z}_4)$ to $C(\mathbf{Z}_2)$ induced by ϕ is a fibration and an opfibration. (If you know about groups, this is an instance of the fact that every surjective group homomorphism is a fibration and an opfibration.)

b. Show that ϕ is not a split fibration or opfibration.

5. Let $\mathcal{C}$ be a category with pullbacks and $\mathcal{A}$ its arrow category. For an arrow $f : A \to B$ (object of $\mathcal{A}$) let $P(f) = B$. For an arrow $(h, k) : f \to g$ (where $g : C \to D$ in $\mathcal{C}$) in $\mathcal{A}$, let $P(h, k) = k$.

a. Show that $P : \mathcal{A} \to \mathcal{C}$ is a functor.

b. Show that P is a fibration.

11.2 The Grothendieck construction

The Grothendieck construction is a way of producing fibrations. It generalizes the semidirect product construction for monoids, which is defined here. Coquand [1988] uses the Grothendieck construction in a study of polymorphism.

The construction can be applied to either set-valued functors or category-valued functors. Given such a functor $F : \mathcal{C} \to \mathbf{Set}$ or $F : \mathcal{C} \to \mathbf{Cat}$, it constructs a category $\mathbf{G}(\mathcal{C}, F)$ and a functor from $\mathbf{G}(\mathcal{C}, F)$ to $\mathcal{C}$. When F is set-valued we will write $\mathbf{G}_0$ instead of $\mathbf{G}$. We will look at the set-valued case first, since it is simpler.

11.2.1 Let $\mathcal{C}$ be a small category and let $F : \mathcal{C} \to \mathbf{Set}$ be a functor. For each object C of $\mathcal{C}$, $F(C)$ is a set, and for each arrow $f : C \to D$, $F(f) : F(C) \to f(D)$ is a set function. There is no set of all sets or set of all set functions, but, since $\mathcal{C}$ is small, there certainly is a set consisting of all the elements of all the sets $F(C)$, and similarly there is a set consisting of all the functions $F(f)$. In other words, although **Set** is large, the description of $F : \mathcal{C} \to \mathbf{Set}$ requires only a small amount of data. ('Small' and 'large' are used here in the technical sense, referring to whether or not a set of data is involved. See 1.3.8.)

By contrast, to describe a functor $G : \mathbf{Set} \rightarrow \mathcal{C}$ would require a large amount of data – an object of $\mathcal{C}$ for each set, and so on.

We will formalize these observations about $F : \mathcal{C} \rightarrow \mathbf{Set}$ by taking the disjoint union of all the sets of the form $F(C)$ for all objects C of $\mathcal{C}$. The elements of this disjoint union can be represented as pairs (x, C) for all objects C of $\mathcal{C}$ and elements $x \in F(C)$. (Thus we construct the disjoint union of sets by labeling the elements. The disjoint union is the construction in **Set** corresponding to the categorical concept of 'sum', discussed in Section 5.4.)

We must do more than this to capture the *functorial nature* of F – what it does to arrows of $\mathcal{C}$. The category $\mathbf{G}_0(\mathcal{C}, F)$ constructed by the Grothendieck construction does capture this structure, and its set of objects is the disjoint union just mentioned.

11.2.2 If we were to draw a picture to explain what F does, the result might be Diagram (11.4), in which $f : C \rightarrow C'$ and $g : C' \rightarrow C''$ are arrows of $\mathcal{C}$ and x and x' are elements of $F(C)$ and $F(C')$ respectively. The box over each object C of $\mathcal{C}$ represents the elements of $F(C)$. The

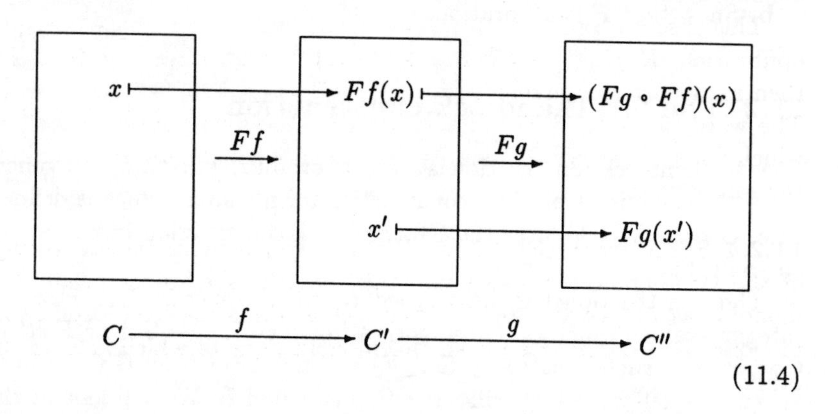

(11.4)

arrows from x to $Ff(x)$ and from x' to $Fg(x')$ are there informally to illustrate what the set functions Ff and Fg do. *These informal arrows become actual arrows in* $\mathbf{G}_0(\mathcal{C}, F)$.

11.2.3 Definition $\mathbf{G}_0(\mathcal{C}, F)$ is the category defined as follows.

GS–1 An object of $\mathbf{G}_0(F)$ is a pair (x, C) where C is an object of $\mathcal{C}$ and x is an element of $F(C)$ (as observed, the C occurs in the pair (x, C) because we want the disjoint union of the values of F).

GS–2 An arrow is a pair of the form $(x, f) : (x, C) \rightarrow (x', C')$ where f is an arrow $f : C \rightarrow C'$ of $\mathcal{C}$ for which $Ff(x) = x'$.

GS–3 If $(x, f) : (x, C) \to (x', C')$ and $(x', g) : (x', C') \to (x'', C'')$, then $(x', g) \circ (x, f) : (x, C) \to (x'', C'')$ is defined by

$$(x', g) \circ (x, f) = (x, g \circ f)$$

Note in GS–3 that indeed $F(g \circ f)(x) = x''$ as required by GS–2.

The reason that we use the notation (x, f) is the requirement that an arrow must determine its source and target. The source of (x, f) is (x, C), where C is the source of f and x is explicit, while its target is (x', C'), where C' is the target of f and $x = Ff(x)$. In the literature, (x, f) is often denoted simply f, so that the same name f may refer to many different arrows – one for each element of $F(C)$.

Projection on the first coordinate defines a functor

$$\mathbf{G}_0(F) : \mathbf{G}_0(\mathcal{C}, F) \to \mathcal{C}$$

$\mathbf{G}_0(\mathcal{C}, F)$ together with $\mathbf{G}_0(F)$ is called the **split discrete opfibration** induced by F, and $\mathcal{C}$ is the **base category** of the opfibration.

If C is an object of $\mathcal{C}$, the inverse image under $\mathbf{G}_0(F)$ of C is simply the set $F(C)$, although its elements are written as pairs so as to form a disjoint union.

This discrete opfibration is indeed an opfibration, in fact a split opfibration. If $f : C \to C'$ in $\mathcal{C}$ and (x, C) is an object of $\mathbf{G}_0(\mathcal{C}, F)$, then an opcartesian arrow is $(x, f) : (x, C) \to (Ff(x), C')$ (Exercise 4). The word 'discrete' refers to the fact that the fibers are categories in which the only arrows are identity arrows; such categories are essentially the same as sets.

11.2.4 Semidirect products We now give a more general version of the Grothendieck construction that has the semidirect product of monoids as a special case. We first define the semidirect product of monoids: it is constructed from two monoids, one of which acts on the other.

11.2.5 Definition If M and T are monoids, an **action** of M on T is a function $\alpha : M \times T \to T$ for which

MA–1 $\alpha(m, 1_T) = 1_T$ for all $m \in M$.

MA–2 $\alpha(m, tu) = \alpha(m, t)\alpha(m, u)$ for all $m \in M$ and $t, u \in T$.

MA–3 $\alpha(1_M, t) = t$ for all $t \in T$.

MA–4 $\alpha((mn), t) = \alpha(m, \alpha(n, t))$ for all $m, n \in M$ and $t \in T$.

If we curry α as in 6.1.2, we get a family of functions $\phi(m) : T \to T$ with the properties listed in MA′–1 through MA′–4 below.

MA′–1 $\phi(m)(1_T) = 1_T$ for all $m \in M$.

MA′–2 $\phi(m)(tu) = \phi(m)(t)\phi(m)(u)$ for all $m \in M$ and $t, u \in T$.

MA′–3 $\phi(1_M)(t) = t$ for all $t \in T$.

MA′–4 $\phi(mn)(t) = \phi(m)[\phi(n)(t)]$ for all $m, n \in M$ and $t \in T$.

Thus we see that an alternative formulation of monoid action is that it is a monoid homomorphism $\phi : M \to \mathbf{End}(T)$ ($\mathbf{End}(T)$ being the monoid of endomorphisms of T). MA′–1 and MA′–2 say that each function $\phi(m)$ is an endomorphism of T, and MA′–3 and MA′–4 say that ϕ is a monoid homomorphism.

11.2.6 Definition The **semidirect product** of M and T with the given action α as just defined is the monoid with underlying set $T \times M$ and multiplication defined by

$$(t, m)(t', m') = (t\alpha(m, t'), mm')$$

To see the connection with the categorical version below you may wish to write this definition using the curried version of α.

11.2.7 The categorical construction corresponding to a monoid acting on a monoid is a functor which takes values in **Cat** rather than in **Set**. A functor $F : \mathcal{C} \to \mathbf{Cat}$ can be regarded as an action of $\mathcal{C}$ on a *variable category* which plays the role of T in the definition just given.

In the case of a monoid action defined by MA–1 through MA–4, the variable category is actually not varying: it is the category $C(T)$ determined by the monoid T. The functor F in that case takes the single object of M to the single object of T, and, given an element $m \in M$, $F(m)$ is the endomorphism of T which takes $t \in T$ to mt: in other words, $F(m)(t) = mt$. Thus F on the arrows is the curried form of the action α.

A set-valued functor is a special case of a category-valued functor, since a set can be regarded as a category with only identity arrows. Note that this is different from the monoid case: an action by a monoid on a set is *not* in general a special case of an action by the monoid on a monoid. It is, however, a special case of the action of a monoid on a category – a discrete category.

11.2.8 Given a functor $F : \mathcal{C} \to \mathbf{Cat}$, the Grothendieck construction in this more general setting constructs the opfibration induced by F, a category $\mathbf{G}(\mathcal{C}, F)$ defined as follows:

GC–1 An object of $\mathbf{G}(\mathcal{C}, F)$ is a pair (x, C) where C is an object of $\mathcal{C}$ and x is an object of $F(C)$.

GC–2 An arrow $(u, f) : (x, C) \to (x', C')$ has $f : C \to C'$ an arrow of $\mathcal{C}$ and $u : Ff(x) \to x'$ an arrow of $F(C')$ (note that by definition $Ff(x)$ is an object of $F(C')$).

GC–3 If $(u, f) : (x, C) \to (x', C')$ and $(v, g) : (x', C') \to (x'', C'')$, then $(v, g) \circ (u, f) : (x, C) \to (x'', C'')$ is defined by

$$(v, g) \circ (u, f) = (v \circ Fg(u), g \circ f)$$

11.2.9 Theorem *Given a functor $F : \mathcal{C} \to \mathbf{Cat}$, $\mathsf{G}(\mathcal{C}, F)$ is a category and the first projection is a functor $P : \mathsf{G}(\mathcal{C}, F) \to \mathcal{C}$ which is a split opfibration with splitting*

$$\kappa(f, X) = (\mathrm{id}_{Ffx}, f) : (x, C) \to (Ffx, C')$$

for any arrow $f : C \to C'$ of $\mathcal{C}$ and object (x, C) of $\mathsf{G}(\mathcal{C}, F)$.

We omit the proof of this theorem. $\mathsf{G}(\mathcal{C}, F)$ is called the **crossed product** $\mathcal{C} \times F$ by some authors.

It is instructive to compare this definition with the discrete opfibration constructed from a set-valued functor. In the case that F is set-valued, the first component u of an arrow $(u, f) : (x, C) \to (x', C')$ has to be an identity arrow and it has to be $\mathrm{id}_{Ff(x)}$. Thus the only arrows are of the form $(\mathrm{id}_{Ff(x)}, f) : (x, C) \to (x', C)$. Such an arrow is denoted (x, f) in GS–1 through GS–3.

To visualize the $\mathbf{Cat}$-valued Grothendieck construction, we can modify the picture in Diagram (11.4) to get Diagram (11.5). The arrows from

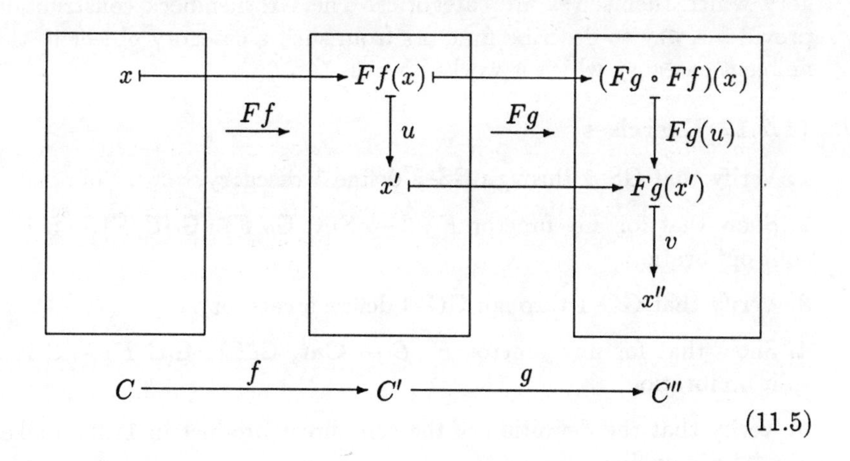

(11.5)

inside one box to inside another, such as the arrow from x to $Ff(x)$, are parts of arrows of $\mathsf{G}(f)$, which are now (in contrast to the discrete case) allowed to miss the target and be rescued by an internal arrow of the codomain category.

Thus in the picture above there is an arrow from x to $Ff(x)$ and $Ff(x)$ is not necessarily x'; the gap is filled by the arrow $u : Ff(x) \to x'$ of $F(C')$. The arrow $(u, f) : (x, C) \to (x', C')$ of $\mathbf{G}(\mathcal{C}, F)$ may be pictured as the arrow from x to $Ff(x)$ followed by u. Observe that the definition of composition says that the square in the picture with corners $Ff(x)$, $(Fg \circ Ff)(x)$, x' and $Fg(x')$ 'commutes'.

As before, one writes (x, C) for x and (x', C') for x' only to ensure that the union of all the categories of the form $F(C)$ is a *disjoint* union.

11.2.10 When the Grothendieck construction is applied to a functor $G : \mathcal{C}^{\mathrm{op}} \to \mathbf{Cat}$, the result is a split fibration $\mathcal{F}(\mathcal{C}, G)$. We spell out the definition for this case for later use.

FC–1 An object of $\mathcal{F}(\mathcal{C}, G)$ is a pair (C, x) where C is an object of $\mathcal{C}$ and x is an object of $G(C)$.

FC–2 An arrow $(f, u) : (C, x) \to (C', x')$ has $f : C \to C'$ an arrow of $\mathcal{C}$ and $u : x \to Gf(x')$ an arrow of $G(C)$.

FC–3 If $(f, u) : (C, x) \to (C', x')$ and $(g, v) : (C', x') \to (C'', x'')$, then $(g, v) \circ (f, u) : (C, x) \to (C'', x'')$ is defined by

$$(g, v) \circ (f, u) = (g \circ f, Gf(v) \circ u)$$

11.2.11 In line with the concept that a category is a mathematical workspace, one could ask to construct objects in a suitably rich category which themselves are categories. The Grothendieck construction provides a way to describe functors from such a category object to the ambient category which is worked out in 14.7.2.

11.2.12 Exercises

1. Verify that GS–1 through GS–3 define a category.

2. Show that for any functor $F : \mathcal{C} \to \mathbf{Set}$, $\mathbf{G}_0(F) : \mathbf{G}_0(\mathcal{C}, F) \to \mathcal{C}$ is a split opfibration.

3. Verify that GC–1 through GC–3 define a category.

4. Show that for any functor $F : \mathcal{C} \to \mathbf{Cat}$, $\mathbf{G}(F) : \mathbf{G}(\mathcal{C}, F) \to \mathcal{C}$ is a split opfibration.

5. Verify that the definition of the semidirect product in 11.2.6 makes $T \times M$ a monoid.

6. Let $F : \mathcal{C} \to \mathbf{Cat}$ be a functor. Show that for each object C of $\mathcal{C}$, the arrows of the form $(u, \mathrm{id}_C) : (x, C) \to (y, C)$ (for all arrows $u : x \to y$ of $F(C)$) (and their sources and targets) form a subcategory of the opfibration $\mathbf{G}(\mathcal{C}, F)$ which is isomorphic to $F(C)$.

11.3 An equivalence of categories

In this section, we describe how the construction of a functor from an opfibration given in Proposition 11.1.8 (in one direction) produces an equivalence of categories (with the Grothendieck construction as inverse) between a category of functors and a suitably defined category of split opfibrations.

11.3.1 Cat-valued functors For a category $\mathcal{C}$, $\mathbf{Fun}(\mathcal{C}, \mathbf{Cat})$ is the category whose objects are functors from $\mathcal{C}$ to the category of categories, and whose arrows are natural transformations between them.

If $F : \mathcal{C} \to \mathbf{Cat}$ is such a functor and $f : C \to D$ is an arrow of $\mathcal{C}$, then $F(C)$ and $G(C)$ are categories and $Ff : F(C) \to F(D)$ is a functor. If also $G : \mathcal{C} \to \mathbf{Cat}$ and $\alpha : F \to G$ is a natural transformation, then for each object of C, $\alpha C : F(C) \to G(C)$ is a functor and the following diagram is a commutative diagram of categories and functors:

$$\begin{array}{ccc} F(C) & \xrightarrow{\alpha C} & G(C) \\ {\scriptstyle Ff}\downarrow & & \downarrow{\scriptstyle Gf} \\ F(D) & \xrightarrow[\alpha D]{} & G(D) \end{array} \qquad (11.6)$$

11.3.2 The category of split opfibrations of $\mathcal{C}$ Let $P : \mathcal{E} \to \mathcal{C}$ and $P' : \mathcal{E}' \to \mathcal{C}$ be two split opfibrations of the same category $\mathcal{C}$ with splittings κ and κ' respectively. A **homomorphism of split opfibrations** is a functor $\zeta : \mathcal{E} \to \mathcal{E}'$ for which

HSO–1 The diagram

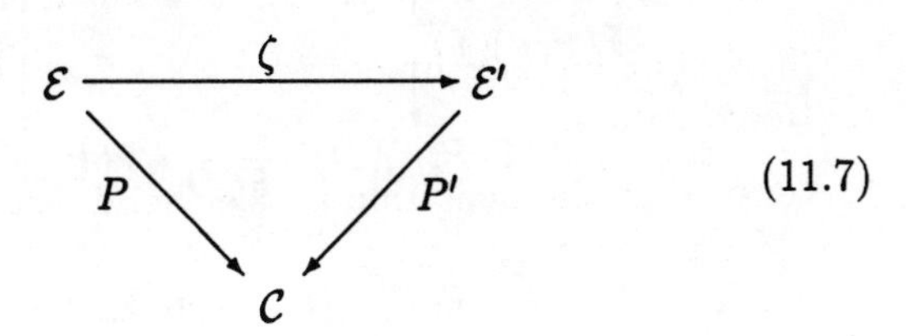

(11.7)

commutes.

HSO–2 For any arrow $f : C \to D$ in $\mathcal{C}$ and object X of $\mathcal{E}$ such that $P(X) = C$,

$$\zeta(\kappa(f, x)) = \kappa'(f, x)$$

Thus a homomorphism of split fibrations 'takes fibers to fibers' and 'preserves the splitting'.

11.3.3 Definition Split opfibrations of $\mathcal{C}$ and homomorphisms between them form a category $\mathbf{SO}(\mathcal{C})$.

We will show that $\mathbf{SO}(\mathcal{C})$ is equivalent to $\mathbf{Fun}(\mathcal{C}, \mathbf{Cat})$. We do this by defining two functors

$$\mathsf{F} : \mathbf{SO}(\mathcal{C}) \to \mathbf{Fun}(\mathcal{C}, \mathbf{Cat})$$

and

$$\mathsf{G} : \mathbf{Fun}(\mathcal{C}, \mathbf{Cat}) \to \mathbf{SO}(\mathcal{C})$$

so that F is an equivalence with pseudo-inverse G as defined in Section 3.4.

11.3.4 Definition For a category $\mathcal{C}$, define the functor

$$\mathsf{F} : \mathbf{SO}(\mathcal{C}) \to \mathbf{Fun}(\mathcal{C}, \mathbf{Cat})$$

as follows:

FI–1 If $P : \mathcal{E} \to \mathbf{Cat}$ is a split opfibration with splitting κ, then $\mathsf{F}(P, \kappa) : \mathcal{C} \to \mathbf{Cat}$ is the functor F satisfying FF–1 through FF–3 defined in 11.1.7.

FI–2 If $\zeta : (P, \kappa) \to (P', \kappa')$ is a homomorphism of opfibrations, $\mathsf{F}\zeta : \mathsf{F}(P, \kappa) \to \mathsf{F}(P', \kappa')$ is the natural transformation whose component at an object C of $\mathcal{C}$ is the functor ζ restricted to $P^{-1}(C)$.

To show that $F\zeta$ is a natural transformation, it is necessary to show that for every $f : C \to D$ in $\mathcal{C}$ the following diagram commutes:

$$\begin{array}{ccc}
\mathsf{F}(P,\kappa)(C) & \xrightarrow{\mathsf{F}\zeta C} & \mathsf{F}(P',\kappa')(C) \\
{\scriptstyle \mathsf{F}(P,\kappa)(f)}\downarrow & & \downarrow{\scriptstyle \mathsf{F}(P',\kappa')(f)} \\
\mathsf{F}(P,\kappa)(D) & \xrightarrow[\mathsf{F}\zeta D]{} & \mathsf{F}(P',\kappa')(D)
\end{array} \tag{11.8}$$

Let $u : X \to X'$ be in $\mathsf{F}(P, \kappa)(C)$. Note that u is an arrow in the inverse image $P^{-1}C$, so $Pu = \mathrm{id}_C$. Moreover, ζu is an arrow for which $P'u = \mathrm{id}_C$.

By definition of cleavage, there are unique arrows $\widetilde{u}$ such that $P(\widetilde{u}) = f$ and $\widehat{u}$ such that $P'(\widehat{u}) = f$, for which

$$\widetilde{u} \circ \kappa(f, X) = \kappa(f, X') \circ u$$

and

$$\widehat{u} \circ \kappa'(f, X) = \kappa'(f, X') \circ \zeta u$$

The top route in Diagram (11.8) takes u to $\widehat{u}$ and the bottom route takes it to $\widetilde{u}$. The following calculation shows that the diagram commutes:

$$\begin{aligned} \widehat{u} \circ \kappa'(f, X) &= \kappa'(f, X') \circ u \\ &= \zeta(\kappa(f, X)) \circ \zeta(u) \\ &= \zeta(\kappa(f, X) \circ u) \\ &= \zeta(\widetilde{u} \circ \kappa(f, X)) \\ &= \zeta(\widetilde{u}) \circ \kappa'(f, X) \end{aligned} \tag{11.9}$$

so that $\widehat{u} = \zeta(\widetilde{u})$ by the uniqueness requirement in the definition of $\widehat{u}$.

11.3.5 The Grothendieck functor To define the functor going the other way we extend the Grothendieck construction.

11.3.6 Definition For a category $\mathcal{C}$, define the functor

$$\mathbf{G} : \mathbf{Fun}(\mathcal{C}, \mathbf{Cat}) \to \mathbf{SO}(\mathcal{C})$$

as follows:

GR–1 For $F : \mathcal{C} \to \mathbf{Cat}$, $\mathbf{G}(F) = \mathbf{G}(\mathcal{C}, F)$.

GR–2 For a natural transformation $\alpha : F \to G : \mathcal{C} \to \mathbf{Cat}$,

$$\mathbf{G}\alpha(x, C) = (\alpha F x, C)$$

for (x, C) an object of $\mathbf{G}(\mathcal{C}, F)$ (so that C is an object of $\mathcal{C}$ and x is an object of FC), and

$$\mathbf{G}\alpha(u, f) = (\alpha C' u, f)$$

for (u, f) an arrow of $\mathbf{G}(\mathcal{C}, F)$ (so that $f : C \to C'$ in $\mathcal{C}$ and $u : Ffx \to x'$ in FC').

Note that in GR–2, $\alpha C' u$ has domain $\alpha C'(Ffx)$, which is $Gf(\alpha Cx)$ because α is a natural transformation. The verification that $\mathbf{G}\alpha$ is a functor is omitted.

11.3.7 Theorem *The functor* $\mathbf{F} : \mathbf{SO}(\mathcal{C}) \to \mathbf{Fun}(\mathcal{C}, \mathbf{Cat})$ *as defined in 11.3.4 is an equivalence of categories with pseudo-inverse* $\mathbf{G}$.

There is a similar equivalence of categories between split fibrations and contravariant functors. The details are in [Nico, 1983]. Moreover, the nonsplit case for both fibrations and opfibrations corresponds in a precise way to 'pseudo-functors', which are like functors except that identities and composites are preserved only up to natural isomorphisms. See [Gray, 1966] (the terminology has evolved since that article).

11.3.8 Exercises

1. Verify that $\mathbf{G}\alpha$ as defined by GR–2 is a functor.
2. Verify that $\mathbf{G}$ as defined by GR–1 and GR–2 is a functor.

11.4 Wreath products

In this section, we introduce the idea of the wreath product of categories (and of functors), based on an old construction originating in group theory. In the monoid case, this construction allows a type of series-parallel decomposition of finite state machines (the Krohn–Rhodes Theorem). This section is not needed later.

11.4.1 Let $\mathcal{A}$ and $\mathcal{B}$ be small categories and $G : \mathcal{A} \to \mathbf{Cat}$ a functor. With these data we define the **shape functor** $S(G, \mathcal{B}) : \mathcal{A}^{\mathrm{op}} \to \mathbf{Cat}$ as follows. If A is an object of $\mathcal{A}$, then $S(G, \mathcal{B})(A)$ is the category of functors from the category $G(A)$ to $\mathcal{B}$ with natural transformations as arrows.

Thus an object of $S(G, \mathcal{B})(A)$ is a functor $P : G(A) \to \mathcal{B}$ and an arrow from $P : G(A) \to \mathcal{B}$ to $P' : G(A) \to \mathcal{B}$ is a natural transformation from P to P'. It is useful to think of $S(G, \mathcal{B})(A)$ as the category of diagrams of shape $G(A)$ (or models of $G(A)$) in $\mathcal{B}$; the arrows between them are homomorphisms of diagrams, in other words natural transformations.

Embedding or modeling a certain shape (diagram, space, structure, etc.) into a certain workspace (category, topological space, etc.) in all possible ways is a tool used all over mathematics. In particular, what we have called the shape functor is very reminiscent of the singular simplex functors in algebraic topology.

We must say what $S(G, \mathcal{B})$ does to arrows of $\mathcal{A}^{\mathrm{op}}$. If $f : A \to A'$ is an arrow of $\mathcal{A}$, then $S(G, \mathcal{B})(f) : G(A') \to G(A)$ takes a functor H :

$G(A') \to \mathcal{B}$ to $H \circ Gf : G(A) \to \mathcal{B}$. This is the usual action of a functor on diagrams in a category.

11.4.2 Since $S(G,\mathcal{B}) : \mathcal{A}^{\text{op}} \to \mathbf{Cat}$ is a category-valued functor, we can use the Grothendieck construction to form the split fibration of $\mathcal{A}$ by $S(G,\mathcal{B})$. This fibration consists of a category denoted $\mathcal{A}\,\mathrm{wr}^G\,\mathcal{B}$, called the **wreath product** of $\mathcal{A}$ by $\mathcal{B}$ with given action G, and a functor $\Pi : \mathcal{A}\,\mathrm{wr}^G\,\mathcal{B} \to \mathcal{A}$.

We now unwind what this implies to give an elementary definition of the wreath product.

11.4.3 Definition Given small categories $\mathcal{A}$ and $\mathcal{B}$ and a functor $G : \mathcal{A} \to \mathbf{Cat}$, the wreath product $\mathcal{A}\,\mathrm{wr}^G\,\mathcal{B}$ is a category defined as follows:

WP–1 The objects of $\mathcal{A}\,\mathrm{wr}^G\,\mathcal{B}$ are pairs (A, P), where A is an object of $\mathcal{A}$ and $P : G(A) \to \mathcal{B}$ is a functor.

WP–2 An arrow $(f, \lambda) : (A, P) \to (A', P')$ of $\mathcal{A}\,\mathrm{wr}^G\,\mathcal{B}$ has $f : A \to A'$ an arrow of $\mathcal{A}$ and $\lambda : P \to P' \circ Gf$ a natural transformation.

WP–3 If $(f, \lambda) : (A, P) \to (A', P')$ and $(g, \mu) : (A', P') \to (A'', P'')$ are arrows of $\mathcal{A}\,\mathrm{wr}^G\,\mathcal{B}$, as in

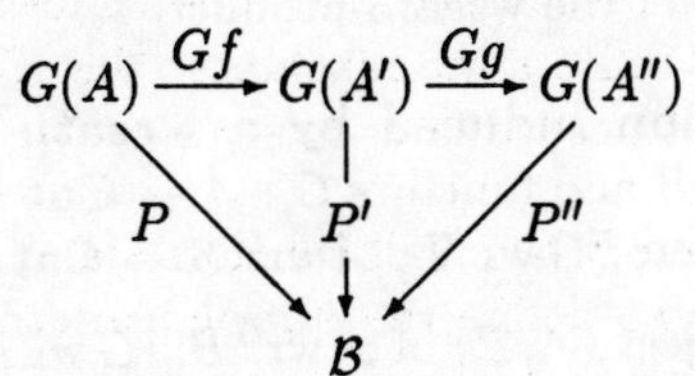

then

$$(g, \mu) \circ (f, \lambda) = (g \circ f, \mu.Gf \circ \lambda) : (A, P) \to (A'', P'')$$

To see the meaning of WP–3, observe that $\lambda : P \to P' \circ Gf$ and $\mu : P' \to P'' \circ Gg$ are natural transformations. Then

$$\mu.Gf : P' \circ Gf \to P'' \circ Gg \circ Gf = P'' \circ G(g \circ f)$$

is the natural transformation whose component at an object x of $G(A)$ is the component of μ at $Gf(x)$ (this was described in Section 4.4). Then

$$\mu.Gf \circ \lambda : P \to P'' \circ G(g \circ f)$$

is the usual composite of natural transformations (see 4.2.11); it is the natural transformation whose component at an object x of $G(A)$ is the composite of the components $(\mu.Gf(x)) \circ \lambda x$.

It follows from WP–3 that there is a **projection functor**

$$\Pi : \mathcal{A}\,\mathrm{wr}^G\,\mathcal{B} \to \mathcal{A}$$

taking (A, P) to A and (f, λ) to f.

11.4.4 Special cases of the wreath product If the functor G in definition 11.4.3 is set-valued, then one obtains the **discrete wreath product** of $\mathcal{A}$ by $\mathcal{B}$ with action G. When $\mathcal{A}$ and $\mathcal{B}$ are both monoids, the *discrete* wreath product is also a monoid. (The general case need not be a monoid.) This case is extensively treated in the literature; see [Wells,1976] for an exposition and bibliography and [Eilenberg, 1976] for a treatment of the Krohn–Rhodes Theorem and other matters in this context.

11.4.5 Definition For any small category $\mathcal{C}$, the **right regular representation** of $\mathcal{C}$ is the functor $R_{\mathcal{C}} : \mathcal{C} \to \mathbf{Set}$ defined as follows:

RR–1 If C is an object of $\mathcal{C}$, then $R_{\mathcal{C}}(C)$ is the set of arrows of $\mathcal{C}$ with codomain C.

RR–2 If $f : C \to C'$ in $\mathcal{C}$ and $g \in R_{\mathcal{C}}(C)$, then $R_{\mathcal{C}}(f)(g) = f \circ g$.

For small categories $\mathcal{A}$ and $\mathcal{B}$, the **standard wreath product** $\mathcal{A} \,\mathrm{wr}\, \mathcal{B}$ is the wreath product $\mathcal{A} \,\mathrm{wr}^{R_{\mathcal{A}}} \mathcal{B}$. This is a generalization of what is called the standard wreath product for groups and monoids. It is the wreath product used in [Rhodes and Tilson, 1989]. They also have a two-sided version of the wreath product.

11.4.6 The action induced by a wreath product Given small categories $\mathcal{A}$ and $\mathcal{B}$ and functors $G : \mathcal{A} \to \mathbf{Cat}$ and $H : \mathcal{B} \to \mathbf{Cat}$, there is an induced functor $G \,\mathrm{wr}\, H : \mathcal{A} \,\mathrm{wr}^G \mathcal{B} \to \mathbf{Cat}$ defined as follows:

WF–1 For an object (A, P) of $\mathcal{A} \,\mathrm{wr}^G \mathcal{B}$, $(G \,\mathrm{wr}\, H)(A, P)$ is the split opfibration induced by $H \circ P : G(A) \to \mathbf{Cat}$.

WF–2 If (h, λ) is an arrow of $\mathcal{A} \,\mathrm{wr}^G \mathcal{B}$ with domain (A, P), and (t, x) is an object of $(G \,\mathrm{wr}\, H)(A, P)$, so that x is an object of $G(A)$ and t is an object of $H(P(x))$, then

$$(G \,\mathrm{wr}\, H)(h, \lambda)(t, x) = (H\lambda x(t), Gh(x))$$

WF–3 If $(u, f) : (t, x) \to (t', x')$ is an arrow of $(G \,\mathrm{wr}\, H)(P, A)$, then

$$(G \,\mathrm{wr}\, H)(h, \lambda)(u, f) = (H(\lambda x')(u), Gh(f))$$

WF–1 can be perceived as saying that $G \,\mathrm{wr}\, H$ is obtained by composing the shapes given by G (see the discussion in 11.4.1) with H. Indeed, G. M. Kelly, who invented this concept ([Kelly, 1974b]) called what we call the wreath product the 'composite' of the categories. That is in some ways a better name: the word 'product' suggests that the two factors are involved in the product in symmetric ways, which is not the case, as the next subsection describes.

11.4.7 The action $G \,\text{wr}\, H$ of $\mathcal{A} \,\text{wr}^G\, \mathcal{B}$ just defined is said to be **triangular** because it is a precise generalization of the action of a triangular matrix. For example, the action

$$\begin{bmatrix} a & b \\ 0 & c \end{bmatrix} \times \begin{bmatrix} x \\ y \end{bmatrix} = \begin{bmatrix} ax + by \\ cy \end{bmatrix}$$

can be described this way: the effect on the first coordinate depends on both the first and second coordinates, but the effect on the second coordinate depends only on the second coordinate.

The dependency of the action on the coordinates given in WF–2 and WF–3 is analogous to the dependency for the matrices in the example just given.

The wreath product can be generalized to many factors, using the following theorem, proved in [Kelly, 1974], Section 7. This theorem allows one to think of the wreath product as generalizing triangular matrices bigger than 2×2.

11.4.8 Proposition *Let $G : \mathcal{A} \to$ **Cat**, $H : \mathcal{B} \to$ **Cat** and $K : \mathcal{C} \to$ **Cat** be functors. Then there is an isomorphism of categories I making this diagram commute.*

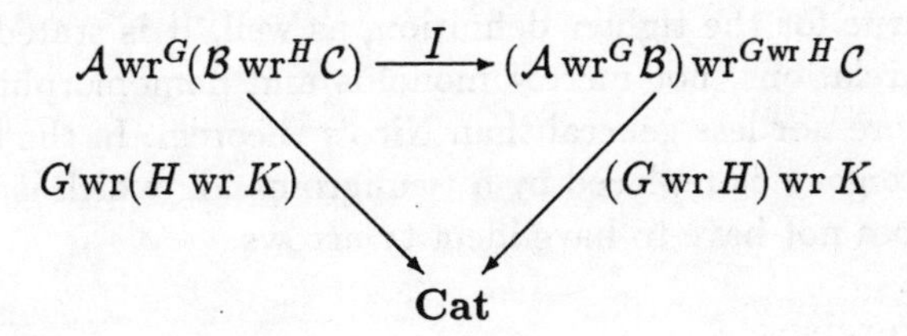

Note that the *standard* wreath product is not associative.

11.4.9 Applications of the wreath product It is natural to wish to simulate complicated state transition systems using systems built up in some way from a small stock of simpler ones. This requires a precise notion of simulation. This is defined in various ways in the literature. In some cases one says a functor $F : \mathcal{C} \to \mathcal{D}$ is a simulation of $\mathcal{D}$ (the word 'cover' is often used) if it has certain special properties. Other authors have the functor going the other way.

The Krohn–Rhodes Theorem for monoids says that every finite monoid action is simulated by an iterated wreath product of finite simple groups and certain very small monoids. The original Krohn–Rhodes Theorem was stated for semigroups. Discussions are in [Wells, 1976] and Eilenberg [1976]; the latter uses a different definition of cover. Wells [1980] proved a generalization of a weak form of the Krohn–Rhodes

Theorem for finite categories and set-valued functors (which generalize the concept of action by a monoid, as discussed in Section 3.2). Wells [1988a] describes how to use some of these decomposition techiques for category-valued functors.

Rhodes and Tilson use the idea of 'division' of categories, which is related to the notion of cover, as the basic idea of an extensive study of varieties of semigroups and complexity. See [Rhodes and Tilson, 1989], [Rhodes and Weil, 1989]

Nico [1983] defines a category induced by any functor called the 'kernel category' of the functor and proves a theorem which embeds the domain of the functor into the standard wreath product of the codomain and the kernel. This generalizes an old theorem of Kaloujnine and Krasner. It follows that every functor factors as a full and faithful functor which is injective on objects, followed by a fibration. Street and Walters [1973] have a related theorem.

In the case of groups, the kernel category of a homomorphism is a category equivalent, but not isomorphic, to the actual kernel of the homomorphism. In the case that $\mathcal{C}$ and $\mathcal{D}$ are monoids, the kernel category (which is not a monoid in general) is called the **derived category** of F. Rhodes and Tilson [1986] have a tighter definition obtained by imposing a congruence on the kernel. A theorem analogous to Nico's theorem is true for the tighter definition, as well. It is stated for semigroups and relations, not merely monoids and homomorphisms, so it is neither more nor less general than Nico's theorem. In the semigroup case, the 'category' is replaced by a 'semigroupoid', which is like a category but does not have to have identity arrows.

11.4.10 Exercises

1. Show that the discrete wreath product of two monoids is a monoid.

2. Show that the wreath product of two groups (monoids in which every element is invertible) regarded as categories is a category in which every arrow is an isomorphism.

12
Adjoints

Adjoints are about the most important idea in category theory, except perhaps for the closely related notion of representable functors. Many constructions made earlier in this book are examples of adjoints, as we will describe. Moreover, toposes are most conveniently described in terms of adjoints.

This chapter develops the basic theory of adjoints enough to use in the last two chapters of the book. Section 12.1 revisits the concept of free monoids, describing them in a way which suggests the general definition of adjoint. Adjoints are described and some basic properties developed in Sections 12.2 and 12.3. Section 12.4 describes locally cartesian closed categories, which are best described using adjoints.

More detail concerning adjoints can be found in [Barr and Wells, 1985] and [Mac Lane, 1971]. [Diers, 1977] describes a generalization which includes the initial families of FD sketches (see Section 7.10). [Hagino, 1986] has a general approach to type constructors for functional programming languages based on adjoints (see also [Chen and Cockett, 1989].)

12.1 Free monoids

In 3.1.14, we gave the universal property of the free monoid. We now state it more carefully than we did there, to pay closer attention to the categories involved. The point is that when we speak of a sub*set* of a monoid, we are mixing two things. A monoid is not just a set and most of its subsets are not submonoids. To describe a monoid, you must give three data: the set of elements, the operation and the identity element. From this point of view the phrase 'subset of a monoid' does not make sense. It is actually a subset of the underlying set of the monoid. Insistence on this is not pointless pedantry; it is the key to understanding the situation.

Let **Mon** denote the category of monoids and let $U : \mathbf{Mon} \to \mathbf{Set}$ denote the underlying set functor. We define a monoid to be a triple $(UM, \cdot, 1)$, where UM is a set called the **underlying set** of M, $\cdot$ is a binary operation on UM and 1 is the identity element for that operation.

A homomorphism of monoids is defined to be a triple (M, f, N) where f is a function $UM \to UN$ that preserves the operation. The underlying functor is defined on homomorphisms by $U(M, f, N) = f$. This very precise notation avoids the abuse of notation mentioned in 2.3.2.

12.1.1 Characterization of the free monoid Given a set X, the free monoid $F(X) = (X^*, \cdot, \langle\rangle)$ is characterized by the property given in the following proposition; the property is called the **universal mapping property** of the free monoid. In the proposition, $\eta X : X \to X^* = UF(X)$ takes $x \in X$ to the string $\langle x \rangle$ of length 1. We systematically distinguish the free monoid $F(X)$ from its underlying set, the Kleene closure $X^* = U(F(X))$.

12.1.2 Proposition *Let X be a set. For any function $u : X \to U(M)$, where M is any monoid, there is a unique monoid homomorphism $g : F(X) \to M$ such that $u = Ug \circ \eta X$.*

Proof. Define $g(\langle\rangle)$ to be the identity element of M, $g(\langle x \rangle) = u(x)$ for $x \in X$, and $g(\langle x_1 \cdots x_n \rangle)$ to be the product $u(x_1) \cdot \cdots \cdot u(x_n)$ in M. That makes g a monoid homomorphism and $u = g \circ \eta X$; the details are left as an exercise. □

Just as the theory of a sketch is the universal model of that sketch, the free monoid generated by a set is universal for functions from that set into the set underlying a monoid.

This proposition says that ηX is a universal element for the functor which takes a monoid M to $\mathrm{Hom}(X, U(M))$ and a monoid homomorphism $f : M \to N$ to $\mathrm{Hom}(X, U(f)) : \mathrm{Hom}(X, U(M)) \to \mathrm{Hom}(X, U(N))$. It follows that the universal mapping property characterizes the free monoid up to a unique isomorphism. This is spelled out more precisely in Exercise 2.

12.1.3 The free monoid functor Each set X generates a free monoid $F(X)$. In 3.1.12, we extended this to a functor $F : \mathbf{Set} \to \mathbf{Mon}$. In this section, we will prove this (as part of the proof of Proposition 12.1.4 below) using only the universal mapping property of free monoids; thus the argument will work in complete generality.

Let ηY denote the arrow from Y into Y^* described in 12.1.2 and let $f : X \to Y$ be a function. Then $\eta Y \circ f : X \to Y^*$ is a function from X into the set underlying a monoid. The universal property of $F(X)$ gives a unique monoid homomorphism $F(f) : F(X) \to F(Y)$ such that if $f^* = UF(f)$, then

$$\begin{array}{ccc} X & \xrightarrow{f} & Y \\ {\scriptstyle \eta X}\downarrow & & \downarrow{\scriptstyle \eta Y} \\ X^* & \xrightarrow[f^*]{} & Y^* \end{array} \qquad (12.1)$$

commutes.

12.1.4 Proposition $F : \mathbf{Set} \to \mathbf{Mon}$ *is a functor and, for any set* X, ηX *is the component at* X *of a natural transformation* $\eta : \mathrm{id}_{\mathbf{Set}} \to U \circ F$.

Proof. Once F is shown to be a functor, that η is a natural transformation will follow from Diagram (12.1).

First note that if $\mathrm{id} : X \to X$ is the identity, then $F(\mathrm{id})$ is the unique arrow $h : F(X) \to F(X)$ such that

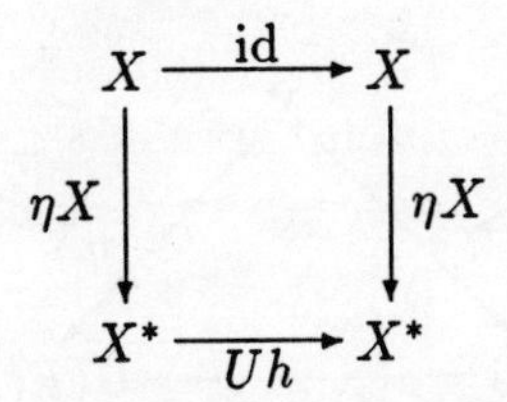

commutes. But if we replace Uh in the diagram above by id_{X^*}, the diagram still commutes. The uniqueness property of the free monoid implies that $Uh = \mathrm{id}_{X^*}$.

Similarly, if $g : Y \to Z$ is a function, the commutativity of both squares in

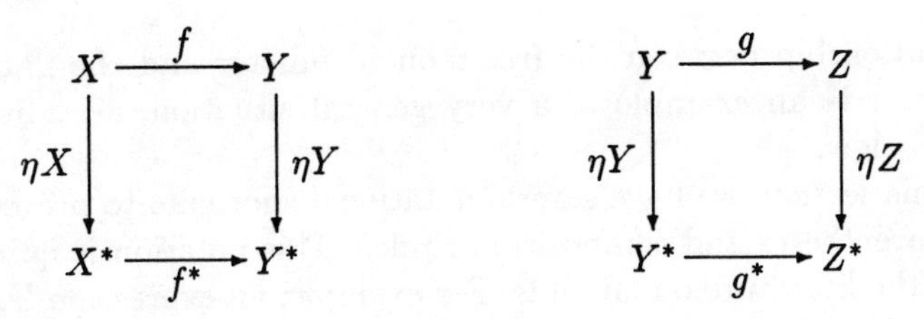

implies that

$$\begin{array}{ccc} X & \xrightarrow{g \circ f} & Z \\ {\scriptstyle \eta X}\downarrow & & \downarrow{\scriptstyle \eta Z} \\ X^* & \xrightarrow[g^* \circ f^*]{} & Z^* \end{array}$$

commutes. But $g^* \circ f^* = UF(g) \circ UF(f) = U(F(g) \circ F(f))$ since U is a functor. This means that the arrow $F(g) \circ F(f)$ satisfies the same equation that characterizes $F(g \circ f)$ uniquely and hence that they are equal. This shows that F is a functor and that η is a natural transformation. □

Another example of functors F and U satisfying this is the free category given by a graph as described in 2.6.11. This very same proof shows that the free category construction is really the object part of a functor.

12.1.5 Exercises

1. Show that the function g defined in the proof of Proposition 12.1.2 is a monoid homomorphism from $F(X)$ to M.

2. Let X be a set and $\gamma : X \to U(E)$ a function to the underlying set of some monoid E with the property that if $u : X \to U(M)$, where M is any monoid, there is a unique monoid homomorphism $f : E \to M$ such that $u = U(f) \circ \gamma$. Prove that there is a unique isomorphism $\phi : E \to F(X)$ for which

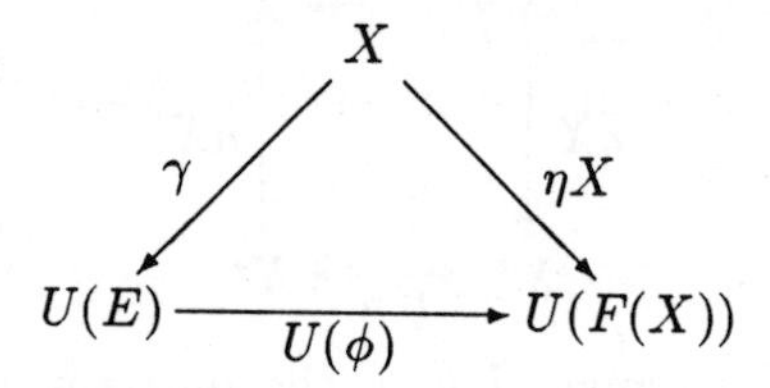

commutes.

12.2 Adjoints

The relationship between the free monoid functor and the underlying set functor is an example of a very general situation, an adjunction, defined below.

In this section, we have several notational shortcuts to prevent clutter of parentheses and composition circles. This notation is quite standard in the literature on adjoints. For example, an expression $UFUA$ is shorthand for $(U \circ F \circ U)(A)$, which is the same as $U(F(U(A)))$.

12.2.1 Definition Let $\mathcal{A}$ and $\mathcal{B}$ be categories. If $F : \mathcal{A} \to \mathcal{B}$ and $U : \mathcal{B} \to \mathcal{A}$ are functors, we say that F is **left adjoint** to U and U is **right adjoint** to F provided there is a natural transformation $\eta : \mathrm{id} \to UF$ such that for any objects A of $\mathcal{A}$ and B of $\mathcal{B}$ and any arrow $f : A \to UB$, there is a unique arrow $g : FA \to B$ such that $f = Ug \circ \eta A$.

This definition asserts that there is a functional way to convert any arrow $f : A \to UB$ to an arrow $g : FA \to B$ in such a way that g solves the equation $f = U(?) \circ \eta A$, and that the solution is unique.

The property of η given in the last sentence of Definition 12.2.1 is called its **universal mapping property**. The existence of the unique arrow $g : FA \to B$ such that $f = Ug \circ \eta A$ is the arrow-lifting property of Section 3.1.14. Just as in the discussion after Proposition 12.1.2, for each object A, the arrow ηA is a universal element for the functor $\mathrm{Hom}(A, U(-))$.

It is customary to write $F \dashv U$ to denote the situation described in the definition. The data (F, U, η) constitute an **adjunction**. The transformation η is called the **unit** of the adjunction.

Theorem 12.3.4 below says in effect that every adjunction arises from data satisfying requirements analogous to the properties of the free monoid as described in Proposition 12.1.2.

In some texts the adjunction is written as a rule of inference, like this:

$$\frac{A \to UB}{FA \to B} \tag{12.2}$$

The definition appears to be asymmetric in F and U. This is remedied in the next proposition.

12.2.2 Proposition *Suppose $F : \mathcal{A} \to \mathcal{B}$ and $U : \mathcal{B} \to \mathcal{A}$ are functors such that $F \dashv U$. Then there is a natural transformation $\epsilon : FU \to \mathrm{id}_{\mathcal{B}}$ such that for any $g : FA \to B$, there is a unique arrow $f : A \to UB$ with $g = \epsilon B \circ Ff$.*

The transformation ϵ is called the **counit** of the adjunction.

It is an immediate consequence of categorical duality that this proposition is reversible and the adjunction is equivalent to the existence of either natural transformation η or ϵ with its appropriate universal mapping property.

We give the proof of Proposition 12.2.2 as an illustration of an argument using natural transformations.

Proof. Take $A = UB$ in the definition of adjoint. Then corresponding to the identity arrow $A = UB \to UB$, there is a unique arrow we call

$\epsilon B : FA = FUB \to B$ such that

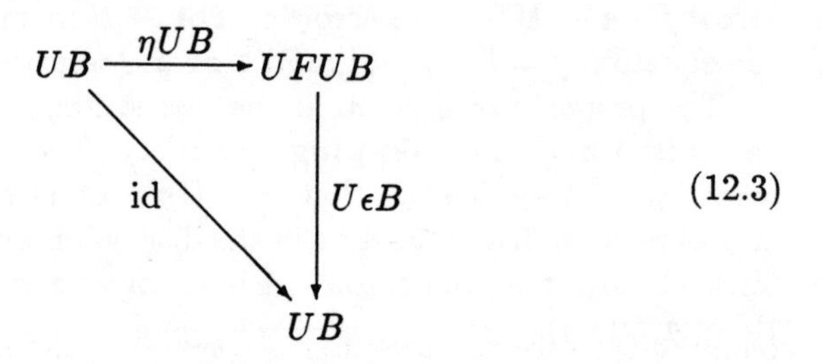

commutes. We first show that ϵB is the component at B of a natural transformation. Let $h : B' \to B$ be an arrow of $\mathcal{B}$. We want to show that the diagram

$$\begin{array}{ccc} FUB' & \xrightarrow{\epsilon B'} & B' \\ {\scriptstyle FUh}\downarrow & & \downarrow{\scriptstyle h} \\ FUB & \xrightarrow[\epsilon B]{} & B \end{array}$$

commutes. First we form the diagram

$$\begin{array}{ccccc} & & \text{id} & & \\ UB' & \xrightarrow[\eta UB']{} & UFUB' & \xrightarrow[U\epsilon B']{} & UB' \\ {\scriptstyle Uh}\downarrow & & \downarrow{\scriptstyle UFUh} & & \downarrow{\scriptstyle Uh} \\ UB & \xrightarrow{\eta UB} & UFUB & \xrightarrow{U\epsilon B} & UB \\ & & \text{id} & & \end{array}$$

whose left hand square commutes because of the naturality of η. Then we have, using (12.3),

$$U(h \circ \epsilon B') \circ \eta UB' = Uh \circ U\epsilon B' \circ \eta UB' = Uh$$

and

$$U(\epsilon B \circ FUh) \circ \eta UB' = U\epsilon B \circ UFUh \circ \eta UB' = U\epsilon B \circ \eta UB \circ Uh = Uh$$

Thus both $h \circ \epsilon B'$ and $\epsilon B \circ FUh$ are solutions to the equation

$$U(?) \circ \eta UB' = Uh$$

Since, by definition of adjointness, this equation has a unique solution, they are equal. This shows that ϵ is natural.

Next we claim that the diagram

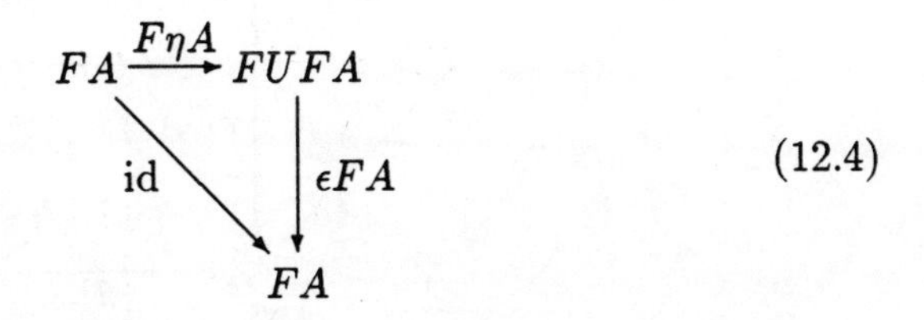

commutes. To see this, we first observe that the naturality of η means that for any arrow $h : A \to A'$ in $\mathcal{A}$, the diagram

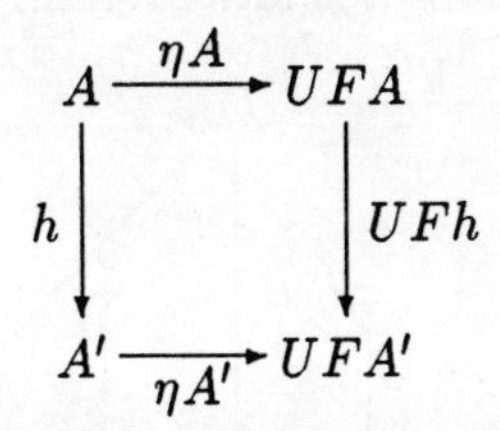

commutes. Letting $A' = UFA$ and $h = \eta A$, this implies that the upper square of the following diagram commutes:

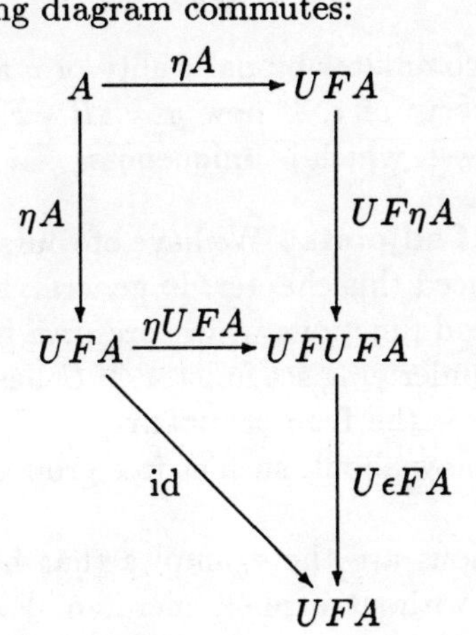

while the commutation of the lower triangle is a special case of (12.3) with $B = FA$. It follows that $U\epsilon FA \circ UF\eta A \circ \eta A = \eta A$, which means that $\epsilon FA \circ F\eta A$ and the identity are both solutions to the equation $U(?) \circ \eta A = \eta A$. The uniqueness implies that they are equal.

Now let $g : FA \to B$. Define $f = Ug \circ \eta A : A \to UB$. In the diagram

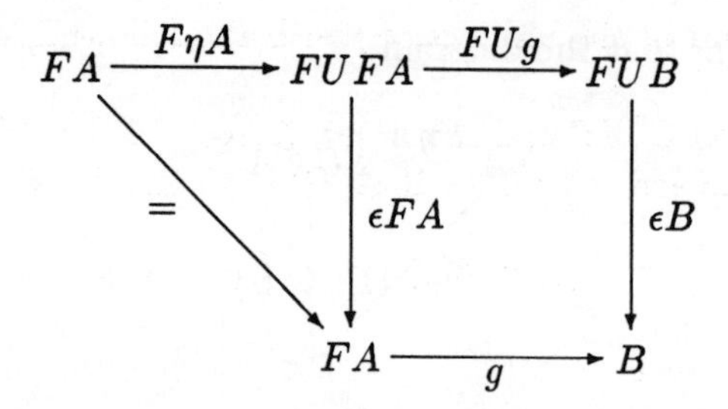

we have just shown that the triangle commutes and the square does by the naturality of ϵ. This shows that f is a solution to $\epsilon B \circ F(?) = g$. Next suppose that h is another solution. Then we have the diagram

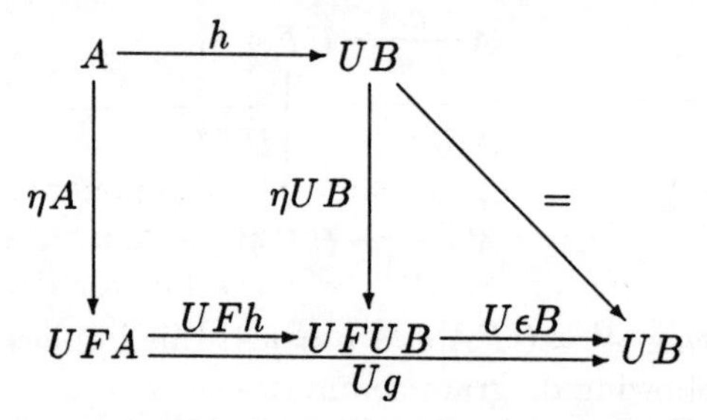

in which the square commutes by naturality of η and the triangle does by the defining property of ϵ. If now $g = \epsilon B \circ Fh$, then the diagram shows that $h = Ug \circ \eta A$, which is uniqueness. □

12.2.3 Examples of adjoints We have of course the example of free monoids that introduced this chapter. In general, let $\mathcal{C}$ be a category of sets with structure and functions which preserve the structure, and let $U : \mathcal{C} \to \mathbf{Set}$ be the underlying set functor. If U has a left adjoint F and S is a set, then $F(S)$ is the **free structure** on S. This description fits structures you may know about, such as free groups, free abelian groups and free rings.

Perhaps less obvious are the examples that have been introduced earlier in this book without explicit mention. For example, let $\mathcal{C}$ be a category. Consider the category $\mathcal{C} \times \mathcal{C}$ which has as objects pairs of objects (A, B) of $\mathcal{C}$ and in which an arrow $(A, B) \to (A', B')$ is a pair (f, g) of arrows $f : A \to A'$ and $g : B \to B'$. There is a functor $\Delta : \mathcal{C} \to \mathcal{C} \times \mathcal{C}$ given by $\Delta(A) = (A, A)$ and $\Delta(f) = (f, f)$. Let us see what a right adjoint to this functor is.

Assuming there is a right adjoint Π to Δ, there should be an arrow we call

$$\langle p_1, p_2 \rangle : \Delta\Pi(A, B) = (\Pi(A, B), \Pi(A, B)) \to (A, B)$$

(this is the counit of the adjunction) with the following universal property.

For any object C of $\mathcal{C}$ and any arrow $\langle q_1, q_2\rangle : \Delta C = (C, C) \to (A, B)$ there is a unique arrow $q : C \to \Pi(A, B)$ such that

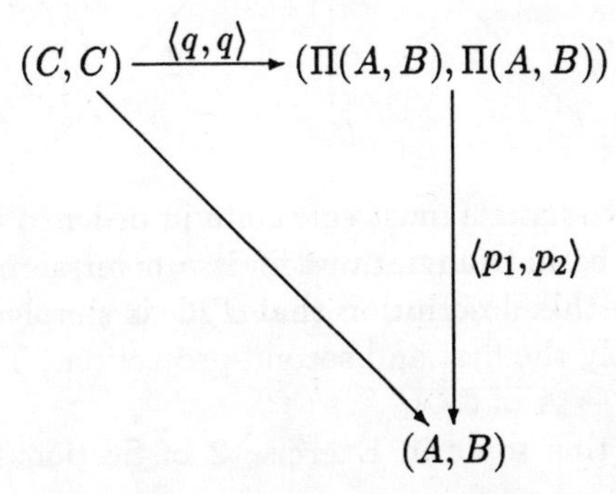

commutes. If you separate the pairs into components and write $A \times B$ instead of $\Pi(A, B)$ you will recover the categorical definition of the product. Thus a right adjoint to Δ is just a functor $\Pi : \mathcal{C} \times \mathcal{C} \to \mathcal{C}$ that chooses a product for each pair of objects of $\mathcal{C}$.

12.2.4 Now let us suppose we are given such a functor Π (which not surprisingly is called a **binary product functor**). We can fix an object A of $\mathcal{C}$ and consider the functor denoted $- \times A : \mathcal{C} \to \mathcal{C}$ whose value at an object B is the object $B \times A$ and at an arrow $f : B \to C$ is the arrow $f \times \mathrm{id}_A : B \times A \to C \times A$ (see 5.2.17).

A right adjoint to the functor $- \times A$, if one exists, can be described as follows. If the value at C of this adjoint is denoted $R_A(C)$, then there is an arrow $e : R_A(C) \times A \to C$ with the universal property that for any arrow $f : B \times A \to C$, there is a unique arrow we may call $\lambda f : B \to R_A(C)$ such that $e \circ (\lambda f \times A) = f$. But this is precisely the universal property that describes the exponential $[A \to C]$ (see 6.1.3). The counit e is the arrow eval defined there. The essential uniqueness of adjoints (see 12.3.3 below) implies that this adjunction property determines the exponential, if it exists, up to isomorphism.

Thus the two main defining properties defining cartesian closed categories can be (and commonly are) described in terms of adjoints.

12.2.5 We can describe this functor $- \times A$ in a slightly different way. For a category $\mathcal{C}$ and object A of $\mathcal{C}$ we described the slice category $\mathcal{C}/A$ in 2.6.9 and functor $U_A : \mathcal{C}/A \to \mathcal{C}$ in 3.1.10. A right adjoint $P_A : \mathcal{C} \to \mathcal{C}/A$ has to associate to each object C of $\mathcal{C}$ an object $\phi_A C = P_A(C) : T_A C \to A$ of $\mathcal{C}/A$ and an arrow (the counit) $\epsilon_A C : T_A C \to C$. This object and arrow must have the universal mapping property that for any other object $f : B \to A$ of $\mathcal{C}/A$, and any arrow $g : B \to C$ there is a unique arrow $h : B \to T_A C$ such that

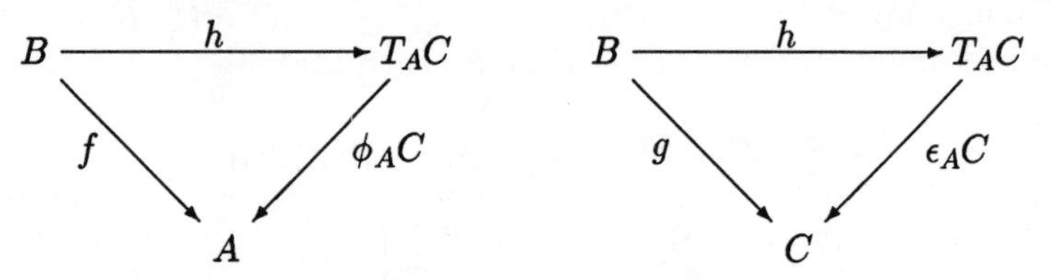

commute. The left triangle must commute in order to have an arrow in $\mathcal{C}/A$ and the right hand triangle must for its universal mapping property. It is evident from this description that $T_A C$ is simply $C \times A$ and $\epsilon_A C$ and $\phi_A C$ are simply the first and second projections. Thus $P_A(C)$ is the object $p_2 : C \times A \to A$ of $\mathcal{C}/A$.

We return to this topic in Exercise 2 of Section 12.3 and in Section 12.4.

12.2.6 Adjoints to the inverse image functor Here is another example of an adjoint. If S is a set, the set of subsets of S is a poset ordered by inclusion. This becomes a category in the usual way. That is, if S_0 and S_1 are subsets of S, then there is exactly one arrow $S_0 \to S_1$ if and only if $S_0 \subseteq S_1$. You should not view this arrow as representing a function, but just a formal arrow that represents the inclusion. We will call this category Sub(S).

If $f : S \to T$ is a function, then with any subset $T_0 \subseteq T$, we get a subset, denoted

$$f^{-1}(T_0) = \{s \in S \mid f(s) \in T_0\}$$

which is called the **inverse image** under f of T_0. If $T_0 \subseteq T_1$, then $f^{-1}(T_0) \subseteq f^{-1}(T_1)$. This means that f^{-1} is a functor from Sub(T) $\to$ Sub(S). This functor turns out to have both left and right adjoints.

The left adjoint is a familiar one (to mathematicians at least), the so-called **direct image.** For $S_0 \subseteq S$, let

$$f_*(S_0) = \{f(s) \mid s \in S_0\}$$

Then it is left as an exercise to show that $f_*(S_0) \subseteq T_0$ if and only if $S_0 \subseteq f^{-1}(T_0)$ which is just the statement that the direct image is left adjoint to the inverse image.

The right adjoint is usually denoted $f_!$ and is defined by saying that $t \in f_!(S_0)$ if and only if $f^{-1}(\{t\}) \subseteq S_0$. Another way of saying this is that *every* element of the inverse image of t is in S_0, while $t \in f_*(S)$ if and only if *some* element of the inverse image of t is in S_0.

We discussed these constructions in 3.1.18 and 3.1.19 for **Set**.

12.2.7 Exercises

1. Show that if $f : S \to T$ is a function between sets, then for $S_0 \subseteq S$ and $T_0 \subseteq T$, $S_0 \subseteq f^{-1}(T_0)$ if and only if $f(S_0) \subseteq T_0$.

2. Show that a left adjoint to the functor Δ of 12.2.3, if it exists, takes the pair of objects A, B to the sum $A + B$.

3. Find the unit of the adjunction $\Delta \dashv \Pi$ described in Section 12.2.3.

4.† Show that the construction of free models of a linear sketch in 4.7.14 gives an example of an adjunction analogous to the free monoid functor.

5. Show that the construction of the path category of a graph in 2.6.11 is left adjoint to the underlying functor from categories to graphs.

6. Show that for any set A with other than one element, the functor $- \times A : \mathbf{Set} \to \mathbf{Set}$ defined in 12.2.4 does not have a left adjoint in **Set**. (Hint: $1 \times A$ is isomorphic to A.)

7. In Section 12.2.4, a right adjoint to the functor $- \times A$ is described. The counit of the adjunction is the transformation e defined there. Describe the unit in general and in **Set**.

12.3 Further topics on adjoints

12.3.1 Hom set adjointness There is an alternative formulation of adjointness which is often found in the categorical literature.

12.3.2 Theorem *If $\mathcal{A}$ and $\mathcal{B}$ are categories and $U : \mathcal{B} \to \mathcal{A}$ and $F : \mathcal{A} \to \mathcal{B}$ are functors, then $F \dashv U$ if and only if* $\mathrm{Hom}(F-,-)$ *and* $\mathrm{Hom}(-,U-)$ *are naturally isomorphic as functors* $\mathcal{A}^{\mathrm{op}} \times \mathcal{B} \to \mathbf{Set}$.

Proof. Let $F \dashv U$, and let A and B be objects of $\mathcal{A}$ and $\mathcal{B}$ respectively. Define $\beta_{A,B} : \mathrm{Hom}(FA, B) \to \mathrm{Hom}(A, UB)$ by $\beta_{A,B}(g) = Ug \circ \eta A$, and $\gamma_{A,B} : \mathrm{Hom}(A, UB) \to \mathrm{Hom}(FA, B)$ by $\gamma_{A,B}(f) = \epsilon B \circ Ff$.

Then $\beta_{A,B}(\gamma_{A,B}(f)) = U\epsilon B \circ UFf \circ \eta A$ by definition and the fact that U is a functor; that is $U\epsilon B \circ \eta UB \circ f$ by naturality of η; but that is f because Diagram (12.3) commutes. It follows similarly, using Diagram (12.4), that $\gamma_{A,B}(\beta_{A,B}(g)) = g$. Thus $\beta_{A,B}$ is an isomorphism with inverse $\gamma_{A,B}$. The naturality is left as an exercise (Exercise 3).

To go in the other direction, suppose we have the natural isomorphism. Then let A be an arbitrary object of $\mathcal{A}$ and $B = FA$. We then get $\mathrm{Hom}(FA, FA) \cong \mathrm{Hom}(A, UFA)$. Let $\eta A \in \mathrm{Hom}(A, UFA)$ be the arrow that corresponds under the isomorphism to the identity arrow of FA. Now for an arrow $f : A \to UB$, that is $f \in \mathrm{Hom}(A, UB)$, let

$g \in \mathrm{Hom}(FA, B)$ be the arrow that corresponds under the isomorphism. Naturality of the isomorphism implies that we have a commutative diagram

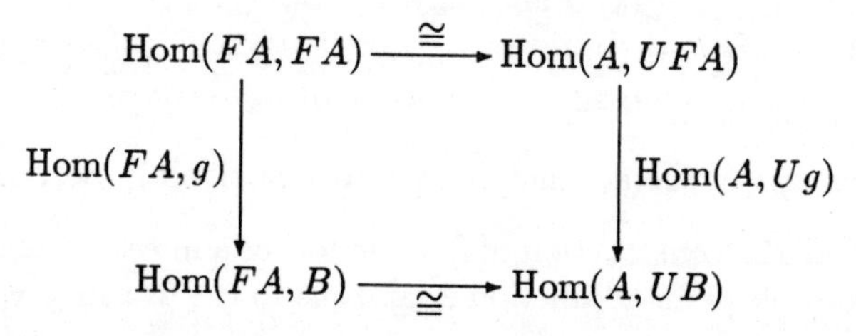

If we follow the identity arrow of FA around the clockwise direction, we get first the arrow ηA by definition, and then $\mathrm{Hom}(A, Ug)(\eta A) = Ug \circ \eta A$. In the other direction, we get $\mathrm{Hom}(FA, g)(\mathrm{id}) = \mathrm{id} \circ g = g$ and that corresponds under the isomorphism to f. Thus we conclude that $f = Ug \circ \eta A$. As for the uniqueness, if $f = Uh \circ \eta A$, then both g and h correspond to f under the isomorphism, so $g = h$. □

12.3.3 Uniqueness of adjoints If $U : \mathcal{B} \to \mathcal{A}$ is a functor, then a left adjoint to U, if one exists, is unique up to natural isomorphism. The reason is that if both F and F' are left adjoint to U, then for any object A of $\mathcal{A}$, the Hom functors $\mathrm{Hom}_{\mathcal{B}}(FA, -)$ and $\mathrm{Hom}_{\mathcal{B}}(F'A, -)$ are each naturally isomorphic to $\mathrm{Hom}_{\mathcal{A}}(A, U-)$ and hence to each other. This follows from the Yoneda embedding, Theorem 4.5.3.

12.3.4 Theorem *Let $\mathcal{A}$ and $\mathcal{B}$ be categories and $U : \mathcal{B} \to \mathcal{A}$ be a functor. Suppose for each object A of $\mathcal{A}$ there is an object FA of $\mathcal{B}$ such that* $\mathrm{Hom}_{\mathcal{B}}(FA, -)$ *is naturally equivalent to* $\mathrm{Hom}_{\mathcal{A}}(A, U-)$ *as a functor from $\mathcal{B}$ to* **Set**. *Then the definition of F on objects can be extended to arrows in such a way that F becomes a functor and is left adjoint to U.*

This theorem is called the **Pointwise Adjointness Theorem**. Its proof generalizes the argument of discussion of 12.1.3 but we omit the details. They can be found in [Barr and Wells, 1985], Section 1.9, Theorem 1, page 52. There is a detailed, general discussion of many equivalent definitions of adjunction in [Mac Lane, 1971], Chapter IV.

One application of this theorem is in showing that the definition of cartesian closed categories given in Chapter 6 is equivalent to the assumptions that the functors Δ of 12.2.3 and $- \times A$ of 12.2.4 have adjoints. In each case, the definition given in Chapter 6 can be input to the Pointwise Adjointness Theorem and the output is the adjoint described in the previous section.

This theorem has a simple formulation in terms of the universal elements of Section 4.5, as follows.

12.3.5 Proposition *A functor $U : \mathcal{B} \to \mathcal{A}$ has a left adjoint if and only if for each object A of $\mathcal{A}$, the functor* $\mathrm{Hom}(A, U-) : \mathcal{B} \to \mathbf{Set}$ *has a universal element.*

If $b : A \to UB$ is the universal element, then $FA = B$ and $b : A \to UB = UFA$ is the component at A of the natural transformation η that appears in 12.2.1.

A detailed discussion of adjoints and universal elements may be found in [Mac Lane, 1971, IV.1].

12.3.6 Theorem *Let $F : \mathcal{A} \to \mathcal{B}$ be left adjoint to $U : \mathcal{B} \to \mathcal{A}$. Then U preserves limits and F preserves colimits.*

Proof. We will not prove this in full generality. We prove, by way of example, that U preserves products. The general case is really no harder; it is just that we are loath to introduce all the notation necessary. The proof that U preserves limits can be turned around using the dual category (Exercise 4) to show that F preserves colimits.

So let $B = \prod_{i \in I} B_i$ with projections $p_i : B \to B_i$. We will show that $\{Up_i : UB \to UB_i\}_{i \in I}$ is a product cone. If for each i, we have an arrow $f_i : A \to UB_i$, we get, by adjointness, a unique arrow $g_i : FA \to B_i$ such that $f_i = Ug_i \circ \eta A$. There is, by definition of product, an arrow $g : FA \to B$ such that $p_i \circ g = g_i$. Then the arrow $f = Ug \circ \eta A$ has the property that

$$Up_i \circ f = Up_i \circ Ug \circ \eta A = U(p_i \circ g) \circ \eta A = Ug_i \circ \eta A = f_i$$

The uniqueness of f follows readily from that of g and from the isomorphism between arrows $FA \to B$ and $A \to UB$. □

The interesting question is the extent to which the converse is true. The basic fact is that the converse is false. First off the category $\mathcal{B}$ may not have enough limits for the condition to be meaningful. The really interesting case is the one in which every (small) diagram in $\mathcal{B}$ has a limit. Even in that case, there is still what can basically be described as a size problem. To go into this in more detail is beyond the scope of this book. The best result is Freyd's Adjoint Functor Theorem. See [Barr and Wells, 1985], Section 1.9, Theorem 3, p. 54 or [Mac Lane, 1971], V.6.

12.3.7 Exercises

1. Suppose that we have categories $\mathcal{A}$ and $\mathcal{B}$ and functors $T : \mathcal{B} \to \mathcal{A}$ and $L, R : \mathcal{A} \to \mathcal{B}$ such that $L \dashv T \dashv R$. Show that $LT \dashv RT$ as endofunctors on $\mathcal{A}$ and $TL \dashv TR$ as endofunctors.

2.† This is a series of exercises designed to show that if the category $\mathcal{C}$ has finite limits, then the existence of a right adjoint R_A to the functor P_A of 12.2.5 is equivalent to the existence of an exponential $[A \to -]$ in $\mathcal{C}$. This is a functor that satisfies the universal mapping property that $\mathrm{Hom}(A \times B, C) \cong \mathrm{Hom}(B, [A \to C])$ for all objects B and C of $\mathcal{C}$. (See Section 6.1.)

a. Let A be an object of $\mathcal{C}$, $P_A : \mathcal{C} \to \mathcal{C}/A$ the functor defined in 12.2.5 which takes an object C to $p_2 : C \times A \to A$, and $L_A : \mathcal{C}/A \to \mathcal{C}$ the underlying object functor which takes $f : B \to A$ to B. Show that L_A is left adjoint to P_A.

b. Show that if $P_A : \mathcal{C} \to \mathcal{C}/A$ has a *right* adjoint R_A, then the functor $A \times - : \mathcal{C} \to \mathcal{C}$ does as well.

c. Suppose that $\mathcal{C}$ is cartesian closed. Show that for each object of the form $P_A(C)$ in $\mathcal{C}/A$, the object $\Phi(P_A(C)) = [A \to C]$ in $\mathcal{C}$ has the basic adjunction property:

$$\mathrm{Hom}(B, \Phi(P_A(C))) \cong \mathrm{Hom}_A(P_A(B), P_A(C))$$

d. Show that every arrow $f : P_A(C) \to P_A(D)$ in $\mathcal{C}/A$ induces an arrow $\Phi(f) : \Phi(C) \to \Phi(D)$ such that

$$\begin{array}{ccc}
\mathrm{Hom}(B, \Phi(C)) & \xrightarrow{\cong} & \mathrm{Hom}(P_A(B), P_A(C)) \\
{\scriptstyle \mathrm{Hom}(B, \Phi(f))} \downarrow & & \downarrow {\scriptstyle \mathrm{Hom}(B, f)} \\
\mathrm{Hom}(B, \Phi(D)) & \xrightarrow[\cong]{} & \mathrm{Hom}(P_A(B), P_A(D))
\end{array}$$

commutes. (Hint: use the Yoneda Lemma.)

e. Show that if $f : C \to A$ is an arbitrary object of $\mathcal{C}/A$, then there is an equalizer diagram

$$f \xrightarrow{d} P_A(C) \underset{d^1}{\overset{d^0}{\rightrightarrows}} P_A L_A P_A(C)$$

(Hint: use elements.)

f. Show that if we let, for an object $f : C \to A$ of $\mathcal{C}/A$,

$$\Phi(f) \longrightarrow \Phi(P_A(C)) \underset{\Phi(d^1)}{\overset{\Phi(d^0)}{\rightrightarrows}} \Phi(P_A(P_A(C))$$

be an equalizer, then for any object B of $\mathcal{C}$,

$$\mathrm{Hom}(B, \Phi(f)) \cong \mathrm{Hom}_A(P_A(B), f)$$

g. Show that P_A has a right adjoint.

3. Show that $\beta_{A,B}$, as defined in the proof of Theorem 12.3.2, is a natural transformation from $\mathrm{Hom}_{\mathcal{B}}(F-,-)$ to $\mathrm{Hom}_{\mathcal{A}}(-,U-)$, both functors from $\mathcal{A}^{\mathrm{op}} \times \mathcal{B}$ to **Set**.

4. For any functor $G : \mathcal{C} \to \mathcal{D}$, let $G^{\mathrm{op}} : \mathcal{C}^{\mathrm{op}} \to \mathcal{D}^{\mathrm{op}}$ be the functor which is the same as G on objects and arrows. Show that, for any functors F and U, if F is left adjoint to U then F^{op} is right adjoint to U^{op}.

12.4 Locally cartesian closed categories

A locally cartesian closed category is a special type of cartesian closed category which has properties desirable for modeling polymorphism. Its definition, which we give here, makes intrinsic use of adjunctions.

12.4.1 The pullback functor Let $\mathcal{C}$ be a category with pullbacks. One can show, using Theorem 12.3.4, that any arrow $f : A \to B$ of $\mathcal{C}$ gives a functor $f^* : \mathcal{C}/B \to \mathcal{C}/A$ between slice categories which we now describe.

If $x : X \to B$ is an object of $\mathcal{C}/B$, $f^*(x) = p_1 : P \to A$, where

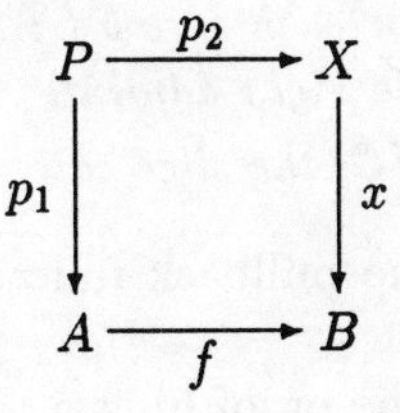

$$\begin{array}{ccc} P & \xrightarrow{p_2} & X \\ {\scriptstyle p_1}\downarrow & & \downarrow{\scriptstyle x} \\ A & \xrightarrow[f]{} & B \end{array} \tag{12.5}$$

is a pullback diagram.

To see what f^* does on arrows, suppose $f^*(x : X \to B) = p_1 : P \to A$ as above, and $f^*(x' : X' \to B) = p_1' : P' \to A$, the latter given by the pullback

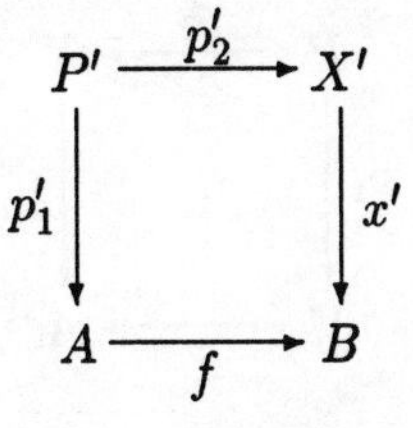

$$\begin{array}{ccc} P' & \xrightarrow{p_2'} & X' \\ {\scriptstyle p_1'}\downarrow & & \downarrow{\scriptstyle x'} \\ A & \xrightarrow[f]{} & B \end{array} \tag{12.6}$$

Suppose $t : x \to x'$ is an arrow of $\mathcal{C}/B$. Then

$$x' \circ t \circ p_2 = x \circ p_2 = f \circ p_1$$

because t is an arrow in $\mathcal{C}/B$ and Diagram (12.5) commutes. Thus P with $t \circ p_2$ and p_1 is a commutative cone over the pullback diagram (12.6), so induces an arrow $s : P \to P'$ such that $p'_1 \circ s = p_1$ and $p'_2 \circ s = t \circ p_2$. The first equation says that s is an arrow from p'_1 to p_1 in $\mathcal{C}/A$, and so $f^*(t) = s$. This shows what f^* must be on arrows. Then Theorem 12.3.4 puts it all together into a functor.

This functor has a left adjoint. Let $\Sigma_f : \mathcal{C}/A \to \mathcal{C}/B$ be the functor for which $\Sigma_f(u : X \to A) = f \circ u$ and, for an arrow $t : u \to u'$ of $\mathcal{C}/A$, $\Sigma_f(t)$ is the arrow $t : f \circ u \to f \circ u'$ of $\mathcal{C}/B$.

12.4.2 Proposition *Let $\mathcal{C}$ be a category with pullbacks and $f : A \to B$ an arrow of $\mathcal{C}$. Then Σ_f as defined above is left adjoint to f^*.*

The proof is omitted. The construction in 12.2.5 is a special case of this construction, at least in the case where $\mathcal{C}$ has a terminal object: To see this, let f be the unique arrow from A to 1 and note that the pullback becomes a product and $\mathcal{C}/1$ is isomorphic to $\mathcal{C}$.

The pullback functor need not have a *right* adjoint. The following theorem characterizes those categories in which it always has a right adjoint.

12.4.3 Theorem *The following are equivalent for any category $\mathcal{C}$.*

(a) *$\mathcal{C}$ has pullbacks and for each arrow $f : A \to B$, the pullback functor $f^* : \mathcal{C}/B \to \mathcal{C}/A$ has a right adjoint.*

(b) *For every object A of $\mathcal{C}$, the slice category $\mathcal{C}/A$ is cartesian closed.*

The right adjoint to the pullback functor f^* is denoted Π_f ($\forall_f$ in some texts).

We sketch the proof. The proof hinges on the following proposition, whose proof we omit.

12.4.4 Proposition *For any arrows $u : X \to A$ and $v : Y \to A$ in any category, the arrow from P to A given by a pullback diagram*

$$\begin{array}{ccc} P & \xrightarrow{p_2} & X \\ {\scriptstyle p_1}\downarrow & & \downarrow{\scriptstyle u} \\ Y & \xrightarrow[v]{} & A \end{array} \qquad (12.7)$$

is the product of u and v in the slice category $\mathcal{C}/A$.

Now we give the proof of Theorem 12.4.3. Suppose the conditions in (a) hold. Let us consider arrows $u : X \to A$, $v : Y \to A$ and $w : Z \to A$ of $\mathcal{C}/A$. It is clear from Proposition 12.4.4 that, in $\mathcal{C}/A$,

$$u \times v = u \circ p_2 = \Sigma_u(u^*(v))$$

By hom set adjointness (Theorem 12.3.2), we then have

$$\begin{aligned}\mathrm{Hom}_{\mathcal{C}/A}(u \times v, w) &= \mathrm{Hom}_{\mathcal{C}/A}(\Sigma_u(u^*(v), w)) \\ &\cong \mathrm{Hom}_{\mathcal{C}/X}(u^*(v), u^*(w)) \\ &\cong \mathrm{Hom}_{\mathcal{C}/A}(v, \Pi_u(u^*(w)))\end{aligned}$$

so that defining $[u \to w]$ to be $\Pi_u(u^*(w))$ gives the cartesian closed structure on $\mathcal{C}/A$.

We leave the converse to Exercise 1.

12.4.5 Definition A category $\mathcal{C}$ which satisfies (a) (or equivalently (b)) of Theorem 12.4.3 is a **locally cartesian closed category**.

If a category $\mathcal{C}$ has a terminal object 1, $\mathcal{C}$ is isomorphic as a category to $\mathcal{C}/1$. We then have the following proposition.

12.4.6 Proposition *A locally cartesian category which has a terminal object is cartesian closed.*

Applications are discussed in [Seely, 1984, 1987]. A 'relatively cartesian closed category' is a category with a special class of arrows having some of the properties discussed in this section. See [Hyland and Pitts, 1989] for the connection between this and polymorphism. See also [Lamarche, 1989].

12.4.7 Exercise

1. This is an exercise designed to prove the converse part of Theorem 12.4.3. For this exercise, $\mathcal{C}$ denotes a category with the property that every slice is a cartesian closed category.

a. Show that every slice of $\mathcal{C}$ has the property that every slice is a cartesian closed category. (Hint: show that a slice of a slice is a slice.)

b. Show that every slice of $\mathcal{C}$ has finite limits.

c. Show that for any $f : A \to B$ in $\mathcal{C}$, the induced $f^* : \mathcal{C}/B \to \mathcal{C}/A$ has a right adjoint.

13
Algebras for endofunctors

In this chapter, we describe some constructions based on an **endofunctor** of a category $\mathcal{C}$, which is a functor from $\mathcal{C}$ to $\mathcal{C}$.

We begin in Section 13.1 to study the concept of the fixed point of an endofunctor. A natural structure to be a fixed point is what is called an algebra for the endofunctor, which is defined there. These algebras are suitable for developing a concept of list object in a category (Section 13.2) and a categorical version of state transition machines (13.2.11).

A triple is a structure which abstracts the idea of adjunction; part of the structure is an endofunctor. Triples have turned out to be an important technical tool in category theory. Section 13.3 defines triples and gives some of their basic properties. In Section 13.4 we develop the idea of an algebra for a triple; it is an algebra for the endofunctor part of the triple with certain properties. The last section describes a technique of Smyth and Plotkin to construct Scott domains, which provide models of computations in cartesian closed categories.

13.1 Fixed points for a functor

13.1.1 What are fixed points for a functor? Let $R : \mathcal{A} \to \mathcal{A}$ be a functor. In this section, we will analyze the notions of a fixed point for a functor and of a least fixed point.

The question of fixed points arises only with an endofunctor. We must understand what sort of structure a fixed point is and what it means to be a least fixed point. For a function on a set, say, we know exactly what a fixed point is. If the set is a total order, we know exactly what a least fixed point is. The relevant structure for fixed points of functors turns out to be the concept of algebra for the functor.

13.1.2 To see why a fixed point for a functor does not have the obvious definition, consider a possible definition of the natural numbers as the least fixed point of the functor R on **Set** defined by $R(S) = 1 + S$ for a set S, and for a function $f : S \to T$, $R(f) = \mathrm{id}_1 + f : 1 + S \to 1 + T$. Here, 1 is a terminal object and '+' denotes disjoint union.

Even supposing that we have chosen a functorial definition for $+$, it is not clear that any set is fixed by this functor. If one is, it is not obvious that it will be countable. Moreover, this fixed point will depend on the surely irrelevant way in which the sums are defined. At first it might seem that what is really wanted is a set $\mathbf{N}$ which is *isomorphic to* $1+\mathbf{N}$. This is consonant with the categorical perception that any use you could make of an object can equally well be made of any isomorphic copy. Thus a fixed point for a functor need only be fixed *up to isomorphism*, thereby avoiding any dependence on arbitrary choices, for example of which particular categorical sum is used.

This leads to a second problem. Although any countably infinite set is now a fixed point, none is *least* in any obvious way. For example, any countably infinite set has a proper countably infinite subset, which is also fixed in the same way. To understand what is going on here, we approach the question from another angle.

13.1.3 Algebras for an endofunctor As above, we let $\mathcal{A}$ be a category and $R : \mathcal{A} \to \mathcal{A}$ be an endofunctor. An **R-algebra** is a pair (A, a) where $a : RA \to A$ is an arrow of $\mathcal{A}$. A homomorphism between R-algebras (A, a) and (B, b) is an arrow $f : A \to B$ of $\mathcal{A}$ such that

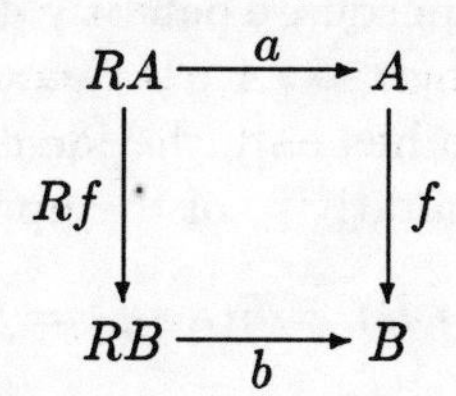

commutes. This construction gives a category $(R : \mathcal{A})$ of R-algebras.

13.1.4 Definition An object (A, a) of $(R : \mathcal{A})$ for which the arrow a is an isomorphism is a **fixed point** for R.

Based on the perception that a category is a generalized poset, the following definition is reasonable.

13.1.5 Definition A **least fixed point** of a functor $R : \mathcal{A} \to \mathcal{A}$ is an initial object of $(R : \mathcal{A})$.

13.1.6 Example In the case of the CPO $\mathcal{P}$ defined in 2.4.8, the function ϕ induces an endofunctor on the category $C(\mathcal{P})$ corresponding to $\mathcal{P}$. One can prove by induction that the only algebra for the functor ϕ is the factorial function itself, which is both A and $\phi(A)$. The algebra map a is the identity map (the factorial function is less than or equal to itself in the CPO $\mathcal{P}$). Thus this algebra is clearly the initial (because it is the only) object of the category of algebras.

If Definition 13.1.5 is to work, we have to show that such an initial object is indeed a fixed point.

13.1.7 Theorem [Lambek, 1970] *Let $R : \mathcal{A} \to \mathcal{A}$ be a functor from a category to itself. If (A, a) is initial in the category $(R : \mathcal{A})$, then a is an isomorphism.*

Proof. Suppose (A, a) is initial. (RA, Ra) is an object of the category $(R : \mathcal{A})$ and hence there is a unique arrow $f : (A, a) \to (RA, Ra)$. This means that the top square of the diagram

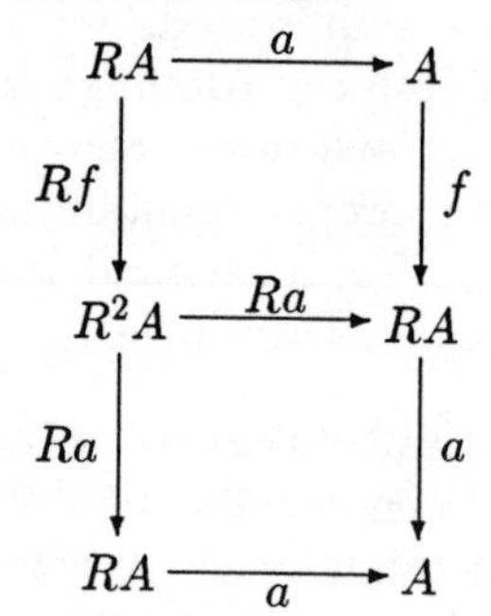

commutes, while the bottom square patently does. Thus the whole rectangle does and so $a \circ f : (A, a) \to (A, a)$ is an arrow between R-algebras. But (A, a) is initial and so has only the identity endomorphism. Thus $a \circ f = \text{id}$. Then the commutativity of the upper square gives us that

$$f \circ a = Ra \circ Rf = R(a \circ f) = R(\text{id}) = \text{id}$$

which means that $f = a^{-1}$ and that a is an isomorphism. □

13.1.8 Example Now let us look again at the functor $R : \mathbf{Set} \to \mathbf{Set}$ which takes S to $1 + S$. The natural numbers $\mathbf{N}$ form an algebra for R, whose R-algebra structure is the function $(0; s) : 1 + \mathbf{N} \to \mathbf{N}$, where 0 is the function picking out 0 and s is the successor function.

This is in fact an initial algebra for R. For suppose $f : 1+S \to S$ is an R-algebra. The required unique R-algebra homomorphism $h : \mathbf{N} \to S$ is the function defined inductively by $h(0) = f(*)$ ($*$ is the unique element of the singleton 1) and $h(n+1) = f(h(n))$ (Exercise 1).

Although $\mathbf{N}$ has many subsets which are fixed up to isomorphism for R, it has no proper subset that is fixed under the given isomorphism $(0; s)$; that is one sense in which it is the 'least' fixed point for R. Now a proper subset fixed under the isomorphism is the same thing as a proper subobject in the category $(R : \mathcal{A})$, and no initial object of any category has a proper subobject; thus least fixed points of functors all have the property that they have no proper subobjects.

However, initiality is strictly stronger than the property of not having proper subobjects; so being the least fixed point of a functor in the sense we have defined is a strong requirement.

Another property of initial objects is that an initial object is determined uniquely up to a unique isomorphism; thus least fixed points in our sense have the desirable property of being uniquely determined in the strongest possible sense of uniqueness consistent with the philosophy of category theory: any two least fixed points are isomorphic in a unique way.

13.1.9 Fixed points of finitary functors Suppose we have a functor R on the category of sets. We say that R is **finitary** if

FF–1 For each set S and each element $x \in RS$, there is a finite subset $S_0 \subseteq S$ and an element $x_0 \in RS_0$ such that $x = Ri_0(x_0)$ where $i_0 : S_0 \to S$ is the inclusion.

FF–2 If also S_1 is a finite subset of S with inclusion i_1 and $x_1 \in S_1$ with $x = Ri_1(x_1)$, then there is a finite subset S_2 containing both S_0 and S_1 with inclusions $j_0 : S_0 \to S_2$ and $j_1 : S_1 \to S_2$ such that $Rj_0(x_0) = Rj_1(x_1)$.

The real meaning of finitary is that everything is determined by what happens on finite subsets. From the point of view of computer science, this seems quite a reasonable hypothesis.

A finitary functor always has a fixed point (Theorem 13.1.10 below). In fact, the underlying functor from (R : **Set**) to **Set** has an adjoint; see [Barr, 1971] for details and generalizations.

We denote by η the unique arrow $\emptyset \to R\emptyset$. We have a sequence

$$0 \xrightarrow{\eta} R(\emptyset) \xrightarrow{R(\eta)} R^2(\emptyset) \xrightarrow{R^2(\eta)} \cdots \xrightarrow{R^{n-1}(\eta)} R^n(\emptyset) \xrightarrow{R^n(\eta)} \cdots \tag{13.1}$$

The colimit of this sequence (which would be the union if the arrows were inclusions) is a set we will call Z.

13.1.10 Theorem *If R is a finitary endofunctor on* **Set** *and Z is the colimit of (13.1), then there is an R-algebra structure $z : RZ \to Z$ which is an initial R-algebra.*

Let $u_n : R^n(\emptyset) \to Z$ be the transition arrow to the colimit. The definition of colimit is that an arrow out of Z can be described by its composite with all the u_n. We let z be the arrow thus defined. The proof that this works is fairly technical and is omitted.

13.1.11 Example This is an example which shows how this construction works in the case of a very simple functor. We let R denote the functor on **Set** given by $RX = 1 + X$. Then the sequence (13.1) comes down to

$$0 \to 1 \to 2 \to \cdots \to n \to \cdots$$

with the arrow from n to $n+1$ given by the injection into the sum. This gives rise to the set $\mathbf{N}$ of natural numbers with the operation successor.

13.1.12 Example Here is an example of a nonfinitary functor. Let $P : \mathbf{Set} \to \mathbf{Set}$ take a set S to the set of subsets of S. If $f : S \to T$ is a function, then Pf takes the subset $S_0 \subseteq S$ to $f(S_0) \subseteq T$. P is not finitary because when S is infinite, not every subset of S (in fact, no infinite subset of S) is a subset of a finite subset of S.

13.1.13 The reader may wonder why it was necessary to introduce the category of all R-algebras when we were interested only in the fixed points. There are two answers to that question.

First off, the underlying functor $U : (R : \mathcal{A}) \to \mathcal{A}$ preserves all limits and therefore may be expected to have a left adjoint F. If, in addition, $\mathcal{A}$ has an initial object 0, then $F0$ is initial in $(R : \mathcal{A})$. The underlying functor from the fixed point category to $\mathcal{A}$ does not generally preserve limits and therefore cannot usually be expected to have an adjoint. Had we begun by constructing the category of fixed points, we would still have wanted to construct $(R : \mathcal{A})$ in order to make it seem reasonable for a least fixed point (or any other) to exist.

The second reason is that the construction of $(R : \mathcal{A})$ is part of the construction of the category of algebras for a triple, which is carried out in 13.4.1.

13.1.14 Exercises

1. Show that the function h defined in 13.1.8 is the unique R-algebra homomorphism from $(0; s) : 1 + \mathbf{N} \to \mathbf{N}$ to $f : 1 + S \to S$.

2. Show that an initial object of a category has no proper subobject.

3. Give an example of a category containing an object with no proper subobject which is nevertheless not an initial object.

4. Let $(S, \leq)$ be a total order and $f : S \to S$ a monotone function. Let $C(S, \leq)$ be the category determined by $(S, \leq)$ as in 2.3.1. Show that f determines an endofunctor, the objects of $(f : S)$ are the $x \in X$ for which $f(x) \leq x$ and the least fixed point is the same as the one defined here.

13.2 Recursive categories

In this section, we discuss two applications to computer science which have been proposed which involve algebras for functors.

13.2.1 In 2.2.1, we described how to represent a functional programming language as a category. The major missing piece from that description was how to produce potentially infinite programs. This is done in traditional languages using WHILE loops and their relatives, but can also be done in a flexible way using recursion, which is available in cartesian closed categories with a natural numbers object (see Section 5.5.) However, the construction of function space objects in cartesian closed categories can lead to noncomputable constructions.

J. R. B. Cockett has proposed axioms on a category which directly allow a limited form of recursion. A consequence of the definition is then a much more extended form of recursion (Proposition 13.2.9 below) while maintaining computability (Theorem 13.2.10 below). This discussion is based on [Cockett, 1987, 1989].

13.2.2 Let $\mathcal{C}$ be a category with finite limits. Suppose that for every object A there is a functor $A \times -$ which takes an object X to $A \times X$ and an arrow $f : X \to Y$ to $\mathrm{id}_A \times f : A \times X \to A \times Y$. An algebra $x : A \times X \to X$ for this functor is an **A-action**. For each object A there is a category $\mathrm{act}(A)$ which is the category of algebras for the functor $A \times -$. There is an underlying functor $U_A : \mathrm{act}(A) \to \mathcal{C}$ which takes $x : A \times X \to X$ to X and an arrow $f : (x : A \times X \to X) \to (y : A \times Y \to Y)$ to f.

13.2.3 Definition A category $\mathcal{C}$ is **recursive** if for every object A the underlying functor U_A has a left adjoint $F_A : \mathcal{C} \to \mathrm{act}(A)$.

For an object B, denote $U_A(F_A(B))$ by $\mathrm{rec}(A, B)$; for $f : B \to C$, we get an arrow

$$\mathrm{rec}(A, f) = U_A(F_A(f)) : \mathrm{rec}(A, B) \to \mathrm{rec}(A, C)$$

The algebra $F_A(B)$ is an algebra structure

$$r(A, B) : A \times \mathrm{rec}(A, B) \to \mathrm{rec}(A, B)$$

The unit of the adjunction has component $r_0(A, B) : B \to \mathrm{rec}(A, B)$ at an object B. In particular, $\mathrm{rec}(1, 1)$ is a natural numbers object (Exercise 2).

The object $\mathrm{rec}(A, B)$ is characterized by the following universal mapping property. If $t_0 : B \to X$ and $t : A \times X \to X$ are arbitrary arrows, then there is a unique $f : \mathrm{rec}(A, B) \to X$ such that the diagram

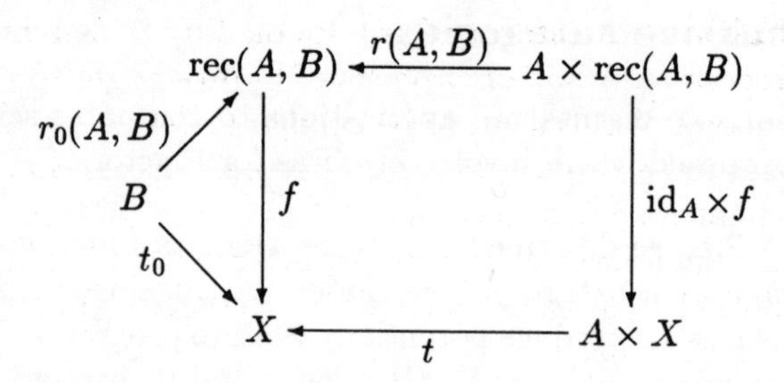

commutes.

13.2.4 **Set** is recursive. You can check that in **Set**, $\mathrm{rec}(A,B)$ is (up to isomorphism) the set of pairs of the form (l,b) where l is a list of elements of A (including the empty list) and $b \in B$, and $r(A,B)$ is the function which takes $(a,(l,b))$ to $(\mathrm{cons}(a,l),b)$, where cons is the function which adjoins a to the list l at the front. The unit $r_0(A,B) : B \to \mathrm{rec}(A,B)$ takes b to $(\langle\rangle, b)$.

In particular, in **Set**, $A^* = \mathrm{rec}(A,1)$ is the set of lists of elements of A. It has the right properties to be regarded as lists of A in any category; for example, Cockett [1987] shows how to define the head and tail of a list in recursive categories. Because of this, it is natural to denote $\mathrm{rec}(A,1)$ by A^* in a recursive category.

13.2.5 Now for objects A and B of a recursive category $\mathcal{C}$, the projection $p_2 : A \times B \to B$ is an algebra for $A \times -$. We have the identity arrow of B and the universal mapping property then produces a unique arrow $a : \mathrm{rec}(A,B) \to B$ with the property that

$$\begin{array}{ccc} & \mathrm{rec}(A,B) & \xleftarrow{\;r(A,B)\;} & A \times \mathrm{rec}(A,B) \\ r_0(A,B) \nearrow & & & \\ B & \Big\downarrow a & & \Big\downarrow \mathrm{id}_A \times a \\ \mathrm{id}_B \searrow & & & \\ & B & \xleftarrow[\;p_2\;]{} & A \times B \end{array}$$

commutes. There is then an arrow

$$c(A,B) = \langle \mathrm{rec}(A,\langle\rangle), a\rangle : \mathrm{rec}(A,B) \to \mathrm{rec}(A,1) \times B$$

13.2.6 Definition Let $\mathcal{C}$ be a recursive category. If for all A and B, $c(A,B)$ is an isomorphism, then $\mathcal{C}$ has **local recursion.** A **locally recursive category** is a recursive category with local recursion for which every slice category $\mathcal{C}/C$ is recursive.

This definition implies that $\mathrm{rec}(A,B) \cong \mathrm{rec}(A,1) \times B = A^* \times B$. The discussion in 13.2.4 shows that this is true in **Set**.

13.2.7 Definition A category $\mathcal{C}$ is a **locos** if it is coherent (see 8.6.8) and locally recursive. A homomorphism of locoses ought to preserve the recursive structure; it turns out to be enough to define a **homomorphism of locoses** to be a coherent functor between locoses which preserves the natural numbers object $\text{rec}(1,1)$.

The axioms on a locos allow many other constructions. Specifically, there is a whole class of functors for which the underlying functor from the algebra category has a left adjoint.

13.2.8 Definition A **polynomial functor** on a category $\mathcal{C}$ with finite sums and products is a functor which is a composite of constant functors, the product functor from $\mathcal{C} \times \mathcal{C} \to \mathcal{C}$ which takes (A, B) to $A \times B$, and the sum functor which takes (A, B) to $A + B$.

13.2.9 Proposition *Let $\mathcal{C}$ be a locos and P a polynomial functor. Then the underlying functor $U : (P : \mathcal{C}) \to \mathcal{C}$ has a left adjoint.*

The proof is in [Cockett, 1987].

This allows the construction of free objects generated by an object. Thus if D is an object and T is the functor for which

$$T(X) = D + X \times D \times X$$

the free object on the terminal object 1 has a structure map

$$[D + F(1) \times D \times F(1)] \to F(1)$$

It can be interpreted as the object of binary trees with data at the root, and the structure map constructs such trees from either a datum or a datum and two given trees. Thus a locos allows a type of initial algebra semantics.

Another property of locoses is that if you start with computable things the constructions of a locos give you computable things in a precise sense. This requires a definition. In any category with finite products, an object D is **decidable** if the diagonal $\Delta : D \to D \times D$ (which is a subobject of $D \times D$) has a complement in $D \times D$ (see 8.6.2). Thus the law of the excluded middle applies internally to equality of elements of a decidable object: two elements are either equal (factor through the diagonal) or not equal (factor through the complement of the diagonal).

13.2.10 Theorem *If $\mathcal{S}$ is a collection of decidable objects in a locos, and no sublocos of the locos contains $\mathcal{S}$, then every object of the locos is decidable.*

For the proof, see [Cockett, 1989]. Locoses may be described as models of an FL sketch, so that there is an initial locos. The theorem above implies that in the initial locos every object is decidable.

13.2.11 Categorical dynamics An **input process** in a category $\mathcal{C}$ is an endofunctor R for which the underlying functor $(R : \mathcal{C}) \to \mathcal{C}$ has a left adjoint. An object of the category $(R : \mathcal{C})$ is called an R**-dynamic**, the category itself is called the category of R**-dynamics** and an arrow in the category is called a **dynamorphism.** A **machine in** $\mathcal{C}$ is a 7-tuple $M = (R, C, c, I, \tau, Y, \beta)$ in which R is an input process, (C, c) is an R-dynamic, the object I of $\mathcal{C}$ is called the initial state object (not to be confused with an initial object of the category) and $\tau : I \to C$ the initial state arrow, and Y is called the output object and $\beta : C \to Y$ is the output arrow.

13.2.12 Example Let A be an object of any recursive category. Then by definition the functor R defined by $R(C) = A \times C$ and $R(f) = \mathrm{id}_A \times f$ is an input process. Thus finite state machines can be defined in an arbitrary recursive category. Many definitions connected with finite state machines can be given in a recursive category. For example, a machine $M = (R, C, c, I, \tau, Y, \beta)$ is reachable if τ factors through no proper subobject of the algebra (C, c).

In particular, let the recursive category be the category of sets and let A be a finite set thought of as an alphabet. An algebra for R is a function $\delta : A \times C \to C$ for some set C. By defining $\delta^*(\langle\rangle, q) = q$ and $\delta^*(aw, q) = \delta(A, \delta^*(w, q))$, one has an action of the free monoid A^* on C. Setting I to be a one-element set, one obtains the usual definition of state transition machine used in the automata theory literature. Reachability here has its usual meaning, that one can reach any state from the start state by a sequence of operations (if not, the states that *can* be so reached form a proper subalgebra containing the image of τ).

The concept of R-dynamic is much more general than this. For example, one can have several 'start' states by choosing I to be a bigger set. More drastically, one could use a different functor R, for example $R(C) = C \times C$.

This discussion of categorical dynamics is from [Arbib and Manes, 1975], which should be consulted for details. See also [Arbib and Manes, 1974].

13.2.13 Exercises

1. Prove the claims in 13.2.4.

2. Prove that in a recursive category, rec(1, 1) is a natural numbers object. (Hint: use the unit of the adjunction for 0.)

13.3 Triples

We now describe a structure based on an endofunctor which has turned out to be an important technical tool in studying toposes and related topics. It is an abstraction of the concept of adjoint and in a sense an abstraction of universal algebra (see the remarks in fine print at the end of 13.4.2 below).

13.3.1 Definition A **triple** $\mathbf{T} = (T, \eta, \mu)$ on a category $\mathcal{A}$ consists of a functor $T : \mathcal{A} \to \mathcal{A}$, together with two natural transformations $\eta : \mathrm{id} \to T$ and $\mu : T^2 \to T$ for which the following diagrams commute.

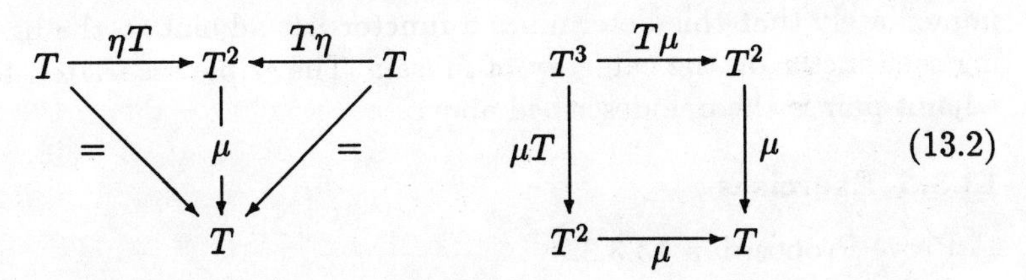

(13.2)

Here, ηT and $T\eta$ are defined as in 4.4.2 and 4.4.3.

The transformation η is the **unit** of the triple and μ is the **multiplication**. The left diagram constitutes the (left and right) **unitary identities** and the right one the **associative identity**. The reason for these names comes from the analogy between triples and monoids. This will be made clear in 13.3.4.

Another widely used name for triple is 'monad'. However, they have nothing to do with the monads of Robinson's theory of infinitesimals.

13.3.2 The triple arising from an adjoint pair An adjoint pair gives rise to a triple on the domain of the left adjoint.

13.3.3 Proposition *Let $U : \mathcal{B} \to \mathcal{A}$ and $F : \mathcal{A} \to \mathcal{B}$ be functors such that $F \dashv U$ with $\eta : \mathrm{id} \to UF$ and $\epsilon : FU \to \mathrm{id}$ the unit and counit. Then $(UF, \eta, U\epsilon F)$ is a triple on $\mathcal{A}$.*

We leave this proof as an exercise. Note that $U\epsilon F : UFUF \to UF$, as required for the multiplication of a triple with functor UF.

Conversely, every triple arises in that way out of some (generally many) adjoint pair. See Section 13.4 for two ways of constructing such adjoints.

13.3.4 Representation triples Let M be a monoid. The representation triple $\mathbf{T} = (T, \eta, \mu)$ on the category of sets is given by letting $T(S) = M \times S$ for a set S. $\eta S : S \to T(S) = M \times S$ takes an element $s \in S$ to the pair $(1, s)$. We define μS by $(\mu S)(m_1, m_2, s) = (m_1 m_2, s)$ for $s \in S$, $m_1, m_2 \in M$. Thus the unit and multiplication of the triple arise directly from that of the monoid. The unitary and associativity equations can easily be shown to follow from the corresponding equations for the monoid.

One way of getting this triple from an adjoint pair is by using M-sets (see 3.2.1). If S and T are M-sets, then a function $f : S \to T$ is said to be **M-equivariant** if $f(ms) = mf(s)$ for $m \in M$, $s \in S$. For a fixed monoid M, the M-sets and the M-equivariant functions form a category, called the category of M-sets.

The **free M-set** generated by the set S is the set $M \times S$ with action given by $m'(m, s) = (m'm, s)$. Using Theorem 12.3.4, one can show immediately that this determines a functor left adjoint to the underlying set functor on the category of M-sets. The triple associated to this adjoint pair is the one described above.

13.3.5 Exercises

1. Prove Proposition 13.3.3.

2. Let $T : \mathbf{Set} \to \mathbf{Set}$ be the functor which takes a set A to the Kleene closure A^* and a function $f : A \to B$ to the function $f^* : A^* \to B^*$ defined in Section 2.5.6. Let $\eta A : A \to A^*$ take an element a to the one-element string (a), and let $\mu A : A^{**} \to A^*$ take a string $(s_1, s_2, \ldots, s_k)$ of strings to the concatenated string $s_1 s_2 \cdots s_n$ in A^* obtained in effect by erasing inner brackets: thus $((a, b), (c, d, e), (), (a, a))$ goes to

$$(a, b)(c, d, e)()(a, a) = (a, b, c, d, e, a, a)$$

In particular, $\mu A((a, b)) = (a, b)$. Show that $\eta : \mathrm{id} \to T$ and $\mu : T \circ T \to T$ are natural transformations, and that (T, η, μ) is a triple.

13.4 Factorizations of a triple

13.4.1 Eilenberg–Moore algebras Let $\mathbf{T} = (T, \eta, \mu)$ be a triple on $\mathcal{A}$. A T-algebra (A, a) is called a $\mathbf{T}$-algebra if two diagrams commute:

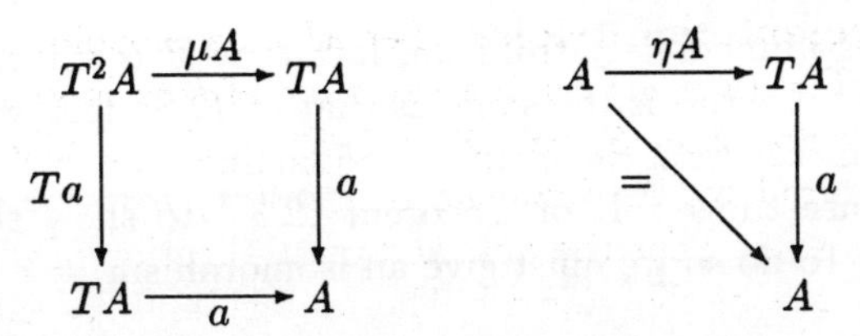

An arrow (homomorphism) between $\mathbf{T}$-algebras is the same as an arrow between the corresponding T-algebras. With these definitions, the $\mathbf{T}$-algebras form a category traditionally denoted $\mathcal{A}^{\mathbf{T}}$ and called the category of $\mathbf{T}$-algebras.

There is an obvious underlying functor $U : \mathcal{A}^{\mathbf{T}} \to \mathcal{A}$ with $U(A, a) = A$ and $Uf = f$. This latter makes sense because an arrow of $\mathcal{A}^{\mathbf{T}}$ *is* an arrow of $\mathcal{A}$ with special properties. There is also a functor $F : \mathcal{A} \to \mathcal{A}^{\mathbf{T}}$ given by $FA = (TA, \mu A)$ and $Ff = Tf$. Some details have to be checked; these are included in the following.

13.4.2 Proposition *The function F above is a functor left adjoint to U. The triple associated to the adjoint pair $F \dashv U$ is precisely $\mathbf{T}$.*

Proof. The first thing to be checked is that $(TA, \mu A)$ is actually a $\mathbf{T}$-algebra. The relevant diagrams that have to be checked to be commutative are

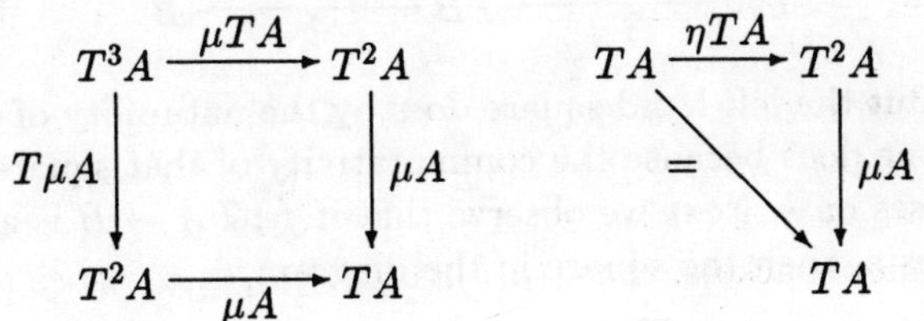

which are two of the three commutative diagrams required for a triple. Next we have to show that for $f : A \to B$, $Ff : FA \to FB$ is an homomorphism in the category of $\mathbf{T}$-algebras. To do this we must show that the diagram

$$\begin{array}{ccc} T^2A & \xrightarrow{T^2 f} & T^2B \\ {\scriptstyle \mu A}\downarrow & & \downarrow{\scriptstyle \mu B} \\ TA & \xrightarrow[Tf]{} & TB \end{array}$$

commutes. But the commutation of this diagram for all f is exactly what is meant by the statement that μ is a natural transformation. The fact that F preserves composition and takes the identity arrows to identity arrows follows immediately from the fact that T is a functor and the value of F on an arrow is the same as that of T. So far, we have shown that F is a functor with the claimed codomain.

Next we use the result of Theorem 12.3.4 to show that F is left adjoint to U. To do so we must give an isomorphism

$$\mathrm{Hom}(FA, (B, b)) \to \mathrm{Hom}(A, B)$$

which is natural in B. The function associates to each arrow

$$f : (TA, \mu A) \to (B, b)$$

the arrow $f \circ \eta A : A \to B$. The naturality of this function is an exercise. It must be shown to be an isomorphism. To do this, we define a function in the other direction that sends $g : A \to B$ to $b \circ Tg : TA \to B$. First we claim that that arrow is an arrow of $\mathbf{T}$-algebras. To see that, we must show that the diagram

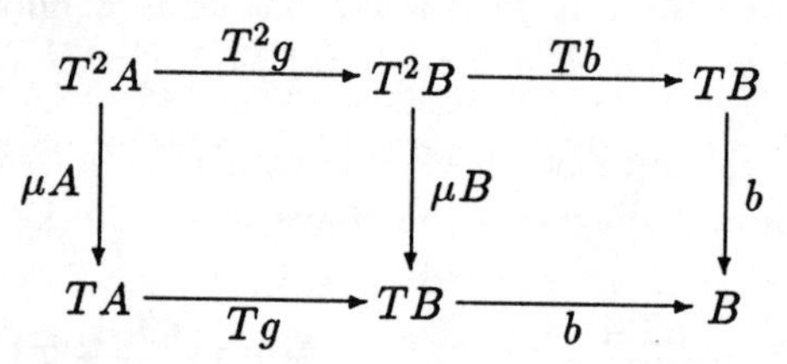

commutes. But the left hand square does by the naturality of μ and the right hand one does because the commutativity of that square is one of the hypotheses on b. Next we observe that if $f : TA \to B$ is an algebra homomorphism, then the square in the diagram

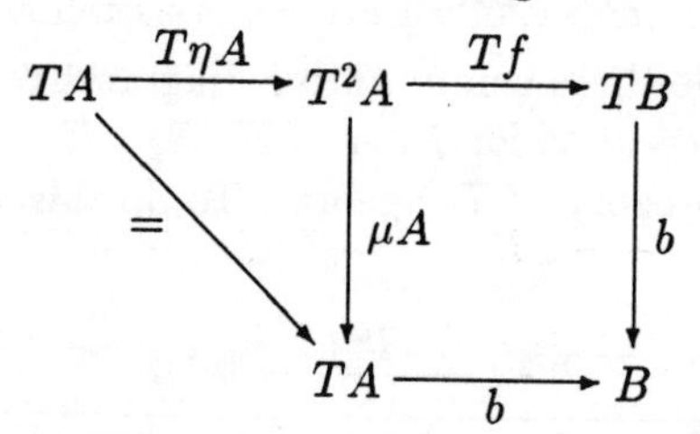

commutes and the triangle does by one of the identities that define a triple. Thus the whole square commutes which shows that one of the composites is the identity. As for the other, that follows from the commutativity of the diagram

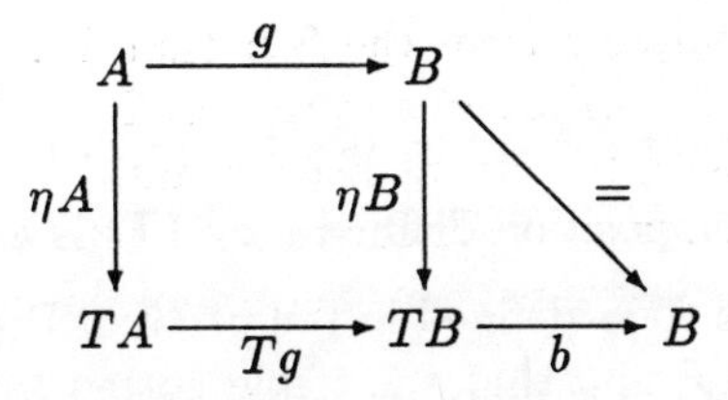

whose square commutes by naturality of η and triangle by one of the hypotheses on b.

It is evident that $T = UF$. For the rest of the proof that $\mathbf{T}$ is the triple associated to the adjoint pair, we refer to [Eilenberg and Moore, 1965]. It is not especially hard, but it does require analyzing the natural transformations associated to an adjunction constructed by Theorem 12.3.4. □

By a theorem of Linton's, every equationally defined category of one-sorted algebraic structures is in fact equivalent the category of Eilenberg–Moore algebras for some triple in **Set** ([Barr and Wells, 1985], Theorem 5 of Section 4.3). (See Exercise 5.) In fact, the converse is true if infinitary operations are allowed (but then a hypothesis has to be added on the direct part of the theorem that there is only a set of operations of any given arity).

13.4.3 The Kleisli category for a triple Let $\mathbf{T} = (T, \eta, \mu)$ be a triple on $\mathcal{C}$. There is another construction which exhibits the triple as coming from an adjoint which is due to Kleisli [1965] and which is sometimes useful.

We define a category $\mathcal{K} = \mathcal{K}(\mathbf{T})$ which has the same objects as $\mathcal{C}$. If A and B are objects of $\mathcal{C}$, then an arrow in $\mathcal{K}$ from A to B is an arrow $A \to TB$ in $\mathcal{C}$. The composition of arrows is as follows. If $f : A \to TB$ and $g : B \to TC$ are arrows in $\mathcal{C}$, we let their composite in $\mathcal{K}$ be the following composite in $\mathcal{C}$:

$$A \xrightarrow{f} TB \xrightarrow{Tg} T^2C \xrightarrow{\mu C} TC$$

The identity of the object A is the arrow $\eta A : A \to TA$. It can be shown that this defines a category. Moreover, there are functors $U : \mathcal{K}(\mathbf{T}) \to \mathcal{C}$ and $F : \mathcal{C} \to \mathcal{K}(\mathbf{T})$ defined by $UA = TA$ and $Uf = \mu A \circ Tf$, where A is the domain of f, and $FA = A$ and for $g : A \to B$, $Fg = Tg \circ \eta A$. Then F is left adjoint to U and $T = U \circ F$. The proof is left as an exercise.

The Kleisli category is equivalent to the full subcategory of free $\mathbf{T}$-algebras of the triple. Its definition makes it clear that the arrows are substitutions. Indeed, in the discussion of unification in Section 8.7, we were essentially working with the Kleisli category of the triple associated with an FP sketch.

13.4.4 Exercises

1. Show that $\eta A : A \to TA$ is the identity on A in $\mathcal{K}(\mathbf{T})$.

2. Show that the composition defined for $\mathcal{K}(\mathbf{T})$ is associative.

3. Show that U and F as defined in the proof of Proposition 13.4.2 are functors, that $U \dashv F$ and that the triple gotten from the adjoint pair is $\mathbf{T}$.

4. Show that the naturality of η implies that the transformation

$$\mathrm{Hom}(FA, (B, b)) \to \mathrm{Hom}(A, B)$$

which takes f to $f \circ \eta A$ in the proof of Theorem 13.4.2 is natural.

5.† Let (T, η, u) be the triple in **Set** defined in Exercise 2 of Section 13.3.

a. Show that an algebra for this triple is a monoid: specifically, if $\alpha : T(A) \to A$ is an algebra, then the definition $ab = \alpha(a, b)$ makes A a monoid, and up to isomorphisms every monoid arises this way.

b. Show that algebra homomorphisms are monoid homomorphisms, and that every monoid homomorphism arises this way.

13.5 Scott domains

In [Scott, 1972], a general construction of a great many models of the untyped λ-calculus is given. (See also [Scott, 1982].) One version of finding such a model involves finding an object D in a cartesian closed category with the property that $D \cong [D \to D]$. This is an example of finding a model for computation in a category other than the category of sets. Another such example involves the category of modest sets (14.8.3).

13.5.1 The trouble with approaching the problem of finding an object $D \cong [D \to D]$ using a fixed point construction is that the operation that takes D to $[D \to D]$ is not a functor since an arrow $D \to E$ induces arrows $[D \to D]$ to $[D \to E]$ and $[E \to D]$ to $[D \to D]$ as well as others, but no arrow in either direction between $[D \to D]$ and $[E \to E]$. There is an ingenious trick due to Smyth and Plotkin [1983] to solve this problem.

Some categories have additional structure on their hom sets; the hom set has the structure of an object from some category $\mathcal{V}$ and the hom functors are arrows of $\mathcal{V}$. Such a category is said to be **enriched over** $\mathcal{V}$. In particular, a cartesian closed category is enriched over **Cat**, since its hom sets are themselves categories (with the arrows as objects and the natural transformations as arrows) and the hom functors preserve the extra structure. (See [Kelly, 1982b].) We will not define the general concept of enriched here, but will concentrate on a particular type of category enriched over the category of posets and monotone maps.

13.5.2 Definition A category $\mathcal{C}$ is a **Smyth–Plotkin category** under the following conditions.

SP–1 $\mathcal{C}$ is cartesian closed.

SP–2 $\mathrm{Hom}(A, B)$ is a poset for each pair of objects A, B of $\mathcal{C}$.

SP–3 If $f : A \to B$, $g, h : B \to C$, $k : C \to D$ and $g \leq h$, then both $g \circ f \leq h \circ f$ and $k \circ g \leq k \circ h$.

SP–4 If $g, h : B \to C$ with $g \leq h$, then for any A, both

$$[g \to A] \leq [h \to E] : [C \to A] \to [B \to A]$$

(note that the order is *not* reversed) and

$$[A \to g] \leq [A \to h] : [A \to B] \to [A \to C]$$

SP–5 $\mathcal{C}$ has limits and colimits along countable chains.

SP–6 If $A = \lim A_i$, then the isomorphism

$$\mathrm{Hom}(B, A) \cong \lim \mathrm{Hom}(B, A_i)$$

for each object B is an order isomorphism.

SP–7 If $A = \mathrm{colim}\, A_i$, then the isomorphism

$$\mathrm{Hom}(A, C) \cong \lim \mathrm{Hom}(A_i, C)$$

for each object C is an order isomorphism.

The constructions in this section work for a more general type of category called 'monoidal closed', which is defined like cartesian closed but with respect to a binary operation on the objects which may be but does not have to be the product structure.

13.5.3 To a Smyth–Plotkin category $\mathcal{C}$ we associate two new categories $\mathrm{LA}(\mathcal{C})$ and $\mathrm{RA}(\mathcal{C})$. $\mathrm{LA}(\mathcal{C})$ (for left adjoint) has the same objects as $\mathcal{C}$. A homomorphism $A \to B$ is a pair of arrows of $\mathcal{C}$, $f : A \to B$ and $g : B \to A$ such that $\mathrm{id}_A \leq g \circ f$ and $f \circ g \leq \mathrm{id}_B$. $\mathrm{RA}(\mathcal{C})$ is defined similarly but with inequalities reversed, so that an arrow $A \to B$ is a pair (f, g) with $g \circ f \leq \mathrm{id}_A$ and $\mathrm{id}_B \leq f \circ g$. These really do mean left and right adjoint, respectively, when we consider the partially ordered sets as categories.

If $(f, g) : A \to B$ is in either $\mathrm{LA}(\mathcal{C})$ or $\mathrm{RA}(\mathcal{C})$, then f and g determine each other. For suppose that both (f, g) and (f, h) are arrows of

LA($\mathcal{C}$), then $h \leq g \circ f \circ h \leq g$ and $g \leq h \circ f \circ g \leq h$, using the definition of homomorphism in LA($\mathcal{C}$) as well as SP–3 above. The proof that g determines f and the proofs in RA($\mathcal{C}$) are similar. A second proof follows from the fact that right and left adjoints determine each other up to isomorphism (see 12.3.3) and isomorphic objects in a poset are equal.

Next we observe that $\mathrm{LA}(\mathcal{C})^{\mathrm{op}} \cong \mathrm{RA}(\mathcal{C})$. In fact, an arrow in LA($\mathcal{C}$) from A to B is a pair (f, g) where $\mathrm{id} \leq g \circ f$ and $f \circ g \leq \mathrm{id}$. This, by definition, is an arrow from $B \to A$ in $\mathrm{LA}(\mathcal{C})^{\mathrm{op}}$. On the other hand an arrow from $B \to A$ in RA($\mathcal{C}$) is a pair (g, f) where $g : B \to A$ and $f : A \to B$ satisfy $f \circ g \leq \mathrm{id}$ and $\mathrm{id} \leq g \circ f$. Thus the isomorphism is the one that takes the pair (f, g) to the pair (g, f).

Now suppose that $\{(f_i, g_i) : A_i \to A\}$ is a cocone in LA($\mathcal{C}$). Purely formally, this is a colimit if and only if the cone $\{(f_i, g_i) : A \to A_i\}$ is a limit cone in $\mathrm{LA}(\mathcal{C})^{\mathrm{op}}$. By the isomorphism above this is true if and only if the cone $\{(g_i, f_i) : A \to A_i\}$ is a limit cone in RA($\mathcal{C}$).

13.5.4 Definition The arrow (f, g) of either LA($\mathcal{C}$) or RA($\mathcal{C}$) is a **retract** if $g \circ f = \mathrm{id}$ and a **coretract** if $f \circ g = \mathrm{id}$.

13.5.5 Theorem *Let $\mathcal{C}$ be a Smyth–Plotkin category. Suppose that the hom sets in $\mathcal{C}$ have least upper bounds along countable increasing chains and that*

$$A_0 \to A_1 \to A_2 \to \cdots \to A_n \to \cdots \tag{13.3}$$

is a countable retract chain in LA($\mathcal{C}$) *with arrows $(h_{ji}, k_{ij}) : A_i \to A_j$ for $i \leq j$. Suppose A is the colimit in $\mathcal{C}$ with transition arrows $f_i : A_i \to A$. Then there are (necessarily unique) arrows $g_i : A \to A_i$ such that (f_i, g_i) is a retract in* LA($\mathcal{C}$) *for each i, and such that A is the colimit of (13.3) in* LA($\mathcal{C}$).

Dually, suppose that

$$\cdots \to A_n \to A_{n-1} \to \cdots \to A_1 \to A_0 \tag{13.4}$$

is a countable coretract chain in RA($\mathcal{C}$) *with arrows $(k_{ji}, h_{ij}) : A_i \to A_j$ for $j \leq i$. Suppose A is the limit in $\mathcal{C}$ with transition arrows $g_i : A \to A_i$. Then there are (necessarily unique) arrows $f_i : A_i \to A$ such that (g_i, f_i) is a coretract in* RA($\mathcal{C}$) *and such that A is the limit of (13.4) in* RA($\mathcal{C}$).

Proof. The two statements are dual to each other, as are their proofs. We sketch the proof of the second. To define for each i an arrow $f_i : A_i \to A$ using the limit property, we have to give a consistent family of arrows $f_{ji} : A_i \to A_j$. We let $f_{ji} = h_{ji}$ for $i < j$, $f_{ji} = k_{ji}$ for $i > j$ and $f_{ii} = \mathrm{id}$. (In point of fact, it is not hard to show that these arrows need be defined only for sufficiently large j so that only the case $i < j$ is important.) To show consistency with the cone, we calculate for $i < j < l$,

$$k_{jl} \circ f_{li} = k_{jl} \circ h_{li} = k_{jl} \circ h_{lj} \circ h_{ji} = \mathrm{id} \circ h_{lj} = f_{lj}$$

The cases where $j < i \leq l$ and $i \leq j < l$ are similar (easier, in fact). The result is an arrow $f_i : A_i \to A$ such that $g_j \circ f_i = f_{ji}$. In particular, $g_i \circ f_i = \mathrm{id}$. Finally we have for $i < j$, $g_j \circ f_i \circ g_i = f_{ji} \circ g_i = h_{ji} \circ g_i = h_{ji} \circ k_{ij} \circ g_j \leq g_j = g_j \circ \mathrm{id}$. At this point, we invoke the fact that the partial order on $\mathrm{Hom}(A, A)$ is the limit of those on $\mathrm{Hom}(A, A_j)$ so that it follows that $f_i \circ g_i \leq \mathrm{id}$.

This defines the arrows f_i. To prove that this is a limit in $\mathrm{RA}(\mathcal{C})$ is tedious, but not hard. If B is an object and (r_i, s_i) is a consistent family of arrows in $\mathrm{RA}(\mathcal{C})$, then since A is the limit in $\mathcal{C}$, there is a unique arrow $r : B \to A$ such that $r_i = g_i \circ r$ for all i. For $i \leq j$, we have $s_i \circ g_i = s_j \circ h_{ji} \circ k_{ij} \circ g_j \leq s_j \circ g_j$. Thus the sequence of arrows $s_i \circ g_i \in \mathrm{Hom}(A, B)$ is a countable increasing sequence and has a sup we call s. To show that $s \circ r \geq \mathrm{id}$ it is sufficient to show that $g_j \circ s \circ r \geq g_j$ for each j. $g_j \circ s \circ r = g_j \circ \sup s_i \circ g_i = \sup_i g_j \circ s_i \circ g_i$. Now to do something with the sup, it sufficient to stick to $j > i$. For such i, $\sup_i g_j \circ s_i \circ g_i = \sup_i k_{ji} \circ s_i \circ r_i \circ g_i \geq \sup_i k_{ji} \circ g_i = \sup_i g_j = g_j$. Also $s \circ r = \sup s_n \circ g_n \circ r = \sup s_n \circ r_n \leq \mathrm{id}$ since every $s_n \circ r_n \leq 1$. We note that if every (r_n, s_n) is a coretract, that is $r_n \circ s_n = \mathrm{id}$, then the same is true of (r, s). The uniqueness of (r, s) follows from the uniqueness of r in $\mathcal{C}$ since r determines s. □

13.5.6 If $(f, g) : D \to E$ is a homomorphism in $\mathrm{LA}(\mathcal{C})$, the diagram

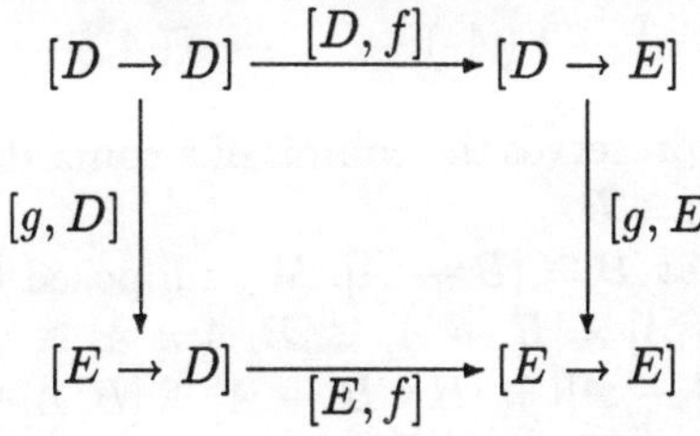

$$\begin{array}{ccc}
[D \to D] & \xrightarrow{[D,f]} & [D \to E] \\
{\scriptstyle [g,D]}\big\downarrow & & \big\downarrow{\scriptstyle [g,E]} \\
[E \to D] & \xrightarrow[{[E,f]}]{} & [E \to E]
\end{array}$$

commutes and determines an arrow $[D, D] \to [E, E]$. This idea can be used directly to find a fixed point for this endoarrow functor. However, it turns out that a slight modification of this idea gives a more useful conclusion.

Now let us begin with an object A of $\mathcal{C}$ with $A \neq 1$ and $A \cong A \times A$. Assume further that there is a retract pair $(h, k) : A \to [A \to A]$ in $\mathrm{LA}(\mathcal{C})$. Define a functor $F : \mathrm{LA}(\mathcal{C}) \to \mathrm{LA}(\mathcal{C})$ that takes B to $[B \to A]$

and $(f,g) : B \to C$ to $([g,A],[f,A])$ from $[B \to a]$ to $[C \to a]$. If (f,g) is a retract pair, so is $F(f,g)$.

Consider the sequence

$$A \to F(A) \to F^2(A) \to \cdots \to F^n(A) \to F^{n+1}(A) \to \cdots$$

with $(h_n,k_n) = F^n(h,k) : F^n(A) \to F^{n+1}(A)$. This is a countable retract chain (the arrow from the ith term to the jth is just the composite of the successive arrows). Let B be the colimit of this chain in $\mathcal{C}$. Then we know that this is also the colimit in $\mathrm{LA}(\mathcal{C})$ and that B is also the limit of the sequence

$$\cdots \to F^n(A) \to \cdots \to F(A) \to A$$

in $\mathcal{C}$. We claim that F commutes with this colimit. In fact, when the object C is the colimit in $\mathcal{C}$ of the sequence

$$C_0 \to C_1 \to \cdots \to C_n \to \cdots$$

$[C,A]$ is the limit of the sequence

$$\cdots \to [C_n \to A] \to \cdots \to [C_1 \to A] \to [C_0 \to A] \qquad (13.5)$$

because $[-,A] : \mathcal{C}^{\mathrm{op}} \to \mathcal{C}$ is a right adjoint and thus preserves limits (see 12.3.6). But (13.5) being a limit is equivalent to the sequence

$$[C_0 \to A] \to [C_1 \to A] \to \cdots \to [C_n \to A] \to \cdots$$

being a colimit. But this is just

$$F(A_0) \to F(A_1) \to \cdots \to F(A_n) \to \cdots$$

which shows that F preserves the colimit of a countable chain of retracts. It follows that $F(B) \cong B$.

Now we have that $B \cong [B \to A]$. We supposed that $A \cong A \times A$ so that $B \times B \cong [B \to A] \times [B \to A] \cong [B, A \times A] \cong [B \to A] \cong B$. Also, $[B \to B] \cong [B \to [B \to A]] \cong [B \times B \to A] \cong [B,A] \cong B$. Thus B is not only a solution to $B \cong [B \to B]$, but simultaneously to $B \cong B \times B$.

13.5.7 Examples There are many examples of this in various categories of posets. Let $\mathcal{C}$ be a (not necessarily full) subcategory of posets and order-preserving maps. The set of order-preserving functions between two objects of $\mathcal{C}$ is partially ordered by saying that $f \leq g$ if $f(x) \leq g(x)$ for all x in the domain of f. Whether $\mathcal{C}$ is cartesian closed and satisfies the other conditions of the theorem we have just proved to be useful is something that has to be proved in each special case.

Here is the example that especially interests us. Let $\mathcal{C}$ be the category whose objects are ω-CPOs and whose arrows are functions that preserve countable sups as described in 2.4.3. It is not hard to see that this category is cartesian closed. In fact, we know that there is a one to one correspondence between functions $A \times B \to C$ and functions $A \to [B \to C]$; the thing to do is to show that this isomorphism remains when everything in sight preserves countable sups and that the isomorphism itself preserves countable sups. We leave this as an exercise. The category has limits computed pointwise. Colimits are not quite so simple. The easiest way to compute a colimit is to first compute the colimit as a poset and then add freely the colimits of countable increasing sequences.

We want to find an object $A \neq 1$ with $A \cong A \times A$ and A having an arrow $A \to [A \to A]$ in $\mathrm{LA}(\mathcal{C})$. Take any object A_0 of $\mathcal{C}$ with $A_0 \neq 1$ and having a least element $\bot$. Then the product A of countably many copies of A_0 will have the property that $A \cong A \times A$. For example, the function that takes the countable sequence $(a_0, a_1, a_2, a_4, \ldots)$ to $((a_0, a_2, a_4, \ldots), (a_1, a_3, a_5, \ldots))$ is an isomorphism.

Since A_0 has a least element, so does A. Let $f : A \to [A \to A]$ be the arrow that takes an element to the constant function at that element which is certainly an order-preserving function. Let $g : [A \to A] \to A$ take an order-preserving function to its value at $\bot$. Then it is immediate that $g \circ f = \mathrm{id}$ and that $f \circ g \leq \mathrm{id}$ so we have an arrow of $\mathrm{LA}(\mathcal{C})$.

We can now apply the construction of 13.5.6 to get an ω-CPO D for which $D \cong D \times D \cong [D \to D]$.

13.5.8 Exercises

1.† Show that the category of ω-CPOs and functions that preserve countable sups is cartesian closed.

2.† Show that every poset freely generates an ω-CPO. (Let P be a poset. Let $\widehat{P}$ denote the set of all countably increasing chains of P. If $c = \{c_n\}$ and $c' = \{c'_n\}$ are two such chains, say that $c \leq c'$ if every element of c is less than or equal to some element of c'. Say that $c = c'$ if both $c \leq c'$ and $c' \leq c$.)

3.† Show that the category of ω-CPOs has limits (computed pointwise) and colimits. Assume that the category of posets has colimits.

4. A subset of a poset is called *directed* or *filtered* if every pair of elements has a common upper bound in the subset. Show that a poset is countably chain complete if and only if every countable directed subset has a least upper bound.

14

Toposes

A topos is a cartesian closed category with some extra structure which produces an object of subobjects for each object. This structure makes toposes more like the category of sets than cartesian closed categories generally are.

Toposes, and certain subcategories of toposes, have proved attractive for the purpose of modeling computation. A particular reason for this is that in a topos, a subobject of an object need not have a complement. One of the fundamental facts of computation is that it may be possible to list the elements of a subset effectively, but not the elements of its complement. Sets which cannot be listed effectively do not exist for computational purposes, and toposes provide a universe of objects and functions which has many nice set-like properties but which does not force complements to exist. We discuss one specific subcategory of a topos, the category of modest sets, which has been of particular interest in the semantics of programming languages.

Toposes have interested mathematicians for other reasons. They are an abstraction of the concept of sheaf, which is important in pure mathematics. They allow the interpretation of second-order statements in the category in an extension of the language associated to cartesian closed categories in Chapter 6. This fact has resulted in toposes being proposed as an alternative to the category of sets for the foundations of mathematics. Toposes can also be interpreted as categories of sets with an internal system of truth values more general than the familiar two-valued system of classical logic; this allows an object in a topos to be thought of as a variable or time-dependent set, or as a set with various degrees of membership. In particular, most ways of defining the category of fuzzy sets lead to a category which can be embedded in a topos.

Sections 14.1 and 14.2 describe the basic properties of toposes, for the most part without proof. Section 14.3 takes a closer look at an aspect of toposes which make many of them a better model of computation than, for example, **Set**.

Sections 14.4 and 14.5 describe a special case of categories of sheaves which makes the connection with sets with degrees of membership clear. The category of graphs is discussed as an example there. Section 14.6 describes the connection with fuzzy sets.

In Section 14.7 we describe category objects in a category, a notion that is needed in Section 14.8, which is a brief description of the realizability topos and modest sets.

This chapter depends on Chapters 1 through 6, Chapter 8, Sections 11.1 and 11.2, and Chapter 12, with some references in proofs to Chapters 7 and 9. Sections 14.1 and 14.2 are needed in all the remaining sections. After that, the chapter consists of four independent units: Section 14.3, Sections 14.4 and 5, Section 14.6 and Sections 14.7 and 8.

We do not discuss the language corresponding to a topos in this book; this is carried out in detail in [Lambek and Scott, 1986] and in [Bell, 1988]. Other discussions of the language and the relation with logic are in [Fourman, 1977], [Fourman and Vickers, 1986], [Boileau and Joyal, 1981] and [Johnstone, 1977]. [Makkai and Reyes, 1977] is concerned with this and also languages corresponding to certain more general classes of categories which include toposes.

The use of toposes specifically as models of programming language semantics is discussed at length in [Hyland and Pitts, 1989].

14.1 Definition of topos

14.1.1 The subobject functor Recall from 2.8.8 that if C is an object of a category, a subobject of C is an equivalence class of monomorphisms $C_0 \rightarrowtail C$ where $f_0 : C_0 \rightarrowtail C$ is equivalent to $f_1 : C_1 \rightarrowtail C$ if and only if there are arrows (necessarily isomorphisms) $g : C_0 \to C_1$ and $h : C_1 \to C_0$ such that $f_1 \circ g = f_0$ and $f_0 \circ h = f_1$.

Assuming the ambient category $\mathcal{C}$ has pullbacks, the 'set of subobjects' function is the object function of a functor $\mathrm{Sub} : \mathcal{C}^{\mathrm{op}} \to \mathbf{Set}$: precisely, for an object C, $\mathrm{Sub}(C)$ is the set of subobjects of C. We must define Sub on arrows.

If $k : C' \to C$ is an arrow and if $f_0 : C_0 \rightarrowtail C$ represents a subobject of C, then in a pullback

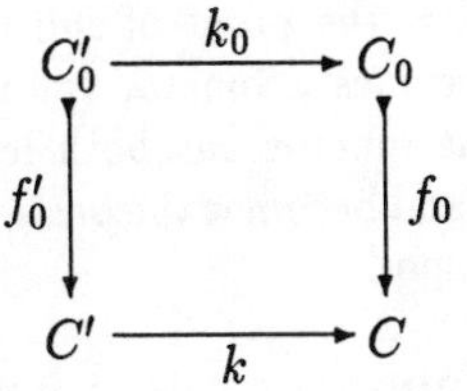

$$\begin{array}{ccc} C_0' & \xrightarrow{k_0} & C_0 \\ {\scriptstyle f_0'}\downarrow & & \downarrow{\scriptstyle f_0} \\ C' & \xrightarrow[k]{} & C \end{array} \tag{14.1}$$

the arrow f_0' is also a monomorphism (see 8.3.4).

It is left as an exercise to prove, using the universal mapping property of pullbacks, that if the monomorphism $f_0 : C_0 \rightarrowtail C$ is equivalent to

$f_1 : C_1 \rightarrowtail C$, then the pullbacks $f_0' : C_0' \rightarrowtail C'$ and $f_1' : C_1' \rightarrowtail C'$ are also equivalent. Thus not only is a pullback of a monomorphism a monomorphism, but also a pullback of a subobject is a subobject.

Thus we can define, for an arrow k as above,

$$\mathrm{Sub}(k) : \mathrm{Sub}(C) \to \mathrm{Sub}(C')$$

to be the function that sends the equivalence class containing f_0 to the equivalence class containing the pullback f_0'.

To show that this is a functor, we must show that the identity arrow induces the identity arrow on subobjects (exercise) and that if $k' : C'' \to C'$, then the diagram

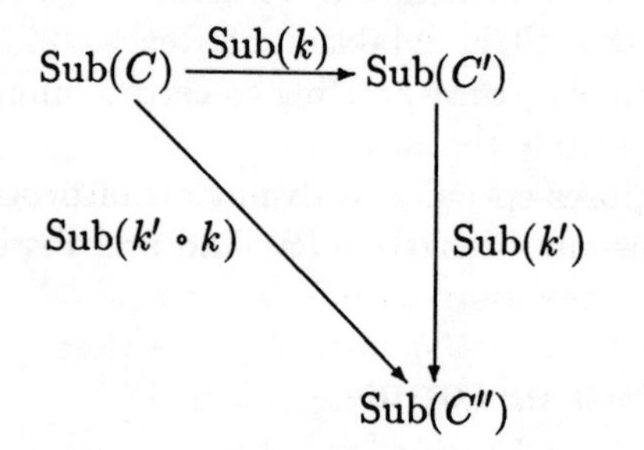

commutes. But the commutativity of this diagram at the subobject represented by f_0 is equivalent to the outer rectangle of the diagram

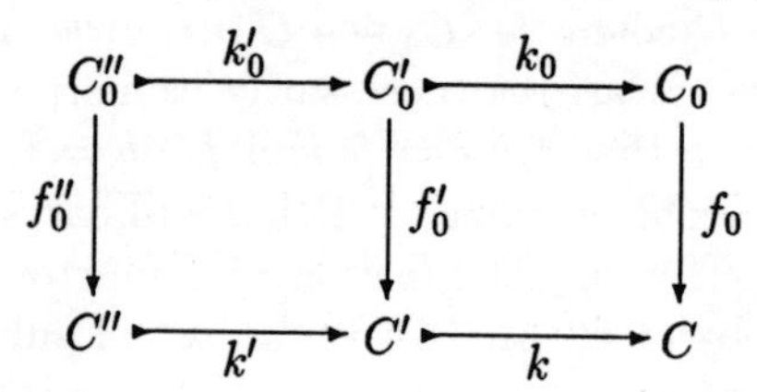

being a pullback when the two smaller squares are, which is true by Exercise 9 of Section 8.3.

In 12.2.6 we saw that when the poset of subsets of a set is viewed as a category, the pullback becomes a functor (monotone function) from one poset to another and that functor has both left and right adjoint. This remains true in any topos and becomes the basis for introducing elementary (first order) logic into a topos.

14.1.2 Definition A **topos** is a category which

TOP–1 has finite limits;

TOP–2 is cartesian closed;

TOP–3 has a representable subobject functor.

We know that a functor is representable if and only if it has a universal element (see 4.5.12). A universal element of the subobject functor is an object, usually called Ω, and a subobject $\Omega_0 \subseteq \Omega$ such that for any object A and subobject $A_0 \subseteq A$, there is a unique arrow $\chi : A \to \Omega$ such that there is a pullback

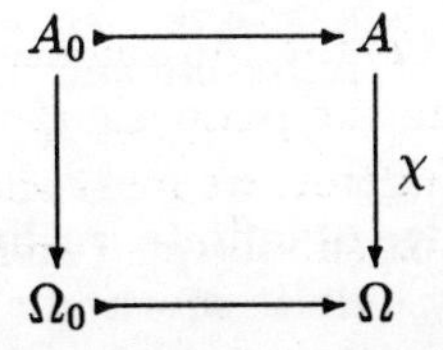

It can be proved that Ω_0 is the terminal object and the left arrow is the unique arrow from A_0 ([Barr and Wells, 1985], Proposition 4 of Section 2.3).

The object Ω is called the **subobject classifier** and the arrow from $\Omega_0 = 1 \to \Omega$ is usually denoted **true**. The arrow χ corresponding to a subobject is called the **characteristic arrow** of the subobject.

The fact that the subobject functor is represented by Ω means precisely that there is a natural isomorphism

$$\phi : \mathrm{Sub}(-) \to \mathrm{Hom}(-, \Omega)$$

which takes a subobject to its characteristic function.

14.1.3 Example The category of sets is a topos. It was shown in 6.1.9 that sets are a cartesian closed category. A two-element set, which we call 2, is a subobject classifier. In fact, call the two elements true and false. Given a set S and subset $S_0 \subseteq S$, define the characteristic function $\chi : S \to \{\text{true}, \text{false}\}$ by

$$\chi(x) = \begin{cases} \text{true} & \text{if } x \in S \\ \text{false} & \text{if } x \notin S \end{cases}$$

Then the square

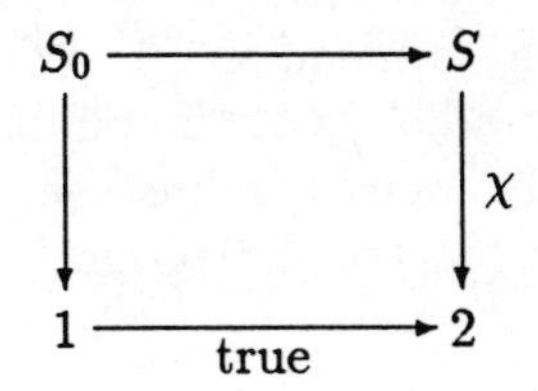

is a pullback. Thus 2 is a subobject classifier.

14.1.4 Exercises

1. Referring to Diagram (14.1), show that if the monomorphism $f_0 : C_0 \rightarrowtail C$ is equivalent to $f_1 : C_1 \rightarrowtail C$, then for any $g : C' \to C$, the pullbacks $f_0' : C_0' \rightarrowtail C'$ and $f_1' : C_1' \rightarrowtail C'$ are also equivalent.

2. Show that the identity arrow $C \to C$ induces the identity arrow $\mathrm{Sub}(C) \to \mathrm{Sub}(C)$.

3. Show that the category of finite sets and all functions between them is a topos.

4.† (Requires some knowledge of infinite cardinals.) Show that the category of finite and countably infinite sets and all functions between them has a subobject classifier but is not cartesian closed. (Hint: the set of subsets of a countable set is not countable.)

14.2 Properties of toposes

We list here some of the properties of toposes, without proof.

14.2.1 In the first place, a topos is not only cartesian closed, it is locally cartesian closed (see Section 12.4). This is Corollary 1.43, p. 36 of [Johnstone, 1977] or Corollary 7, p. 182 of [Barr and Wells, 1985].

14.2.2 Power objects In any topos, the object $[A \to \Omega]$ has the property that

$$\mathrm{Hom}(B, [A \to \Omega]) \cong \mathrm{Hom}(A \times B, \Omega) \cong \mathrm{Sub}(A \times B) \tag{14.2}$$

These isomorphisms are natural when the functors are regarded as functors of either A or of B. The object $[A \to \Omega]$ is often called the **power object** of A and denoted $\mathcal{P}A$. It is the topos theoretic version of the powerset of a set. Theorem 1 of Section 5.4 of [Barr and Wells, 1985] implies that a category with finite limits is a topos if for each object A there is a power object that satisfies (14.2).

The inverse image and universal image constructions in 12.2.6 for the powerset of a set can be made on $[A \to \Omega]$ for any object A in a topos. The left and right adjoints of the pullback functors are related to these images via the diagram in [Johnstone, 1977], Proposition 5.29; this diagram is called the 'doctrinal diagram'.

14.2.3 Effective equivalence relations Let $d, e : E \rightrightarrows A$ be two arrows in a category. For any object B we have a single function

$$\langle \mathrm{Hom}(B,d), \mathrm{Hom}(B,e) \rangle : \mathrm{Hom}(B,E) \to \mathrm{Hom}(B,A) \times \mathrm{Hom}(B,A)$$

which sends f to the pair $(d \circ f, e \circ f)$. If this function is, for each object B, an isomorphism of $\mathrm{Hom}(B,E)$ with an equivalence relation on the set $\mathrm{Hom}(B,A)$, then we say that E is an **equivalence relation** on the object A. This means that the image of $\langle \mathrm{Hom}(B,d), \mathrm{Hom}(B,e) \rangle$ is actually an equivalence relation on $\mathrm{Hom}(B,A)$, if a relation is interpreted as being a set of ordered pairs.

This can be thought of as embodying two separate conditions. First off, the function $\langle \mathrm{Hom}(B,d), \mathrm{Hom}(B,e) \rangle$ must be an injection, because we are supposing that it maps $\mathrm{Hom}(B,E)$ isomorphically to a subset of $\mathrm{Hom}(B,A) \times \mathrm{Hom}(B,A)$. Secondly, that subset must satisfy the usual reflexivity, symmetry and transitivity conditions of equivalence relations.

14.2.4 Kernel pairs Here is one case in which this condition is automatic. If $g : A \to C$ is an arrow, the pullback of the square

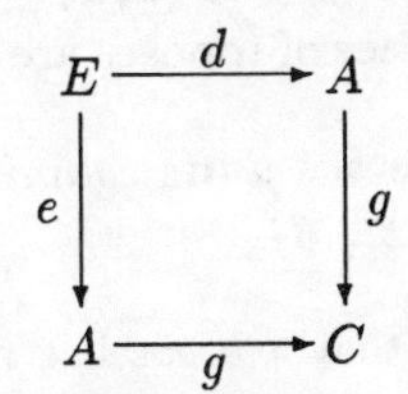

is called a **kernel pair** of g. Notation: we will write that

$$E \overset{d}{\underset{e}{\rightrightarrows}} A \xrightarrow{g} C$$

is a kernel pair. For an object B of $\mathcal{C}$, the definition of this limit is that there is a one to one correspondence between arrows $B \to E$ and pairs of arrows (h, k) from B to A such that $g \circ h = g \circ k$ and that the correspondence is got by composing an arrow from B to E with d and e, resp. To put it in other words, $\mathrm{Hom}(B,E)$ is isomorphic to

$$\{(h,k) \in \mathrm{Hom}(B,A) \times \mathrm{Hom}(B,A) \mid g \circ h = g \circ k\}$$

which is an equivalence relation.

14.2.5 Suppose that $E \rightrightarrows A$ describes an equivalence relation. We say that the equivalence relation is **effective** if it is a kernel pair of some arrow from A. We say that a category has **effective equivalence relations** if every equivalence relation is effective. We give the following without proof. The interested reader may find the proof in [Barr and Wells, 1985], Theorem 7 of Section 2.3 or [Johnstone, 1977], Proposition 1.23, p. 27.

14.2.6 Theorem *In a topos, every equivalence relation is effective.*

14.2.7 Example Equivalence relations in **Set** and **Mon** are effective. An equivalence relation in **Set** is simply an equivalence relation, and the class map to the quotient set is the function with the equivalence relation as kernel pair. An equivalence relation on a monoid M in **Mon** is a congruence relation on M (see Exercise 6 of Section 3.5).

There are many categories which lack effective equivalence relations. One is the category of partially ordered sets and monotone maps. Here is a simple example. Let C be a two-element chain $x < y$. Consider the subset E of $C \times C$ consisting of all four pairs (x, x), (x, y), (y, x) and (y, y). The only ordering is that $(x, x) \leq (y, y)$. Then E is an equivalence relation, but is not the kernel pair of any arrow out of C. The least kernel pair that includes E has the same elements as E, but has the additional orderings $(x, x) \leq (x, y) \leq (y, y)$ and $(x, x) \leq (y, x) \leq (y, y)$.

Other important properties of toposes are contained in the following.

14.2.8 Theorem *A topos has finite colimits, every epi is regular and the category is regular (see 8.4.3).*

An early proof of the fact that a topos has finite colimits ([Mikkelson, 1976]) mimicked the construction in sets. For example, the coequalizer of two arrows was constructed as a set of subsets, which can be identified as the set of equivalence classes mod the equivalence relation generated by the two arrows. However, the argument is difficult. The modern proof (due to Paré) is much easier, but it involves some legerdemain with triples and it is hard to see what it is actually doing. See [Barr and Wells, 1985], Corollary 2 of 5.1 for the latter proof. The rest is found in 5.5 of the same source.

14.2.9 The initial topos There is an FL sketch whose models are toposes. (See [Barr and Wells, 1985], Section 4.4. In that book, FL sketches are called LE sketches.) It follows that there is an initial model of the topos axioms. This topos lacks a natural numbers object. It might be an interesting model for a rigidly finitistic model of computation, but would not allow the modeling of such things as recursion.

In fact, the phrase 'initial topos' is usually reserved for the initial model of the axioms for toposes with a natural numbers object (Section 5.5). This category provides an interesting model for computation. The arrows from N to N in that category are, not surprisingly, the total recursive functions. In fact, all partial recursive functions are modeled in there as well, but as partial arrows, which we now describe.

14.2.10 Partial arrows In 2.1.13 we discussed partial functions between sets. This concept can be extended to any category. Let A and B be objects. A partial arrow A to B consists of a subobject $A_0 \subseteq A$ and an arrow $f : A_0 \to B$. This defines the partial arrow f in terms of a particular representative $A_0 \to A$ of the subobject, but given any other representative $A'_0 \to A$, there is a unique arrow from A_0 to A'_0 commuting with the inclusions which determines an arrow from A'_0 to B by composition with f. The subobject determined by A_0 is called the **domain** of the partial arrow. If $g : A_1 \to B$ is another partial arrow on A we say the $f \leq g$ if $A_0 \subseteq A_1$ and the restriction of g to A_0 is f. If we let $i : A_0 \to A_1$ denote the inclusion arrow, then the second condition means simply that $g \circ i = f$. We will say that f and g are the same partial arrow if both $f \leq g$ and $g \leq f$. This means that the domains of f and g are the same subobject of A and that f and g are equal on that domain.

We say that **partial arrows to B are representable** if there is an object $\widetilde{B}$ and an embedding $B \rightarrowtail \widetilde{B}$ such that there is a one to one correspondence between arrows $A \to \widetilde{B}$ and partial arrows A to B, the correspondence given by pulling back:

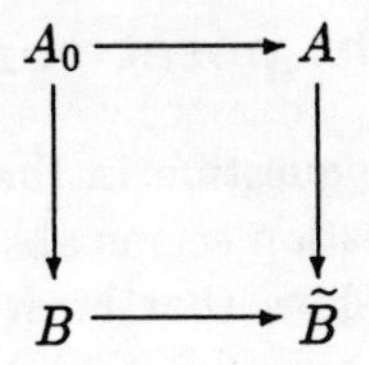

In a topos, the arrow true: $1 \to \Omega$ represents partial functions to 1. The reason is that since each object has a unique arrow to 1, a partial arrow from A to 1 is equivalent to a subobject of A.

14.2.11 Theorem *In a topos, partial arrows into every object are representable.*

See [Johnstone, 1977], 1.26 for the proof.

14.2.12 Exercises

1. Verify the isomorphisms of (14.2) for **Set**.

2. Let S be a set and E be an equivalence relation on S. Then there are two arrows E to S, being the inclusion of E into $S \times S$, followed by the two projections on S. Show that E, together with these two arrows, is the kernel pair of the function that takes each element of S to the equivalence class containing it.

3. a. Let $d, e : E \rightrightarrows M$ be two monoid homomorphisms. Show that they form an equivalence relation in **Mon** if and only if the arrow

$$E \xrightarrow{\langle d, e \rangle} M \times M$$

is a monomorphism and the image of $\langle d, e \rangle$ is a congruence on M (see Exercise 6 of Section 3.5 with d, e the projections).

b. Show that equivalence relations in **Mon** are effective.

4. Show that in any category with kernel pairs, if $f : C \to D$ is a coequalizer, then it is the coequalizer of its kernel pair.

5. Give an example in **Set** of a function with a kernel pair which is not the coequalizer of its kernel pair.

6. Show that in the category of sets, $\widetilde{S}$ can be taken to be $S \cup \{*\}$, where $*$ is an element not in S.

7. Show that in a topos, the subobject classifier is $\widetilde{1}$, the object that represents partial arrows into 1.

14.3 Is a two-element poset complete?

The word 'complete' in the question in the heading means that it has sups of all subsets. The question seems absurd at first, but it improves with age. In fact, we will show that both in a topos and in realistic models of computation, the answer is 'no'. Moreover the answer is no in both cases for the same reason.

The claim we want to make, and for which this discussion is evidence, is that topos semantics is an appropriate model for computation, or at least a more appropriate one than set theory. This will not mean that topos theory will tell you how to compute something that you could not compute without it. What happens is that in many toposes certain computationally meaningless constructions cannot be made at all, although they are all possible in **Set**.

Although we raise the question about the two-element set, everything we say is equally true for flat CPOs.

14.3.1 Computation models Of course, in a most naive sense, 2 is certainly complete as a poset, but we want to look at this in a more sophisticated way. What we really want to be true if we say that a poset P is complete is that for any other object A, the poset $[A \to P]$, with the pointwise order, is complete. In the case of 2, this means that $[A \to 2]$ is complete.

In the case of ordinary set theory, it is still true that 2 is complete in this expanded sense. For computational semantics, the situation is different. Consider the case $A = \mathbf{N}$, although any infinite set would do as well. A function $\mathbf{N} \to 2$ is determined by and determines a subset of $\mathbf{N}$. However, a computable function $\mathbf{N} \to 2$ is determined by and determines a *recursive* subset of $\mathbf{N}$. And it is well known that the set of recursive subsets of $\mathbf{N}$ is not complete. It is not even countably complete; in fact, the recursively enumerable subsets are characterized as the countable unions of recursive subsets. Of course, arbitrary unions of recursive subsets will be even worse.

14.3.2 Topos models The situation in an arbitrary topos is similar. A topos has an object Ω with the property that for any object A, $\mathrm{Hom}(A, \Omega)$ is the set of subobjects of A. In a certain sense, the subobject lattice of any object of a topos is always complete. However, this is not the issue. A topos always has an object $2 = 1 + 1$ and an arrow $A \to 2$ does not represent an arbitrary subobject, but rather a complemented subobject (see 8.6.2). Any arrow $A \to 2$ gives a decomposition of A as a sum by letting $\mathrm{true}, \mathrm{false} : 1 \to 2$ be the two injections and then letting A_0 and A_1 be the pullbacks in

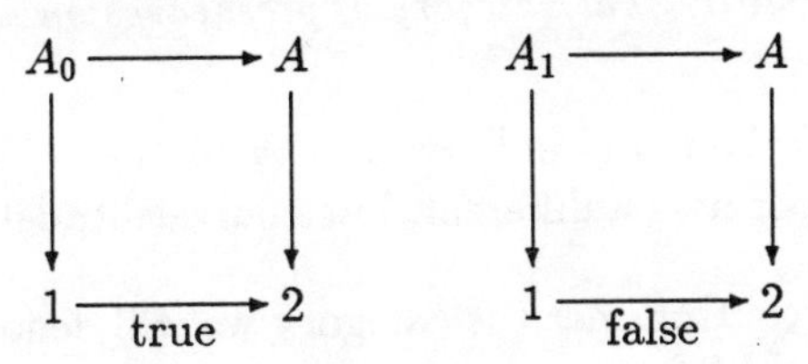

The universality of sums in a topos implies that from $2 = 1 + 1$ we can conclude that $A = A_0 + A_1$. On the other hand, if $A = A_0 + A_1$, there is a unique arrow $A \to 2$ whose restriction to A_0 is $\mathrm{false} \circ \langle\rangle$ and to A_1 is $\mathrm{true} \circ \langle\rangle$. Thus the internal completeness of 2 in a topos is equivalent to the supremum of complemented sets being complemented. It is not hard to show that this is not true in general. In fact, there is a topos in which the arrows from $\mathbf{N}$ to 2 are just the total recursive two-valued functions (and in fact, the arrows from $\mathbf{N}$ to itself are also the total recursive functions). In that topos a complemented subset of $\mathbf{N}$ is exactly a recursive subset (one which, with its complement, has a recursive characteristic function) and the fact that a union of recursive subsets is not recursive finishes the argument.

In 6.5.4 we constructed a semantics for an `If` loop by a construction of a supremum in a flat CPO (of which 2 is an example). However, the last paragraph shows that in general this semantics does not make computational sense. The reason is that, although you can write down the sup as a formal thing, it will not be guaranteed to give a terminating program. The sup of partial functions exists, but the domain of such a partial function is not generally computable.

In a topos model of computational semantics, in which all arrows $\mathbf{N} \to \mathbf{N}$ are given by recursive functions, a subobject of $\mathbf{N}$ is determined by an arrow into Ω. The maps into Ω cannot usually be computed. The one special case in which they can is that of an arrow that factors through the subobject $\langle \mathrm{true}; \mathrm{false} \rangle : 2 \to \Omega$. This corresponds to the traditional distinction between recursively enumerable and recursive subsets. In classical set theory, all subsets are classified by arrows into 2, but here only recursive subsets are.

14.4 Presheaves

14.4.1 Definition Let $\mathcal{C}$ be a category. A functor $E : \mathcal{C}^{\mathrm{op}} \to \mathbf{Set}$ is called a **presheaf** on $\mathcal{C}$. Note that a presheaf on $\mathcal{C}$ is a contravariant functor. The presheaves on $\mathcal{C}$ with natural transformations as arrows forms a category denoted $\mathrm{Psh}(\mathcal{C})$.

14.4.2 Proposition *The category of presheaves on a category $\mathcal{C}$ form a topos.*

The proof may be found in [Barr and Wells, 1985], Section 2.1, Theorem 4. That proof uses a different, but equivalent, definition of topos.

14.4.3 Example Consider the category we will denote by $0 \rightrightarrows 1$. It has two objects and four arrows, two being the identities. A contravariant set-valued functor on this category is a pair of objects G_0 and G_1 and a pair of arrows we will call source, target : $G_1 \to G_0$. The two identities are sent to identities. Thus the category of presheaves on this category is the category of graphs, which is thereby a topos. (This category is the theory of the sketch for graphs given in 4.6.6.)

We described the exponential object for graphs in 6.1.12. It is instructive to see what the subobject classifier is. We have to find a graph Ω and an arrow true : $1 \to \Omega$ such that for any graph $\mathcal{G}$ and subgraph

$\mathcal{G}_0 \subseteq \mathcal{G}$, there is a unique graph homomorphism $\chi : \mathcal{G} \to \Omega$ such that the diagram

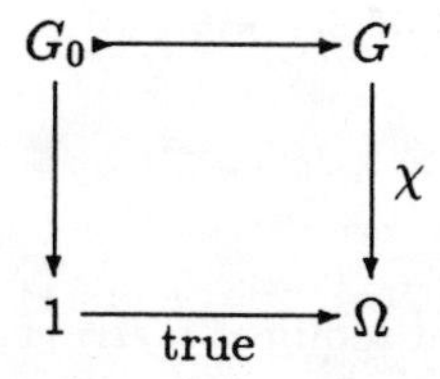

commutes.

We define the graph Ω as follows. It has five arrows we will call 'all', 'source', 'target', 'both' and 'neither'. The reason for these names should become clear shortly. It has two nodes we will call 'in' and 'out'. The arrows 'all' and 'both' go from 'in' to itself, 'neither' goes from 'out' to itself. The arrow 'source' goes from 'in' to 'out' and 'target' from 'out' to 'in'. The terminal graph, which has one arrow and one node, is embedded by the function true that takes the arrow to 'all' and the node to 'in'.

Now given a graph $\mathcal{G}$ and a subgraph $\mathcal{G}_0$ we define a function $\chi : \mathcal{G} \to \Omega$ as follows. For a node n of $\mathcal{G}$, $\chi(n)$ is 'in' or 'out', according to whether n is in $\mathcal{G}_0$ or not. For an arrow a, we let $\chi(a)$ be 'all' if $a \in \mathcal{G}_0$ (whence its source and target are as well). If not, there are several possibilities. If the source but not the target of a belongs to $\mathcal{G}_0$, then $\chi(a) =$'source'. Similarly, if the target but not the source is in $\mathcal{G}_0$, it goes to 'target'. If both the source and target are in it, then $\chi(a) =$'both' and if neither is, then it goes to 'neither'.

14.4.4 Exercises

1. Show that the graphs No and Ar discussed in 6.1.12 are actually the contravariant set-valued functors on $0 \rightrightarrows 1$ represented by the objects 0 and 1, respectively.

2. Show that the object Ω in the category of graphs can be described as follows. The nodes are the subgraphs of No and the arrows are the subgraphs of Ar and the source and target are induced by s and t (defined in 6.1.12), respectively.

3. Let $\mathcal{C}$ be a category, and let M and N be two objects of $\mathrm{Psh}(\mathcal{C})$. Use the Yoneda Lemma and the adjunction defining cartesian closed categories to show that for any object X of $\mathcal{C}$, $[M \to N](X)$ must be the set of natural transformations from $\mathrm{Hom}(-, X) \times M$ to N, up to isomorphism. Note that this does not prove that $\mathrm{Psh}(\mathcal{C})$ *is* cartesian closed. That is true, but requires ideas not given here.

4. Let $\mathcal{C}$ be a category and X an object of $\mathcal{C}$. Let Ω be the subobject classifier of Psh($\mathcal{C}$). Show that $\Omega(X)$ is, up to isomorphism, the set of subfunctors (see Exercise 4 of Section 4.3) of $\mathrm{Hom}(-, X)$.

14.5 Sheaves

To define sheaves in general requires a structure on the category called a **Grothendieck topology**. The theory is developed in full generality in [Barr and Wells, 1985], Chapter 6, especially 6.1 and 6.2 in which it is shown that every sheaf category (including as a special case, every presheaf category) is a topos. Here we will discuss a special case of sheaves in which the category is a partial order.

14.5.1 Let P be a partially ordered set. From the preceding section, a presheaf E on P assigns to each element $x \in P$ a set $E(x)$ and whenever $x \leq y$ assigns a function we will denote $E(x,y) : E(y) \to E(x)$ (note the order; *x precedes y, but the arrow is from $E(y)$ to $E(x)$*). This is subject to two conditions. First, that $E(x,x)$ be the identity function on $E(x)$ and second that when $x \leq y \leq z$, that $E(x,y) \circ E(y,z) = E(x,z)$. The arrows $E(x,y)$ are called **restriction functions.**

14.5.2 Heyting algebras We make the following supposition about P.

HA–1 There is a top element, denoted 1, in P.

HA–2 Each pair of elements x, $y \in P$ has an infimum, denoted $x \wedge y$.

HA–3 Every subset $\{x_i\}$ of elements of P has a supremum, denoted $\bigvee x_i$.

HA–4 For every element $x \in P$ and every subset $\{x_i\} \subseteq P$, $x \wedge (\bigvee x_i) = \bigvee (x \wedge x_i)$.

A poset that satisfies these conditions is called a **complete Heyting algebra.**

14.5.3 If $\{x_i\}$ is a subset with supremum x, and E is a presheaf, there is given a restriction function $e_i : E(x) \to E(x_i)$ for each i. The universal property of product gives a unique function $e : E(x) \to \prod_i E(x_i)$ such that $p_i \circ e = e_i$. In addition, for each pair of indices i and j, there are functions $c_{ij} : E(x_i) \to E(x_i \wedge x_j)$ and $d_{ij} : E(x_j) \to E(x_i \wedge x_j)$ induced by the relations $x_i \geq x_i \wedge x_j$ and $x_j \geq x_i \wedge x_j$. This gives two functions $c, d : \prod_i E(x_i) \to \prod_{ij} E(x_i \wedge x_j)$ such that

$$\begin{array}{ccc}
\prod_i E(x_i) & \xrightarrow{c} & \prod_{ij} E(x_i \wedge x_j) \\
\downarrow p_i & & \downarrow p_{ij} \\
E(x_i) & \xrightarrow[c_{ij}]{} & E(x_i \wedge x_j)
\end{array}$$

and

$$\begin{array}{ccc}
\prod_i E(x_i) & \xrightarrow{d} & \prod_{ij} E(x_i \wedge x_j) \\
\downarrow p_i & & \downarrow p_{ij} \\
E(x_i) & \xrightarrow[d_{ij}]{} & E(x_i \wedge x_j)
\end{array}$$

commute.

14.5.4 Definition A presheaf is called a **sheaf** if it satisfies the following additional condition:

$$x = \bigvee x_i$$

implies

$$E(x) \xrightarrow{e} \prod_i E(x_i) \overset{c}{\underset{d}{\rightrightarrows}} \prod_{ij} E(x_i \wedge x_j)$$

is an equalizer.

14.5.5 Theorem *The category of sheaves on a Heyting algebra is a topos.*

As a matter of fact, the category of sheaves for any Grothendieck topology is a topos. This is developed in Chapter 6 of [Barr and Wells, 1985].

14.5.6 Constant sheaves A presheaf E is called **constant** if for all $x \in P$, $E(x)$ is always the same set and for all $x \leq y$, the function $E(y, x)$ is the identity function on that set.

The constant presheaf at a one-element set is always a sheaf. This is because the sheaf condition comes down to a diagram

$$1 \to 1 \rightrightarrows 1$$

which is certainly an equalizer. No constant presheaf whose value is a set with other than one element can be a sheaf. In fact, the 0 (bottom) element of P is the supremum of the empty set and the product of the empty set of sets is a one-element set (see 5.3.6). Hence the sheaf condition on a presheaf E is that

$$E(0) \to \prod_{\emptyset} \rightrightarrows \prod_{\emptyset}$$

which is

$$E(0) \to 1 \rightrightarrows 1$$

and this is an equalizer if and only if $E(0) = 1$.

14.5.7 A **nearly constant** presheaf is a presheaf which has the same value at every $x \in P$ except that $E(0) = 1$. It is interesting to inquire when a nearly constant presheaf is a sheaf. It turns out to be so if and only if the meet of two nonzero elements of P is nonzero. Thus the condition is a condition on P rather than on E.

To see this, suppose E is a nearly constant presheaf whose value at any $x \neq 0$ is S and that $x = \bigvee x_i$. In the diagram

$$E(x) \to \prod E(x_i) \rightrightarrows \prod E(x_i \wedge x_j)$$

every term in which $x_i = 0$ contributes nothing to the product since $1 \times Y \cong Y$. An element of the product is a string $\{s_i\}$ such that $s_i \in S$. The condition of being an element of the equalizer is the condition that the image of s_i under the induced function $E(x_i) \to E(x_i) \wedge E(x_j)$ is the same as the image of s_j under $E(x_j) \to E(x_i \wedge x_j)$. But in a nearly constant sheaf, all these sets are the same and all the functions are the identity, so this condition is simply that $s_i = s_j$. But this means that an element of the equalizer must be the same in every coordinate, hence that diagram is an equalizer.

14.5.8 Interpretation of sheaves Let E be a sheaf on P. The reader will want to know how E is to be interpreted. Basically one should think of P as representing an algebra of truth values and $E(x)$ as representing the set of all elements of P whose degree of membership in E is at least x. In fact, in the case that P is a Boolean algebra, sheaves are equivalent to P-valued sets.

What about the restriction functions? Well, any element that belongs to E with degree at least x will certainly belong with degree at least y, for any $y \leq x$. This gives the arrow $E(y,x) : E(x) \to E(y)$. This arrow is not in general surjective, since if $y \leq x$ there may be elements of degree y but not degree x. The restriction functions are not in general injective, either, as we will see in the next section.

Sheaves were originally invented to study things defined 'locally' but not necessarily 'globally' (the latter can be interpreted here as being defined over the top of the algebra). From that point of view, one would not expect surjectivity in the restriction functions. In fact, sheaves in which the restrictions are always surjective are a kind of degenerate case.

One should not expect the restrictions to be injective, either. The reason is that not only membership but all predicates, including equality, have truth values indexed by P. Two elements may differ when looked at at level x, but not at level y. In the next section, we will contrast this behavior with that of fuzzy set theory, another attempt at introducing into mathematics sets with more general kinds of truth values.

14.5.9 Time sheaves, I Here is a good example of a topos in which one can see that the restriction arrows should not be expected to be injective. Consider the partially ordered set whose elements are intervals on the real line with inclusion as ordering. It is helpful to think of these as time intervals. And in fact, if the reader prefers to think of these as intervals with positive endpoints nothing will change.

Now consider any definition of a naive set. Some possible definitions will be time invariant, such as the set of mathematical statements about, say, natural numbers, that are true. Others of these 'sets' change with time; one example might be the set of all books that are currently in print; another the set of statements currently known to be true about the natural numbers. These may conveniently be thought of as the presheaves whose value on some time interval is the set of books that were in print over the entire interval and the set of statements about natural numbers known to be true during that entire interval. The restriction to a subinterval is simply the inclusion of the set of books in print over the whole interval to that (larger) set of those in print over that subinterval or the restriction of the knowledge from the interval to the subinterval. In this example, the restrictions are injective.

Instead of books in print, we could take the example of businesses in operation. Because of the possibility of merger, divestment and the like, two businesses which are actually distinct over a large interval might coincide over a smaller subinterval.

Another situation in which the restriction functions are not necessarily injective arises from the set of variables in a block-structured programming language. The presheaf is this: the set for a certain time interval during the running of the program is the quotient of the set of variables which exist over the whole time interval, where the equivalence relation is that of having the same value over the whole interval. Two variables may not be equivalent over a large interval, whereas they may be equivalent over a smaller one; in that case the restriction function would not be injective.

In general, any property describes the set of all entities that have that property over the entire interval. The restriction to a subinterval arises out of the observation that any entity that possesses a property over an interval possesses it over any subinterval.

The sheaf condition in this case reduces to this: if the interval I is a union of subintervals I_k (where k ranges over an index set which need not be countable) and an entity possesses that property over every one of the subintervals I_k, then it does over the whole interval. This condition puts a definite restriction on the kinds of properties that can be used. For example, the property of being **grue**, that is blue over the first half of the interval and green on the second half, is not allowed. The properties that are allowed are what might be called **local**, that is true when they are true in each sufficiently small interval surrounding the point in time. This statement is essentially a tautology, but it does give an idea of what the sheaf condition means.

14.5.10 Time sheaves, II Here is another topos, rather different from the one above, that might also be considered to be time sheaves. Unlike the one above which is symmetric to forward and reverse time, this one definitely depends on which way time is flowing. This is not to say that one is better or worse, but they are different and have different purposes. In this one, the elements of the Heyting algebra are the times themselves. In other words, we are looking at time indexed sets, as opposed to sets indexed by time intervals.

We order this set by the reverse of the usual order. So a presheaf in this model is a family $\{X(t)\}$ of sets, t a real number, together with functions $f(s,t) : X(t) \to X(s)$ for $t \leq s$, subject to the condition that $f(t,t)$ is the identity and $f(r,s) \circ f(s,t) = f(r,t)$ for $t \leq s \leq r$. The sheaf condition is a bit technical, but can easily be understood in the special case of a presheaf all of whose transition arrows $f(s,t)$ are inclusions. In that case, the condition is that when $t = \bigwedge t_i$ (so that t is the greatest lower bound of the t_i), then $X(t) = \bigcap X(t_i)$.

An example, which might be thought typical, of such a sheaf might be described as the 'sheaf of states of knowledge'. At each time t we let $X(T)$ denote the set of all things known to the human race. On the hypothesis (which might be disputed) that knowledge is not lost, we have a function $X(t) \to X(s)$ for $t \leq s$. In common parlance, we might consider this function to be an inclusion, or at least injective, but it is possible to modify it in various ways that will render it not so. We might introduce an equivalence relation on knowledge, so that two bits of knowledge might be considered the same. In that case, if at some future time we discover two bits of knowledge the same, then bits not equal at one time become equal at a later time and the transition arrow is not injective.

For example, consider our knowledge of the set of complex numbers. There was a time in our history when all the numbers e, i, π and -1 were all known, but it was not known that $e^{i\pi} = -1$. In that case, the number $e^{i\pi}$ and -1 were known separately, but not the fact that they were equal. See [Barr, McLarty and Wells, 1985]. The sheaf condition is this: if $\{t_i\}$ is a set of times and t is their infimum, then anything known at time t_i for every i is known at time t.

14.5.11 Exercise

1. Let $\mathcal{H}$ be a complete Heyting algebra. Define the binary operation $\Rightarrow: H \times H \to H$ by requiring that $a \Rightarrow b$ is the join of all elements c for which $a \wedge c \leq b$.

a. Prove that $a \wedge c \leq b$ if and only if $c \leq a \Rightarrow b$.

b. Prove that when $\mathcal{H}$ is regarded as a category in the usual way, it is cartesian closed with $\Rightarrow$ as internal hom.

14.6 Fuzzy sets

Fuzzy set theory is a more-or-less categorical idea that some claim has application to computer modeling. It appears to be closely related to topos theory. In fact, it appears to us that the interesting core of the subject is already implicit in topos theory. We will not go deeply into the subject, but only give a few definitions and point out the connections. More details can be found in [Barr, 1986c] and [Pitts, 1982].

14.6.1 At this point, it would be appropriate to give some definitions. One of the problems with fuzzy sets is that the meaning of the term has been left vague (one might say fuzzy). Rather than attempt to give all possible definitions, we content ourselves with a definition that is common and for which the connection with topos theory is as simple as possible.

14.6.2 Definition Let P be a complete Heyting algebra. A **P-valued set** is a pair (S, σ) consisting of a set S and a function $\sigma : S \to P$. A **fuzzy set** is a P-valued set for some complete Heyting algebra P.

Think of $\sigma(s)$ as being the degree of membership of s in the fuzzy set. If $\sigma(s) = 1$, then s is fully in the fuzzy set, while if $\sigma(s) = 0$, then s is not in the fuzzy set at all.

Actually that last statement is not quite true; we will return to this point later; pretend for the moment that it is.

14.6.3 Let (S, σ) and (T, τ) be fuzzy sets. An **arrow** $f : (S, \sigma) \to (T, \tau)$ is an arrow $f : S \to T$ such that $\sigma \leq \tau \circ f$. Thus the degree of membership of s in (S, σ) cannot exceed that of $f(s)$ in (T, τ). With this definition and the obvious identity arrows, the fuzzy sets based on P form a category Fuzz(P).

The hypothesis actually made on P was that both P and the opposite order P^{op} were Heyting algebras ([Goguen, 1974]). The hypothesis on P^{op} plays no role in the theory and so we have omitted it.

14.6.4 Once we have defined the category of fuzzy sets, the definition of subset of a fuzzy set emerges. For $f : (S, \sigma) \rightarrowtail (T, \tau)$ to be monic it is necessary that f be injective. In particular, we can think of a subset of (T, τ) as being a fuzzy set (T_0, τ_0) where $T_0 \subseteq T$ and $\tau | T_0 \leq \tau_0$.

14.6.5 More ado about nothing Consider the following two fuzzy subsets of (S, σ). The first is the set $(\emptyset, \langle\rangle)$ and the second is the set $(S, 0)$ where $\langle\rangle$ is the unique function of $\emptyset$ to P and 0 stands for the function that is constantly zero. One is the empty set and the other is the set in which every element is not there. There is seemingly no difference between these two sets as neither actually contains any elements. In fact, in fuzzy set theory, these two sets (and sets in between) are not considered to be equal. This results in the class of fuzzy set theories being curiously restricted (see 14.6.10).

14.6.6 Fuzzy sets and sheaves The reader may suspect (from the title of this section, if nothing else) that there is a connection between fuzzy sets and toposes. Both are generalizations of set theory to introduce lattices more general than the two-element lattice as truth values.

One of the two differences has just been mentioned; the different treatment of the null set. Actually, this difference is relatively minor. The second one is not. Suppose (S, σ) is a fuzzy set. We can define a presheaf E by letting

$$E(x) = \{s \in S \mid \sigma(s) \geq x\}$$

as suggested in our informal discussion. Clearly, if $y \leq x$, then $E(x) \subseteq E(y)$ and using these inclusions, we get a presheaf on P. It is almost never a sheaf, however. The essential reason for this is that $E(0) = S$, while we have seen in 14.5.6 that $E(0) = 1$ when E is a sheaf.

14.6.7 It turns out there is a very simple way to make E into a sheaf, but not on P. Let P^+ denote the poset constructed from P by adding a new bottom element. Let us call the new bottom element $\perp$ to distinguish it from the old one we called 0. Now given a P indexed fuzzy set, define a presheaf on P^+ by letting $E(x)$ be defined as above for $x \in P$ and $E(\perp) = 1$.

14.6.8 Proposition *The presheaf E just defined is a sheaf. It is a subsheaf of the near constant sheaf C defined by $C(x) = S$ for $x \neq \perp$ and $C(\perp) = 1$.*

Proof. We first observe that P^+ obviously has the property that the meet of two nonzero elements is nonzero because P has finite meets. Thus C is a sheaf. A diagram chase shows that if C is a sheaf and E a subpresheaf, then E is a sheaf if and only if for each $x = \bigvee x_i$, the diagram

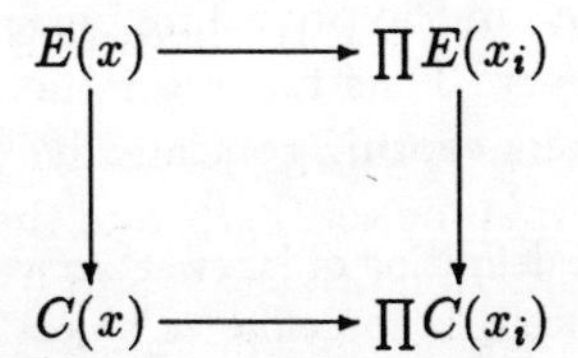

is a pullback. The vertical arrows are just the inclusions. As we saw in 14.5.7, the lower horizontal function is just the inclusion of $C(x)$ into the set of constant strings. It follows that this is essentially what the upper horizontal arrow is. Now in order that a string of elements $\{s_i\} \in \prod E(x_i)$ be constant, it is necessary and sufficient that all the s_i be the same element s and that $s \in E(x_i)$ for all i which means that $\sigma(s) \geq x_i$ for all i. But this is just what is required to have $\sigma(s) \geq x$ and $s \in E(x)$. □

Continuing in this vein, it is possible to show the following.

14.6.9 Theorem *For any Heyting algebra P, the category of fuzzy sets based on P is equivalent to the full subcategory of the category of P^+ sheaves consisting of the sheaves that are subsheaves of the near constant sheaves.*

14.6.10 The introduction of P^+ instead of P is directly traceable to the failure the two kinds of empty sets as mentioned in 14.6.5 to be the same. The fact that the sheaves are subsheaves of the near constant sheaves is really a reflection of the fact that in fuzzy set theory only one of the two predicates of set theory is made to take values in P (or P^+).

This shows up in the fact that in fuzzy set theory there is no fuzzy set of fuzzy subsets of a fuzzy set. In other words, the $\mathcal{P}$ construction is missing. Here's why. Suppose S is a set, considered as a fuzzy set with $\sigma(s) = 1$ for all $s \in S$. (Such a fuzzy set is called a **crisp** set.) Let $x < y$ be two elements of P and consider the subsets $S_x = (S, \sigma_x)$ and $S_y = (S, \sigma_y)$, with σ_x and σ_y being the functions which are constant at x and y respectively. Then of course, $S_x \neq S_y$ (actually S_x is a proper subset of S_y), but it is clear that when looking only at degrees of membership at level x or below, the two subsets are equal. In fact, in the topos, the degree to which S_x equals S_y is just x. But this predicate cannot be stated in the language of fuzzy sets and the result is that there are not and cannot be (unless P has just one element) power objects.

The point is that there are two predicates in set theory, membership and equality. In topos theory, both may be fuzzy, but in fuzzy set theory, only membership is allowed to be. But $\mathcal{P}$ converts membership into equality as explained in the preceding paragraph and so cannot be defined in fuzzy set theory. Thus fuzzy set theory, as currently implemented, lacks a certain conceptual consistency.

One can try to refine the definition of fuzzy set so as to allow fuzzy equality. The obvious way to proceed is to define as objects triplets (S, σ, η), with (S, σ) as above and $\eta : S \times S \to P$, interpreted as fuzzy equality. These must be subject to the condition that the degree to which two elements are equal cannot exceed the degree to which either one is defined. The resultant category is equivalent to the topos of sheaves on P^+.

14.7 External functors

14.7.1 Category objects in a topos One of the tools proposed for programming language semantics is the category of modest sets, which we will describe in the next section. The category of modest sets is not a category in the sense we have been using the word up until now: it is a category *object* in another category, called the effective topos.

Recall the sketch for categories that was described in detail in 9.1.4. A model of this sketch in the category of sets is, of course, a category. A model in a category $\mathcal{C}$ is called a **category object** in $\mathcal{C}$. A homomorphism between such category objects is called an **internal functor** between those category objects.

Referring to the sketch, we see that a category object consists of four objects C_0, C_1, C_2 and C_3 such that $C_2 \cong C_1 \times_{C_0} C_1$ and $C_3 \cong C_1 \times_{C_0} C_1 \times_{C_0} C_1$. There are arrows in $\mathcal{C}$ corresponding to unit, source, target and composition. The crucial commutative diagrams are:

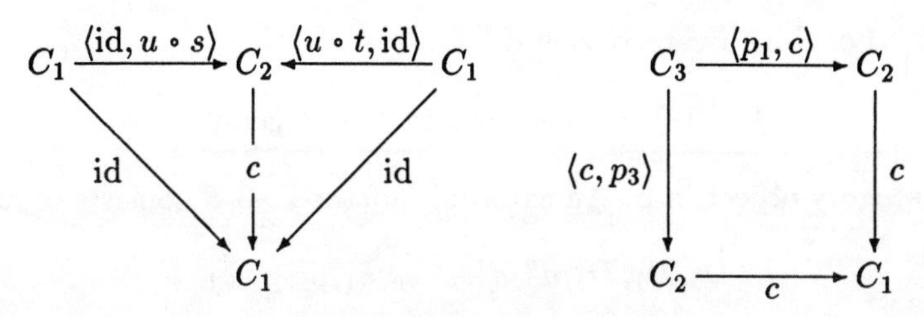

We may denote such a category object by $\mathbf{C} = \langle C_0, C_1, u_{\mathbf{C}}, s_{\mathbf{C}}, t_{\mathbf{C}}, c_{\mathbf{C}} \rangle$ since the remaining data are determined by these. As a matter of convenience, we usually omit the indexes unless they are necessary for comprehension. If $\mathbf{D} = \langle D_0, D_1, u, s, t, c \rangle$ is another category object, then an internal functor $f : \mathbf{D} \to \mathbf{C}$ is given by homomorphisms $f_0 : D_0 \to C_0$ and $f_1 : D_1 \to C_1$ such that the following diagrams commute, where $f_2 : D_2 \to C_2$ is the unique arrow for which $p_i \circ f_2 = f_1 \circ p_i, \quad i = 1, 2.$

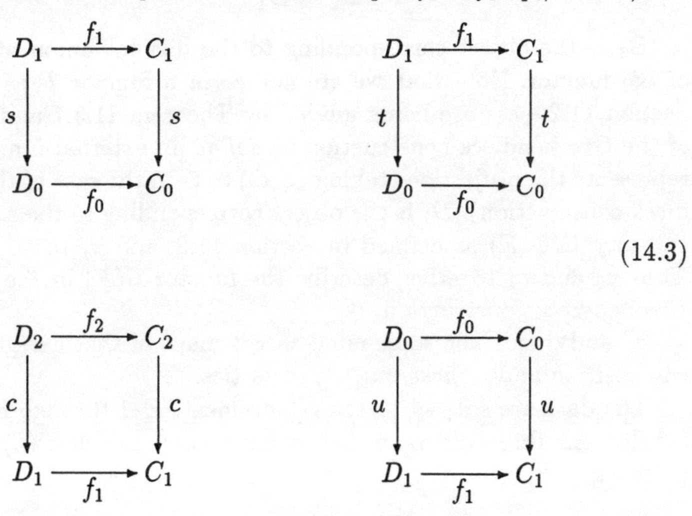

14.7.2 External functors The notion of a functor from a category into the category of sets can be extended to describe a functor from a category in $\mathcal{E}$ to $\mathcal{E}$ itself. Such a functor is called an **external functor**. This construction is based on the construction in Section 11.2 for **Set**. Theorem 11.3.7 gives an equivalence of categories between external functors and the split opfibrations which are produced by the Grothendieck construction, and so justifies the representation of external functors defined on a category object by split opfibrations.

Essentially, what we will do is use objects and arrows representing sets and functions involved in the Grothendieck construction, find enough cones and commutative diagrams to characterize it, and take that as the definition of external functor in an arbitrary category.

14.7.3 Let $\mathcal{E}$ be a category, and let

$$\mathbf{C} = (C_0, C_1, \text{source}, \text{target}, \text{unit}, \text{comp})$$

be a category object in $\mathcal{E}$. An external functor $\mathbf{C} \to \mathcal{E}$ consists of data

$$(D_0, D_1, D_2, d^0, d^1, u, \pi_0, \pi_1, p_1, p_2, c)$$

for which the D_i are objects of $\mathcal{E}$ and the arrows have sources and targets as indicated:

$$\begin{aligned} d^i &: D_1 \to D_0, \quad i = 1, 2 \\ u &: D_0 \to D_1 \\ \pi_i &: D_i \to C_i, \quad i = 0, 1 \\ p_i &: D_2 \to D_1, \quad i = 1, 2 \\ c &: D_2 \to D_1. \end{aligned}$$

D_0 is the object corresponding to the disjoint union of the values of the functor. Note that we are not *given* a functor F as we were in Section 11.2 – we are being guided by Theorem 11.3.7 and the details of the Grothendieck construction to *define* an external functor, and π_0 represents the projection (taking (x, C) to C in the case of the Grothendieck construction). D_1 is the object corresponding to the arrows of the category $\mathbf{G}(C, F)$ as defined in Section 11.2, and π_1 takes (x, f) to f. Thus π_0 and π_1 together describe the functor $\mathbf{G}(F)$ in the case of the Grothendieck construction.

d^0 and d^1 are the source and target maps of that category, c is the composition and u picks out the identities.

The data are subject to the requirements E–1 through E–4 below.

E–1 All three diagrams below must commute and (a) must be a pullback:

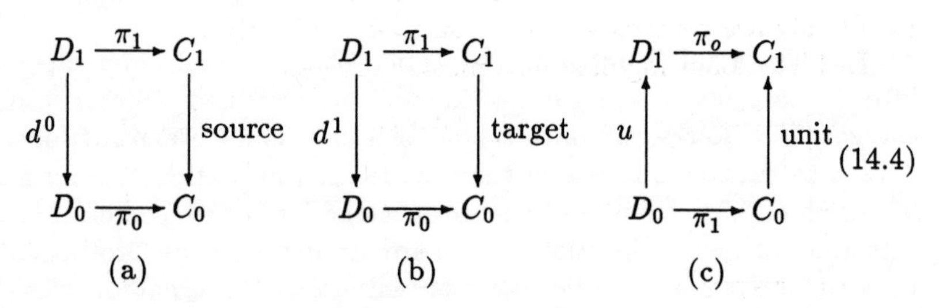

That (a) is a pullback says in the case of the Grothendieck construction that, up to unique isomorphism, D_1 consists of elements of the form (x, f) with f an arrow of $\mathcal{E}$ and $x \in F(C)$ where C is the source of f. This is part of GS–2 (see Section 11.2). The commutation of (b) says that $d_1(x, f) \in F(C')$ where C' is the target of F; that follows from GS–1 and GS–2 (there, x' must be in $F(C')$). That of (C) says that $u(x, C)$ must be (id_C). In the case of the Grothendieck construction that follows from the fact that F is given as a functor.

E–2 The following diagram is a pullback:

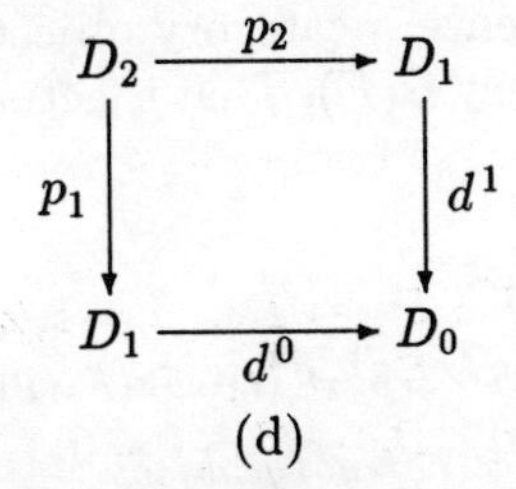

(d)

In the case of the Grothendieck construction this forces D_2 to be the set of composable pairs of arrows of $\mathbf{G}(C, F)$.

E–3 The following diagram must commute:

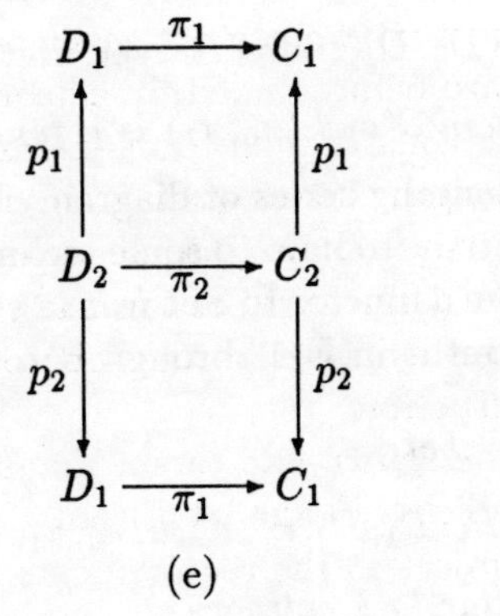

(e)

π_2 is defined by Diagram (14.3) (called f_2 there). In the case of the Grothendieck construction this follows from GS–3: (x, g') composes with (x, f) only if g composes with f (but not conversely!).

E–4 The following diagrams must commute.

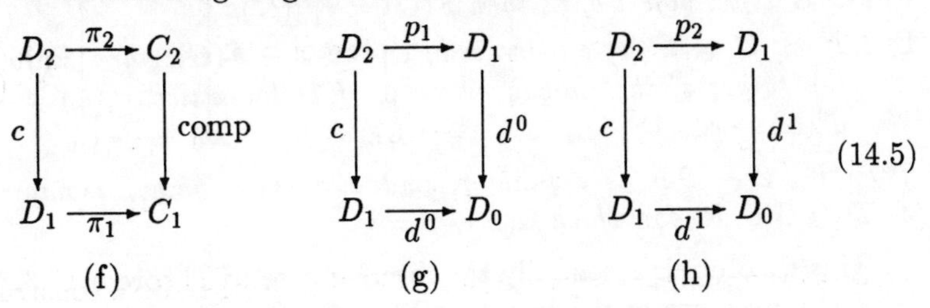

(f) (g) (h) (14.5)

In the case of the Grothendieck construction, (f) says that π_1 preserves composition and (g) and (h) say that the composite has the correct source and target.

As we have presented these diagrams, we have observed that they are all true in the case of the Grothendieck construction in **Set**. In the case of the Grothendieck construction,

$$(D_0, D_1, \text{source}, \text{target}, \text{unit}, \text{comp})$$

is actually a category, hence a category object in **Set**. Moreover, $\pi = (\pi_0, \pi_1)$ is a functor, namely $\mathbf{G}(F)$. That is actually true in any category, as seen in the following.

14.7.4 Proposition *Let*

$$(D_0, D_1, D_2, d^0, d^1, u, \pi_0, \pi_1, p_1, p_2, c)$$

be an external functor to a category object

$$\mathcal{C} = (C_0, C_1, \text{source}, \text{target}, \text{unit}, \text{comp})$$

in a category $\mathcal{E}$. Then

$$(D_0, D_1, \text{source}, \text{target}, \text{unit}, \text{comp})$$

is a category object in $\mathcal{E}$ and (π_0, π_1) is a functor in $\mathcal{E}$.

The proof is a lengthy series of diagram chases.

The converse is true, too: up to some technicalities, an external functor in **Set** arises from a functor to **Set** from a small category. That means we said enough about it in E–1 through E–4 to characterize it.

14.7.5 Theorem *Let*

$$(\mathcal{D}_0, \mathcal{D}_1, \mathcal{D}_2, d^0, d^1, u, \pi_0, \pi_1, p_1, p_2, c)$$

be an external functor to a category

$$\mathcal{C} = (\mathcal{C}_0, \mathcal{C}_1, \text{source}, \text{target}, \text{unit}, \text{comp})$$

in **Set**. *Define $F : \mathcal{C} \to$* **Set** *by*

F–1 *If C is an object of $\mathcal{C}$, then $F(C) = \pi_1^{-1}(C)$.*

F–2 *If $f : C \to C'$ is an arrow of $\mathcal{C}$ and $x \in F(C)$, then $F(f)(x)$ is the target of the unique arrow α of $\mathcal{D}$ for which $\pi_1(\alpha) = f$ and $d^0(\alpha) = x$.*

Then $F : \mathcal{C} \to$ **Set** *is a functor, and there is a unique isomorphism $\beta : \mathcal{D} \to \mathbf{G}(\mathcal{C}, F)$ for which $\mathbf{G}(F) \circ \beta = \pi$.*

This theorem is essentially the discrete case of Theorem 11.3.7.

F–2 makes use of the fact that Diagram (14.4)(a) is a pullback. A seasoned categorist will simply *name* $\alpha \in F^2$ as (x, f), since he knows that between any two pullbacks defined by (14.4)(a) there is a unique isomorphism respecting the projections π_1 and d^0.

14.8 The realizability topos

The category of modest sets has been proposed as a suitable model for the polymorphic λ-calculus. It is a subcategory of a specific topos, the realizability topos. Space limitations prevent us from giving a full exposition of this topic, which in any case is still under development at the time of writing. Here we describe how the realizability topos and the subcategory of modest sets are constructed.

14.8.1 Realizability sets In the discussion below, we will be writing $f(x)$ where f is a partial function. It will be understood that $f(x)$ is defined when we write this.

A **realizability set** is a pair $\mathbf{A} = (A, =_{\mathbf{A}})$ where A is a set called the **carrier** of $\mathbf{A}$ and $=_{\mathbf{A}}$: $A \times A \to \mathcal{P}(N)$ is a function to sets of natural numbers (thought of as reasons that two elements are equal; in fact, they can be thought of as the set of Gödel numbers of proofs that they are). We denote the value of this function at $(a_1, a_2) \in A \times A$ by $[\![a_1 =_{\mathbf{A}} a_2]\!]$, often abbreviated to $[\![a_1 = a_2]\!]$. This is subject to the following conditions:

REAL–1 There is a partial recursive function f such that for any a_1, $a_2 \in A$ and $n \in [\![a_1 = a_2]\!]$, $f(n) \in [\![a_2 = a_1]\!]$.

REAL–2 There is a partial recursive function of two variables g with the property that if $n \in [\![a_1 = a_2]\!]$ and $m \in [\![a_2 = a_3]\!]$, then $g(n, m) \in [\![a_1 = a_3]\!]$.

This is made into a category by defining an arrow $\mathbf{A} = (A, =_{\mathbf{A}}) \to \mathbf{B} = (B, =_{\mathbf{B}})$ to be an equivalence class of partial functions $\phi : A \to B$ such that there is a partial recursive function f such that $n \in [\![a_1 = a_2]\!]$ implies that $f(n) \in [\![fa_1 = fa_2]\!]$. Two such arrows ϕ and ψ are equal if there is a partial recursive f such that $n \in [\![a = a]\!]$ implies that $f(n) \in [\![\phi a = \psi a]\!]$.

Note that the last definition implies that f is defined on all elements $a \in A$ for which $[\![a = a]\!] \neq \emptyset$. The other elements of A are irrelevant.

14.8.2 This category is a topos. We will not prove this, but simply describe some of the constructions required. To find products, for example, first choose a recursive bijection $f : \mathbf{N} \times \mathbf{N} \to \mathbf{N}$. If $\mathbf{A} = (A, =_{\mathbf{A}})$ and $\mathbf{B} = (B, =_{\mathbf{B}})$ are realizability sets, the product is $(A \times B, =_{\mathbf{A}} \times =_{\mathbf{B}})$, where the latter is defined by

$$[\![(a_1, b_1) = (a_2, b_2)]\!] = \{f(n, m) \mid n \in [\![a_1 = a_2]\!], m \in [\![b_1 = b_2]\!]\}$$

Notice that a choice of f is required to show that products exist. The products thereby exist, but do not depend in any way on the choice of the function.

Equalizers can be defined as follows. If $\phi, \psi : \mathbf{A} \to \mathbf{B}$ are two arrows their equalizer is given by $(A, =_{\phi,\psi})$ where $[\![a_1 =_{\phi,\psi} a_2]\!] = [\![\phi a_1 = \psi a_2]\!]$.

Finally power objects can be constructed as follows. Let f denote a pairing as above and also choose an enumeration g_e of all the partial recursive functions. The carrier of $\mathcal{P}\mathbf{A}$ is $[A \to \mathcal{P}\mathbf{N}]$, all functions from A to sets of natural numbers. If P and Q are two such functions, then we will say that $f(d, e) \in [\![P = Q]\!]$ if for all $n \in P(a)$ and $m \in [\![a = b]\!]$, we have $g_d(f(n, m)) \in Q(b)$ and for all $n \in Q(A)$ and $m \in [\![a = b]\!]$, we have $g_e(f(n, m)) \in P(b)$.

This topos is called the **realizability** or **effective** topos. It is due to Martin Hyland [1982], although the basic idea goes back to S. Kleene. See also [Rosolini, 1987].

14.8.3 Modest sets 'Modest sets' is used more-or-less interchangeably to denote either a certain full subcategory of the category of realizability sets or an internal category object of that topos. A great deal of effort has been put into describing the connection between the two. We will begin with the former. The category of modest sets can be described directly and then embedded into the realizability topos.

A **modest set** consists of $\mathbf{A} = (\mathbf{N}, =_{\mathbf{A}})$ where as usual $\mathbf{N}$ denotes the natural numbers and $=_{\mathbf{A}}$ is a **partial equivalence relation** or **PER** on $\mathbf{N}$, which means it is symmetric and transitive, but not necessarily reflexive. We think of $\mathbf{A}$ as a quotient of a subobject of $\mathbf{N}$; the subobject is the set of n for which $n =_{\mathbf{A}} n$ modulo the relation $=_{\mathbf{A}}$.

The category of modest sets has arrows from $\mathbf{A}$ to $\mathbf{B}$ defined as partial recursive functions f defined on all n such that $n =_{\mathbf{A}} n$ and such that $n =_{\mathbf{A}} m$ implies that $f(n) =_{\mathbf{A}} f(m)$. Note that since the relation is symmetric and transitive, as soon as there is some m with $n =_{\mathbf{A}} m$, it is also the case that $n =_{\mathbf{A}} n$ and so $f(n)$ and $f(m)$ are defined.

The modest sets form a cartesian closed category. Choose, as above, an enumeration g_e of the partial recursive functions. Then $[\mathbf{A} \to \mathbf{B}]$ is the PER with relation given by $d =_{[\mathbf{A} \to \mathbf{B}]} e$ if and only if whenever $n =_{\mathbf{A}} m$ then $g_d(n) =_{\mathbf{B}} g_e(m)$. The modest sets do not form a topos.

We embed the modest sets into the realizability sets by choosing a pairing f and associating to the modest set $\mathbf{A} = (\mathbf{N}, =_{\mathbf{A}})$ the realizability set with carrier $\mathbf{N}$ and $e \in [\![n = m]\!]$ if and only if $n =_{\mathbf{A}} m$ and $e = f(n, m)$. Although this appears to depend on the choice of a pairing, it is easy to see that up to isomorphism it does not.

In fact it cannot. The pairing is used only to show that products *exist*. But their properties – in particular their isomorphism class – are independent of the particular construction used to prove their existence. Similar remarks apply to the cartesian closed structure, which does not depend on a particular enumeration of the partial recursive functions.

14.8.4 The internal category of modest sets This is a very brief glance at how one might internalize the description of modest sets to produce a category object inside the category of realizability sets. Except by way of motivation, this category object has no real connection with the category of modest sets. Nevertheless, we will call it the **internal category of modest sets.**

This internal category has the remarkable property that *every* internal diagram in it has a limit. This is not possible for ordinary categories (unless they are just posets), but the impossibility proof requires a property of set theory which is not valid for the realizability sets (namely the axiom of choice).

The construction is an exercise in internal expression. A modest set is a relation on $\mathbf{N}$ with certain properties. The natural numbers object in the category of realizability sets is the object $(\mathbf{N}, =_{\mathbf{N}})$, where $e \in [\![n =_{\mathbf{N}} m]\!]$ if and only if $n = m = e$. That is $[\![n =_{\mathbf{N}} m]\!] = \{n\}$ if $n = m$ and is empty otherwise. We will denote it by $\mathbf{N}$. Then a modest set is a subset of $\mathbf{N} \times \mathbf{N}$ with certain properties.

The set of objects of the internal category of modest sets is then a certain set of subsets of $\mathbf{N} \times \mathbf{N}$, which is the same as a subset of $\mathcal{P}(\mathbf{N} \times \mathbf{N})$. We want to describe this subset as consisting of those relations that are symmetric and transitive. The trick is to define endoarrows s and t of $\mathcal{P}(\mathbf{N} \times \mathbf{N})$ that associate to a relation R the relations R^{op} and $R \circ R$, respectively. For s, this is easy. There is an arrow $\langle p_2, p_1 \rangle : \mathbf{N} \times \mathbf{N} \to \mathbf{N} \times \mathbf{N}$ that switches the coordinates and we let $s = \mathcal{P}(\langle p_2, p_1 \rangle)$.

For t, it is a little harder. By Yoneda, an arrow $\mathcal{P}(\mathbf{N} \times \mathbf{N})$ to itself can be defined by describing a natural transformation $\mathrm{Hom}(-, \mathcal{P}(\mathbf{N} \times \mathbf{N}))$ to itself. An element of $\mathrm{Hom}(A, \mathcal{P}(\mathbf{N} \times \mathbf{N}))$ is, by the defining property of $\mathcal{P}$, a subobject $R \subseteq A \times \mathbf{N} \times \mathbf{N}$. Given such an R, we define R' to be the pullback in the following diagram:

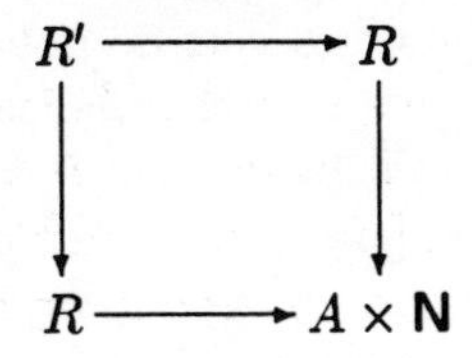

The two maps from $R \to A \times \mathbf{N}$ are the inclusion into $A \times \mathbf{N} \times \mathbf{N}$ followed by $\langle p_1, p_2 \rangle$ and $\langle p_1, p_3 \rangle$, respectively. You should think of R as being a set of triples (a, n_1, n_2) and then $R' \subseteq A \times \mathbf{N} \times \mathbf{N} \times \mathbf{N}$ is the set of 4-tuples (a, n_1, n_2, n_3) such that $(a, n_1, n_2) \in R$ and $(a, n_2, n_3) \in R$. Then the inclusion of R' into $A \times \mathbf{N} \times \mathbf{N} \times \mathbf{N}$ followed by $\langle p_1, p_2, p_4 \rangle$ gives us an arrow $R' \to A \times \mathbf{N} \times \mathbf{N}$ whose image we denote by $t(R)$. It is interpreted as the set of (a, n_1, n_3) such that there is an n_2 with both $(a, n_1, n_2) \in R$ and $(a, n_2, n_3) \in R$. We omit the proof that this construction from R to $t(R)$ is natural in A, but assuming that, it follows from the Yoneda Lemma that this results from an endomorphism t of $\mathcal{P}(\mathbf{N} \times \mathbf{N})$.

Finally, the object M_0 of objects is defined as the limit of

$$\mathcal{P}(\mathbf{N} \times \mathbf{N}) \overset{s}{\underset{t}{\Rrightarrow}} \mathcal{P}(\mathbf{N} \times \mathbf{N})$$

where the middle arrow is the identity. This is to be interpreted as the set of relations that are fixed under both s and t.

The object M_0 is the object of objects of the internal category of modest sets. The object of arrows is defined as a subobject of the object $[\mathbf{N} \to \mathbf{N}]^{\sim}$ of partial arrows of $\mathbf{N}$ to $\mathbf{N}$; namely those that satisfy the internal version of the definition of the ordinary of modest sets. This construction is tedious, but not difficult, and we omit it.

Appendix: Solutions to exercises

Solutions for Chapter 1

Section 1.2

1. Suppose f is injective and that $\mathrm{Hom}(S,f)(g) = \mathrm{Hom}(S,f)(h)$. Then $f \circ g = f \circ h$. Let x be any element of S. Then $f(g(x)) = f(h(x))$ and f is injective, so $g(x) = h(x)$. Since x is arbitrary, $g = h$. Conversely, suppose f is not injective. Then for some $t, u \in T$ with $t \neq u$, $f(t) = f(u)$. Define functions $g : S \to T$ and $h : S \to T$ to be the constant functions with values t and u respectively. Because S is nonempty, $g \neq h$. For any $x \in S$, $f(g(x)) = f(t) = f(u) = f(h(x))$, so $f \circ g = f \circ h$. Hence $\mathrm{Hom}(S,f)$ is not injective.

2. Suppose h is surjective, and suppose $\mathrm{Hom}(h,T)(f) = \mathrm{Hom}(h,T)(g)$. Then $f \circ h = g \circ h$. Let $x \in S$, and let $w \in W$ satisfy $h(w) = x$ (using surjectivity of h). Then $f(x) = f(h(w)) = g(h(w)) = g(x)$. Since x was arbitrary, this shows that $f = g$, so that $\mathrm{Hom}(h,T)$ is injective. Conversely, suppose that h is not surjective. This means there is a particular $x \in S$ which is not $h(w)$ for any $w \in W$. Let $t, u \in T$ be two distinct elements of T. Let $f : S \to T$ be the constant function with value t. Let $g : S \to T$ take x to u and all other elements to t. Then $\mathrm{Hom}(h,T)(f) = \mathrm{Hom}(h,T)(g)$, so $\mathrm{Hom}(h,T)$ is not injective.

3.a. To show that the mapping is injective, suppose $\langle f, g\rangle = \langle f', g'\rangle$. Then for any $x \in X$,

$$(f(x), g(x)) = \langle f, g\rangle(x) = \langle f', g'\rangle(x) = (f'(x), g'(x))$$

The coordinates of these pairs must be the same, so $f(x) = f'(x)$ and $g(x) = g'(x)$ for all $x \in X$. Hence $f = f'$ and $g = g'$, so $(f, g) = (f', g')$. To get surjectivity, suppose that $q : X \to S \times T$ is any function. Define $f : X \to S$ by requiring that $f(x)$ be the first coordinate of the pair $q(x)$ for all $x \in X$; in other words, $f(x) = \mathrm{proj}_1(q(x))$. Similarly set $g(x) = \mathrm{proj}_2(q(x))$. Then the mapping of the problem takes (f, g) to q, so it is surjective.

b. The pair $(\mathrm{proj}_1, \mathrm{proj}_2)$.

Section 1.3

1. The graph

$$a \overset{s}{\underset{t}{\rightrightarrows}} n \xrightarrow{\text{label}} l$$

expresses the concept. Interpret n as nodes and l as labels; then the arrow called 'label' takes a node to its label.

2. By definition, $\mathcal{G}$ is simple if and only if for any two distinct arrows f and g, either source$(f) \neq$ source(g) or target$(f) \neq$ target(g). Since for any arrow x,

$$\langle \text{source}, \text{target} \rangle(x) = (\text{source}(x), \text{target}(x))$$

$\mathcal{G}$ is simple if and only if for any two distinct arrows f and g,

$$\langle \text{source}, \text{target} \rangle(f) \neq \langle \text{source}, \text{target} \rangle(g)$$

But that says that $\langle$source, target$\rangle$ is injective by definition.

Section 1.4

1. Let $\phi : \mathcal{G} \to \mathcal{H}$ be a graph homomorphism with $\mathcal{H}$ simple. Let $f : a \to b$ be an arrow of $\mathcal{G}$. Then $\phi_1(f)$ has to be an arrow from $\phi_0(a)$ to $\phi_0(b)$. But there is at most one such arrow because $\mathcal{H}$ is simple. Thus $\phi_1(f)$ is determined by $\phi_0(a)$ and $\phi_0(b)$.

2. We must show that if $u : m \to n$ is an arrow of $\mathcal{G}$, then

$$\psi_1(\phi_1(u)) : \psi_0(\phi_0(m)) \to \psi_0(\phi_0(n))$$

in $\mathcal{K}$. Since ϕ is a graph homomorphism, $\phi_1(u) : \phi_0(m) \to \phi_0(n)$ in $\mathcal{H}$. The required equation then follows because ψ is a graph homomorphism.

3.a. Let $h : c \to d$ be a graph homomorphism in $\mathcal{H}$. We must show that $\psi_1(h) : \psi_0(c) \to \psi_0(d)$ in $\mathcal{G}$. Suppose $\psi_1(h) : a \to b$. Then, by definition of ψ and the fact that ϕ is a graph homomorphism,

$$h = \phi_1(\psi_1(h)) : \phi_0(a) \to \phi_0(b)$$

in $\mathcal{H}$. Thus $\phi_0(a) = c$ and $\phi_0(b) = d$, so $\psi_0(c) = a$ and $\psi_0(d) = b$ as required.

Here is a second proof that fits a pattern we use over and over in this book in showing that if an invertible arrow has a certain property, then its inverse does too.

$$\begin{aligned} \text{source}(\psi_1(h)) &= \psi_0(\phi_0(\text{source}(\psi_1(h)))) \\ &= \psi_0(\text{source}(\phi_1(\psi_1(h)))) \quad = \quad \psi_0(\text{source}(h)) \end{aligned}$$

and similarly for target.

b. By definition, we must prove the following equations:

$$\begin{aligned} \psi_0 \circ \phi_0 &= (\text{id}_{\mathcal{G}})_0 \\ \psi_1 \circ \phi_1 &= (\text{id}_{\mathcal{G}})_1 \\ \phi_0 \circ \psi_0 &= (\text{id}_{\mathcal{H}})_0 \\ \phi_1 \circ \psi_1 &= (\text{id}_{\mathcal{H}})_1 \end{aligned}$$

For any graph $\mathcal{F}$, $(\text{id}_{\mathcal{F}})_0 = \text{id}_{F_0}$ and $(\text{id}_{\mathcal{F}})_1 = \text{id}_{F_1}$. The result then follows from the fact that $\psi_i = (\phi_i)^{-1}$ for $i = 1, 2$.

Solutions for Chapter 2

Section 2.1

1. Functional composition is associative, and the identity functions satisfy C–3 and C–4, so it is only necessary to show that the identity functions are injective and that composite of injective functions is injective. If S is a set and $x, y \in S$ with $x \neq y$, then $\mathrm{id}_S(x) = x \neq y = \mathrm{id}_S(y)$, so id_S is injective. Let $f : S \to T$ and $g : T \to V$. If $x, y \in S$ and $x \neq y$, then $f(x) \neq f(y)$ because f is injective. But then $g(f(x)) \neq g(f(y))$ because g is injective. Hence $g \circ f$ is injective.

2. It is necessary to show that id_S is surjective for each set S and that if $f : S \to T$ and $g : T \to V$ are surjective then so is $g \circ f$. If $x \in S$, then $\mathrm{id}_S(x) = x$ so id_S is surjective. If $v \in V$, then there is a $t \in T$ for which $g(t) = v$ since g is surjective. There is $x \in S$ for which $f(x) = t$ since f is surjective. Then $g(f(x)) = v$, so $g \circ f$ is surjective.

3.a. $\mathrm{id}_A \circ u = u$ by definition of id_A, but $\mathrm{id}_A \circ u = \mathrm{id}_A$ by assumption. Hence $u = \mathrm{id}_A$. This is an example of a useful heuristic: when a property is given in terms of all arrows and you want to prove something about the property, see what it says for identity arrows.

b. $u \circ \mathrm{id}_A = u$ by definition of id_A, but $u \circ \mathrm{id}_A = \mathrm{id}_A$ by assumption. Hence $u = \mathrm{id}_A$.

Section 2.2

1. This requires an arrow **nonzero** : **NAT** $\to$ **BOOLEAN** and the following equations: **nonzero** $\circ$ 0 = **false** and **nonzero** $\circ$ **succ** = **true**. Since an arrow to **NAT** has to be a composite ending in 0 or **succ**, this takes care of all possibilities.

Section 2.3

1. A must be empty or have only one element.

2. Composition is certainly a binary operation on $\mathrm{Hom}(A, A)$ since all the arrows involved have the same source and target, so that every pair of arrows is composable. It is associative by C–2 and id_A is an identity by C–4.

3. If a semigroup S has two identity elements e and e', then $e = ee' = e'$. The first equation is true because e' is an identity and the second one is true because e is an identity.

Section 2.4

1.a. If (S, α) is a set with relation, then $\mathrm{id}_S : (S, \alpha) \to (S, \alpha)$ is a homomorphism and satisfies C–3 and C–4. We must show that the composite of homomorphisms is a homomorphism; since they are set functions, C–1 and C–2 will follow. Suppose $f : (S, \alpha) \to (T, \beta)$ and $g : (T, \beta) \to (V, \gamma)$ are homomorphisms. If $x\alpha y$, then $f(x)\beta f(y)$ because f is a homomorphism. Then $g(f(x))\gamma g(f(y))$ since g is a homomorphism. Hence $g \circ f$ is a homomorphism.

b. This is an immediate consequence of the definition of monotone in 2.4.2.

2. The identity functions are clearly continuous and strict (if relevant), so we need only show that the composite of continuous functions is continuous and the composite of strict functions is strict. Let $f : S \to T$ and $g : T \to V$ be continuous functions. Let s be the supremum of a chain $C = (c_0, c_1, c_2, \ldots)$ in S. Then the image $f(C) = \{f(c_i) \mid i \in \mathbf{N}\}$ is a chain in T because a continuous function is monotone. Moreover, $f(s)$ is the supremum of $f(C)$. Since g is continuous, $g(f(s))$ is the supremum of $g(f(C))$, so that $g \circ f$ is continuous. Finally, if f and g are strict, so is $g \circ f$, since then $g(f(\bot)) = g(\bot) = \bot$.

3. The nonnegative integers form a chain in $\mathbf{R}^+$. There is no element in $\mathbf{R}^+$ satisfying SUP–1, since such an element would be a real number which is bigger than any integer.

4. Let $C = (C_0, C_1, C_2, \ldots)$ be a chain in $(\mathcal{P}(S), \subseteq)$. Let V be the union $\bigcup_{i=1}^{\infty} C_i$. Then for every i, $C_i \subseteq V$, so SUP–1 is satisfied. Let $C_i \subseteq W$ for every i and suppose $v \in V$. Then $v \in C_i$ for at least one i since V is the union of all the C_i. Thus $v \in W$, which proves that $V \subseteq W$ so that SUP–2 holds.

5. Let $S = \mathbf{Z} \cup \{\top, \hat{\top}\}$, where $\top$ and $\hat{\top}$ are distinct and are not integers. Let the ordering $\leq$ on S be the usual ordering on $\mathbf{Z}$, with $\top \leq \hat{\top}$ and for any integer n, $n \leq \top$ and $n \leq \hat{\top}$. Then $(S, \leq)$ is an ω-CPO. The function $f : (S, \leq) \to (S, \leq)$ for which $n \mapsto n$ for $n \in \mathbf{Z}$, $\top \mapsto \hat{\top}$ and $\hat{\top} \mapsto \hat{\top}$ is monotone, but not continuous, since it does not preserve the sup of the chain $\mathbf{Z}$: that sup is $\top$ but $f(\top) = \hat{\top}$. The bottom element is $\emptyset$.

6. If h and k are partial functions, then $h \leq k$ means that whenever h is defined at n then so is k and moreover $h(n) = k(n)$. Suppose $\mathcal{H} = (h_0, h_1, \ldots)$ is a chain with supremum h. Then h is the union of all the partial functions h_i as sets of ordered pairs. To show that $\phi(h)$ is the supremum of $\phi(\mathcal{H}) = (\phi(h_0), \phi(h_1), \ldots)$, we must show that for any n, $\phi(h)(n)$ is defined if and only if $\phi(h_i)(n)$ is defined for some i and then $\phi(h)(n) = \phi(h_i)(n)$; that says that $\phi(h)$ is the union of all the partial functions $\phi(h_i)$ and so is the supremum of $\phi(\mathcal{H})$.

Suppose $\phi(h_i)(n)$ is defined. If $n = 0$ then $\phi(h_i)(0) = \phi(h)(0) = 1$ by definition of ϕ. Otherwise, $h_i(n-1)$ is defined and $\phi(h_i)(n) = nh_i(n-1)$. Then $h(n-1) = h_i(n-1)$ since h is the union of the h_i, so $\phi(h)(n) = nh(n-1) = nh_i(n-1) = \phi(h_i)(n)$.

Suppose $\phi(h)(n)$ is defined. Then $h(n-1)$ is defined and $\phi(h)(n) = nh(n-1)$. Since h is the union of the h_i there must be some i for which $h_i(n-1)$ is defined and $h(n-1) = h_i(n-1)$. But then by definition of ϕ, $\phi(h_i)(n) = nh_i(n-1) = nh(n-1) = \phi(h)(n)$ as required.

Now we must show that the factorial function f is the unique fixed point of ϕ. In the first place, $f(0) = 1 = \phi(f)(0)$. Also, f is a total function and $f(n) = nf(n-1)$, so that $\phi(f)$ is defined and by definition of ϕ, $\phi(f)(n) = nf(n-1) = f(n)$, so $\phi(f) = f$. If also $\phi(g) = g$, then $g(0) = 1$ and $g(n) = \phi(g)(n) = ng(n-1)$ so by induction g is the factorial function.

7. For a partial function h, define $\psi(h)$ by requiring that $\psi(h)(0) = \psi(h)(1) = 1$ and that for $n > 1$, if $h(n-1)$ and $h(n-2)$ are defined, then $\psi(h)(n)$ is defined and $\psi(h)(n) = h(n-1) + h(n-2)$. Then the Fibonacci function is the unique fixed point of ψ.

8. Let p be the supremum of the chain C. Since f is continuous, $f(p)$ is the supremum of the chain $f(C) = (f(\bot), f^2(\bot), \ldots)$. But p is an upper bound of $f(C)$ so $f(p) \le p$. On the other hand, the only element in C not in $f(C)$ is $\bot$, which is less than anything, so $f(p)$ is an upper bound of C. Thus $p \le f(p)$. Hence $p = f(p)$.

If $f(q) = q$, then $\bot \le q$, $f(\bot) \le f(q) = q$, ..., $f^n(\bot) \le f^n(q) = q$, and so on, so that q is an upper bound of C. Hence $p \le q$.

Section 2.5

1. Let $f : S \to T$ be a bijective semigroup homomorphism with inverse g. Let $t, t' \in T$. Then

$$g(tt') = g(f(g(t))f(g(t'))) = g(f(g(t)g(t'))) = g(t)g(t')$$

2. Let $f : S \to T$ and $g : T \to V$ be semigroup homomorphisms. Then for any two elements $s, s' \in S$, $g(f(s))g(f(s')) = g(f(s)f(s')) = g(f(ss'))$, the first equation because g is a homomorphism and the second because f is a homomorphism. Thus $g \circ f$ is a homomorphism.

3. For $m \in \mathbf{Z}_k$, $m = m + 0 = 0 \cdot k + m$, so $m +_k 0 = m$ and similarly $0 +_k m = m$. To verify the associative law let m, n and $p \in \mathbf{Z}_k$. Define r_i, q_i, $i = 1, 2, 3, 4$ by

$$\begin{aligned} m + n &= q_1 k + r_1 \\ n + p &= q_2 k + r_2 \\ r_1 + p &= q_3 k + r_3 \\ m + r_2 &= q_4 k + r_4 \end{aligned}$$

and $0 \le r_i < k$ for $i = 1, 2, 3, 4$. We must show that $r_3 = r_4$. It follows from the equations just given that $r_3 - r_4 = (q_2 + q_4 - q_1 - q_3)k$. Since $0 \le r_i < k$ for each i, the difference $r_3 - r_4$ has to be between $-k$ and k not inclusive. Since it is a multiple of k, it must be 0; hence $r_3 = r_4$.

4. Call the function ϕ, so that for some q, $n = qk + \phi(n)$ and $0 \le \phi(n) < k$. Since $0 = 0 \cdot k + 0$, $\phi(0) = 0$. For arbitrary m and n in $\mathbf{Z}$, let q_1 and q_2 be the integers for which $m + n = q_1 k + \phi(m + n)$ and $\phi(m) + \phi(n) = q_2 k + \phi(m) +_k \phi(n)$. Then

$$\phi(m + n) - (\phi(m) +_k \phi(n)) = (q_1 - q_2)k + (m - \phi(m)) + (n - \phi(n))$$

so that the left hand side of that equation is a multiple of k. Since it is the difference of two numbers whose absolute values are less than k, the difference must be 0, so that $\phi(m) +_k \phi(n) = \phi(m + n)$.

5. It is easy to show that there is a one to one correspondence between homomorphisms ϕ from that first monoid to any given monoid M and elements $x \in M$ such that $x^4 = 1$. The correspondence takes such an element x to the homomorphism ϕ which take the elements 0, 1, 2 and 3 to 1, x, x^2 and x^3, respectively. This clearly preserves the identity and the equation $\phi(m+n) = \phi(m)\phi(n)$ is interesting only when there is a carry mod 4, and in that case reduces to the requirement that $x^4 = 1$. A ϕ constructed this way is injective if and only if $x^n \neq 1$ for $n = 1$, 2 or 3. The second monoid is easily seen to have two such elements with that property, namely 2 and 3.

6. Given $f : A \to M$ and a string $a = (a_1, a_2, \ldots, a_m)$ in $F(A)$, let

$$\widehat{f}(a_1, a_2, \ldots, a_m) = f(a_1)f(a_2)\cdots f(a_m), \text{ if } n > 0$$

the product of the elements $f(a_i)$ in M and the identity of M otherwise. A monoid homomorphism such that $\widehat{f}(a) = f(a)$ for all $a \in A$ clearly must satisfy this equation; but we must prove that $\widehat{f}$ is a monoid homomorphism. For the empty string, which is the identity of $F(A)$, $\widehat{f}()$ is the identity of M. Thus $\widehat{f}$ preserves the identity. If $a' = (a'_1, a'_2, \ldots, a'_n)$ is another string, then

$$\begin{aligned}
\widehat{f}(aa') &= \widehat{f}\Big((a_1, a_2, \ldots, a_m)(a'_1, a'_2, \ldots, a'_n)\Big) \\
&= \widehat{f}(a_1, a_2, \ldots, a_m, a'_1, a'_2, \ldots, a'_n) \\
&= f(a_1)f(a_2)\cdots f(a_m)f(a'_1)f(a'_2)\cdots f(a'_n) \\
&= \Big(f(a_1)f(a_2)\cdots f(a_m)\Big)\Big(f(a'_1)f(a'_2)\cdots f(a'_n)\Big) \\
&= \widehat{f}(a)\widehat{f}(a')
\end{aligned}$$

so that $\widehat{f}$ preserves multiplication. Note that the fourth equality is an instance of the general associative law.

7. f^* is the unique function $\widehat{f}$ of the previous problem and so preserves concatenation because that is what the operation in the free monoid $F(B)$ is.

8. Since f^* is a homomorphism by the preceding problem, it is necessary only to show that it is bijective. We are given that f is a bijection (isomorphisms in **Set** are bijections). If a and a' are lists in A^* and they are different, then they differ in some coordinate; suppose the ith coordinates a_i and a'_i are different. The ith coordinates of $f^*(a)$ and $f^*(a')$ are then $f(a_i)$ and $f(a'_i)$, which must be different because f is injective. Thus f^* is injective. If b is a list in B, let a be a list in A of the same length for which $f(a_i) = b_i$ for each coordinate i. There is such a list since f is surjective, and then $f^*(a) = b$, so f^* is surjective.

Section 2.6

1. $C(M)^{\mathrm{op}}$ has exactly one object $*$ because $C(M)$ does. We define M^{op} to be $\mathrm{Hom}_{C(M)^{\mathrm{op}}}(*, *)$. If x, y, $z \in M$ and $xy = z$ in M then $yx = z$ in M^{op}. Clearly $C(M^{\mathrm{op}}) = C(M)^{\mathrm{op}}$.

2. The ordering in all posets in this answer will be written '$\leq$'. If P is a poset then between any two objects in $C(P)^{\text{op}}$ there is at most one arrow because that is true in $C(P)$. Thus $C(P)^{\text{op}}$ is the category determined by a poset called P^{op}; $x \leq y$ in P^{op} if and only if $y \leq x$ in P.

3. Let P be the set $\{1,2\}$ with discrete ordering (no two different elements are related). Let Q be the same set with the usual ordering $(1 \leq 2)$. Let R be the set $\{1,2,3\}$ with the discrete ordering. $C(P)$ is wide in $C(Q)$ (they have the same elements) but not full ($1 \leq 2$ in Q but not in P). $C(P)$ is full in $C(R)$ because in both $x \leq y$ holds only when $x = y$, but it is not wide since they do not have the same elements.

4. Let M be the two-element monoid $M = \{1, e\}$ with identity element 1 and $ee = e$. Then in $C(M)$, the only object is $*$ and the arrows are $1 : * \to *$ and $e : * \to *$. Consider the category $\mathcal{D}$ with one object $*$ and one arrow e. Note that e is the identity arrow for $*$ in $\mathcal{D}$ but not in $\mathcal{C}$. $\mathcal{D}$ satisfies requirements S–1, S–2 and S–4 but not S–3. Note that $\mathcal{D}$ is even a category; it is just not a subcategory.

5. Let $h : f \to f'$ and $h' : f' \to f''$. This means $f' \circ h = f$ and $f'' \circ h' = f'$. To show that $h' \circ h : f \to f''$ is an arrow of $\mathcal{C}/A$, we must show that $f'' \circ (h' \circ h) = f$. That follows from this calculation:

$$f'' \circ (h' \circ h) = (f'' \circ h') \circ h = f' \circ h = f$$

6. The free category generated by the graph (1.2) has the following arrows

(a) An arrow $\text{id}_1 : 1 \to 1$.

(b) For each nonnegative integer k, the arrow $\text{succ}^k : n \to n$. This is the path $(\text{succ}, \text{succ}, \ldots, \text{succ})$ (k occurrences of succ). This includes $k = 0$ which gives id_n.

(c) For each nonnegative integer k, the arrow $\text{succ}^k \circ 0 : 1 \to n$. Here $k = 0$ gives $0 : 1 \to n$.

Composition obeys the rule $\text{succ}^k \circ \text{succ}^m = \text{succ}^{k+m}$.

As for the graph (1.3), in the free category there are two objects a and n and four arrows id_a, id_n and $s, t : a \to n$.

Section 2.7

1. Let $f : A \to B$, $g : B \to A$ and $h : B \to A$ be arrows in a category with $g \circ f = h \circ f = \text{id}_A$ and $f \circ g = f \circ h = \text{id}_B$. Then

$$g = g \circ \text{id}_B = g \circ (f \circ h) = (g \circ f) \circ h = \text{id}_A \circ h = h$$

Note that this does not use the full power of the hypothesis.

2. We have $f^{-1} \circ g^{-1} \circ g \circ f = f^{-1} \circ f = \text{id}_A$ and similarly $g \circ f \circ f^{-1} \circ g^{-1} = \text{id}_C$. Thus $(g \circ f)^{-1} = f^{-1} \circ g^{-1}$ because the inverse is unique (preceding exercise).

3. (a) $M = M^{op}$: any monoid with commutative operation, for example the one-element monoid $\{e\}$ with $ee = e$.

(b) $M \neq M^{op}$ but M isomorphic to M^{op}: let $A = \{x, y\}$ and $M = A^*$. Then in M, $(x)(y) = (x, y)$, while in M^{op}, $(x)(y) = (y, x)$ so that $M \neq M^{op}$. On the other hand the function $\phi : M \to M^{op}$ defined by $\phi(a_1, \ldots, a_n) = (a_n, \ldots, a_1)$ is easily seen to be an isomorphism of M with M^{op}.

(c) M not isomorphic to M^{op}. A useful general construction for any set S is the 'right monoid' which is defined on the set $S \cup \{1\}$ (1 is some element not in S) by $xy = y$ for x, $y \in S$ and $x1 = 1x = x$ for $x \in S \cup \{1\}$. M^{op} (the 'left monoid') is not isomorphic to M so long as S has two or more elements.

4. Let P be the set $\{1, 2\}$ with discrete ordering (no two different elements are related). Then $P = P^{op}$ since $x \leq y$ if and only if $x = y$. Let Q be the same set with the usual ordering $(1 \leq 2)$. Then $Q \neq Q^{op}$ since $1 \leq 2$ in Q but not in Q^{op}. However the map $1 \mapsto 2$, $2 \mapsto 1$ is an isomorphism from Q to Q^{op}. Finally, let $S = \{\{1\}, \{2\}, \{1, 2\}\}$ ordered by inclusion. Then S is not isomorphic to S^{op} since S has an upper bound, namely $\{1, 2\}$ but S^{op} has no upper bound.

5. Let $(P, \leq)$ and $(Q, \leq)$ be posets and $\phi : P \to Q$ an isomorphism. Suppose P is totally ordered. Let q, $q' \in Q$. Since P is totally ordered, we may assume (after perhaps renaming) that $\phi^{-1}(q) \leq \phi^{-1}(q')$. Then $q = \phi(\phi^{-1}(q) \leq \phi(\phi^{-1}(q'))) = q'$ so Q is totally ordered.

6. If x and y are isomorphic objects, then there is an arrow from x to y and one from y to x (the isomorphism and its inverse). Then $x \leq y$ and $y \leq x$, so $x = y$ by antisymmetry.

7. Let $\phi : S \to T$ be an isomorphism of semigroups. Since it a set function with an inverse, it is a bijection by Proposition 2.7.7. Conversely, suppose that $\phi : S \to T$ is a bijective homomorphism. Then its inverse is a homomorphism by Exercise 1 of Section 2.5.

8.a. There is exactly one *set* function from the empty set to any set. It preserves the multiplication of the empty semigroup vacuously; thus the empty semigroup is the initial object. Let E be a one-element semigroup and S any semigroup. If e is the only element of E, the multiplication must be given by $ee = e$. The unique function $f : S \to E$ taking everything in S to e preserves the multiplication: if x, $y \in S$, then $f(x)f(y) = ee = e = f(xy)$. Thus E is a terminal object.

b. Let U be the semigroup with two elements 1 and e with $1 \cdot e = e \cdot 1 = e$, $1 \cdot 1 = 1$ and $e \cdot e = e$. Then any nonempty semigroup S has *two* homomorphisms to U: the constant function taking everything to 1 and the one taking everything to e.

9. Let $f : A \to B$. If $A = \emptyset$, let b be any element of B. If not, let $f(x) = f(y) = b$ for all x, $y \in A$. Define $k : 1 \to A$ to be $* \mapsto b$ where $*$ is the unique element of 1. Then $f = k \circ \langle\rangle$. Conversely, if $k : 1 \to A$ and $f = k \circ \langle\rangle$, then for any x, $y \in A$, $f(x) = k(*) = f(y)$.

10. Let $\mathcal{E}$ be the graph with one node e and one arrow $f : e \to e$. Let $\mathcal{G}$ be any graph. Define $\phi : \mathcal{G} \to \mathcal{E}$ by $\phi_0(g) = e$ for any node g and $\phi_1(u) = f$ for any arrow $f : g \to h$. Then $\phi_1(u) : \phi_0(g) \to \phi_0(h)$, so ϕ is a graph homomorphism. It is clearly the only possible one.

11. Suppose $f : S \to S$ is an idempotent set function. Let x be an element of the image, so there is some $s \in S$ such that $f(s) = x$. Then $f(x) = f(f(s)) = f(s) = x$, so x is a fixed point. Conversely, if $f(x) = x$, then x is in the image of f by definition.

Conversely, suppose the image of f is the same as its set of fixed points. Then for any $x \in S$, $f(f(x)) = f(x)$, so that $f \circ f = f$.

12.a. Let $f : A \to A$ be an idempotent set function. Let B be the image of f and $g : A \to B$ the corestriction of f. Let $h : B \to A$ be the inclusion of b in A. Then for any $x \in A$, $h(g(x)) = g(x) = f(x)$, so $h \circ g = f$. And for any $b \in B$, $g(h(b)) = g(b) = f(b) = b$ by the result of the preceding problem, so $g \circ h = \mathrm{id}_B$.

b. Let $\mathcal{E}$ be the category with one object $*$ and two arrows $1, e$ with $1 = \mathrm{id}_*$ and $e \circ e = e$. There are no arrows g, h for which $h \circ g = e$ and $g \circ h = 1$ since in this category, composition is commutative. (Note that 1 is split. The identity is split in any category.)

13. (i) is false. The category with two distinct objects and only identity arrows is a counterexample.

(ii) is false. Take the category of the preceding example and add one arrow u going from one of the objects, say A, to itself, with $u \circ u = \mathrm{id}_A$.

(iii) is true. Suppose $f : A \to B$ is an arrow of $C(P)$. Its inverse goes from B to A so $A \leq B$ and $B \leq A$ in P. Hence $A = B$. Since there is never more than one arrow with the same source and target in $C(P)$, f must be id_A.

Section 2.8

1. Suppose $f \circ x = f \circ y$. Then

$$x = \mathrm{id}_A \circ x = f^{-1} \circ f \circ x = f^{-1} \circ f \circ y = \mathrm{id}_A \circ y = y$$

hence f is a monomorphism.

2. Let $f : S \to T$ be a semigroup homomorphism. Suppose it is a monomorphism. Let $x, y \in S$ be distinct. Let $\mathbf{N}^+$ denote the semigroup of positive integers on addition. Let $p_x : \mathbf{N}^+ \to S$ take k to x^k and similarly define p_y; p_x and p_y are homomorphisms since for all x, $x^{k+n} = x^k x^n$. Since $x \neq y$, p_x and p_y are *distinct* homomorphisms. If $f(x) = f(y)$ then for all positive integers k,

$$f(p_x(k)) = f(x^k) = f(x)^k = f(y)^k = f(y^k) = f(p_y(k))$$

so that $f \circ p_x = f \circ p_y$ which would mean that f is not a monomorphism. Thus we must have $f(x) \neq f(y)$ so that f is injective. The same trick works with monoids, taking the nonnegative integers $\mathbf{N}$ on addition and setting $p_x(0)$ to be the identity.

Conversely, suppose that f is injective. Let $g, h : V \to S$ be homomorphisms for which $f \circ g = f \circ h$. For any $v \in V$, $f(g(v)) = f(h(v))$, so $g(v) = h(v)$ since f is injective. Hence $g = h$. Thus f is a monomorphism.

3.a. Let $f : A \to B$ be a monomorphism in $\mathcal{C}$ and let $\mathcal{D}$ be a subcategory of $\mathcal{C}$. Then if $f \circ g = f \circ h$ in $\mathcal{D}$ the same equation is true in $\mathcal{C}$, so that $g = h$ in $\mathcal{C}$. Hence f is a monomorphism in $\mathcal{D}$.

b. Let $\mathcal{C}$ be the full subcategory of **Set** determined by the sets $A = \{1\}$ and $B = \{1,2\}$. Let $k : B \to A$ be the (only possible) function. Let $\mathcal{D}$ be the subcategory of $\mathcal{C}$ consisting of id_A, id_B, and k. Then k is a monomorphism in $\mathcal{D}$ since there are not two distinct arrows from B to B, but it is not a monomorphism in $\mathcal{C}$ since k composed with any of the four functions from B to B gives k again.

4. In this answer we use these facts repeatedly:

(a) An isomorphism in the category of sets is a bijection and conversely.

(b) If $i : X \to S$ and $j : Y \to S$ are injective functions and $\beta : X \to Y$ is a bijection for which $j \circ \beta = i$ then $i \circ \beta^{-1} = j$.

(a)(i) Let $s = m(a)$ be in the image of m and let $\beta : A \to B$ be the bijection for which $m = n \circ \beta$. Then $s = n(\beta(a))$ so s is in the image of n. A symmetric argument using β^{-1} shows that the image of n is included in the image of m, so the images are equal.

(a)(ii) Let $m : A \to S$ be an injection in $\mathcal{O}$. Define $\beta : A \to I$ to be the corestriction of m to I. β is injective because m is and surjective because I is the image of m. Hence it is bijective. Clearly $i \circ \beta = m$.

(a)(iii) Let $\beta : J \to I$ be the bijection for which $i \circ \beta = j$. Since i and j are both inclusions, for any $x \in J$, $i(x) = x = j(x) = i(\beta(x))$, so $i \circ \beta = i$. Since i is injective, $\beta(x) = x$ for all $x \in J$. Since β is surjective, $I = J$ and $\beta = \mathrm{id}_I$.

(a)(iv) Immediate from (i), (ii) and (iii).

(b)(i) Immediate from the definition of subobject.

(b)(ii) Immediate from properties of equivalence relations.

(b)(iii) Immediate from (i) and (ii).

5. Let $m : A \to 1$ be monic and let $f, g : B \to A$. Then $m \circ f = m \circ g : B \to 1$ by definition of terminal object, and m is monic, so $f = g$. In particular, for any arrow $h : A \to C$, $h \circ f = h \circ g$ implies that $f = g$; hence h is monic.

6.a. The terminal object is a one-element set. By the remarks in 2.8.9, such a set has two subobjects because it has two subsets: the set itself and the empty set. (Note in connection with the preceding exercise that there is at most one function from any set B to the empty set: none at all if B is nonempty, and the identity function if B is empty.)

b. The terminal graph has one node and one arrow. It has three subgraphs: itself, the graph with one node and no arrows, and the graph with no nodes or arrows. These are in one to one correspondence with the subobjects.

c. The terminal monoid is the monoid with one element. It has only itself as a submonoid.

Section 2.9

1. Let $f : S \to T$ be a surjective monoid homomorphism and let $g, h : T \to V$ be monoid homomorphisms. Suppose that $g \circ f = h \circ f$. Let t be any element of T. Then there is an element $v \in V$ for which $f(v) = t$. Then $g(t) = g(f(v)) = h(f(v)) = h(t)$. Since t was arbitrary, $g = h$ so f is an epimorphism.

2. That ϕ is a homomorphism requires verifying that $\phi(0) = 0$ (which is true by definition) and that $\phi(m +_4 n) = \phi(m) +_2 \phi(n)$ for all $m, n \in \mathbf{Z}_4$: for example, $\phi(2 +_4 3) = \phi(1) = 1$ and $\phi(2) +_2 \phi(3) = 0 +_2 1 = 1$. (The comments in the answer to Exercise 5 of Section 2.5 apply here too.) It is surjective since it has both 0 and 1 as values.

Now suppose $\psi : \mathbf{Z}_2 \to \mathbf{Z}_4$ is a monoid homomorphism satisfying $\phi \circ \psi = \mathrm{id}_{\mathbf{Z}_2}$. Since $\psi(0) = 0$ necessarily, there are only four possibilities for ψ, determined by what it does to 1. If $\psi(1) = 0$ then $\phi(\psi(1)) = \phi(0) = 0 \neq 1$. If $\psi(1) = 1$ then $\psi(1 +_2 1) = \psi(0) = 0$ but $\psi(1) +_4 \psi(1) = 2$, so ψ is not a homomorphism. If $\psi(1) = 2$ then $\phi(\psi(1)) \neq 1$. Finally, if $\psi(1) = 3$, then again ψ is not a homomorphism because $\psi(1 +_2 1) = 0 \neq 2 = \psi(1) +_4 \psi(1)$.

3.a. The statement that $m \in M$ is a monomorphism says that $mn = mp$ implies $n = p$ (m is left cancellable). That it is an epimorphism says that $nm = pm$ implies $n = p$ (m is right cancellable) If m is an isomorphism then m is a monomorphism and an epimorphism by Proposition 2.9.9. To see the converse, let m be a monomorphism. Let $M = \{m_1, m_2, \ldots, m_k\}$ be finite. The elements $mm_1, mm_2, \ldots, mm_k$ are all different, so there are k of them, so one of them, call it mn, is 1. But then $mnm = m = m \cdot 1$ and m is a monomorphism, so $nm = 1$. Hence m is an isomorphism. A symmetric argument shows that an epimorphism is an isomorphism.

Warning: In a finite *semigroup*, a left cancellable element need not be right cancellable.

b. Every nonzero element of the nonnegative integers with addition as operation is both a monomorphism and an epimorphism but not an isomorphism. (In other words, $m + n = m + p$ implies $n = p$ and similarly on the other side, but if $m \neq 0$ then m has no inverse.)

4. If $f : A \to B$ has a splitting $g : B \to A$, then in P, $A \leq B$ and $B \leq A$ so $A = B$. The only arrow from A to itself is id_A.

5. See the answer to Exercise 3(b).

6. $h \circ g$ is an idempotent because $h \circ g \circ h \circ g = h \circ \mathrm{id} \circ g = h \circ g$. It is split because h and g fit the requirements for the h and g of the definition in the exercise mentioned: $h \circ g = h \circ g$ and $g \circ h = \mathrm{id}$.

7. (i) $\mathrm{Hom}(A,f)(g) = \mathrm{Hom}(A,f)(h)$ if and only if $f \circ g = f \circ h$ by definition of $\mathrm{Hom}(A,f)$. The result is the immediate from the definitions of injective and monomorphism.

(ii) $\mathrm{Hom}(f,D)(u) = \mathrm{Hom}(f,D)(v)$ if and only if $u \circ f = v \circ f$. Again, the result follows from the definition of injective and epimorphism.

(iii) Suppose f is a split monomorphism with splitting g, so $g \circ f = \mathrm{id}_B$. Suppose $u : B \to D$. Then $\mathrm{Hom}(f,D)(u \circ g) = u \circ g \circ f = u$, so $\mathrm{Hom}(f,D)$ is surjective. Conversely suppose $\mathrm{Hom}(f,D)$ is surjective for every object D. Then there is a $g : C \to B$ such that $\mathrm{hom}(f,B)(g) = \mathrm{id}_B$; that is, $g \circ f = \mathrm{id}_B$, so f is split.

(iv) Suppose f is a split epimorphism, with $f \circ g = \mathrm{id}_C$. Let $u : A \to C$. Then $\mathrm{Hom}(A,f)(g \circ u) = f \circ g \circ u = u$ so $\mathrm{Hom}(A,f)$ is surjective. Conversely, suppose $\mathrm{Hom}(A,f)$ is surjective for every A. Let $g : C \to B$ satisfy $\mathrm{Hom}(B,f)(g) = \mathrm{id}_C$, that is, $f \circ g = \mathrm{id}_C$, so f is a split epimorphism.

(v) An isomorphism is a split epimorphism and a split monomorphism because it is split by its inverse. Thus an isomorphism satisfies (a) and (b). In particular an isomorphism is an epimorphism and a monomorphism. It then follows from (i) through (iv) that an isomorphism satisfies (c) and (d). Conversely, suppose f is a split epi, split by $g : C \to B$, so $f \circ g = \mathrm{id}_C$. Then $f \circ g \circ f = f = f \circ \mathrm{id}_B$. If f is also mono then $g \circ f = \mathrm{id}_B$. Hence (a) implies that f is an isomorphism. That also follows from (b) by doing the same proof in the opposite category (note that the dual of the concept of isomorphism is isomorphism). Finally, (c) implies (a) by (i) and (iv) and (d) implies (b) by (ii) and (iii).

8. (i) Suppose an arrow $x : m \to n$ in $\mathcal{H}$ is not in the image of f_1. Let $\mathcal{U}$ be the graph with one node a and two arrows $u, v : a \to a$. Let $g : \mathcal{H} \to \mathcal{U}$ take all nodes to a and all arrows to u, and let h take all nodes to a and all arrows except x to u, with $h(x) = v$. Both g and h are graph homomorphisms. Then $g \circ f = h \circ f$ but $g \neq h$, showing that f is not epi.

If all the arrows of $\mathcal{H}$ are in the image of f_1 but there is a node n not in the image of f_0, then there can be no arrows with n as source or target. Let $\mathcal{V}$ be the graph with two nodes a and b and one arrow $u : a \to a$. Let $g : \mathcal{H} \to \mathcal{V}$ take all nodes to a and all arrows to u; let $h : \mathcal{H} \to \mathcal{V}$ take all nodes except n to a, n to b and all arrows to u. Again, g and h are graph homomorphisms and $g \neq h$ but $g \circ f = h \circ f$.

Conversely, suppose that f_0 and f_1 are both surjective. Let $g, h : \mathcal{H} \to \mathcal{K}$ satisfy $g \circ f = h \circ f$. Then for every node *or* arrow x of $\mathcal{H}$ there is a node or arrow m of $\mathcal{G}$ for which $f_i(m) = x$ ($i = 0$ if x is a node, $i = 1$ if x is an arrow). Then $g_i(x) = g_i(f_i(m)) = h_i(f_i(m)) = h_i(x)$, so $g = h$ and f is epi.

(ii) Suppose f is monic and suppose u and v are arrows of $\mathcal{G}$ for which $f_1(u) = f_1(v)$. Let $\mathcal{F}$ denote the graph with two nodes a and b and one arrow $x : a \to b$, and define $g : \mathcal{F} \to \mathcal{G}$ to take x to u and $h : \mathcal{F} \to \mathcal{G}$ to take x to v. What g and h do on nodes is then forced. Then $f \circ g = f \circ h$ but $g \neq h$, so f is not monic. If f is monic and f_1 is injective but there are nodes m and n of $\mathcal{G}$ for which $f_0(m) = f_0(n)$, then we carry out the same trick except that we take $\mathcal{F}$ to be the graph with one node and no arrows and let g and h take that node to m and n respectively.

Conversely, suppose f_0 and f_1 are injective. Suppose $g : \mathcal{F} \to \mathcal{G}$ and $h : \mathcal{F} \to \mathcal{G}$ are graph homomorphisms such that $f \circ g = f \circ h$. Let x be any node or arrow of $\mathcal{F}$. Then $f_i(g_i(x)) = f_i(h_i(x))$ ($i = 0$ if x is a node, $i = 1$ if x is an arrow), so because f_i is injective, $g_i(x) = h_i(x)$. Hence $g = h$ so f is monic.

(iii) This is an immediate consequence of Exercise 3 of Section 1.4 and the fact that an isomorphism in the category of sets is a bijection.

Solutions for Chapter 3

Section 3.1

1. Since functors preserve the operations of domain and codomain, the fact that $g \circ f$ is defined implies that the domain of g is the codomain of f. But then the domain of $F_1(g)$, which is F_1 applied to the domain of g, is the same as the codomain of $F_1(f)$. Hence $F_1(g) \circ F_1(f)$ is defined in $\mathcal{D}$.

2. The initial category has no objects and, therefore, no arrows. It clearly has exactly one functor to every other category. The terminal category has just one object and the identity arrow of that object. To any category $\mathcal{C}$ there is just one functor that takes every object to that single object and every arrow to that one arrow.

3. If $f : A \to B$ is a function between sets, the existential powerset functor takes f to f_* defined by $f_*(A_0) = \{f(a) \mid a \in A_0\}$. If $g : B \to C$, then $c \in (g_* \circ f_*)(A_0)$ if and only if there is an $a \in A_0$ such that $c = g(f(a))$. This is the same condition that $c \in (g \circ f)_*(A_0)$. It is evident that when f is the identity, so is f_*.

The universal powerset functor $f_!$ is defined by $b \in f_!(A_0)$ if and only if $b = f(a)$ implies that $a \in A_0$. If f is the identity, then $b = f(a)$ if and only if $b = a$ so that $f_!(A_0) = A_0$ (the identity arrow). If $f : A \to B$ and $g : B \to C$, then $c \in (g_! \circ f_!)(A_0)$ if and only if $c = g(b)$ implies that $b \in f_!(A_0)$, which is true if and only if $b = f(a)$ implies $a \in A_0$. Putting these together, we see that $c \in (g_! \circ f_!)(A_0)$ if and only if $c = f(g(a))$ implies $a \in A_0$. But this is the condition that $c \in (g \circ f)_!(A_0)$.

4. Let $\mathcal{C} = \mathbf{2} + \mathbf{2}$, the category that can be pictured as

$$0 \longrightarrow 1 \qquad 1' \longrightarrow 2$$

with no other nonidentity arrows, and $\mathcal{D}$ the category $\mathbf{3}$ that looks like

$$0 \longrightarrow 1 \longrightarrow 2$$

The functor $F : \mathcal{C} \to \mathcal{D}$ identifies 1 and $1'$. The image of F includes all of $\mathcal{D}$ except the composite arrow from $0 \to 2$. Thus the image is not closed under composition and is not a subcategory.

5.a. Since a functor is determined by its value on objects and arrows, a functor that is injective on objects and arrows is certainly a monomorphism. (Compare Exercise 8 of Section 2.9.) The other way is less obvious. If F is not injective on objects, say $C \neq C'$ but $F(C) = F(C')$ then let 1 be the category with one object and only the identity arrow. Let $G : 1 \to \mathcal{C}$ take that object to C, while $G' : 1 \to \mathcal{C}$ takes it to C'. Then $G \neq G'$, while $F \circ G = F \circ G'$, whence F is not a monomorphism. Next suppose F is injective on objects, but not on arrows. If $f \neq g$ are two arrows with $F(f) = F(g)$, then $F(f)$ and $F(g)$ have the same domain and the same codomain. Since F is injective on objects, it must be that f and g have the same domain and the same codomain. Say that $f, g : C \to C'$. Now let $\mathbf{2}$ be the category with two objects and one nonidentity arrow between them:

$$0 \longrightarrow 1$$

Let $G : \mathbf{2} \to \mathcal{C}$ take 0 to C, 1 to C' and the nonidentity arrow to f, while G' is the same on objects, but takes the nonidentity arrow to g. Clearly $G \neq G'$, while $F \circ G = F \circ G'$.

b. The functor simply forgets the existence of the identity element. Two distinct monoids must differ in either their sets of elements or multiplication and in either case must differ as semigroups. Thus the functor is injective on objects. For similar reasons, it is injective on arrows. We have just seen that a functor that is injective on objects and arrows is a monomorphism.

6.a. It being evident that e is a two-sided identity, it is necessary only to show that the multiplication is associative. In any equation $x(yz) = (xy)z$ as soon as any of the variables is e, both sides reduce to a binary product of the remaining two terms (one or both of which might also be e). If none of them is e, then this is an equation involving terms from S and so is valid because it is in S.

b. If we denote by e_S and e_T the elements added to S and T respectively, then we define $F(f) = f^1 : S^1 \to T^1$ by

$$f^1(x) = \begin{cases} f(x) & \text{if } x \in S \\ e_T & \text{if } x = e_S \end{cases}$$

In verifying that $f^1(xy) = f^1(x)f^1(y)$ it is necessary to consider cases. If neither x nor y is e_S, then it follows from the fact that f is a homomorphism of semigroups. If, say, $x = e_S$, then both sides reduce to $f^1(y)$ and similarly if $y = e_S$. Finally, we must show that F is a functor. It is clear that if $f : S \to S$ is the identity, then $f^1 : S^1 \to S^1$ is the identity as well. It is also clear that if $g : T \to R$ is another monoid homomorphism, then

$$(g^1 \circ f^1)(x) = \begin{cases} (g \circ f)(x) & \text{if } x \in S \\ e_R & \text{if } x = e_S \end{cases} = (g \circ f)^1(x)$$

c. Since F is injective on objects and arrows, it is a monomorphism (see preceding exercise).

7. Define $\alpha : \text{Hom}_{\textbf{Mon}}(F(A), M) \to \text{Hom}_{\textbf{Set}}(A, U(M))$ by $\alpha(g)(a) = g(a)$. Then $(\alpha(\beta(f)))(a) = f(a)$ is clear. On the other hand, if $g : F(A) \to M$ is a monoid homomorphism, then

$$\begin{aligned}(\beta(\alpha(g)))(a_1, a_2, \cdots, a_n) &= \alpha(g)(a_1)\alpha(g)(a_2)\cdots\alpha(g)(a_n)\\ &= g(a_1)g(a_2)\cdots g(a_n) = g(a_1, a_2, \cdots, a_n)\end{aligned}$$

the last equality coming from the fact that g is a monoid homomorphism. Thus β is invertible and hence bijective.

8. In Exercise 6, we showed that $\gamma(h)$ (which would have been called h^1 there) is a monoid homomorphism. To see that γ is a bijection, we define

$$\delta : \text{Hom}_{\textbf{Mon}}(F(S), M) \to \text{Hom}_{\textbf{Sem}}(S, U(M))$$

by $\delta(g)(x) = g(x)$ for $x \in S$. Since a monoid homomorphism is also a semigroup homomorphism, this is well defined. Clearly $\delta(\gamma(h))(x) = h(x)$ for $x \in S$ and $h : S \to U(M)$. To go the other way, suppose $g : F(S) \to M$. For $x \in S$, $\gamma(\delta(g))(x) = \delta(g)(x) = g(x)$. For e_S, we have $\gamma(\delta(g))(e_S) = 1$ by definition of γ, but $g(e_S) = 1$ since g is a monoid homomorphism. Thus $\delta = \gamma^{-1}$ and γ is a bijection.

9.a. This is an immediate consequence of Exercises 5.b and 6.c.

b. It is obvious that **Set** is a subcategory of the category of partial functions. To get a functor in the other direction, let $F(S) = S \cup \{S\}$. If $f : S \to T$ is a partial function, let $F(f) : F(S) \to F(T)$ be the total function defined by

$$F(f)(x) = \begin{cases} f(x) & \text{if } x \in S \text{ and } f(x) \text{ is defined} \\ T & \text{if } x = S \text{ or } f \text{ is not defined at } x \end{cases}$$

It is clear that if $f \neq g$, then $F(f) \neq F(g)$.

10. On one hand, if $\mathcal{A}$ is discrete, for any function $F_0 : \mathcal{A}_0 \to \mathcal{B}_0$ there is a unique functor $F : \mathcal{A} \to \mathcal{B}$ whose value at the identity of some object A of $\mathcal{A}$ is the identity of $F_0(A)$. To go the other way, suppose $\mathcal{A}$ is not discrete. Suppose first that there is an arrow $f : A \to B$ in $\mathcal{A}$ with $A \neq B$. Let $\mathcal{B}$ be the category with the same objects as $\mathcal{A}$, but with no nonidentity arrows. Then the identity function $F_0 : \mathcal{A}_0 \to \mathcal{B}_0$ cannot be extended to a functor since there is nowhere to send f. Now suppose that all arrows of $\mathcal{A}$ are endoarrows. Let $\mathcal{B}$ have the same objects as $\mathcal{A}$ and let

$$\text{Hom}_{\mathcal{B}}(A, A) = \text{Hom}_{\mathcal{A}}(A, A) \times \text{Hom}_{\mathcal{A}}(A, A)$$

for an object A of $\mathcal{A}$. The identity function $F_0 : \mathcal{A}_0 \to \mathcal{B}_0$ can be extended in at least two ways to a functor. The first way is to take $F(f) = (f, \text{id}_A)$ for $f : A \to A$ in $\mathcal{A}$ and the second is $F'(f) = (\text{id}_A, f)$.

11. We claim that a category is indiscrete if and only if there is exactly one arrow between any two objects. It is obvious that such a category is indiscrete. To go the other way, first suppose that $\mathcal{A}$ has two objects A and B with no arrows between them, then $\mathcal{A}$ cannot be indiscrete. For if **2** denotes the category with two objects 0 and 1 and one arrow between them, then the object function that takes 0 to A and 1 to B cannot be extended. If there is more than one arrow from A to B, then the same object function on **2** has more than one extension to a functor.

Section 3.2

1. If M acts on S, let $\phi : M \to FT(S)$ be defined by $\phi(a)(x) = ax$ for $a \in M$ and $x \in S$. We have $\phi(1)(a) = 1a = a = \mathrm{id}(a)$ so that $\phi(1) = \mathrm{id}$. Also, for $a, b \in M$,

$$\phi(ab)(x) = (ab)x = a(bx) = \phi(a)(\phi(b)(a))$$

so that $\phi(ab) = \phi(a) \circ \phi(b)$. Conversely, if $\phi : M \to FT(S)$, then let M act on S by letting $ax = \phi(a)(x)$. The computations above can be reversed to show that since ϕ is a monoid homomorphism, the M-set identities are satisfied. It is clear that these processes are inverse to each other.

2. The identity of A^* is the empty word () and the definition of ϕ^* is that $\phi^*((),a) = a$ for all $a \in A$, so that A–1 is satisfied. Also, if we assume that

$$\phi^*(wv, m) = \phi^*(w, \phi^*(v, m))$$

for words w of length k, then

$$\begin{aligned}\phi^*((a)wv, m) &= \phi(a, \phi^*(wv, m)) \\ &= \phi(a, \phi^*(w, \phi^*(v, m))) = \phi^*((a)w, \phi^*(v, m))\end{aligned}$$

The first and third equality are from the definition of ϕ, while the second is from the inductive hypothesis.

Section 3.3

1.a. Since a functor is faithful if it is injective between hom sets and a monoid has only one hom set, such a functor is faithful if and only if it is injective.

b. Since hom sets are either singleton or empty and a function on such a set is always injective, such a functor is always faithful.

2. A functor is full if it is surjective on hom sets so a functor between monoids is full if and only if it is surjective. As for posets, a functor $f : P \to Q$ between posets is full if and only if whenever $f(x) \leq f(y)$, then $x \leq y$.

3. No. For example let $(\mathbf{N}, +)$ and $(\mathbf{N}, *)$ denote the monoids of integers with the operations of addition and multiplication, respectively. The function $f : (\mathbf{N}, +) \to (\mathbf{N}, *)$ that is constantly 0 is a semigroup homomorphism that is not a monoid homomorphism since it does not preserve the identity.

4. It is faithful, but not full. Certainly, if $f \neq g : S \rightarrow T$, then $F(f) \neq F(g) : S^* \rightarrow T^*$ since on strings of length 1, $F(f)$ is essentially the same as f. On the other hand, there are infinitely many homomorphisms from $\mathbf{N} = F(1)$ to itself (take the generating element to any power of itself), but only one function from $\{1\}$ to itself.

5. It is faithful because f can be recovered from $\mathcal{P}(f)$ by its actions on singletons. On the other hand, it cannot be full since, for example, there is only one function from 1 to 1 and four from $\mathcal{P}(1)$ to $\mathcal{P}(1)$.

6. A functor from a monoid to a category takes the single object of the monoid to some object of the monoid and is a monoid homomorphism from the monoid to the monoid of endoarrows of that object. In other words, its arrow part is a monoid homomorphism. Therefore, if $i : \mathbf{N} \rightarrow \mathbf{Z}$ is the inclusion, the nonexistence of monoid homomorphisms $g \neq h$ such that $g \circ i = h \circ i$ implies that there also cannot exist functors.

7. Let $\mathcal{C}$ be the category with three objects and three nonidentity arrows as shown:

$$A \underset{g}{\overset{f}{\rightrightarrows}} B \xrightarrow{h} C$$

plus the single composite arrow $h \circ f = h \circ g$. Then h is not a monomorphism in $\mathcal{C}$ because it composes on the left with two distinct arrows to give the same arrow. Let $\mathcal{C}_0$ be the subcategory consisting of $h : B \rightarrow C$, together with identity arrows of B and C. Then h is a monomorphism in $\mathcal{C}_0$ because there is no pair of distinct arrows to compose it with.

8. Suppose $f : A \rightarrow B$ is a split mono in a category. Then there is a $g : B \rightarrow A$ with $g \circ f = \mathrm{id}_A$. For any functor F, $F(g) \circ F(f) = F(g \circ f) = F(\mathrm{id}_A) = \mathrm{id}_{F(A)}$, since functors preserve composition and identities.

9.a. There is exactly one semigroup structure on the empty set and also on the one point set and these are the initial, respectively terminal semigroups. Thus the underlying functor preserves, reflects and creates both initial and terminal objects.

b. There is exactly one category structure on the empty graph and that is the initial category, so the underlying functor preserves, reflects and creates the initial object. The terminal category has one object and one arrow, the identity. This is the unique category structure on the graph with one object and one arrow and that is the terminal graph. Thus the terminal object is also preserved, reflected and created.

Section 3.4

1. Let $F : \mathcal{C} \rightarrow \mathcal{D}$ be an isomorphism with inverse G. Then G is a pseudo-inverse for F. We define the arrows u_C required by E–2 to be id_C for each object C, and similarly $v_C = \mathrm{id}_C$. Since $G(F(f)) = f$ and $F(G(g)) = g$ for all arrows f of $\mathcal{C}$ and g of $\mathcal{D}$, requirements E–2 and E–3 are satisfied.

2. Let $\mathcal{C} = \mathbf{1}$ be the category with one object and its identity arrow. Let $\mathcal{D}$ be the category with two objects, their identities, and two other arrows that are inverse isomorphisms between the objects. The unique functor from $\mathcal{D}$ to $\mathcal{C}$ has two pseudo-inverses, each taking the unique object of $\mathcal{C}$ to one of the two isomorphic objects of $\mathcal{D}$.

3. Define $F : \mathcal{PF} \to \mathcal{PS}$ by $F(S) = (S \cup \{S\}, S)$. S is chosen as the additional element to guarantee that it is not already an element of S. If $f : S \to T$ is a partial function, let $F(f)$ be defined by

$$F(f)(x) = \begin{cases} f(x) & \text{if } x \in S \text{ and } f \text{ is defined at } x \\ T & \text{otherwise} \end{cases}$$

It is obvious that F preserves identities. As for composition, if $g : T \to R$ is a partial function, then $g(f(x))$ is defined if and only if f is defined at x and g is defined at $f(x)$, which is exactly when $g \circ f$ is defined at x. From this, it is immediate that F preserves composition.

We define a pseudo-inverse $G : \mathcal{PS} \to \mathcal{PF}$ by $G(S, s) = S - \{s\}$ and if $f : (S, s) \to (T, t)$ is an arrow, then

$$G(f)(x) = \begin{cases} f(x) & \text{if } f(x) \neq t \\ \text{undefined} & \text{if } f(x) = t \end{cases}$$

It is easy to show that G is a functor. Now if S is a set, it is clear that $G(F(S)) = S$ and $(S, s) \cong F(G(S, s))$ by an isomorphism v_S which is the identity on S and takes s to S (the latter being the added element of $F(S - \{s\})$). What has to be shown is that for any partial function $f : S \to T$ and any function $g : (S, s) \to (T, t)$, E–2 and E–3 hold. These are a simple matter of considering cases and we omit them.

4. Let $\mathcal{P}$ be the category of preordered sets and $\mathcal{Q}$ the category of small categories as described. We let $F : \mathcal{P} \to \mathcal{Q}$ take a preordered set $(P, \leq)$ to the category $C(P, \leq)$ as described in 2.3.1. If $f : P \to P'$ is a monotone function, then $F(f)$ agrees with f on objects and when $a \leq b$, then we let $F(f)(b, a) = (f(b), f(a))$. In the other direction, let Q be a category in $\mathcal{Q}$ and let $G(Q)$ be the preordered set whose elements are the objects of Q with the preorder that $a \leq b$ if there is an arrow $a \to b$. The composition law in the category makes this relation transitive and the identity makes it reflexive. If $f : Q \to Q'$ is a functor in $\mathcal{Q}$, let $G(f)$ be the object function of f. $G(f)$ is monotone because if $x \leq y$ in Q then f must take the corresponding arrow to an arrow in Q'. It is clear that $G \circ F$ is the identity, so that E–2 is satisfied with $u_C = \mathrm{id}_C$. $F \circ G$ is the identity on the objects and that there will be an arrow $a \to b$ in Q if and only if $a \leq b$ in $G(Q)$ if and only if there is an arrow $a \to b$ in $F(G(Q))$. Thus there is an isomorphism $v_C : Q \to F(G(Q))$ which is the identity on objects and which takes an arrow $a \to b$ in Q to (the only) arrow $a \to b$ in $F(G(Q))$. These isomorphisms must satisfy E–3 because for any arrow g there is only one arrow that $v_{D'} \circ g \circ v_D^{-1}$ can be.

5. There is an functor $F : \mathcal{M} \to \mathcal{L}$ that takes n to the space of n-rowed column vectors and a matrix $A : m \to n$ to the linear transformation of multiplying on the left by A. In case one of the numbers is 0, there is only the 0 linear transformation between them. It is well known that this is a functor (matrix multiplication corresponds to composition of linear transformations) which is full and faithful and that every finite dimensional vector space is isomorphic to a space of column vectors. A pseudo-inverse G is found by choosing, for each space V, a basis B and then letting $G(V)$ be the number of elements of B. If V' is another space with chosen basis B' and $T : V \to V'$ is a linear transformation, then $G(T)$ is the matrix of T with respect to B and B'. Although G is uniquely defined on objects, its value on arrows depends completely on the choice, for each vector space V, of a basis for that space. In this case u_m is the $m \times m$ identity matrix, so E–2 is satisfied, and E–3 follows from the definition of the linear transformation determined by a matrix, given a basis.

6.a. Let $F : \mathbf{Mon} \to \mathbf{Ooc}$ take the monoid M to the category with one object and with M as its set of endoarrows. If $f : M \to N$ is a monoid homomorphism, then $F(f)(M) = N$ on the object and $F(f)(a) = f(a)$ for $a \in M$. It is immediate that F is a functor. Let $G : \mathbf{Ooc} \to \mathbf{Mon}$ take the one-object category C to the object of endomorphisms of that single object. It is clear that $G \circ F$ is the identity, so E–2 holds. $F \circ G$ is the identity on the arrows of the category and is evidently the only possible isomorphism on the singleton set of objects, so E–3 holds because there is no choice possible for the arrow.

b. Suppose $f, g : M \to N$ are homomorphisms of monoids. Then since $F(f)$ agrees with f on the arrows of $F(M)$, $F(f) = F(g)$ implies $f = g$. This shows directly that F is faithful. To see it is full, let $h : F(M) \to F(N)$ be a functor. Then the arrow function of h is a monoid homomorphism f from the monoid of endomorphisms of the object of $F(M)$ to the object of $F(N)$. $F(f)$ agrees with h on arrows and certainly does on objects since there is no choice.

Section 3.5

1. Suppose that both CR–1 and CR–2 of 3.5.1 are satisfied. Then special cases result from letting h or k be an identity. Using these special cases, we have that

$$g_1 \circ f_1 \sim g_2 \circ f_1 \sim g_2 \circ f_2$$

On the other hand, special cases of the diagram here result by setting $f_1 = f_2$ or $g_1 = g_2$. Using them, we have that $f \sim g$ implies that $k \circ f \sim k \circ g$ and that $f \circ g \sim f \circ h$.

2. Suppose that $\sim_1$ and $\sim_2$ are congruences. The intersection of two equivalence relations is an equivalence relation on any set. Also if $f_1 \sim_1 f_2$ and $g_1 \sim_1 g_2$ implies that $f_1 \circ g_1 \sim_1 f_2 \circ g_2$ and if $f_1 \sim_2 f_2$ and $g_1 \sim_2 g_2$ implies that $f_1 \circ g_1 \sim_2 f_2 \circ g_2$, then for $\sim = \sim_1 \cap \sim_2$, $f_1 \sim f_2$ and $g_1 \sim g_2$ imply that $f_1 \circ g_1 \sim f_2 \circ g_2$.

3. A functor F is full if every arrow in $\mathrm{Hom}(F(A), F(B))$ has the form $F(f)$ for some $f : A \to B$. A quotient as described here is surjective on arrows, so the condition of fullness is certainly satisfied.

4.a. It is an equivalence because $F(f) = F(f)$, $F(f) = F(g)$ implies $F(g) = F(f)$ and $F(f) = F(g)$ and $F(g) = F(h)$ implies $F(f) = F(h)$. It is a congruence because $F(f_1) = F(f_2)$ and $F(g_1) = F(g_2)$ implies that

$$F(f_1 \circ g_1) = F(f_1) \circ F(g_1) = F(f_2) \circ F(g_2) = F(f_2 \circ g_2)$$

b. Suppose $[f], [g] : A \to B$ are arrows of $\mathcal{C}/{\sim}$ such that $F_0([f]) = F_0([g])$. Then $F(f) = F(g)$, whence $f \sim g$ and $[f] = [g]$.

c. $F = F_0 \circ Q$ and we just seen that F_0 is faithful and that Q is full (previous exercise).

5. It follows from 2.2.5(ii) that $\neg\mathbf{true} = \mathbf{false}$ and $\neg\mathbf{false} = \mathbf{true}$. Thus $\widehat{F}(\neg\mathbf{true}) = \widehat{F}(\mathbf{false})$ and $\widehat{F}(\neg\mathbf{false}) = \widehat{F}(\mathbf{true})$. Similarly, $\widehat{F}(\mathbf{chr} \circ \mathbf{ord}) = \mathrm{id}$. Thus $f \sim g$ implies $\widehat{F}(f) = \widehat{F}(g)$ for the generators of the congruence. The set of pairs $\{(f, g)\}$ for which $\widehat{F}(f) = \widehat{F}(g)$ thus includes the generators and is closed under composition since functors preserve composition. Thus it includes all pairs for which $f \sim g$.

6.a. Such a relation satisfies Definition 3.5.1(a) automatically, since all elements of M are arrows of $C(M)$ with the same domain and codomain. The definition forces it to satisfy (b).

b. Suppose K is a submonoid and $n \sim n'$. Then (m, m) and (n, n') are both in K and $(m, m)(n, n') = (mn, mn')$ so $mn \sim mn'$. Similarly, $nm \sim n'm$. Conversely, suppose $\sim$ is a congruence on M. Let (m, m') and (n, n') be elements of K. Then $m \sim m'$ and $n \sim n'$, so by Exercise 1, $(mn, m'n') \in K$. Finally, $(1, 1) \in K$ because the relation is reflexive. Thus K is a submonoid of $M \times M$.

Solutions for Chapter 4

Section 4.1

1.

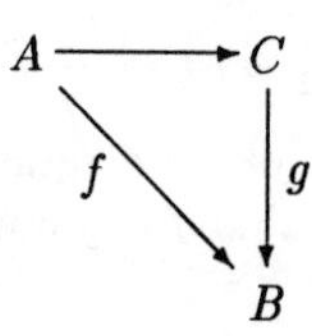

2. Let s be the function $(x, y) \mapsto (y, x)$. Then the diagram is

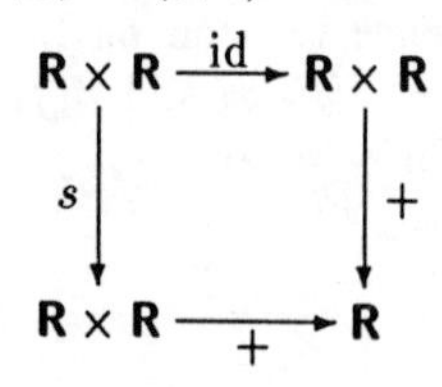

3.

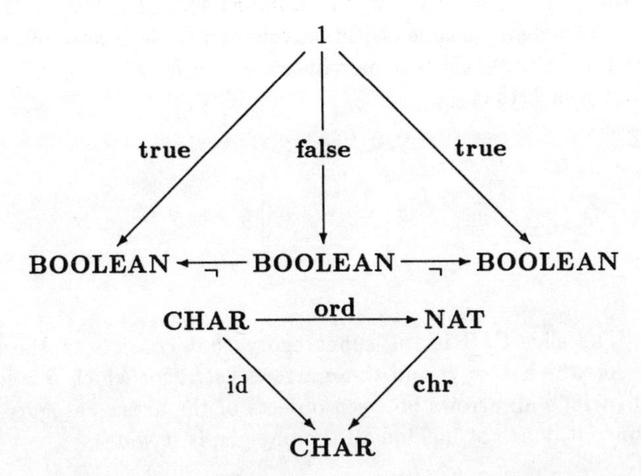

4. Let $\mathcal{C}$ and $\mathcal{D}$ be categories with sets of objects $\mathcal{C}_0$ and $\mathcal{D}_0$, sets of arrows $\mathcal{C}_1$ and $\mathcal{D}_1$, and sets of composable pairs of arrows $\mathcal{C}_2$ and $\mathcal{D}_2$, respectively. A functor $F : \mathcal{C} \to \mathcal{D}$ consists of functions $F_0 : \mathcal{C}_0 \to \mathcal{D}_0$, $F_1 : \mathcal{C}_1 \to \mathcal{D}_1$ along with the uniquely determined function $F_2 : \mathcal{C}_2 \to \mathcal{D}_2$ such that

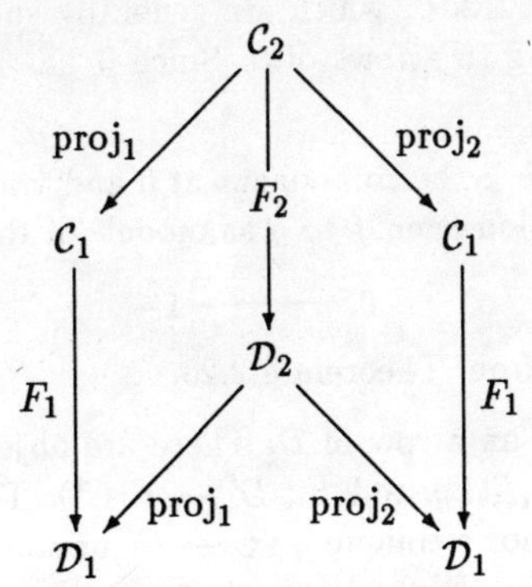

commutes. In addition, the following diagrams must commute:

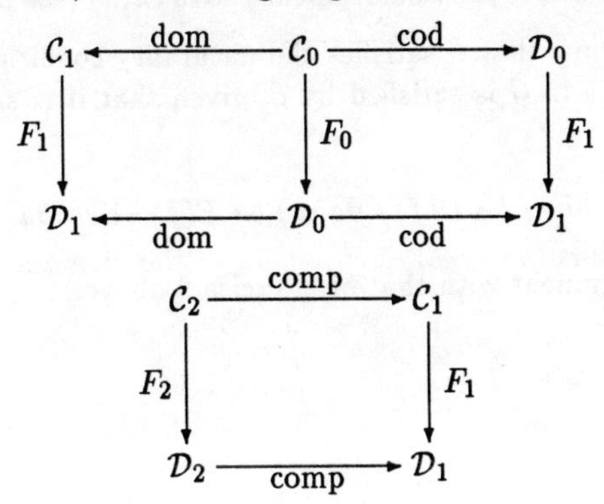

Section 4.2

1. The arrow category has as objects arrows $f : C \to D$ and an arrow from $f : C \to D$ to $f' : C' \to D'$ is a pair of arrows (g, h) where $g : C \to C'$ and $h : D \to D'$ are such that

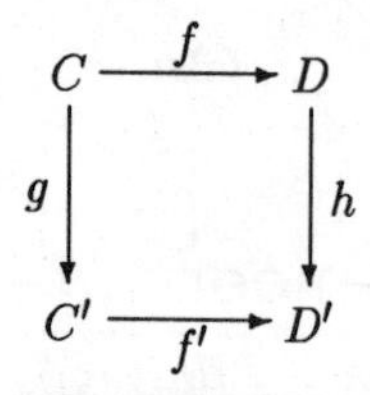

commutes. The slice $\mathcal{C}/B$ is the subcategory that consists of those objects $f : C \to D$ for which $D = B$ and those arrows (g, h) for which $h = \mathrm{id}_B$. Since it does not include all arrows between objects of the arrow category, it is not full (nor, since it does not include all the objects, is it wide).

2. An object of $\mathbf{Mod}(\mathcal{G},\mathcal{C})$ is given by a graph homomorphism $\mathcal{G} \to \mathcal{C}$. Since $\mathcal{G}$ has two nodes such a homomorphism is given by a pair of objects of $\mathcal{C}$. Since there are no arrows in $\mathcal{G}$, such a pair of arrows is exactly a graph homomorphism so the objects of $\mathbf{Mod}(\mathcal{G},\mathcal{C})$ are the pairs. An arrow from one pair to another is a pair of arrows as in $\mathcal{C} \times \mathcal{C}$, which are generally subject to commutativity conditions corresponding to arrows of $\mathcal{G}$. Since $\mathcal{G}$ has no arrows, there are no conditions in this case.

3. In diagram (4.17), h is the component at 0 and k is the component at 1 of the natural transformation from f to g as models of the graph

$$0 \longrightarrow 1$$

The result now follows from Theorem 4.2.20.

4. Let $f : D \to D'$ be an arrow of $\mathcal{D}$. There are objects C and C' of $\mathcal{C}$ and isomorphisms $h : D \to F(C)$ and $k : D' \to F(C')$. The arrow $k \circ f \circ h^{-1} : F(C) \to F(C')$ is $F(g)$ for a unique $g : C \to C'$ because an equivalence is full and faithful. Then $k \circ f = F(g) \circ h$, which means that (h, k) is an arrow from f to $F(g)$. Since h and k are isomorphisms, so is (h, k) (see previous exercise).

5. We have to show that β satisfies the naturality condition engendered by an arrow $f : a \to b$ in $\mathcal{G}$ is satisfied by β, given that it is satisfied by α. We have

$$D(f) \circ \beta a = \beta b \circ \alpha b \circ D(f) \circ \beta a = \beta b \circ E(f) \circ \alpha a \circ \beta a = \beta b \circ E(f)$$

(Compare this argument with that of Exercise 3 above.)

Section 4.3

1. Recall that if $\mathcal{G}$ is graph, $\eta\mathcal{G}$ is the identity on objects and is defined on arrows by $\eta\mathcal{G}[u] = (u)$, where we use square brackets to denote application to distinguish it from the parentheses used for lists. If $f : \mathcal{G} \to \mathcal{H}$ is a graph homomorphism, we have to show that

$$\begin{array}{ccc} \mathcal{G} & \xrightarrow{\eta\mathcal{G}} & U(F(\mathcal{G})) \\ {\scriptstyle f}\downarrow & & \downarrow{\scriptstyle U(F(f))} \\ \mathcal{H} & \xrightarrow[\eta\mathcal{H}]{} & U(F(\mathcal{H})) \end{array}$$

commutes. Applied to objects, we get $\eta\mathcal{H} \circ f[a] = f[a]$, while

$$U(F(f)) \circ \eta\mathcal{G}[a] = U(F(f))[a] = f[a]$$

If $u : a \to b$ is an arrow, $\eta\mathcal{H} \circ f[u] = (f[u])$, while

$$U(F(f)) \circ \eta\mathcal{G}[u] = U(F(f))[(u)] = (f[u])$$

2. Conditions E–2 and E–3 of 3.4.2 can be recast as $G(F(f)) \circ u_C = u_{C'} \circ f$ and that $F(G(g)) \circ v_D = v_{D'} \circ g$. In this form, they are the statements that u is a natural transformation from $\mathrm{id}_{\mathcal{C}}$ to $G \circ F$ and that v is a natural transformation from $\mathrm{id}_{\mathcal{D}} \to F \circ G$. Since each component of u and each component of v is an isomorphism, u and v are natural equivalences.

3. We must show that if $f : \mathcal{G} \to \mathcal{H}$ is a graph homomorphism,

$$\begin{array}{ccc} \mathcal{G} & \xrightarrow{\beta\mathcal{G}} & W(\mathcal{G}) \\ {\scriptstyle f}\downarrow & & \downarrow{\scriptstyle W(f)} \\ \mathcal{H} & \xrightarrow[\beta\mathcal{H}]{} & W(\mathcal{H}) \end{array}$$

commutes. But for a node a of $\mathcal{G}$, $W(f)(\beta\mathcal{G}(a))$ is calculated as the component of $\mathcal{H}$ containing $f(a)$, which is exactly what $\beta\mathcal{H}(f(a))$ is. In other words, the naturality is the very definition of $W(f)$.

4. We must show that for each arrow $f : C \to C'$ of $\mathcal{C}$,

$$\begin{array}{ccc} G(C) & \xrightarrow{i_C} & F(C) \\ {\scriptstyle G(f)}\downarrow & & \downarrow{\scriptstyle F(f)} \\ G(C') & \xrightarrow[i_{C'}]{} & F(C') \end{array}$$

commutes. We have, for $x \in G(C)$,

$$F(f)((i_C)(x)) = F(f)(x) = G(f)(x) = i_{C'}(G(f)(x))$$

5. We must show that for any graph homomorphism $f : \mathcal{G} \to \mathcal{H}$, the square

$$\begin{array}{ccc} A(\mathcal{G}) & \xrightarrow{\text{source}\,\mathcal{G}} & N(\mathcal{G}) \\ {\scriptstyle A(f)}\downarrow & & \downarrow{\scriptstyle N(f)} \\ A(\mathcal{H}) & \xrightarrow[\text{source}\,\mathcal{H}]{} & N(\mathcal{H}) \end{array}$$

commutes. We have for $u : a \to b$ in $\mathcal{G}$, $N(f)(\text{source}\,\mathcal{G}(u)) = N(f)(a) = f(a)$, while $\text{source}\,\mathcal{H}(A(f)(u)) = \text{source}(f(u)) = f(a)$. The argument for target is similar.

6. A functor $F : \mathcal{C} \to \mathbf{Set}$ is determined uniquely by giving, for each $C \in \mathcal{C}_0$, a set $F(C)$. The disjoint union of these sets can be modeled as $W(F) = \bigcup\{(x, C) \mid x \in F(C)\}$. The function $w(F) : W(F) \to \mathcal{C}_0$ defined by $w(F)(x, C) = C$ makes $w(F)$ an object of $\mathbf{Set}/\mathcal{C}_0$. Given a natural transformation $\alpha : F \to G$, define $W(\alpha) : W(F) \to W(G)$ by $W(\alpha)(x, C) = (\alpha C(x), C)$. This makes $W : \mathbf{Fun}(\mathcal{C}, \mathbf{Set}) \to \mathbf{Set}/\mathcal{C}_0$ a functor.

Conversely, given an object $f : S \to \mathcal{C}_0$ of $\mathbf{Set}/\mathcal{C}_0$, define a functor $V(f) : \mathcal{C} \to \mathbf{Set}$ by $V(f)(C) = \{x \in S \mid f(x) = C\}$ for an object C of $\mathcal{C}$. If $u : f \to g : S' \to \mathcal{C}_0$ is an arrow of $\mathbf{Set}/\mathcal{C}_0$, define a natural transformation $V(u) : V(f) \to V(g)$ by $(V(u)C)(x) = u(x)$ for $x \in S'$. Then $V : \mathbf{Set}/\mathcal{C}_0 \to \mathbf{Fun}(\mathcal{C}, \mathbf{Set})$ is a functor.

We have that $V(W(F))(C) = \{(x, C) \mid x \in F(C)\}$. Let $\beta C(x, C) = x$ for $x \in F(C)$. Then β is a natural isomorphism from $V \circ W$ to the identity functor on $\mathbf{Fun}(\mathcal{C}, \mathbf{Set})$. In the other direction, for $f : S \to \mathcal{C}_0$,

$$W(V(f)) = \bigcup\{(x, C) \mid x \in V(f)(C)\} = \{(x, f(x)) \mid x \in S\}$$

which is isomorphic to S. The diagram

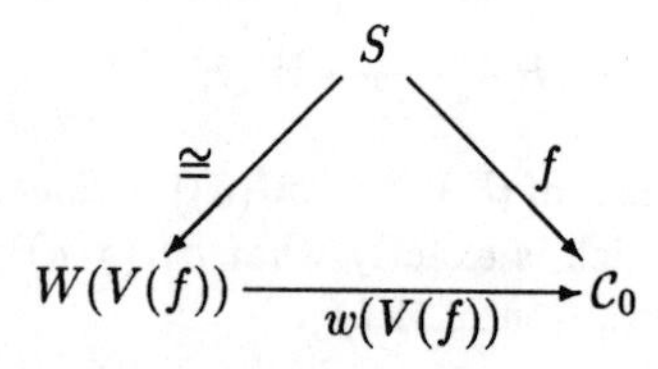

commutes and the naturality of the isomorphism is straightforward to verify.

7.a. We must show that if $f : S \to T$ is a function, then

$$\begin{array}{ccc} S & \xrightarrow{\{\}S} & \mathcal{P}S \\ {\scriptstyle f}\downarrow & & \downarrow{\scriptstyle f_*} \\ T & \xrightarrow[\{\}T]{} & \mathcal{P}T \end{array}$$

commutes. But each path applied to an $x \in S$ produces $\{f(x)\}$.

b. The diagram that would have to commute is

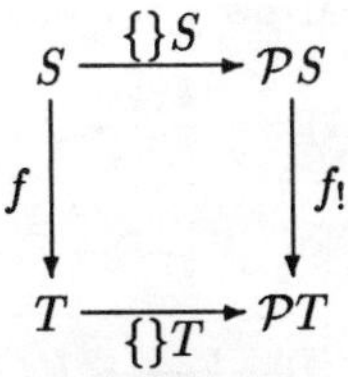

$$\begin{array}{ccc} S & \xrightarrow{\{\}S} & \mathcal{P}S \\ {\scriptstyle f}\downarrow & & \downarrow{\scriptstyle f_!} \\ T & \xrightarrow[\{\}T]{} & \mathcal{P}T \end{array}$$

for an $f : S \to T$. The reader may want to verify that it does if and only if f is injective. When, for example, $f : 2 \to 1$ is the unique function, going around counter-clockwise gives, at 0, the element $\{0\}$. Going the other way, we get $f_!(\{0\})$ which is the set of $x \in 1$ whose *every* inverse image is included in $\{0\}$. Since no element of 1 has this property, $f_!\{0\} = \emptyset$.

c. It makes no sense to ask for a natural transformation between a contravariant functor and a covariant functor.

8. A natural transformation from $F \to G$ is a function f from $S = F(M)$ to $T = G(M)$ and it must satisfy the naturality condition that for any $a \in M$,

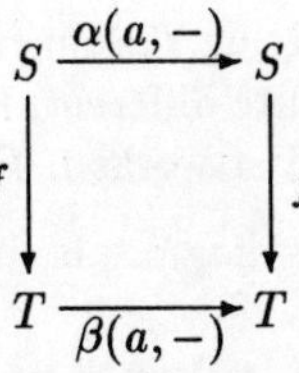

$$\begin{array}{ccc} S & \xrightarrow{\alpha(a,-)} & S \\ {\scriptstyle f}\downarrow & & \downarrow{\scriptstyle f} \\ T & \xrightarrow[\beta(a,-)]{} & T \end{array}$$

commutes. This is the definition of an equivariant function.

9. If $F, G : \mathcal{E} \to \mathcal{G}$ is such that $\alpha(F) = \alpha(G)$, then the sole vertex of $\alpha(F)$ is the same as that of $\alpha(G)$. But that is all there is to a homomorphism on $\mathcal{E}$. Thus α is injective. Similarly, every node of $\mathcal{G}$ does give graph homomorphism on $\mathcal{E}$ so α is surjective.

10.a. If $f : \mathcal{G} \to \mathcal{H}$ is a graph homomorphism, we define

$$P_k(f)((u_n, u_{n-1}, \dots, u_1)) = (f(u_n), f(u_{n-1}), \dots, f(u_1))$$

This makes sense since f preserves source and target. The functoriality is clear.

b. Since a path of length 0 is a node, the object functions are the same. The arrow functions are the same by definition.

c. Since a path of length 1 is an arrow, the object functions are the same. The arrow functions are the same by definition.

Section 4.4

1. We must show that for an arrow $f : A \to A'$ of $\mathcal{A}$, the diagram

$$\begin{array}{ccc} H(F(A)) & \xrightarrow{\beta FA} & K(F(A)) \\ {\scriptstyle H(F(f))}\downarrow & & \downarrow{\scriptstyle K(F(f))} \\ H(F(A')) & \xrightarrow[\beta FA']{} & K(F(A')) \end{array}$$

commutes. This is just the naturality diagram of β applied to the arrow $F(f) : F(A) \to F(A')$. We must also show that the diagram

$$\begin{array}{ccc} H(F(A)) & \xrightarrow{H(\alpha A)} & H(G(A)) \\ {\scriptstyle H(F(f))}\downarrow & & \downarrow{\scriptstyle H(G(f))} \\ H(F(A')) & \xrightarrow[H(\alpha A')]{} & H(G(A')) \end{array}$$

commutes. This is done by applying the functor F to the naturality diagram of α. Note that the proofs are quite different. For example, the second requires that H be a functor, while the first works if F is merely an object function.

2.a. The commutativity of this diagram is simply the naturality of β applied to the arrow $\alpha A : FA \to GA$.

b. The transformation $\beta * \alpha$ is defined as the composite $K\alpha \circ \beta F$ and is therefore the vertical composite of two natural transformations.

c. In the situation,

$$\mathcal{A} \underset{G}{\overset{F}{\rightrightarrows}} \Downarrow\alpha \ \mathcal{B} \underset{K}{\overset{H}{\rightrightarrows}} \Downarrow\beta \ \mathcal{C} \underset{M}{\overset{L}{\rightrightarrows}} \Downarrow\gamma \ \mathcal{D}$$

We have that

$$\gamma * (\beta * \alpha) = \gamma KG \circ L(\beta G \circ H\alpha) = \gamma KG \circ L\beta G \circ LH\alpha$$

while

$$(\gamma * \beta) * \alpha = (\gamma K \circ L\beta)G \circ LH\alpha = \gamma KG \circ L\beta G \circ LH\alpha$$

3. By Definition 4.4.2 the component of $(G_1\alpha)E$ at an object A of $\mathcal{A}$ is $(G_1\alpha)E(A)$. This is $G_1(\alpha E(A))$ by Definition 4.4.3. By Definition 4.4.3, the component of $G_1(\alpha E)$ at A is $G_1((\alpha E)A)$. This is $G_1(\alpha E(A))$ by Definition 4.4.2.

4. The horizontal composites $\beta * \mathrm{id}_F$ and $\mathrm{id}_H * \alpha$ are defined to be the two equal composites in each of these special cases of Diagram (4.20):

$$\begin{array}{ccc} (H \circ F)A & \xrightarrow{H(\mathrm{id}_F)A} & (H \circ F)A \\ {\scriptstyle (\beta F)A}\downarrow & & \downarrow{\scriptstyle (\beta F)A} \\ (K \circ F)A & \xrightarrow[H(\mathrm{id}_F)A]{} & (K \circ F)A \end{array}$$

$$\begin{array}{ccc} (H \circ F)A & \xrightarrow{(H\alpha)A} & (H \circ G)A \\ {\scriptstyle (\mathrm{id}_H)FA}\downarrow & & \downarrow{\scriptstyle (\mathrm{id}_H)FA} \\ (H \circ F)A & \xrightarrow[(H\alpha)A]{} & (H \circ G)A \end{array}$$

Since

$$H(\mathrm{id}_F)A = H(\mathrm{id}_{FA}) = \mathrm{id}_{(H \circ F)A}$$

(the first equality by the definition of identity natural transformation in 4.2.13), we have that

$$(\beta * \mathrm{id}_F)A = (\beta F)A \circ \mathrm{id}_{H \circ FA} = (\beta F)A$$

as required. Similarly $(\mathrm{id}_H)GA = \mathrm{id}_{(H \circ G)A}$, so that

$$(\mathrm{id}_H * \alpha)A = (\mathrm{id}_H)GA \circ (H\alpha)A = (H\alpha)A$$

5.a. The first equation in Godement's fifth rule follows from this calculation, using the Interchange Law and Exercise 4.

$$\begin{aligned} (\gamma F_2) \circ (G_1 \alpha) &= (\gamma * \mathrm{id}_{F_2}) \circ (\mathrm{id}_{G_1} * \alpha) \\ &= (\gamma \circ \mathrm{id}_{G_1}) * (\mathrm{id}_{F_2} \circ \alpha) = \gamma * \alpha \end{aligned}$$

A similar argument shows that $\gamma\alpha = (G_2\alpha) \circ (\gamma F_1)$

b. This is shown by the following calculation, using (in order) Exercise 4, the Interchange Law, the definition of the (vertical) composite of natural transformations and Exercise 4 again:

$$\begin{aligned} (G_1 \beta E) \circ (G_1 \alpha E) &= (\mathrm{id}_{G_1} * \beta E) \circ (\mathrm{id}_{G_1} * \alpha E) \\ &= (\mathrm{id}_{G_1} \circ \mathrm{id}_{G_1}) * (\beta E \circ \alpha E) \\ &= \mathrm{id}_{G_1} * (\beta \circ \alpha)E = G_1(\beta \circ \alpha)E \end{aligned}$$

Section 4.5

1. Let $H = \mathrm{Hom}_{\mathbf{Set}}(1,-)$ and I the identity functor. Let $\alpha : H \to I$ assign to each function from 1 to I the element which is the image of that function. This is clearly bijective. If $f : S \to T$ is a function, we have to show that the diagram

$$\begin{array}{ccc} H(S) & \xrightarrow{\alpha S} & I(S) \\ {\scriptstyle H(f)}\downarrow & & \downarrow{\scriptstyle I(f)} \\ H(T) & \xrightarrow[\alpha T]{} & I(T) \end{array}$$

commutes. If we take a function $g : 1 \to S$ whose value is $x \in S$, then

$$I(f)(\alpha S(g)) = I(f)(x) = f(x) = \alpha T(f \circ g) = \alpha T(H(f)(g))$$

Note that this says that every element of any set is a universal element for H.

2. A graph homomorphism $\mathbf{2} \to \mathcal{G}$ takes the single arrow of $\mathbf{2}$ to an arrow u of $\mathcal{G}$ and takes 0 to the source of u and 1 to the target of u. In other words, it is completely determined by u. Thus $A(\mathcal{G}) \cong \mathrm{Hom}(\mathbf{2}, \mathcal{G})$. It remains only to show that the isomorphism is natural in $\mathcal{G}$. If $f : \mathcal{G} \to \mathcal{H}$, we must show that

$$\begin{array}{ccc} A(\mathcal{G}) & \xrightarrow{\cong} & \mathrm{Hom}(\mathbf{2}, \mathcal{G}) \\ {\scriptstyle A(f)}\downarrow & & \downarrow{\scriptstyle \mathrm{Hom}(\mathbf{2}, f)} \\ A(\mathcal{H}) & \xrightarrow[\cong]{} & \mathrm{Hom}(\mathbf{2}, \mathcal{H}) \end{array}$$

commutes. The argument is trivial.

3. Global elements are arrows from the terminal object. In the case of categories, the terminal object is the category $\mathbf{1}$ with one object and one arrow, the identity of that object. A functor from that object is determined by where it sends that object, which can be to any arbitrary object. The identity is sent to the identity of that chosen object. Thus $\mathrm{Hom}_{\mathbf{Cat}}(\mathbf{1}, \mathcal{C})$ is isomorphic to the set of objects of $\mathcal{C}$. The set of objects is a functor $\mathcal{O} : \mathbf{Cat} \to \mathbf{Set}$: if $F : \mathcal{C} \to \mathcal{D}$ is a functor then $\mathcal{O}(F)$ is F_0, the object part of the functor. To verify naturality amounts to showing that if $G : \mathbf{1} \to \mathcal{C}$ is a functor, then $\mathrm{Hom}(\mathbf{1}, F)(G(*)) = F(G(*))$ which is immediate from the definition. (This says that the global elements of a category are its objects. Note that the global elements of a graph are its *loops*.)

4. Yes and yes. The representing object is the category $\mathbf{2}$ which is the graph $\mathbf{2}$ with the addition of two identity arrows. The argument is virtually identical to the argument for graphs.

5. Set $F^*(C) = \{(x, C) \mid x \in F(C)\}$ and for $f : C \to D$, $F^*(f)(x, C) = (F(f)(x), D)$. The function $\beta C = x \mapsto (x, C)$ is clearly an isomorphism. To be natural requires $F^*(f)(\beta C(x))$ be the same as $\beta D(F^*(f)(x))$ which is immediate from the definition.

6. Let $\mathcal{D} = \mathcal{C}^{\mathrm{op}}$. Then the ordinary Yoneda embedding

$$\mathcal{D}^{\mathrm{op}} \to \mathbf{Fun}(\mathcal{D}, \mathbf{Set})$$

is full and faithful. But $\mathcal{C}^{\mathrm{op}} = \mathcal{D}$ means $\mathcal{D}^{\mathrm{op}} = \mathcal{C}$ so this is just 4.5.5.

7. Let $\alpha C' : \mathrm{Hom}(C, C') \to F(C')$ take f to $F(f)(c)$ for $c \in F(C)$. We must show that for any $g : C' \to B$,

$$\begin{array}{ccc} \mathrm{Hom}(C, C') & \xrightarrow{\alpha C'} & F(C') \\ {\scriptstyle \mathrm{Hom}(C,g)}\downarrow & & \downarrow{\scriptstyle F(g)} \\ \mathrm{Hom}(C, B) & \xrightarrow[\alpha B]{} & F(B) \end{array}$$

We have, for $f \in \mathrm{Hom}(C, C')$,

$$\begin{aligned} \alpha B(\mathrm{Hom}(C, g)(f)) &= \alpha B(g \circ f) \\ &= F(g \circ f)(c) = F(g)(F(f)(c)) = F(g)(\alpha C'(f)) \end{aligned}$$

8. Suppose $\beta : \mathrm{Hom}(C, -) \to F$ is a natural transformation. Then $\beta C : \mathrm{Hom}(C, C) \to F(C)$. Let $c = \beta C(\mathrm{id}_C) \in F(C)$. For any $f : C \to C'$, the naturality diagram (see solution to preceding problem) gives that

$$\beta C'(\mathrm{Hom}(C, f)(\mathrm{id}_{C'})) = F(f)(\beta C)(\mathrm{id}_C)$$

But the left hand side is $\beta C'(f)$ and the right hand side is $F(f)(c)$.

9. In the preceding exercise we constructed, for each $c \in F(C)$, a natural transformation we may call $\beta(C, F)(c) : \mathrm{Hom}(C, -) \to F$. This is the value at c of a function we call $\beta(C, F) : F(C) \to \mathrm{NT}(\mathrm{Hom}(C, -), F)$ where $\mathrm{NT}(G, F)$ stands for the set of natural transformations between functors G and F. It is easy to see that $\mathrm{NT}(G, F)$ is contravariant in G and covariant in F. It is, in fact, the hom functor in the category **Cat**. Then the formulation of naturality is that if $f : C \to C'$ is an arrow and $\alpha : F \to F'$ is a natural transformation, the square

$$\begin{array}{ccc} F(C) & \xrightarrow{\beta(C,F)} & \mathrm{NT}(\mathrm{Hom}(C, -), F) \\ {\scriptstyle \alpha f}\downarrow & & \downarrow{\scriptstyle \mathrm{NT}(\mathrm{Hom}(f, -), \alpha)} \\ F'(C') & \xrightarrow[\beta(C',F')]{} & \mathrm{NT}(\mathrm{Hom}(C', -), F') \end{array}$$

commutes. Here αf is defined to be $F'f \circ \alpha C = \alpha C' \circ Ff$, equal by naturality. Note the double contravariance. From $f : C \to C'$, we get

$$\mathrm{Hom}(f, -) : \mathrm{Hom}(C', -) \to \mathrm{Hom}(C, -)$$

and then

$$\mathrm{NT}(\mathrm{Hom}(C, -), F) \to \mathrm{Hom}(\mathrm{Hom}(C', -), F)$$

Now to prove this, we must apply it to a $c \in F(C)$. Going around clockwise gives us the natural transformation from $\mathrm{Hom}(C', -) \to F$ whose value at an object A takes an arrow $g : C' \to A$ to $\alpha A(F(g \circ f)(c))$. Going the other way we get the natural transformation whose value at an object A is $F'g(\alpha C'(Ff(c)))$. From the functoriality of F and naturality of α, we have

$$\alpha A(F(g \circ f)(c)) = \alpha A(Fg(Ff(c))) = F'g(\alpha C'(Ff(c)))$$

10. If $c \in F(C)$ is a universal element of F, then there is a unique $f : C \to C'$ such that $Ff(c) = c'$. Symmetrically, there is a unique $g : C' \to C$ such that $Fg(c') = c$. Then $Fg(Ff(c)) = Fg(c') = c$. But the universality of c says that there is a unique arrow $h : C \to C$ such that $Fh(c) = c$. Clearly, $g = \mathrm{id}_C$ is one such; it follows that $Fg \circ Ff = \mathrm{id}_C$. Symmetrically, $Ff \circ Fg = \mathrm{id}_{C'}$.

Section 4.6

1. The sketch has one node, call it s, and two arrows $u, v : s \to s$. There is one diagram $D : \mathcal{I} \to \mathcal{S}$ based on the shape

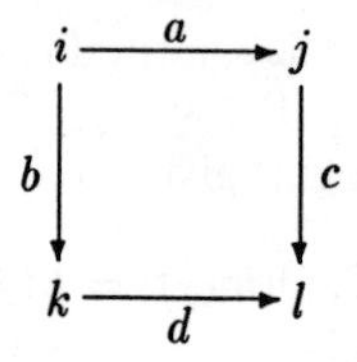

defined by $D(i) = D(j) = D(k) = D(l) = s$, $D(a) = D(d) = u$ and $D(b) = D(c) = v$.

2. Since the sketch underlying a category has the same objects as the category and since the theory of a linear sketch also has the same objects as the sketch, the objects are same. For each arrow $f : A \to B$, there is an arrow in the sketch. Whenever $(f_1, f_2, \ldots, f_n)$ is a path in the graph underlying $\mathcal{C}$ there is a diagram $(f_1, f_2, \ldots, f_n) = (f_1 \circ f_2 \circ \cdots \circ f_n)$ so that in the theory category every path is equal to a single arrow and the obvious functor is full. It is also faithful since there is no relation among paths in the theory that does not come from a commutative diagram in $\mathcal{C}$.

3. In general a homomorphism must commute in that way with every arrow in the sketch. However, an arrow that commutes in this way with an invertible arrow also commutes with the inverse:

$$\begin{aligned} f \circ M(v) &= N(v) \circ N(u) \circ f \circ M(v) \\ &= N(v) \circ f \circ M(u) \circ M(v) = N(v) \circ f \end{aligned}$$

Section 4.7

1. If two terms are forced to be equal by the equivalence relation, then they are certainly equal in every model since the relations are valid in every model. On the other hand, the theory is a model and if the two terms are not forced to be equal by the equivalence relation, they are not equal in the theory.

Solutions for Chapter 5

Section 5.1

1. Let $\mathcal{C}$, $\mathcal{D}$ and $\mathcal{E}$ be categories. The category $\mathcal{C} \times \mathcal{D}$ has functors $P_1 : \mathcal{C} \times \mathcal{D} \to \mathcal{C}$ and $P_2 : \mathcal{C} \times \mathcal{D} \to \mathcal{D}$ defined by $P_1(C, D) = C$, $P_2(C, D) = D$, $P_1(f, g) = f$ and $P_2(f, g) = g$ for C and D objects and f and g arrows of $\mathcal{C}$ and $\mathcal{D}$ respectively. That these are functors follows immediately from the fact that source, target and composition in the product category are defined coordinatewise (see 2.6.6). Now let $F : \mathcal{E} \to \mathcal{C}$ and $G : \mathcal{E} \to \mathcal{D}$ be functors. Define $\langle F, G\rangle : \mathcal{E} \to \mathcal{C} \times \mathcal{D}$ by $\langle F, G\rangle(E) = (F(E), G(E))$ and $\langle F, G\rangle(h) = (F(h), G(h))$ for E an object and h an arrow of $\mathcal{E}$. The proof that this is a functor is immediate. Then $P_1 \circ \langle F, G\rangle(E) = P_1(F(E), G(E)) = F(E)$ and similarly for arrows. And similarly, $P_2 \circ \langle F, G\rangle = G$. If $H : \mathcal{E} \to \mathcal{C} \times \mathcal{D}$ is any functor with $P_1 \circ H = F$ and $P_2 \circ H = G$, let $H(E) = (H_1(E), H_2(E))$. Then $F(E) = P_1 \circ H(E) = H_1(E)$ and similarly for arrows. Thus $F = H_1$ and similarly $G = H_2$, which proves uniqueness.

2. Let M and N be monoids. The product is the product of the underlying sets with multiplication $(m_1, n_1)(m_2, n_2) = (m_1m_2, n_1n_2)$. The identity element is $(1, 1)$.

3. If P and Q are posets, their product is the product of the underlying sets with $(p_1, q_1) \leq (p_2, q_2)$ if and only if $p_1 \leq p_2$ and $q_1 \leq q_2$.

4. This is essentially the same as for categories.

5. Use the notation from the text. Then

$$\begin{aligned}\mathrm{proj}_1((s_1, t_1)(s_2, t_2)) &= \mathrm{proj}_1(s_1s_2, t_1t_2) = s_1s_2 \\ &= \mathrm{proj}_1(s_1, t_1)\,\mathrm{proj}_1(s_2, t_2)\end{aligned}$$

so that proj_1 is a semigroup homomorphism. Similarly, so is proj_2. Now let R be another semigroup and $q_1 : R \to S$ and $q_2 : R \to T$ be homomorphisms. Then

$$\begin{aligned}\langle q_1, q_2\rangle(r_1r_2) &= (q_1(r_1r_2), q_2(r_1r_2)) = (q_1(r_1)q_1(r_2), q_2(r_1)q_2(r_2)) \\ &= (q_1(r_1), q_2(r_1))(q_1(r_2), q_2(r_2)) \\ &= \langle q_1, q_2\rangle(r_1)\langle q_1, q_2\rangle(r_2)\end{aligned}$$

6. Given any object B and arrows $f : B \to 1$ and $g : B \to A$, then $g : B \to A$ is evidently the unique arrow such that $\mathrm{id}_A g = g$. But also $\langle\rangle \circ g = f$ since that is the only arrow from B to 1.

7. In the category of sets the product of any set A with the empty set is the empty set. If A is nonempty, the projection onto A is not surjective, hence not an epimorphism.

Section 5.2

1. The isomorphism is given by

$$f(x) = \begin{cases} (2,1) & \text{if } x = 1 \\ (1,1) & \text{if } x = 2 \\ (3,1) & \text{if } x = 3 \\ (2,2) & \text{if } x = 4 \\ (1,2) & \text{if } x = 5 \\ (3,2) & \text{if } x = 6 \end{cases}$$

2. The isomorphism of the preceding exercise can be composed with any of the $6! = 720$ permutations of 6 to give another one.

3. Let $f_i : A_i \to B_i$ and $g_i : B_i \to C_i$ for $i = 1, 2$. Then

$$(g_1 \circ f_1) \times (g_2 \circ f_2) : A_1 \times A_2 \to B_1 \times B_2$$

is the unique arrow such that

$$\text{proj}_1 \circ ((g_1 \circ f_1) \times (g_2 \circ f_2)) = (g_1 \circ f_1) \circ \text{proj}_1$$

and

$$\text{proj}_2 \circ ((g_1 \circ f_1) \times (g_2 \circ f_2)) = (g_2 \circ f_2) \circ \text{proj}_2$$

We have

$$\text{proj}_1 \circ (g_1 \times g_2) \circ (f_1 \times f_2) = g_1 \circ \text{proj}_1 \circ (f_1 \times f_2) = g_1 \circ f_1 \circ \text{proj}_1$$

and similarly

$$\text{proj}_2 \circ (g_1 \times g_2) \circ (f_1 \times f_2) = g_2 \circ f_2 \circ \text{proj}_2$$

The result follows from the uniqueness of $(g_1 \circ f_1) \times (g_2 \circ f_2)$.

Define the functor $- \times - : \mathcal{C} \times \mathcal{C} \to \mathcal{C}$ by choosing, for each pair A and B of $\mathcal{C}$ a product object $A \times B$ and letting $(- \times -)(A, B) = A \times B$ and $(- \times -)(f, g) = f \times g$. The identities above are what is necessary to show that this mapping preserves composition. That it preserves identities follows immediately from them as well.

4. By 5.2.13, the left vertical arrow in (5.10) takes the pair of arrows (q_1, q_2), where $q_1 : W \to A$ and $q_2 : W \to B$, to $(q_1 \circ f, q_2 \circ f)$, and the right vertical arrow takes $q : W \to A \times B$ to $q \circ f : V \to A \times B$. Therefore, by Definition 5.2.8, if you start at lower left with (q_1, q_2) and go north and then east, you get the unique arrow q' which makes

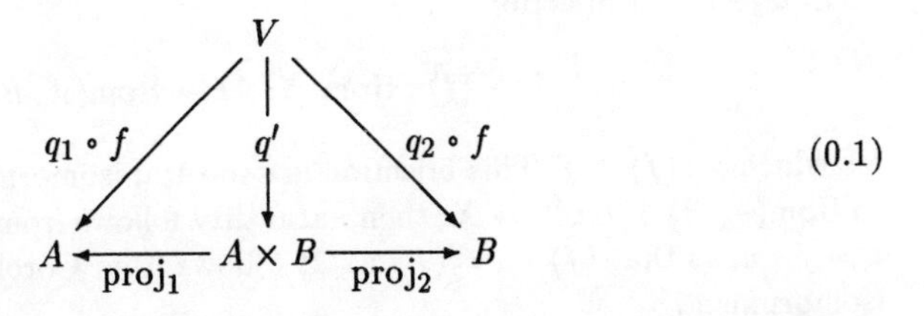

commute. If you go east and north, you get $q \circ f$, where q is the unique arrow which makes

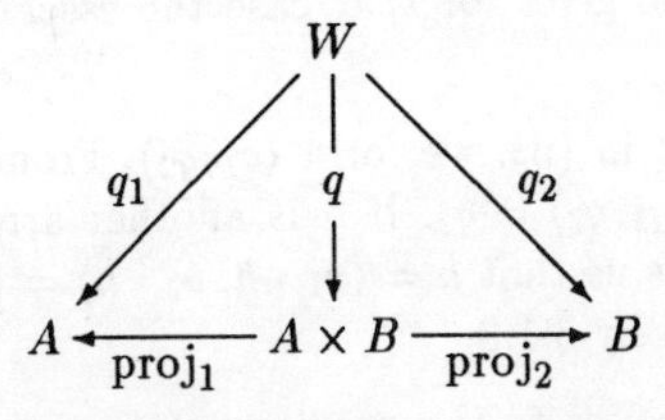

commute. Because $\mathrm{proj}_i \circ (q \circ f) = (\mathrm{proj}_i \circ q) \circ f = q_i \circ f$ for $i = 1, 2$, it follows that $q \circ f = q'$. This is a consequence of the fact that q' is the *unique* arrow making (0.1) commute. Hence (5.10) commutes.

5. Let $f : C \to A$ and $g : C \to B$. Then $i^{-1} \circ \langle f, g \rangle : C \to V$ is an arrow such that

$$\mathrm{proj}_1 \circ i \circ i^{-1} \circ \langle f, g \rangle = \mathrm{proj}_1 \circ \mathrm{id}_U \circ \langle f, g \rangle = \mathrm{proj}_1 \circ \langle f, g \rangle = f$$

and similarly $\mathrm{proj}_2 \circ i \circ i^{-1} \circ \langle f, g \rangle = g$. Moreover if $h : C \to V$ is such that $\mathrm{proj}_1 \circ i \circ h = f$ and $\mathrm{proj}_2 \circ i \circ h = g$, then we have $p_1 \circ i \circ h = p_1 \circ i \circ i^{-1} \circ \langle f, g \rangle$ and $p_2 \circ i \circ h = p_2 \circ i \circ i^{-1} \circ \langle f, g \rangle$. But since arrows to U are uniquely determined by their projections to A and B, we conclude that $i \circ h = i \circ i^{-1} \circ \langle f, g \rangle$ from which the isomorphism i can be cancelled to give $h = i^{-1} \circ \langle f, g \rangle$.

Section 5.3

1.a. Define $q_1 = \mathrm{proj}_1 : A \times (B \times C) \to A$, $q_2 = \mathrm{proj}_1 \circ \mathrm{proj}_2 : A \times (B \times C) \to B$ and $q_3 = \mathrm{proj}_2 \circ \mathrm{proj}_2 : A \times (B \times C) \to C$. The meaning of the last, for example, is the second projection to $B \times C$, followed by the second projection from the latter to C. Now suppose D is an object and $f_1 : D \to A$, $f_2 : D \to B$ and $f_3 : D \to C$ are given. Then there is a unique arrow $\langle f_2, f_3 \rangle : D \to B \times C$ such that $\mathrm{proj}_1 \circ \langle f_2, f_3 \rangle = f_2$ and $\mathrm{proj}_2 \circ \langle f_2, f_3 \rangle = f_3$. It follows that there is a unique arrow $\langle f_1, \langle f_2, f_3 \rangle \rangle : D \to A \times (B \times C)$ such that $\mathrm{proj}_1 \circ \langle f_1, \langle f_2, f_3 \rangle \rangle = f_1$ and $\mathrm{proj}_2 \circ \langle f_1, \langle f_2, f_3 \rangle \rangle = \langle f_2, f_3 \rangle$ from which the required identities follow immediately. If $g : D \to A \times (B \times C)$ is another arrow with $q_i \circ g = f_i$ for $i = 1$, 2, 3, then $\text{proj}_1 \circ \text{proj}_2 \circ g = f_2$ and $\text{proj}_2 \circ \text{proj}_2 \circ g = f_3$, from which it follows from the uniqueness of arrows into a product, that $\text{proj}_2 \circ g = \langle f_2, f_3 \rangle$. Also, $p_1 \circ g = f_1$ so that $g = \langle f_1, \langle f_2, f_3 \rangle \rangle$.

b. If $p : B \to A$ is a unary product diagram, then by definition there is for each object X a bijection

$$f \mapsto \langle f \rangle : \text{Hom}(X, A) \to \text{Hom}(X, B)$$

for which $p \circ \langle f \rangle = f$. This bijection is a natural isomorphism from $\text{Hom}(-, A)$ to $\text{Hom}(-, B)$: if $u : Y \to X$, then naturality follows from the fact that $p \circ \langle f \rangle \circ u = f \circ u$, so that $\langle f \rangle \circ u = \langle f \circ u \rangle$. It follows from Corollary 4.5.4 that p is an isomorphism.

c. The preceding part shows that every category has unary products, so such a category has n-ary products for $n = 0$, 1 and 2. The first part shows the same for $n = 3$ and also gives for that case the essential step for an obvious induction on n.

2. Given q_1 and q_2 as in (ii), we form $\langle q_1, q_2 \rangle$. From (iii), we see that $p_1 \circ \langle q_1, q_2 \rangle = q_1$ and $p_2 \circ \langle q_1, q_2 \rangle = q_2$. If h is another arrow satisfying the same identities, then (iv) tells us that $h = \langle p_1 \circ h, p_2 \circ h \rangle = \langle q_1, q_2 \rangle$ so that we have the uniqueness required by 5.1.3.

3. We saw in the exercises to the previous section that the product in each category took as objects of the product category the product of the objects and as arrows of the product the product of the arrows. Thus the product is constructed in the same way in both categories.

4. Let $\mathcal{C} = \mathcal{D}$ be the category of countably infinite sets and all functions between them. Let $F : \mathcal{C} \to \mathcal{D}$ be defined by $F(X) = 1 + X$ for such a set X and if $f : X \to Y$, $F(f) = 1 + f$ is the function whose restriction to X is f and which takes the added element of X to the added element of Y. Then although $F(X \times Y) \cong F(X) \times F(Y)$ since any two countable sets are isomorphic, the cone

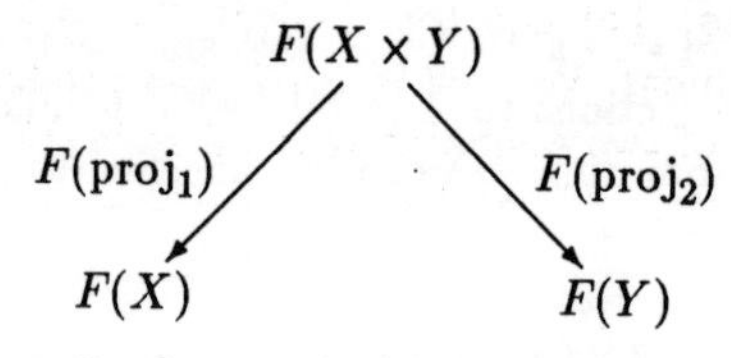

is not a product cone. In fact, $F(X \times Y)$ contains no point u such that $f(\text{proj}_1)(u)$ is the added point of X but $f(\text{proj}_2)(u)$ is *not* the added point of Y or vice versa.

5.a. Since a unary product diagram is an isomorphism and every functor preserves isomorphisms, every functor preserves unary products. Now for $n \geq 3$, n-ary products are defined by induction. Assuming that a functor F preserves $n-1$-ary products, then the product

$$A_1 \times A_2 \times \cdots \times A_n = A_1 \times (A_2 \times \cdots \times A_n)$$

Then

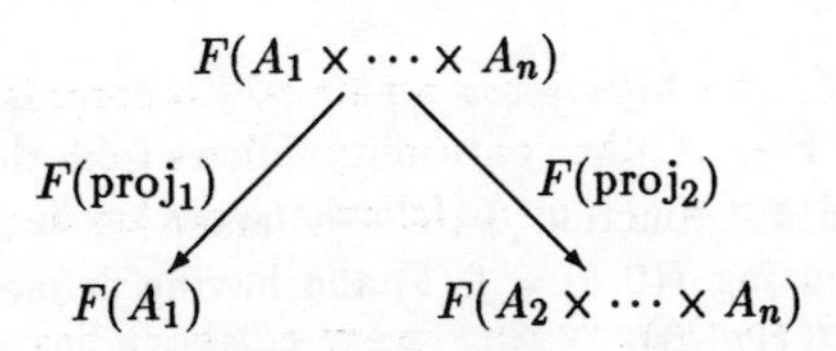

is a product cone. By the inductive assumption so is

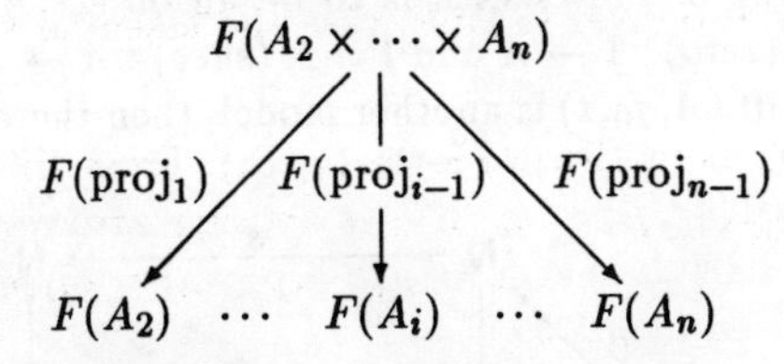

from which the conclusion follows.

b. Let $\mathcal{C} = \mathbf{1}$, the category with one object called 0 and its identity arrow. Let $\mathcal{D} = \mathbf{2}$, the category with two objects, called 0 and 1, their identities and one arrow $0 \to 1$. Then 0 is the terminal object of $\mathcal{C}$ and 1 is the terminal object of $\mathcal{D}$. Both categories have finite products, with $0^n = 0$ in both and $1^n = 1$ and $0 \times 1 = 0$ in $\mathcal{D}$. The products are canonical since there is only one possible choice. Then the functor $F : \mathcal{C} \to \mathcal{D}$ given by $F(0) = 0$ preserves n-ary products for $n \geq 1$, but not nullary products.

6. A terminal object in a category is an object that every other object has an arrow to (unique, of course). In a poset, that is an element that every element is less than or equal to, that is a top element. There is no largest integer so $\mathbf{N}$ has no terminal element. We saw in 5.1.7 that products in posets are just meets. Since the meet of two nonnegative integers is the smaller of the two, $\mathbf{N}$ has binary products.

Section 5.4

1. Using the version of the disjoint sum of 5.4.4, define $(f+g)(s,0) = (f(s),0)$ and $(f+g)(t,1) = (g(t),1)$.

2. Let P be a poset and x, $y \in P$. The sum $x+y$ is characterized by the fact that there is an arrow $x + y \to z$ corresponding to every pair consisting of an arrow $x \to z$ and an arrow $y \to z$. In a poset, there is an arrow $x \to z$ and only one if and only if $x \leq z$ and similarly for y. Thus $x + y \leq z$ if and only if $x \leq z$ and $y \leq z$. But this property characterizes the join $x \vee y$.

3. Let P and Q be two posets and $P+Q$ denote the disjoint sum of the sets P and Q as described in 5.4.4. Define $(x,i) \le (y,i)$, $i = 0, 1$ if and only if $x \le y$, while $(x,0) \not\le (y,1)$ and $(x,1) \not\le (y,0)$ for all $x, y \in P+Q$. The proof that this is the sum is essentially the same as the sum for the category of sets, augmented by the observation that an arrow from $P+Q \to R$ preserves the partial order just defined if and only if its restrictions to P and Q do.

4. Use the example given in the answer to Exercise 7 of Section 5.1 in the dual category.

Section 5.5

1. Given sets A and B, a function $f_0 : B \to A$ and $t : A \to A$, define a function $f : B \times \mathbf{N} \to A$ by letting $f(0,b) = f_0(b)$ and having defined $f(i,b)$ for $i \le n$, define $f(n+1,b) = t(f(n,b))$.

2. To be a model M of that sketch is to be an object $A = M(n)$ together with arrows $f_0 = M(\text{zero}) : 1 \to A$ and $t = M(\text{succ}) : A \to A$, so that $(\mathbf{N}, 0, s)$ is certainly a model. If (A, f_0, t) is another model, then there is a unique arrow $f : \mathbf{N} \to A$ such that

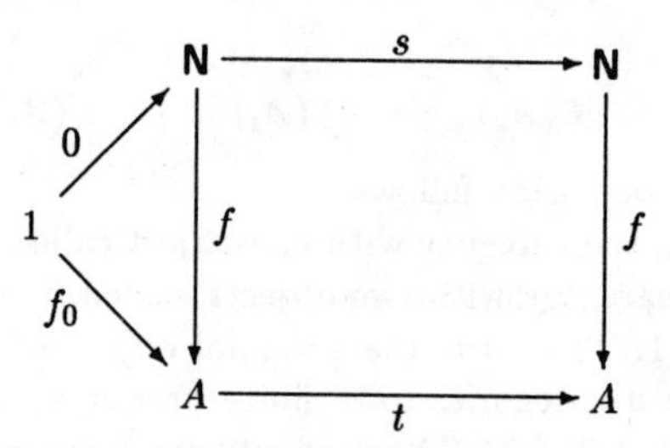

commutes. But the commutation of the two parts of that diagram express the fact that f is an arrow in the category of models of the sketch. Since f is unique, this shows that there is exactly one such arrow from $(\mathbf{N}, 0, s)$ to any other model of the sketch, whence that is the initial model.

3. Define $t : B \times \mathbf{N} \times A \to B \times \mathbf{N} \times A$ by $t(b,n,a) = h(b,n,f(b,n))$. Define $k_0 : B \to B \times \mathbf{N} \times A$ by $k_0(b) = (b, 0, g(b))$. Then the induction property gives an arrow $k : B \times \mathbf{N} \to B \times \mathbf{N} \times A$ such that $k(b,0) = k_0(b) = (b,0,g(b))$ and $k(b,s(n)) = t(k(b,n))$. If we let $k_i = \text{proj}_i \circ k$, $i = 1, 2, 3$, then $k(b,n) = (k_1(b,n), k_2(b,n), k_3(b,n))$. Next we claim that $k_1(b,n) = b$ and $k_2(b,n) = n$. For consider the diagram (in which we have abbreviated proj as p)

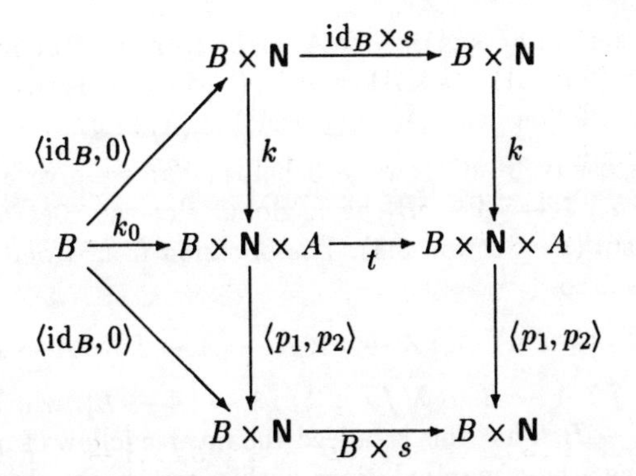

Compare this to the diagram

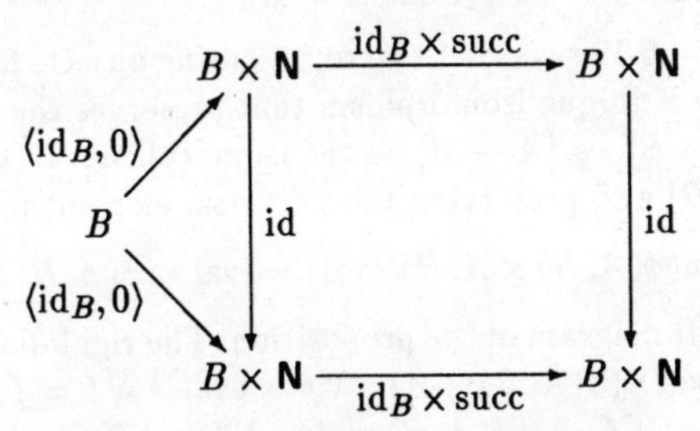

and the uniqueness of recursively defined arrows implies that $\langle p_1, p_2\rangle \circ k = \langle \mathrm{id}_B, \mathrm{id}_{\mathbf{N}}\rangle$. Then $f = k_3$ has the required properties.

Section 5.6

1. There is, by CC–1 and CC–2, a proof of $A \Rightarrow$ **true** and only one for each object A of $\mathcal{C}$ so that **true** is the terminal object. If $q_1 : X \to A$ and $q_2 : X \to B$ are arrows in $\mathcal{C}$, then from CC–3, there is an arrow $\langle q_1, q_2\rangle : X \to A \wedge B$. According to CC–4, $p_1 \circ \langle q_1, q_2\rangle = q_1$ and $p_2 \circ \langle q_1, q_2\rangle = q_2$ and by CC–5, the arrow $\langle q_1, q_2\rangle$ is unique with this property, so that $A \wedge B$ together with p_1 and p_2 is a product of A and B in $\mathcal{C}$.

2.a. Either projection is a proof.
b. $\langle \mathrm{id}_A, \mathrm{id}_A\rangle$ is a proof.
c. $\langle \mathrm{proj}_2, \mathrm{proj}_1\rangle$ is a proof.
d. $\langle \mathrm{proj}_1 \circ \mathrm{proj}_1, \langle \mathrm{proj}_2 \circ \mathrm{proj}_1, \mathrm{proj}_2\rangle\rangle$ is a proof.

Solutions for Chapter 6

Section 6.1

1. By CCC–3, $\mathrm{eval} \circ ((\lambda \circ \mathrm{eval}) \times A) = \mathrm{eval} : [A \to B] \times A \to B$. By the uniqueness requirement of CCC-3, $(\lambda \circ \mathrm{eval}) \times A = [A \to B] \times A$ (the identity arrow) so the first component of $(\lambda \circ \mathrm{eval}) \times A$ must be the identity.

2. By CCC–3, $\text{eval} \circ (\lambda f \times A) \circ (g \times A) = f \circ (g \times A)$. But by Exercise 3 of Section 5.2, $\text{eval} \circ (\lambda f \times A) \circ (g \times A) = \text{eval} \circ ((\lambda f \circ g) \times A)$. By the uniqueness property of eval, it follows that $\lambda(f \circ (g \times A)) = (\lambda f) \circ g$.

3. This follows easily from Exercise 6 below, but we give an independent proof here. Let $h : 1 \to [A \to B]$ be a global element. Let $u : A \to 1 \times A$ be an isomorphism (see Section 5.3). The bijection from $\text{Hom}(1, [A \to B])$ to $\text{Hom}(A, B)$ takes h to

$$\text{eval} \circ (h \times A) \circ u : A \to 1 \times A \to [A \to B] \times A \to B$$

Its inverse takes $f : A \to B$ to $\lambda(f \circ u^{-1}) : 1 \to [A \to B]$, which fits since $f \circ u^{-1} : 1 \times A \to A \to B$. That this is indeed the inverse follows from CCC–3. (In categorical writing, these manipulations with u are nearly always suppressed by assuming that $A = A \times 1$ and $u = \text{id}_A$.)

4. Corollary 4.5.12 says that two representing objects for the same functor are isomorphic by a unique isomorphism that preserves the universal element. Let $\phi(A, B) : [A \to B]' \to [A \to B]$ be the isomorphism. In this case the functor is $\text{Hom}(- \times A, B)$ and preserving the universal element means that

$$\text{Hom}(\phi(A, B) \times A, B)(\text{eval}) = \text{eval} \circ (\phi(A, B) \times A) = \text{eval}'$$

which is the left diagram of the proposition. The right diagram follows from this calculation: $\text{eval} \circ (\phi(A, B) \times A) \circ \lambda' f = \text{eval}' \circ \lambda' f = f$, so by the uniqueness requirement of CCC–3, $(\phi(A, B) \times A) \circ \lambda' f = \lambda f$.

5. Define $\text{eval} : [\mathcal{C} \to \mathcal{D}] \times \mathcal{C} \to \mathcal{D}$ on objects as follows. Let $F : \mathcal{C} \to \mathcal{D}$ and let C be an object of $\mathcal{C}$. Then $\text{eval}(F, C) = F(C)$. On arrows, suppose $\alpha : F \to F'$ is a natural transformation and $f : C \to C'$ is an arrow of $\mathcal{C}$. Then set $\text{eval}(\alpha, f) = F'f \circ \alpha C = \alpha C' \circ Ff : F(C) \to F'(C')$ (they are the same by naturality).

Since (α, f) is an arrow from (F, C) to (F', C'), eval preserves source and target because $\text{eval}(F, C) = F(C)$ and $\text{eval}(F', C') = F'(C')$. Preservation of identities is easy to verify. As for composition, let $\beta : F' \to F''$ and $g : C' \to C''$. Then

$$\begin{aligned}
\text{eval}((\beta, g) \circ (\alpha, f)) &= \text{eval}(\beta \circ \alpha, g \circ f) \\
&= F''(g \circ f) \circ (\beta \circ \alpha)C \\
&= F''(g) \circ F''(f) \circ \beta C \circ \alpha C \\
&= F''(g) \circ \beta C' \circ F'(f) \circ \alpha C \\
&= \text{eval}(\beta, g) \circ \text{eval}(\alpha, f)
\end{aligned}$$

The third line follows from the functoriality of F'' and the definition of composition of natural transformations; the fourth line is by naturality of β.

For a functor $F : \mathcal{B} \times \mathcal{C} \to \mathcal{D}$ and an object (B, C) of $\mathcal{B} \times \mathcal{C}$,

$$\begin{aligned}
\text{eval} \circ (\lambda F \times \mathcal{C})(B, C) &= \text{eval} \circ (\lambda F(B), C) \\
&= \lambda F(B)(C) = F(B, C)
\end{aligned}$$

Thus $\text{eval} \circ (\lambda F \times \mathcal{C}) = F$, as required.

6. In this exercise we regard λ as binding more tightly than composition or product, so that for example $\lambda f \circ g$ means $(\lambda f) \circ g$ and $\lambda f \times g$ means $(\lambda f) \times g$.

6.a. Define $\Gamma B : \mathrm{Hom}(C, [A \to B]) \to \mathrm{Hom}(C \times A, B)$ by $\Gamma B(g) = \mathrm{eval} \circ (g \times A)$. Then by CCC–3, $\Gamma B \circ \Lambda B(f) = \mathrm{eval} \circ (\lambda f \times A) = f$ for $f : C \times A \to B$, and by the uniqueness requirement for eval,

$$\Lambda B \circ \Gamma B(g) = \lambda(\mathrm{eval} \circ (g \times A)) = g$$

for $g : C \to [A \to B]$, so ΓB is the inverse of ΛB which is therefore bijective.

b. Let $g : C' \to C$ and $h : B \to B'$. Naturality requires that this diagram commute:

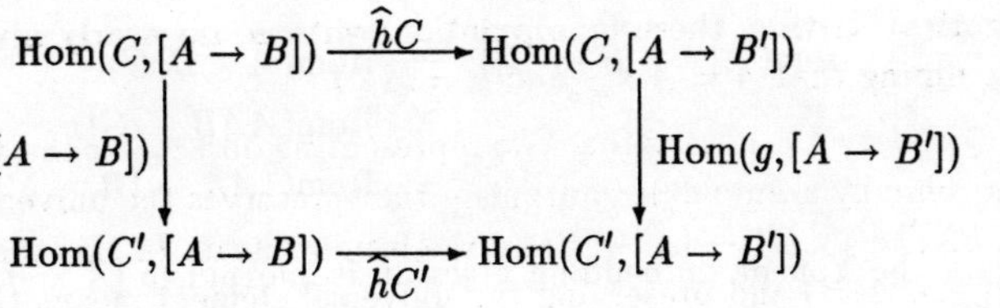

$$\begin{array}{ccc} \mathrm{Hom}(C, [A \to B]) & \xrightarrow{\widehat{h}C} & \mathrm{Hom}(C, [A \to B']) \\ {\scriptstyle \mathrm{Hom}(g, [A \to B])} \downarrow & & \downarrow {\scriptstyle \mathrm{Hom}(g, [A \to B'])} \\ \mathrm{Hom}(C', [A \to B]) & \xrightarrow[\widehat{h}C']{} & \mathrm{Hom}(C', [A \to B']) \end{array}$$

Because ΛB is a bijection, an arbitrary arrow of $\mathrm{Hom}(C, [A \to B])$ can be taken to be λf for some $f : C \times A \to B$. The upper route around the diagram takes λf to

$$\lambda(h \circ \mathrm{eval} \circ (\lambda f \times A)) \circ g = \lambda(h \circ f) \circ g$$

whereas the lower route takes it to

$$\begin{aligned} \lambda(h \circ \mathrm{eval} \circ ((\lambda f \circ g) \times A)) &= \lambda(h \circ \mathrm{eval} \circ (\lambda f \times A) \circ (g \times A)) \\ &= \lambda(h \circ f \circ (g \times A)) \end{aligned}$$

which is the same thing by Exercise 2.

c. Let $f : B \to B'$. The required diagram for naturality, namely

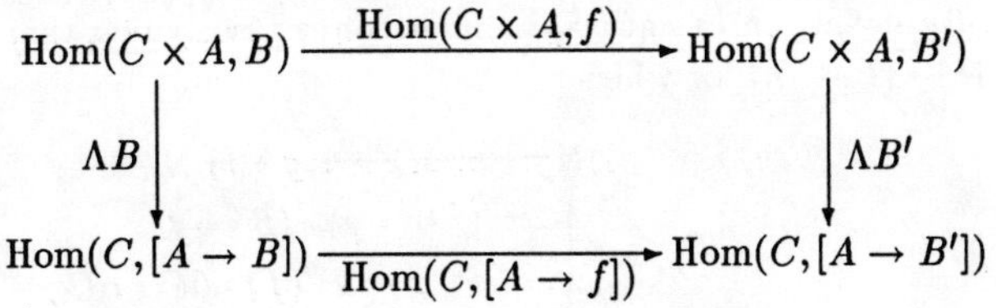

$$\begin{array}{ccc} \mathrm{Hom}(C \times A, B) & \xrightarrow{\mathrm{Hom}(C \times A, f)} & \mathrm{Hom}(C \times A, B') \\ {\scriptstyle \Lambda B} \downarrow & & \downarrow {\scriptstyle \Lambda B'} \\ \mathrm{Hom}(C, [A \to B]) & \xrightarrow[\mathrm{Hom}(C, [A \to f])]{} & \mathrm{Hom}(C, [A \to B']) \end{array}$$

commutes by definition of $[A \to f]$. It is a natural isomorphism because each component is a bijection.

d. Let $C = [A \to B]$ in the preceding diagram and start with eval in the upper right corner. The upper route gives $\lambda(f \circ \mathrm{eval})$ and the lower route gives $[A \to f] \circ \lambda(\mathrm{eval})$, which is $[A \to f]$ by Exercise 1.

7.a. This is shown by the following calculation, in which each isomorphism is a consequence either of Exercise 6.c or of Proposition 5.2.14.

$$\begin{aligned}\mathrm{Hom}(1,[A \to B \times C]) &\cong \mathrm{Hom}(A, B \times C)\\ &\cong \mathrm{Hom}(A,B) \times \mathrm{Hom}(A,C)\\ &\cong \mathrm{Hom}(1,[A \to B]) \times \mathrm{Hom}(1,[A \to C])\\ &\cong \mathrm{Hom}(1,[A \to B] \times [A \to C])\end{aligned}$$

By the fullness of the Yoneda embedding (Theorem 4.5.3), this corresponds to an isomorphism $[A \to B] \times [A \to C] \to [A \to B \times C]$.

b.

$$\begin{aligned}\mathrm{Hom}(1,[A \times B \to C]) &\cong \mathrm{Hom}(A \times B, C)\\ &\cong \mathrm{Hom}(A,[B \to C])\\ &\cong \mathrm{Hom}(1,[A \to [B \to C]])\end{aligned}$$

so again the Yoneda embedding gives an isomorphism $[A \times B \to C] \to [A \to [B \to C]]$.

8. Let $\mathbf{N}$ be a natural numbers object and suppose $f_0 : B \to A$ and $t : A \to A$. We must find a unique $f : B \times \mathbf{N} \to A$ for which

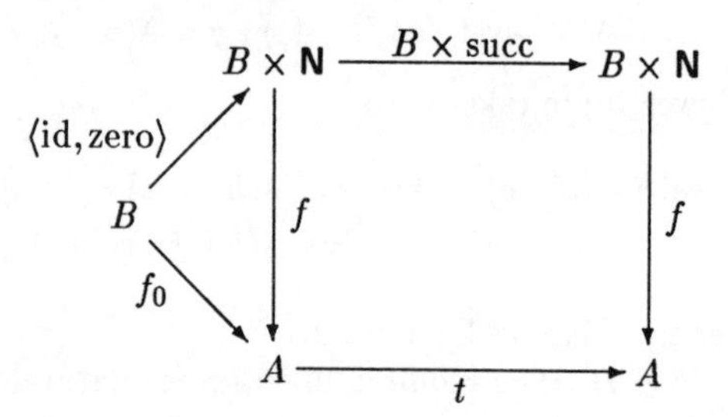

commutes. By definition of natural numbers object, we know that there is a unique $g : \mathbf{N} \to [B \to A]$ for which

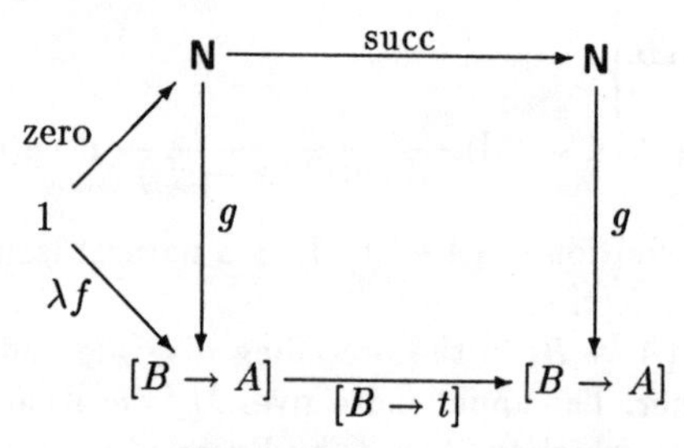

commutes, where $[B \to t]$ is the arrow defined in Exercise 6(c). Then by CCC–3 and the second diagram,

$$\mathrm{eval} \circ (B \times g) \circ (B \times \mathrm{zero}) = \mathrm{eval} \circ B \times (g \circ \mathrm{zero}) = \mathrm{eval} \circ (B \times \lambda f_0)$$

which is f_0, so that defining $f = \text{eval} \circ B \times g$ will make the triangle in the first diagram commute. (Note that we are using Proposition 6.1.6 here to evaluate on the first coordinate instead of the second.) As for the square, $f \circ (B \times \text{succ}) = \text{eval} \circ (B \times (g \circ \text{succ})) = \text{eval} \circ (B \times ([B \to t] \circ g)) = \text{eval} \circ ((B \times [B \to t]) \circ (B \times g))$ because the square in the second diagram above commutes. By Exercise 6.d, this is

$$\text{eval} \circ (B \times \lambda(t \circ \text{eval})) \circ (B \times g) = t \circ \text{eval} \circ (B \times g) = t \circ f$$

as required. By the uniqueness property of eval, f is unique.

9. Let $\mathcal{G}_0$ be the set of nodes of $\mathcal{G}$ regarded as a graph with no arrows. We must show that

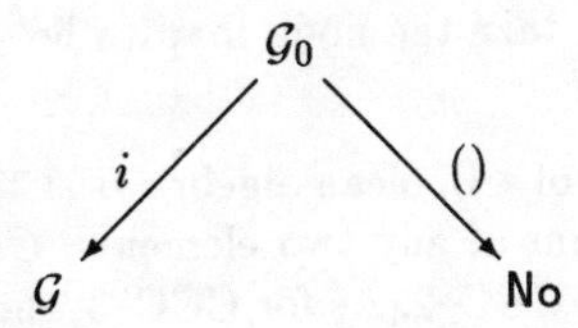

is a product diagram, where $()$ is the unique function and $i : \mathcal{G}_0 \to \mathcal{G}$ is the identity function on nodes and (necessarily) the empty function on arrows. Let $\mathcal{T}$ be a graph and $f : \mathcal{T} \to \mathcal{G}$, $g : \mathcal{T} \to \mathsf{No}$ be graph homomorphisms. We must find $u : \mathcal{T} \to \mathcal{G}_0$ for which $i \circ u = f$ and $() \circ u = ()$. The second equation is automatic. The first equation requires that u be the same as f on nodes. The fact that there is a homomorphism from $\mathcal{T}$ to No (which has no arrows) means that $\mathcal{T}$ has no arrows. Thus u has to be the empty function on arrows; since f must also be the empty function on arrows it follows that $i \circ u = f$.

A node of $\mathcal{G} \times \mathsf{No}$ as constructed in Exercise 4 of Section 5.1 is a pair (g, n) where g is a node of $\mathcal{G}$ and n is the unique node of No. (The construction in that exercise shows that $\mathcal{G} \times \mathsf{No}$ has no arrows since No has none.) Thus the function $(g, n) \mapsto g$ is clearly a bijection between $\mathcal{G} \times \mathsf{No}$ and the set of nodes of $\mathcal{G}$. What we have constructed here is more than that: it is a natural isomorphism in the category of graphs.

10. Given $f : \mathcal{C} \times \mathcal{G} \to \mathcal{H}$, we must show that $\text{eval} \circ (\lambda f \times \mathcal{G}) = f$. First, suppose that (c, n) is a node of $\mathcal{C} \times \mathcal{G}$. Then

$$\begin{aligned} \text{eval}((\lambda f \times \mathcal{G})(c, n)) &= \text{eval}(\lambda f(c), n) \\ &= \lambda f(c)(n) = f(c, n) \end{aligned}$$

For an arrow $(a : c \to d, w)$ of $\mathcal{C} \times \mathcal{G}$, let

$$\begin{aligned} \text{eval}((\lambda f \times \mathcal{G})(a, w)) &= \text{eval}(\lambda f(a), w) \\ &= f_a(w) = f(a, w) \end{aligned}$$

as required.

11. We use the formulation of $[\mathcal{G} \to \mathcal{H}]$ in Exercise 10. Let $h : 1 \to [\mathcal{G} \to \mathcal{H}]$ be a global element. Remember that in the category of graphs and graph homomorphisms, 1 is the graph with one node n and one arrow e. The node n goes to a node of $[\mathcal{G} \to \mathcal{H}]$, which is a function $f_0 : G_0 \to H_0$. The arrow e goes to an arrow of $[\mathcal{G} \to \mathcal{H}]$, which is an ordered triple $(f_1, f_2, f_3) : f_1 \to f_2$ as described in Exercise 10. The fact that 1 has only one node means that, since h is a graph homomorphism, necessarily $f_1 = f_2 = f_0$. Then the conditions in the description in Exercise 10 say that for any arrow g of $\mathcal{G}$, $\text{source}(f_3(g)) = f_0(\text{source}(g))$ and $\text{target}(f_3(g)) = f_0(\text{target}(g))$. Thus f_0 must be the node map and f_3 the arrow map of a graph homomorphism from $\mathcal{G}$ to $\mathcal{H}$. It follows from Exercise 3 that the *loops* of $[\mathcal{G} \to \mathcal{H}]$ are in one to one correspondence with the graph homomorphisms from $\mathcal{G}$ to $\mathcal{H}$: to recover the homomorphism from the loop $(f_1, f_1, f_3) : f_1 \to f_1$, take the node map to be f_1 and the arrow map to be f_3.

12. The terminal object of a Boolean algebra B is $\top$, so $C(B)$ satisfies CCC–1. Since B has the infimum of any two elements, $C(B)$ has binary products (see 5.1.7) and so satisfies CCC–2. As for CCC–3, showing eval exists requires showing that $[a \to b] \wedge a \leq b$. This follows from this calculation that uses the distributive law:

$$\begin{aligned}
[a \to b] \wedge a &= (\neg a \vee b) \wedge a \\
&= (\neg a \wedge a) \vee (b \wedge a) \\
&= \bot \vee (b \wedge a) = b \wedge a \leq b
\end{aligned}$$

The existence of the λ function requires showing that if $c \wedge a \leq b$ then $c \leq [a \to b]$ (the uniqueness of λf and the fact that $\text{eval} \circ \lambda f \times A = f$ are automatic since no hom set in the category determined by a poset has more than one element). This follows from this calculation, assuming $c \wedge a \leq b$:

$$\begin{aligned}
c &= c \wedge \top = c \wedge (a \vee \neg a) \\
&= (c \wedge a) \vee (c \wedge \neg a) \\
&\leq b \vee (c \wedge \neg a) \leq b \vee \neg a = [a \to b]
\end{aligned}$$

Section 6.2

1. For $i = 1, 2$,

$$\begin{aligned}
\text{proj}_i(c) &=_X \lambda_{x \in A \times B}(\text{proj}_i(x))\text{`}c && \text{(TL–16)} \\
&=_X \lambda_{x \in A \times B}(\text{proj}_i(x))\text{`}c' && \text{(TL–12)} \\
&=_X \text{proj}_i(c') && \text{(TL–16)}
\end{aligned}$$

where in the applications of TL–16, we judiciously choose a variable x of type $A \times B$ that does not occur freely in c or c'.

2.

$$\begin{aligned}(a,b) &=_X \lambda_{x\in A}(x,b)`a \quad \text{(TL–16)}\\ &=_X \lambda_{x\in A}(x,b)`a' \quad \text{(TL–12)}\\ &=_X (a',b) \quad \text{(TL–16)}\end{aligned}$$

and similarly $(a',b) =_X (a',b')$. The result follows from TL–9.

Section 6.3

1. We must show that for $f : C \times A \to B$, $\text{eval} \circ (\lambda f \times A) = f$. First note that $\lambda f \times A = \langle \lambda f \circ p_1, p_2\rangle : C \times A \to [A \to B] \times A$. (See 5.2.17.) Now let z be a variable of type C, y a variable of type A which is not in X, and suppose f is determined by a term $\phi(z,y)$ of type B. Then $\lambda f \times A$ is represented by $(\lambda_y\phi(z,y) \circ z, y) =_X (\lambda_y\phi(z,y), y)$ by definition of composition in $C(\mathcal{L})$. Then $\text{eval} \circ (\lambda f \times A)$ is

$$(p_1(\lambda_y\phi(z,y),y))`p_2(\lambda_y\phi(z,y),y) =_X \lambda_y\phi(z,y)`y$$

by TL–14 and Exercise 2 of Section 6.2. Since $y \notin X$, this is $\lambda_y\phi(z,y)$ by TL-17.

Section 6.4

1.a. $\mathbf{N}\times\mathbf{N} \xrightarrow{\langle p_1,p_1,p_2,p_2\rangle} \mathbf{N}\times\mathbf{N}\times\mathbf{N}\times\mathbf{N} \xrightarrow{*\times *} \mathbf{N}\times\mathbf{N} \xrightarrow{+} \mathbf{N}$

b. $\mathbf{N}\times\mathbf{N}\times\mathbf{N} \xrightarrow{\langle p_1,p_1,p_2,p_2\rangle} \mathbf{N}\times\mathbf{N}\times\mathbf{N}\times\mathbf{N} \xrightarrow{*\times *} \mathbf{N}\times\mathbf{N} \xrightarrow{+} \mathbf{N}$

c. $\mathbf{N}\times\mathbf{N} \xrightarrow{()} 1 \xrightarrow{5} \mathbf{N}$

Solutions for Chapter 7

Section 7.1

1. Let us temporarily denote the usual addition by $\oplus$. The fact that $0+m = m$ implies that $k+m = k\oplus m$ when $k=0$. Assuming that equation for some k, we have that

$$\text{succ}(k)+m = \text{succ}(k+m) = \text{succ}(k\oplus m) = \text{succ}(k)\oplus m$$

2. What we must do is to add an operation and equations to implement the standard inductive definition of multiplication: $0*m = 0$ and $\text{succ}(k)*m = m + k*m$. We do this by adding one operation $*: n\times n \to n$ and diagrams

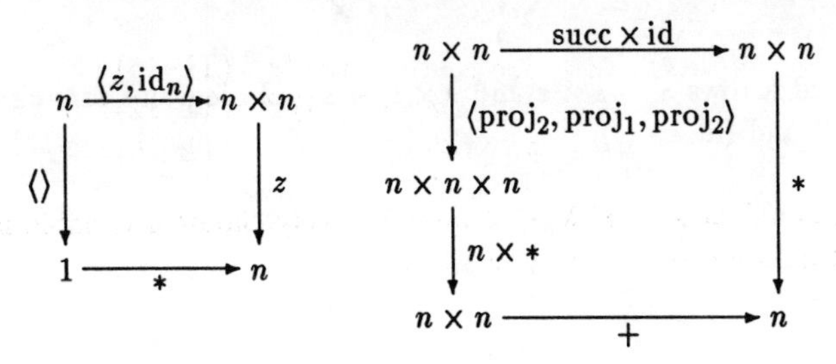

In order to make these diagrams, we also have to add arrows $n \to n \times n$, $n \times n \to n \times n$, $n \times n \to n \times n \times n$ and $n \times n \times n \to n \times n$ and, following the pattern of similar constructions in 7.1.7, diagrams forcing them to be $\langle z, \mathrm{id}_n \rangle$, $\mathrm{succ} \times \mathrm{id}$, $\langle \mathrm{proj}_2, \mathrm{proj}_1, \mathrm{proj}_2 \rangle$ and $n \times *$ respectively.

Section 7.4

1. What has to be proved is that if M and N are models of a sketch and if there is a homomorphism from M to N whose component at the node s is $h : M(s) \to N(s)$, then the component at $s \times s$ is $h \times h$ and at $s \times s \times s$ is $h \times h \times h$. More generally, if the homomorphism has component h_i at s_i, $i = 1, \ldots, n$, then the component at $s_1 \times \cdots \times s_n$ is $h_1 \times \cdots \times h_n$. Of course, it has to be well understood that the 'products' nodes in a sketch are purely symbolic, while that of homomorphisms in a model is their categorical product. We will do this for $n = 2$, but the extension to larger n poses no problem. So suppose s and t are nodes and M and N are models and we have a homomorphism $M \to N$ whose components at s and t are h and k respectively. Then the unique arrow l that makes

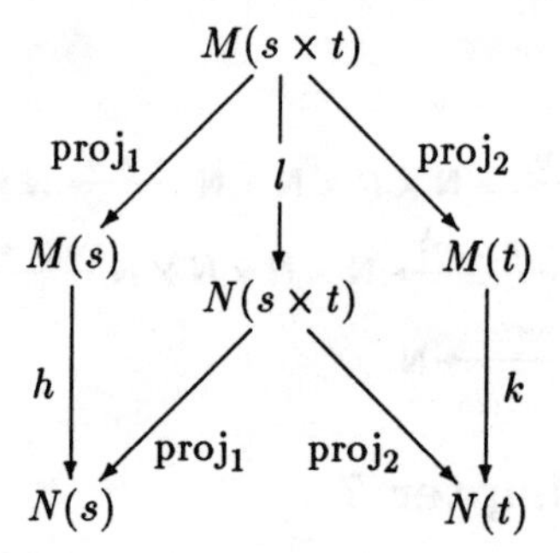

commute is $h \times k$.

2. Besides the additions suggested in the statement of the exercise, you have to add an operation $e : 1 \to s$ and the diagrams

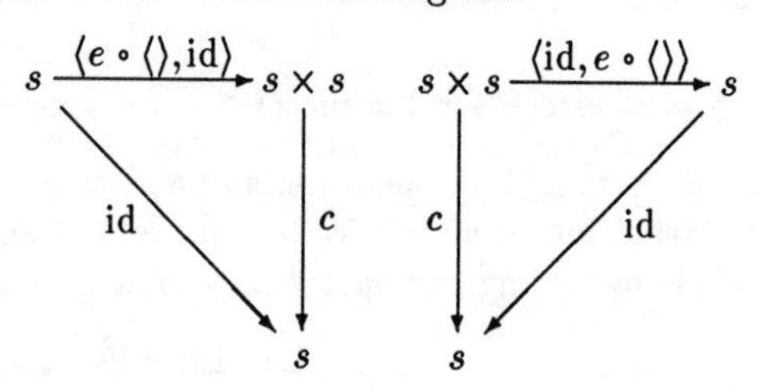

You also need arrows $s \to s \times s$ and $s \times s \to s$ and diagrams forcing them to be $\langle e \circ \langle\rangle, \mathrm{id} \rangle$ and $\langle \mathrm{id}, e \circ \langle\rangle \rangle$.

3. We give the answer for the case of real vector spaces; the other is similar. We assume the real number field **R** as a given structure. We suppose, for each $r \in R$, a unary operation we will denote $r^* : s \to s$. We also suppose a second binary operation $p : s \times s \to s$ to stand for addition. We require a unit element $z : 1 \to s$ for the operation p and a diagram similar to the previous exercise to say that z is the unit element for the operation. We have to say that p is associative and commutative. The diagram for the latter is

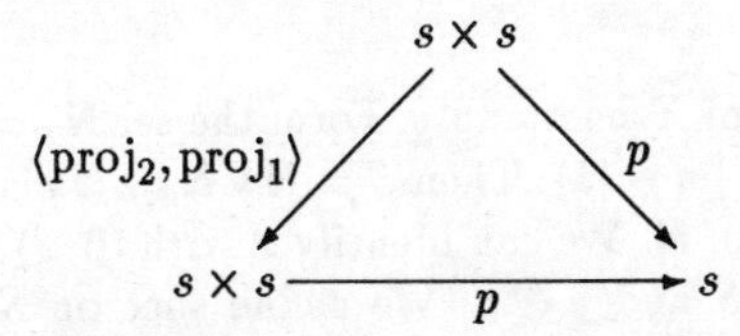

We have to say that p distributes over c which is done via the diagram in which we have abbreviated by w the arrow $\langle \mathrm{proj}_1, \mathrm{proj}_2, \mathrm{proj}_1, \mathrm{proj}_3 \rangle$

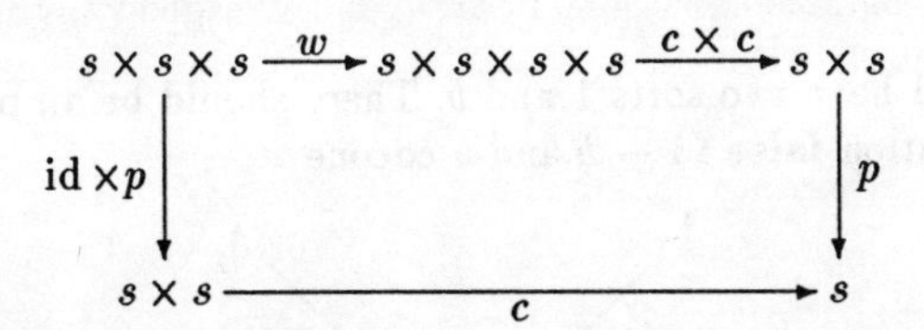

We have to add a unary negation operator $n : s \to s$ together with a diagram to say it is the negation operator:

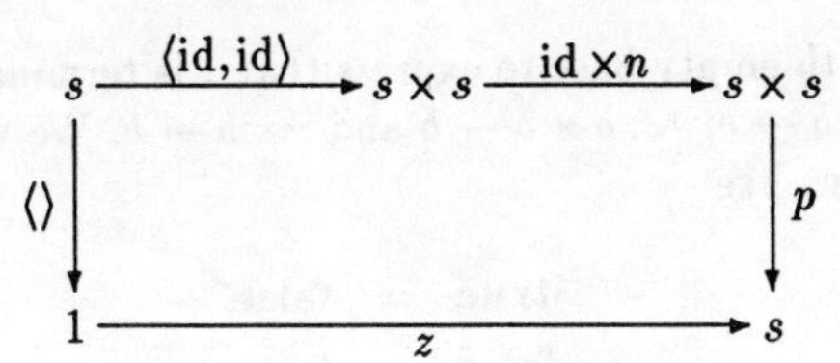

In addition, we need diagrams that express the following identities:

$$
\begin{aligned}
0^*(x) &= z \\
r^*(p(x, y)) &= p(r^*(x), r^*(y)) \\
(r + s)^*(x) &= p(r^*(x), s^*(x)) \\
1^*(x) &= x \\
r^*(s^*(x)) &= (rs)^*(x)
\end{aligned}
$$

We give, for example, the diagram required to express the third of the equations above:

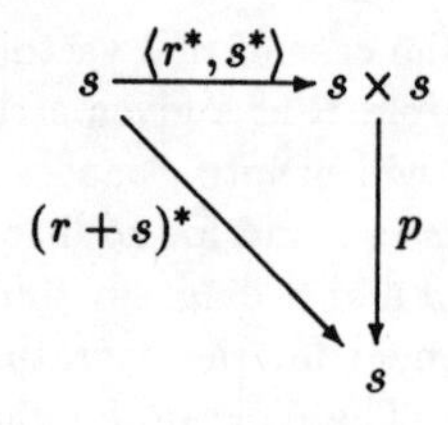

Section 7.6

1. Let the elements of X be x and y. Form the set $\mathbf{N}_x = \{(n,x) \mid n \in \mathbf{N}\}$ and similarly $\mathbf{N}_y = \{(n,y) \mid n \in \mathbf{N}\}$. Then $S = \mathbf{N} \vee \mathbf{N}_x \vee \mathbf{N}_y$ is certainly the disjoint union of three copies of $\mathbf{N}$. We can identify x with $(0,x)$ and $(0,y)$ so that, up to isomorphism, $x \in S$ and $y \in S$. We define succ on S by $\text{succ}(n) = n+1$, $\text{succ}(n,x) = (n+1,x)$ and $\text{succ}(n,y) = (n+1,y)$. With these definitions and no relations, this is the initial term model.

Section 7.8

1. This should have two sorts 1 and b. There should be an operation **true** : $1 \to b$, an operation **false** : $1 \to b$ and a cocone

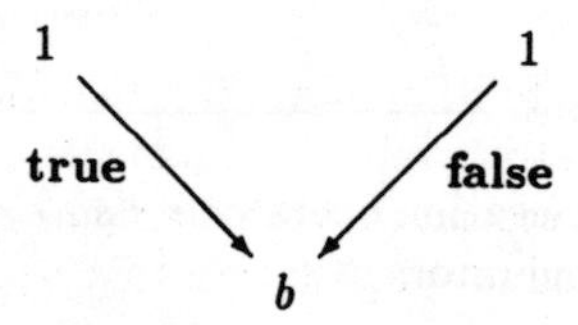

We need a cone with empty base to express that 1 is terminal. We should have operations $\vee : b \times b \to b$, $\wedge : b \times b \to b$ and $\neg : b \to b$. We will need diagrams to express equations like

$$\begin{aligned}
\neg\mathbf{true} &= \mathbf{false}\\
\neg\mathbf{false} &= \mathbf{true}\\
\neg \wedge (x,y) &= \vee(x,y)\\
\vee(0,0) &= 0\\
\vee(0,1) &= 1\\
\vee(1,0) &= 1\\
\vee(1,1) &= 1
\end{aligned}$$

There are other equations, but they all follow from these. In particular, we do not, in this case, have to express the distributive laws since they follow, there being so few elements.

Section 7.9

1. This is most readily done using elements, although it can all be done with diagrams. If $x * x^{-1} = 1$ and $y * y^{-1} = 1$, then

$$(x * y) * (y^{-1} * x^{-1}) = x * (y * (y *^{-1} * x^{-1})) = x * ((y * y^{-1}) * x^{-1})$$
$$= x * (1 * x^{-1}) = x * x^{-1} = 1$$

If $2 \neq 0$ in a field, then 2^{-1} exists, whence $(2 * 2)^{-1} = 4^{-1}$ also exists. Hence $4 \neq 0$.

2. Let $\mathbf{Q}$ and F_2 denote the fields of rational numbers and the two-element field respectively. In order to have a product of $\mathbf{Q}$ and F_2, we have to have a cone

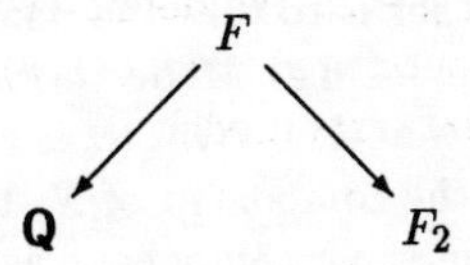

which is exactly what we do not have in the category of fields.

Section 7.10

1.a. Several things have to be shown. First that the image $N_0 \subseteq N$ is closed under the operations (that is if all the arguments of the operation are in the image, so does the value); second that the diagrams continue to commute; third that the cones remain products; and fourth that the cocones remain sums.

Let $u : s_1 \times s_2 \times \cdots \times s_n \to s$ be an operation. For $i = 1, \ldots, n$ let $x_i \in M(s_i)$ and $x = u(x_1, \ldots, x_n)$. Then $f(x) = u(f(x_1), \ldots, f(x_n))$, which demonstrates the first point. Since the operations in N_0 are the restrictions of those in N any two paths that agree on N do so on N_0 (actually, whether or not they do on M). For the cones, we illustrate on binary cones. Suppose $s \leftarrow r \rightarrow t$ is a cone in the sketch. Then in the diagram

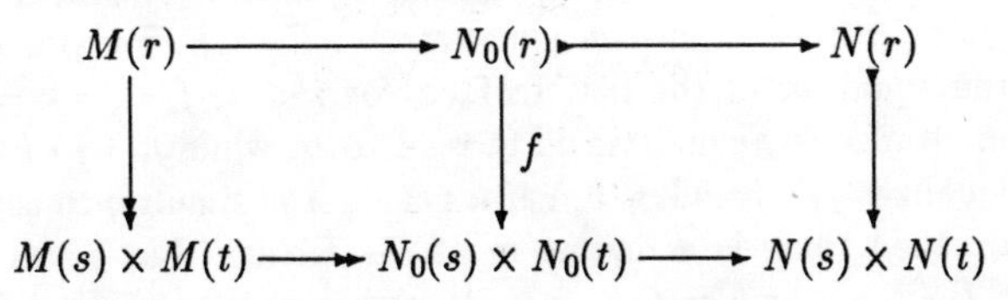

certain functions have been labeled as being surjective or injective (although others are and some are bijective). These use the facts that the product of two injective functions is injective and the product of two surjective functions is surjective. That f is surjective follows from the fact that the composite of two surjective functions is surjective and if the composite of two functions is surjective, so is the second one. Dually, the fact that f is injective uses the facts that the composite of two injective functions is injective and that if the composite of two functions is injective, the first one is.

The dual argument, using the fact that a sum of injectives is injective and a sum of surjectives is surjective, gives the corresponding result for discrete cocones.

It is worth noting that these arguments fail if either the cones or cocones fail to be discrete or if the category in which the models are taken is other than the category of sets. The reason is that even in the category of sets, the arrow induced between equalizers of epis is not necessarily epic and between coequalizers of monos will not be monic. (Equalizers and coequalizers are defined in chapter 8.) In categories other than sets, even the sum of epics will not generally be epic, nor the sums of monics monic.

b. There are two ways of doing this. One is to show that the intersection of submodels is a submodel. Then the intersection of all submodels is clearly the smallest submodel. Another is to take the image of the initial term model in the component of that model. Let M be the initial term model, N is the given model and N_0 this submodel. If $N_1 \subseteq N$ is any other submodel, there is an initial term model M' in the component of N_1 that has an arrow $M' \to N_1$. But $N_1 \subseteq N$ and so is in the same component as N. Thus $M' = M$ and the map $M \to N_1 \subseteq N$ is the original map $M \to N$. Since that factors through N_1, it follows that $N_0 \subseteq N_1$.

Solutions for Chapter 8

Section 8.1

1. Suppose $h, k : C \to E$ with $j \circ h = j \circ k = l$. Then $f \circ l = f \circ j \circ k = g \circ j \circ k$ so there is a unique arrow $m : C \to E$ with $j \circ m = l$. But both h and k are such arrows and so $h = k$.

2. Suppose by induction that this is true for every list of $n-1$ parallel arrows and $f_1, \ldots, f_n : A \to B$ is a list of n parallel arrows. Let $j : E \to A$ be the equalizer of $f_1, \ldots, f_{n-1}$. We may clearly suppose that $n \geq 3$. Then the parallel pair $f_1 \circ j, f_n \circ j : E \to B$ has an equalizer $k : F \to E$. I claim that $j \circ k : F \to A$ is the equalizer of the list. In fact, for $i < n$, $f_i \circ j \circ k = f_1 \circ j \circ k$, because j simultaneously equalizes all those arrows, while $f_n \circ j \circ k = f_1 \circ j \circ k$ because k equalizes $f_1 \circ j$ and $f_n \circ j$. If $h : C \to A$ is simultaneously equalized by $f_1, \cdots, f_n$, then there is a unique $m : C \to E$ such that $j \circ m = h$. Since $f_1 \circ j \circ m = f_1 \circ h = f_n \circ h = f_n \circ j \circ m$, there is a unique $g : C \to F$ such that $k \circ g = m$. Then $j \circ k \circ g = j \circ m = h$. If $j \circ k \circ g' = h$, then $g' = g$ because $j \circ k$ is monic; hence g has the required uniqueness property.

3. Let A and B be monoids and $f, g : A \to B$ be monoid homomorphisms. Let $E = \{x \in A \mid f(a) = g(a)\}$. Then $f(1) = 1 = g(1)$ so that $1 \in E$. Also if $x, y \in E$, then $f(xy) = f(x)f(y) = g(x)g(y) = g(xy)$ so that $xy \in E$ and E is a submonoid of A. Let $j : E \to A$ be the inclusion homomorphism. Now let $h : C \to A$ be a monoid homomorphism with $f \circ h = g \circ h$. Then for all $x \in C$, $f(h(x)) = g(h(x))$ so that $h(x) \in E$. Thus $h(C) \subseteq E$ so there is a function

$k : C \to E$ with $j \circ k = h$. It has to be shown that k is a homomorphism, but $k(x) = h(x)$ for all $x \in C$, so that is trivial. The fact that j is injective makes the uniqueness of k evident.

4.a. For x and y to have an equalizer, we would need, at least, an element z with $xz = yz$ and this never happens in a free monoid.

b. Suppose the element x is the equalizer of y and z. Then x is monic, which means, according to the cited exercise, that x is invertible. But then $yx = zx$ implies that $y = yxx^{-1} = zxx^{-1} = z$.

5.a. A monomorphism in **Set** is an injective function (see Theorem 2.8.3). so let $f : A \to B$ be an injective function. Let C be the set of all pairs

$$\{(b, i) \mid b \in B,\ i = 0, 1\}$$

and impose an equivalence relation on these pairs forcing $(b, 0) = (b, 1)$ if and only if there is an $a \in A$ with $f(a) = b$ (and not forcing $(b, i) = (c, j)$ if b and c are distinct). Since f is injective, if such an a exists, there is only one. Let $g : B \to C$ by $g(b) = (b, 0)$ and $h : B \to C$ by $h(b) = (b, 1)$. Then clearly $g(b) = h(b)$ if and only if there is an $a \in A$ with $f(a) = b$. Now let $k : D \to B$ with $g \circ k = h \circ k$. It must be that for all $x \in D$, there is an $a \in A$, and only one, such that $k(x) = f(a)$. If we let $l(x) = a$, then $l : D \to A$ is the unique arrow with $f \circ l = k$.

b. Let $f : A \to B$ be both an epimorphism and the equalizer of $g, h : B \to C$. Since $g \circ f = h \circ f$ and f is epi, $g = h$. Then $g \circ \mathrm{id}_B = h \circ \mathrm{id}_B$ so there is a $k : B \to A$ such that $f \circ k = \mathrm{id}_B$. But then $f \circ k \circ f = f = f \circ \mathrm{id}_A$. But f being mono can be cancelled from the left to conclude that $k \circ f = \mathrm{id}_A$.

c. The reason v and w are inverse to each other is that there is no other arrow for $v \circ u$ to be but id_C and similarly $u \circ v = \mathrm{id}_D$. Let $F : \mathcal{A} \to \mathcal{B}$ be the inclusion. For any functors $G, H : \mathcal{B} \to \mathcal{C}$, if $G \circ F = H \circ F$, then $G(u) = H(u)$. But then

$$G(v) = G(u^{-1}) = G(u)^{-1} = H(u)^{-1} = H(u^{-1}) = H(v)$$

Since the objects of $\mathcal{B}$ and the other arrows are in the image of F, it follows that $G = H$. Since F is not an isomorphism, it cannot be a regular monomorphism by (b).

Section 8.2

1. If D is the base of Diagram (8.1), then cone(E, D) is the set of all commutative cones of that shape. The equalizer of f and g is the universal element of the functor F of the proof of Proposition 8.1.4. These two functors are naturally isomorphic by an isomorphism $\phi : F \to \mathrm{cone}(-, D)$: if X is any object and $u : X \to A$ is in $F(X)$ (so $f \circ u = g \circ u$), then $\phi A(u)$ is the cone with components $u : X \to A$ and $f \circ u : X \to B$. By its very definition this is a commutative cone, so an element of cone(X, D). Since u determines $f \circ u$ uniquely, this function ϕA is a bijection, so ϕ is a natural isomorphism. Since the two functors are isomorphic, so are the objects that represent them, by the uniqueness of universal elements.

2. The limits are as indicated:

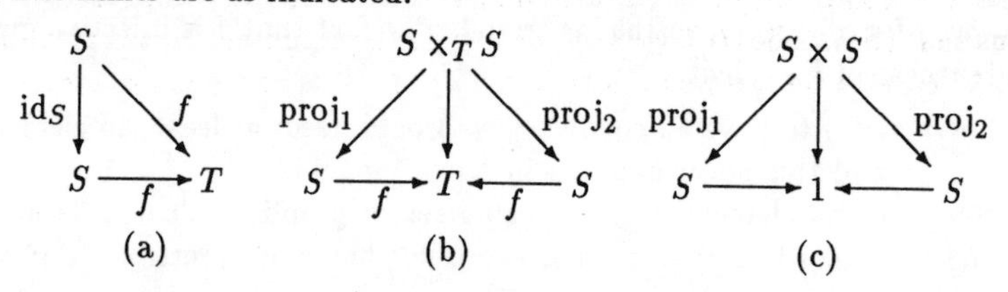

where $S \times_T S$ is the subset $\{(s, s') \mid f(s) = f(s')\}$ of $S \times S$. (This is standard notation to be introduced in Section 8.3.)

3.a. An arrow $f : C \to \prod_{i \in \mathcal{I}} Di$ is simply a collection of arrows $f_i : C \to Di$, one for each object of $\mathcal{I}$. In order that it be a cone on the diagram D it is necessary and sufficient that it satisfy the additional condition that $Da \circ f_j = f_k$. This is equivalent to

$$\mathrm{proj}_k \circ f = f_k = Da \circ f_j = Da \circ \mathrm{proj}_j \circ f$$

which is a necessary and sufficient condition that f factor through $E \to \prod Di$. Thus a cone over D is equivalent to an arrow to E, which means that E is a limit of D.

b. An arrow $f : C \to \prod Di$ is a collection of arrows $f_i : C \to Di$. In order that it be a cone on D it must simultaneously satisfy the conditions $Da \circ f_j = f_k$ and $Db \circ f_l = f_m$. This is equivalent to $\langle f_i, f_l \rangle : C \to Di \times Dl$ and $\langle f_k, f_m \rangle : C \to Dk \times D_m$ satisfying $(Da \times Db) \circ \langle f_j, f_l \rangle = \langle f_k, f_m \rangle$, in turn equivalent to $r \circ f = s \circ f$, that is to factoring through $E \to \prod Di$.

c. Again an arrow $f : C \to A$ is a family of arrows $f_i : C \to Di$. In order to be a cone on D it must satisfy $Da \circ f_{\mathrm{source}(a)} = f_{\mathrm{target}(a)}$ for every arrow $a \in \mathcal{I}$. This is exactly the condition that $r \circ f = s \circ f$, which means that f is a cone over D if and only if it factors through $E \to A$. Hence a cone over D is equivalent to an arrow to E, which means that E is the limit.

Section 8.3

1. Let $f : R \to S$ and $k : R \to A$ be functions such that $g \circ f = i \circ k$. Then for each $x \in R$, $g(f(x)) \in A$ which means that $f(x) \in g^{-1}(A)$. Thus f factors through $g^{-1}(A)$ by a unique map $m : R \to g^{-1}(A)$, so $f = j \circ m$. Also $i \circ h \circ m = i \circ k$, so $h \circ m = k$ because i is monic.

2. Suppose that $f : A \rightarrowtail B$ is monic. Then the only way you can have $g, h : C \to A$ such that $f \circ g = f \circ h$ is if $g = h$. In that case $g : C \to A$ is the unique arrow such that $\mathrm{id} \circ g = g$ and $\mathrm{id} \circ g = h$ so that

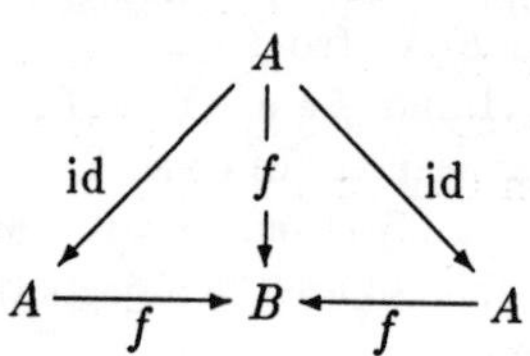

satisfies the condition of being a pullback cone. Thus (a) implies (b). It is obvious that (b) implies (c). Now suppose that the diagram in (c) is a pullback. Let h, $k : C \to A$ be arrows such that $f \circ h = f \circ k$. Then we have a cone

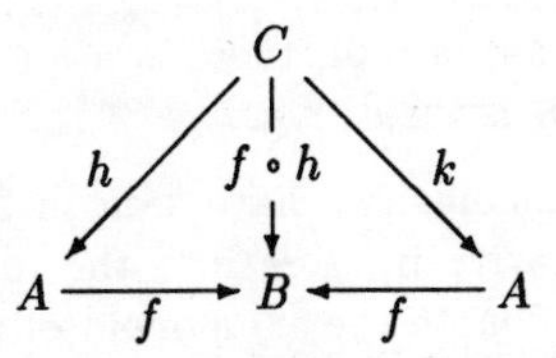

so that there is an arrow $l : C \to P$ such that $h = g \circ l = k$. Thus (c) implies (a).

3. Suppose $h, k : D \to P$ with $p_2 \circ h = p_2 \circ k$. We have

$$f \circ p_1 \circ h = g \circ p_2 \circ h = g \circ p_2 \circ k = f \circ p_1 \circ k$$

and f is monic by assumption so that $p_1 \circ h = p_1 \circ k$. Thus $x = h$ and $x = k$ are solutions to the equation $p_1 \circ x = p_1 \circ h$ and $p_2 \circ x = p_2 \circ h$. But the definition of pullback requires that solution to be unique so that $h = k$.

4. If $f, g : A \to B$, then the equalizer is the arrow $j : E \to A$ in the pullback

$$\begin{array}{ccc} E & \xrightarrow{\quad j \quad} & A \\ \big\downarrow & & \big\downarrow{\scriptstyle f \times g} \\ A \times A & \xrightarrow[\langle \mathrm{id}, \mathrm{id} \rangle]{} & B \times B \end{array}$$

The proof of this fact comes down to showing that the pullback of this square and the equalizer of f and g represent equivalent functors: the functor G representing the pullback of the square has value

$$\begin{aligned} G(X) &= \{(u : X \to A, v : X \to A \times A) \mid \langle \mathrm{id}, \mathrm{id} \rangle \circ v = (f \times g) \circ u\} \\ &= \{(u, v) \mid \langle v, v \rangle = \langle f \circ u, g \circ u \rangle\} \\ &= \{(u, v) \mid f \circ u = v = g \circ u\} \\ &\cong \{u \mid f \circ u = g \circ u\} \end{aligned}$$

which is the value of the functor representing the equalizer of f and g. (Compare the proof of Exercise 1 of Section 8.2.)

5. A pullback of a diagram $A \to 1 \leftarrow B$ is just a product $A \times B$. We know from the preceding exercise that a category with products and pullbacks has equalizers. Hence such a category has finite products and and equalizers. The conclusion now follows from Corollary 8.2.10.

6. In **Set**, the pullback is

$$P = \{(a, b) \mid f(a) = g(b)\}$$

Since f is surjective, for any $a \in A$, there is a $b \in B$ such that $g(a) = f(b)$. Then $p_1(a, b) = a$. Thus p_1 is surjective.

7. In the category of monoids, we have seen in 2.9.3 that the inclusion of the monoid $\mathbf{N}$ of nonnegative integers into the monoid $\mathbf{Z}$ of integers is an epimorphism. Let $A \subseteq \mathbf{Z}$ denote the submonoid of nonpositive integers. Both inclusions are epic, but the pullback, which is the intersection, is the submonoid $\{0\}$ and the maps to both $\mathbf{N}$ and A are not epic. In fact all pairs of arrows on $\mathbf{N}$ (or on A) agree on $\{0\}$.

8. Suppose the arrow $g : A \to C$ in Diagram (8.2) is a split epi, so that there is an arrow $h : C \to A$ so that $g \circ h = \mathrm{id}_C$. Then we have $\mathrm{id} : B \to B$ and $h \circ f : B \to A$ satisfy $f \circ \mathrm{id} = g \circ h \circ f$ so there is an arrow $k : B \to P$ such that $p_1 \circ k = h \circ f$ and $p_2 \circ k = \mathrm{id}_B$. This latter identity is what we want.

9. Suppose we have arrows $A \to B$ and $A \to C$ giving a commutative cone. By composing the latter with $C \to D$, we get a pair of arrows $A \to B$ and $A \to D$ giving a commutative cone and this leads to a unique arrow $A \to Q$. We now have $A \to C$ and $A \to Q$ giving a commutative cone and this leads to a unique arrow $A \to P$ making the left hand cone commute. Clearly, the outer rectangle commutes as well.

It says that the weakest precondition of the composite of two procedures can be calculated as the weakest precondition under the second procedure of the weakest precondition under the first.

Section 8.4

1. In Exercise 1 of Section 8.1, we showed that any equalizer is a monomorphism. Interpreted in the dual category, it is the result of this exercise.

2. By Proposition 2.9.2, every epimorphism in **Set** is surjective. We claim that every surjective function is a regular epimorphism. In fact, let $f : S \to T$ be surjective and suppose $E = \{(x, y) \in S \times S \mid f(x) = f(y)\}$. Let $p, q : E \to S$ be the first and second projections. We claim that f is the coequalizer of p and q. Clearly $f \circ p = f \circ q$. Let $h : S \to R$ such that $h \circ p = h \circ q$. Define $k : T \to R$ as follows. For $t \in T$, there is an $s \in S$ with $f(s) = t$. We would like to define $k(t) = h(s)$. If s' is another element of S with $f(s') = t$, then $(s, s') \in E$ and so $h(s) = (h \circ p)(s, s') = (h \circ q)(s, s') = h(s')$. Thus k is well defined and clearly $k \circ f = h$. The uniqueness of k follows from the fact that f is epic.

3. In Exercise 5 of Section 8.1, it is shown that a regular mono that is an epimorphism is an isomorphism. The dual says that a monomorphism that is a regular epimorphism is an isomorphism. In 2.9.3 the inclusion was shown to be epic. Since the inclusion is evidently a monomorphism and not an isomorphism, it cannot be a regular epimorphism.

4. We use the terminology of Exercise 2, just assuming that S and T are monoids and f a monoid homomorphism. First we have to say that if f is not surjective, then it still may be an epimorphism, as the inclusion of $\mathbf{N}$ into $\mathbf{Z}$ shows, but it cannot be regular. The reason is the easily verified fact that the image of a monoid homomorphism $f : S \to T$ is a submonoid $T_0 \subseteq T$ and if it is not all of T, the properties of regular epimorphism and the fact that the inclusion is a monomorphism combine to provide an arrow $T \to T_0$ such that the composite $T \to T_0 \to T$ is the identity of T. This is possible if and only if $T_0 = T$.

To go the other way, we need add to the construction of Exercise 2 only the facts that $E \subseteq S \times S$ is a submonoid and that all the arrows constructed are monoid homomorphism. Of all of these, only the fact that k is a monoid homomorphism is interesting. In fact, if t and t' are elements of T and if $s, s' \in S$ are such that $f(s) = t$ and $f(s') = t'$, then since f is a monoid homomorphism, $f(ss') = f(s)f(s') = f(t)f(t') = f(tt')$ so that $k(tt') = h(ss') = h(s)h(s') = k(t)k(t')$. Similarly, since f is a monoid homomorphism, $f(1) = 1$, so $k(1) = h(1) = 1$.

5. A kernel pair of f is characterized by the following mapping property: $f \circ d^0 = f \circ d^1$ and if $e^0, e^1 : C \to A$ satisfies $f \circ e^0 = f \circ e^1$, then there is a unique $g : C \to K$ such that $d^i \circ g = e^i$, $i = 0, 1$. Now if f is the coequalizer of e^0 and e^1 and the kernel pair exists, let $g : C \to K$ as described. Suppose $h : A \to D$ is an arrow such that $h \circ d^0 = h \circ d^1$. Then $h \circ e^0 = h \circ d^0 \circ g = h \circ d^1 \circ g = h \circ e^1$ so that there is a unique $k : B \to D$ with $k \circ f = h$. Thus f is the coequalizer of d^0 and d^1.

Section 8.5

1. Suppose E is some object and $\langle l; l' \rangle : C+C' \to E$ and $\langle m; m' \rangle : B+B' \to E$ satisfy $\langle l; l' \rangle \circ (g + g') = \langle m; m' \rangle \circ (f + f')$. If i and j represent the inclusions of the components, then by definition, $\langle l; l' \rangle \circ (g + g') \circ i = \langle l; l' \rangle \circ i \circ g = l \circ g$ and similarly $\langle m; m' \rangle \circ (f + f') \circ i = m \circ f$ so these equations imply that $l \circ g = m \circ f$. By using j we similarly conclude that $l' \circ g' = m' \circ f'$. The mapping properties of the pushout imply the existence of $n : D \to E$ and $n' : D' \to E$ such that $n \circ k = l$, $n \circ h = m$ $n' \circ k' = l'$ and $n' \circ h' = m'$. Then $\langle n, n' \rangle$ is the required arrow. Uniqueness follows from the uniqueness of the components, together with the uniqueness of an arrow from a sum, given its components.

2.a. Since e is injective, $e(C) \cong C$. Up to isomorphism, the diagram is the sum of the following two:

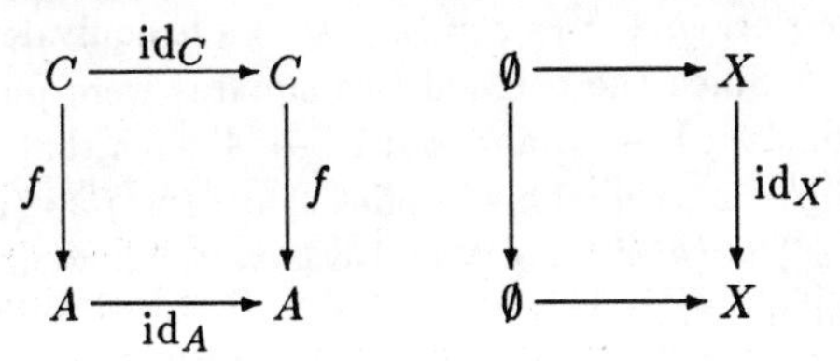

and the preceding exercise completes the argument.

b. e is an arbitrary monomorphism and the function i is injective.

3.a. As suggested by the hint, we take pairs of natural numbers with coordinatewise addition and subject to the relation that for any $a \in \mathbf{N}$, $(b,c) = (a+b, a+c)$. This relation is not an equivalence relation (it is not symmetric because a cannot be negative), but the symmetric closure is an equivalence relation. We will show that the quotient is isomorphic to $\mathbf{Z}$. To do this we define a function $f : \mathbf{N} \times \mathbf{N} \to \mathbf{Z}$ by $f(b,c) = b - c$. Clearly, $f(a+b, a+c) = f(b,c)$ so that $f \circ g = f \circ h$. On the other hand, if $f(b,c) = f(b',c')$, then $c - b = c' - b'$ or $c - c' = b - b'$. If $c \leq c'$, then $(b,c) = g(c' - c, b, c)$ while $(b',c') = h(c' - c, b, c)$. If $c' \leq c$, their roles are reversed. In either case, f is the quotient by the generated equivalence relation. The function f is also surjective, since every $n \geq 0$ is $f(n,0)$, while every $n < 0$ is $f(0,-n)$. Since $f(0,0) = 0$ and $f(b+b', c+c') = b + b' - (c+c') = b - c + b' - c' = f(b,c) + f(b',c')$ so that f is a monoid homomorphism and the coequalizer is the coequalizer in the category of monoids.

b. Let us denote this subset by A. The function $f : A \to \mathbf{Z}$ defined by

$$f(n,m) = \begin{cases} n & \text{if } m = 0 \\ -m & \text{if } n = 0 \end{cases}$$

is obviously bijective. It is a matter of consideration of cases to see that it is additive if addition is defined in A as suggested above.

4. Define $h : \mathbf{Z} \times \mathbf{N}^+ \to \mathbf{Q}$ by $h(b,c) = b/c$. Since $b/c = (ab)/(ac)$, $h \circ f = h \circ g$. Moreover, if $h(b,c) = h(b',c')$, then $bc' = b'c$. We have the equations

$$\begin{aligned} f(c',b,c) &= (b,c) \\ g(c',b,c) &= (c'b, c'c) \\ f(c,b',c') &= (b',c') \\ g(c,b',c') &= (cb', cc') \end{aligned}$$

Thus the coequalizer of f and g must render (b,c) and (b',c') equal. Thus h is injective. It is clearly surjective.

5. Let X be a set, $r : X \to B + B'$ and $s : X \to C + C'$ functions such that $(h + h') \circ r = (k + k') \circ s = v$. If we let $Y = v^{-1}(D)$ and $Y' = v^{-1}(D')$, then $X = Y + Y'$. Moreover, it is clear that $r(Y) \subseteq B$, $r(Y') \subseteq B'$, $s(Y) \subseteq C$ and $s(Y') \subseteq C'$. Let $t : Y \to B$ and $t' : Y' \to B'$ be the restrictions of r to Y and Y', respectively. Then $r = \langle t; t' \rangle$. Similarly, we have $s = \langle u; u' \rangle$ for $u : Y \to C$ and $u' : Y' \to C'$. Moreover, $(h + h') \circ r = (k + k') \circ s$ is equivalent to $h \circ t = k \circ u$ and $h' \circ t' = k' \circ u'$. Since the original two squares were pullbacks, it follows that there are arrows $w : Y \to A$ and $w' : Y' \to A'$ such that $f \circ w = t$, $g \circ w = u$, $f' \circ w' = t'$ and $g' \circ w' = u'$. This implies that $(f + f') \circ \langle w; w' \rangle = \langle t; t' \rangle = r$ and $(g + g') \circ \langle w; w' \rangle = \langle u; u' \rangle = s$. We also have to show uniqueness, but the arguments are similar.

6.a. Let $e^0 : \mathbf{1} \to \mathbf{2}$ take the single object of $\mathbf{1}$ to one object of $\mathbf{2}$ and let e^1 take it to the other. Then to say of a functor $u : \mathbf{2} \to \mathcal{A}$ that $u \circ e^0 = u \circ e^1$ is simply to say that u takes the two objects to the same one. If a is the single arrow in $\mathbf{2}$, then $\text{source}(u(a)) = \text{target}(u(a))$ so that not only will there be $u(a)$, but also $u(a) \circ u(a)$ and $u(a) \circ u(a) \circ u(a)$ and so on. Now let $\mathbf{N}$ also denote the category with one object and the natural numbers as arrows as defined in the exercise. Then $q(a) \circ q(a) = 2$, $q(a) \circ q(a) \circ q(a) = 3$ and so on. Given a functor u as above, we can define a functor $v : \mathbf{N} \to \mathcal{A}$ by $v(0) = \text{id}_{\text{source}(u(a))}$, $v(1) = u(a)$, $v(2) = u(a) \circ u(a)$ and so on. Clearly v is unique such that $v \circ q = u$. Thus q is the coequalizer of u and v.

b. Let $v : \mathcal{A} \to \mathbf{2}$ and $w : \mathcal{A} \to \mathbf{N}$ be such that $q \circ v = t \circ w$. If f is an arrow in $\mathcal{A}$ then $q(v(f)) = t(w(f))$. But the only integer that is in the image of $q \circ v$ and $t \circ w$ is 0. Thus $q(v(f)) = 0$, which means that $v(f)$ is an identity. Thus v takes every arrow to an identity arrow, that is it factors through $\mathbf{1} + \mathbf{1}$, and the factorization is clearly unique.

The arrow $s : \mathbf{1} + \mathbf{1} \to \mathbf{N}$ is not even epic, let alone regular. In fact, $t \circ s = s = \text{id}_{\mathbf{N}} \circ s$ without $t = \text{id}_{N}$.

c. Let $\mathcal{D}$ be the category whose nonidentity arrows can be pictured as:

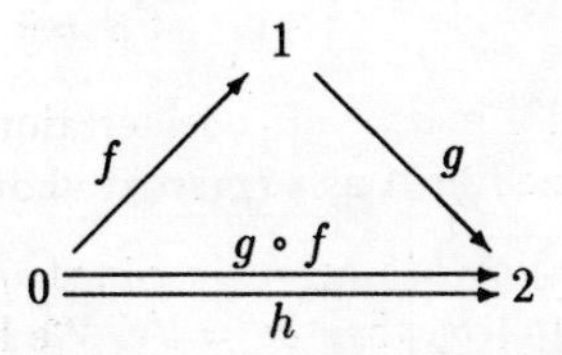

There is a functor $G : \mathcal{C} \to \mathcal{D}$ that is like F except that $G(h) = h$. This functor cannot factor through F because under any $H : \mathbf{3} \to \mathcal{D}$, $H(F(h)) = H(g \circ f) = H(g) \circ H(f) \neq h$.

d. A complete answer is too long, but we give enough details that the reader should have no difficulty filling in the rest. The first thing to observe is that being surjective on composable pairs of arrows includes, as special cases, being surjective on objects and on arrows. Suppose $T : \mathcal{A} \to \mathcal{B}$ is a functor which fails to be surjective on composable pairs, but is surjective on arrows and objects. Let f_1 and g_1 be a composable pair of arrows such that whenever $T(f_2) = f_1$ and $T(g_2) = g_1$, then f_2 and g_2 are not composable. Then let $S : \mathbf{3} \to \mathcal{B}$ be the unique functor such that $S(f) = f_1$ and $S(g) = g_1$. The pullback of S along T will be a category that includes subcategories like $\mathcal{C}$ and other pieces. It will have a functor to the category $\mathcal{D}$ of the preceding part that takes every arrow lying above f to f, every arrow lying above g to g and every arrow lying above $g \circ f$ to h, which is not the composite $g \circ f$. Then just as in the preceding part, this functor makes all the identifications made by the functor to $\mathbf{3}$, but does not factor through that functor. Thus that functor is not a regular epi. If T fails to be epi on arrows or on objects, even easier arguments suffice.

The other direction is to show that a functor that is surjective on composable pairs of arrows is a stable regular epi. It is immediate that the condition of being surjective on composable pairs is stable under pullback, so it is sufficient to show that such an arrow is a regular epi. The argument is similar to that for monoids and we omit it.

Section 8.6

1. The functor that assigns to each object the set of commutative cocones to that object on the base $A \leftarrow 0 \rightarrow B$ is naturally isomorphic to $\mathrm{Hom}(A,-) \times \mathrm{Hom}(B,-)$, which is the functor which assigns the set of cocones on the discrete base consisting of A and B. Indeed, the functors are actually identical, since any pair of arrows $f : A \rightarrow C$, $g : B \rightarrow C$ is a commutative cocone on the base $A \leftarrow 0 \rightarrow B$. Thus the universal elements are isomorphic.

2. Use exactly the same argument as for the preceding exercise.

3. We have to show that if

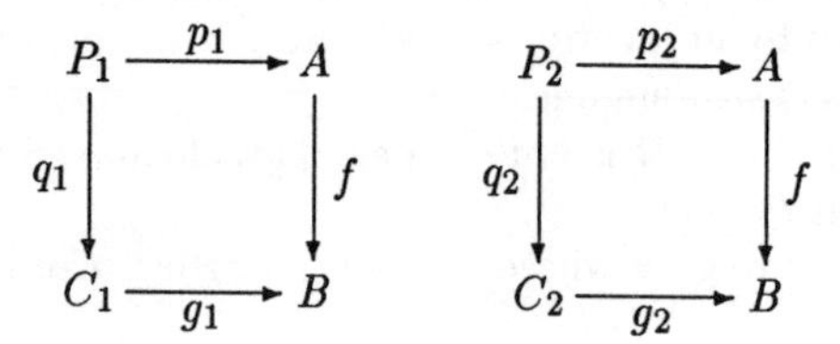

are both pullbacks, then so is

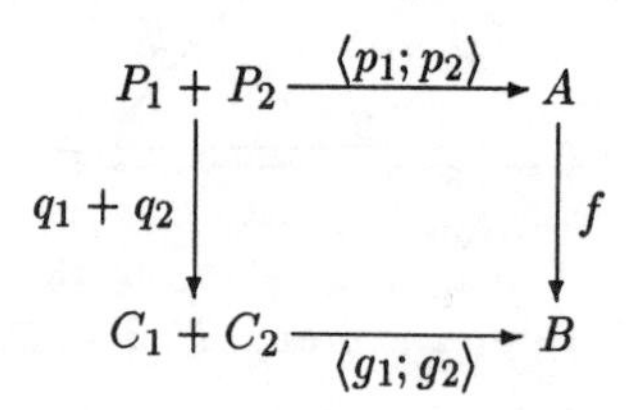

Let $i_1 : C_1 \rightarrow C_1 + C_2$, $i_2 : C_2 \rightarrow C_1 + C_2$, $j_1 : P_1 \rightarrow P_1 + P_2$ and $j_2 : P_2 \rightarrow P_1 + P_2$ be the coproduct injections. If $a \in A$ and $c \in C_1 + C_2$ such that $f(a) = \langle g_1; g_2\rangle(c)$, then either $c = i_1(c_1)$ for some $c_1 \in C_1$ and $f(a) = g_1(c)$ or $c = i_2(c_2)$ for some $c_2 \in C_2$ and $f(a) = g_2(c)$. In the first case, there is a unique $x_1 \in P_1$ such that $p_1(x_1) = a$ and $q_1(x_1) = c_1$. Then also $(q_1 + q_2)(j_1(x_1)) = j_1(q_1(x_1)) = j_1(c_1) = c$ and $\langle p_1; p_2\rangle(j_1(x_1)) = p_1(x_1) = a$. We get a similar conclusion if $c = j_2(c_2)$. This shows that $P_1 + P_2$ satisfies the existence condition of pullback. The uniqueness condition is similar.

4. Let us deal first with the monoid case. If M is a monoid and $f, g : E \rightarrow M$ are monoid homomorphisms, then $f = g$ is the function taking the unique element of $e \in E$ to the identity of M. Hence $f : E \rightarrow M$ is the unique monoid homomorphism such that $f \circ \mathrm{id}_E = f$ and $f \circ \mathrm{id}_E = g$. Thus the cocone

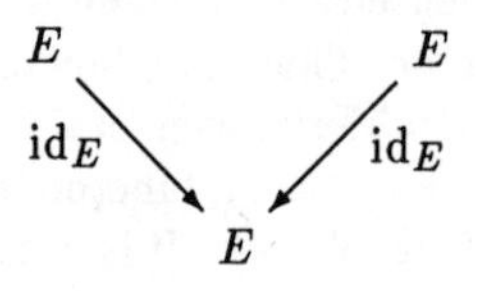

is a colimit.

In the category of semigroups, the situation is quite different. If $f : E \to S$ is a semigroup homomorphism, $f(e)$ need not be an identity arrow, if any exists. Since $e^2 = e$, it must be that $f(e)^2 = f(e)$ (such an element of S is called an **idempotent**). Other than that, there is no restriction on what $f(e)$ can be. Now let S be the semigroup of endofunctions on $\mathbf{N}$. The elements u and v defined in the problem are idempotents such that for all n, $(u \circ v)^n$ are distinct. In fact, $(u \circ v)^n$ adds $2n$ to odd integers and $2n - 2$ to even ones (and $(v \circ u)^n$ does the opposite). Now let $i : E \to F$ and $j : E \to F$ give a sum $E+E$. Let $i(e) = a$ and $j(e) = b$. Since there are arrows $f, g : E \to S$ defined by $f(e) = u$ and $g(e) = v$, there is an $h : F \to S$ such that $h \circ i = f$ and $h \circ j = g$. In F there are all the elements $(ab)^n$ and since $h((ab)^n) = (u \circ v)^n$ in order to be a semigroup homomorphism, all the $(ab)^n$ must also be distinct. Thus F is infinite.

5. This follows because they are each isomorphic to E in the previous example.

6.a. The recursive functions are defined as the smallest set of functions including successor and projections and closed under certain operations, of which the simplest is composition. Since the identity is also recursive (it is, for one thing, the projection from the unary product), it follows that this defines a category.

b. The first thing we do is to choose a bijective pairing with the additional property that it reflects. This means we choose a pair of functions $l, r : \mathbf{N} \to \mathbf{N}$ such that the function $\langle l, r \rangle : \mathbf{N} \to \mathbf{N} \times \mathbf{N}$ is bijective and such that if $l(m) \leq l(m')$ and $r(m) \leq r(m')$, then $m \leq m'$. It follows that if $l(m) = a$, then the number of $m' \leq m$ for which $l(m') = a$ is $r(m)$. A pairing that works is given by $l(m) = \exp_2(m)$, the number of powers of 2 that divide m and

$$r(m) = \frac{m + 1 - 2^l}{2^{l+1}}$$

where $l = l(m)$. Let M_i be the ith Turing machine in some enumeration of them. Let M be the process that at step n runs $M_{l(n)}$ one step. Because of our assumption on the bijective pairing, this is actually the $r(n)$th step that this machine has executed. Define a recursive function $f : \mathbf{N} \to \mathbf{N}$ by letting $f(n) = r(n)$ if machine $M_{l(n)}$ halts at the nth step and 0 otherwise. This is not injective (takes the value 0 more than once) but the composite $\langle r, l \rangle^{-1} \circ \langle \mathrm{id}, f \rangle$ is and defines an injective recursive function $\mathbf{N} \to \mathbf{N}$ whose image is not decidable (else the halting problem would be). But then there cannot be any recursive function whose image is the complement. Thus the subobject defined by that injection has no complement in the category.

7. Let N denote the sum of countably many copies of 1 and $i_n : 1 \to N$ be the nth injection to the sum. Let $z = n_0 : 1 \to N$ and let $s : N \to N$ be defined by $s \circ i_n = i_{n+1}$. Given an object A, an arrow $f_0 : 1 \to A$ and an arrow $t : A \to A$, we use ordinary induction to define a sequence of arrows $f_n : 1 \to A$, for $n \geq 1$

by $f_{n+1} = t \circ f_n$. Then the arrow $f = \langle f_0, f_1, \ldots, \rangle : N \to A$ clearly satisfies the recursion. The uniqueness is shown similarly. If the countable sums are stable under pullbacks, then this construction is clearly stable.

Section 8.7

1. The definition of epimorphism implies that for any $h : B \to A$, the condition $h \circ f = h \circ g$ is equivalent to the condition $h \circ f \circ e = h \circ g \circ e$ so that the condition to be satisfied by a coequalizer of f and g is identical to that to be satisfied by a coequalizer of $f \circ e$ and $g \circ e$. Thus the cocone functors are equivalent and are therefore the universal elements.

2. If $h = \langle h_1; h_2 \rangle : B_1 + B_2 \to D$, then the condition $\langle h_1; h_2 \rangle \circ (f_1 + f_2) = \langle h_1; h_2 \rangle \circ (g_1 + g_2)$ is equivalent to the two conditions $h_1 \circ f_1 = h_1 \circ g_1$ and $h_2 \circ f_2 = h_2 \circ g_2$. There results unique arrows $k_1 : C_1 \to D$ and $k_2 : C_2 \to D$ such that $k_1 \circ c_1 = h_1$ and $k_2 \circ c_2 = h_2$. This gives $\langle k_1; k_2 \rangle \circ (c_1 + c_2) = \langle h_1; h_2 \rangle$.

3. For $x \in X$, $h(x)$ is a polynomial in W. That is, we have either $h(x) = w \in W$ or $h(x) = f(u_1, \ldots, u_n)$ where f is an n-ary operation in the theory and each u_i is either an element of W or such a polynomial of lower nesting depth than $h(x)$. At the bottom is some finite subset of $W_0 \subseteq W$ consisting of all the elements of W that are constituents of h. It is even possible that $W_0 = \emptyset$ in which case $h(x)$ is constant. Now if $k, l : F(W) \to F(T)$ are homomorphisms and if there is a $w \in W_0$ with $k(w) \neq l(w)$, then by induction on the depth, we show that $(k \circ h)(x) \neq (l \circ h)(x)$. If, say, $w = u_1$, then $f(k(w), \ldots) \neq f(l(w), \ldots)$ because $F(T)$ is free and such an equality contradicts the fact that there are no equations. This argument works for one x. If every element of W is involved in some $h(x)$, then this argument would show that $k \circ h = l \circ h$ implies that $k = l$. If not, then there is a subset $W' \subseteq W$, consisting of all elements of W that are involved in at least one $h(x)$, that h factors through. In the application, W consists of the variables used in defining h and so h is epimorphic.

Solutions for Chapter 9

Section 9.1

1. Let (C_0, C_1, s, t, u, c) and (D_0, D_1, s, t, u, c) be two models of the sketch for categories. Note that we have used the common convention of using the same letter to stand for the arrow in the sketch and in the model (in every model, in fact). A homomorphism consists of an arrow $F_0 : C_0 \to D_0$ and an arrow $F_1 : C_1 \to D_1$ that satisfy some conditions forced by the fact that a homomorphism is a natural transformation. First off, we have $s \circ F_1 = F_0 \circ s$, $t \circ F_1 = F_0 \circ t$ and $u \circ F_0 = F_1 \circ u$. These mean that the homomorphism preserves source, target and identity arrows. These identities induce a unique arrow $F_2 : C_2 \to D_2$ such that $p_1 \circ F_2 = F_1 \circ p_1$ and $p_2 \circ F_2 = F_1 \circ p_2$. We further suppose that $c \circ F_2 = F_1 \circ c$. These conditions mean that the homomorphism preserves composition; thus the homomorphism is a functor.

Section 9.2

1. Referring to Diagram (9.1), we see that incl is the arrow opposite not $\circ$ true : $1 \to b$ in a cone, which means that in a model it is the arrow opposite an arrow from 1. But every arrow from 1 is monic (see Section 2.8, Exercise 5). Hence incl is monic.

Section 9.3

1. For each object C of $\mathcal{C}$, let $M'(c) = \{(x,c) \mid x \in M(c)\}$. If $t : c \to d$ is an arrow in $\mathcal{S}$, let $t(x,c) = (t(x),d)$. Then the first projection is an isomorphism $M' \to M$ and it is clear that $M(c) \cap M(d) = \emptyset$ for $c \neq d$.

Section 9.4

1. There is a sort c, a sort $l \subseteq c \times c$, an operation $r : c \to l$ and an operation $t : l \times_c l \to l$. There are diagrams to express the fact that r is the reflexive law and t the transitive law, but these are routine. There is an object $l + l$ and a cocone to express that $l + l$ is the sum of two copies of l. Finally, if $q_1, q_2 : l \to c$ are the operations corresponding to the inclusion followed by the projections, there is an arrow $\langle\langle q_1, q_2\rangle; \langle q_2, q_1\rangle\rangle : l + l \to c \times c$ and we need a cone and cocone to express that this arrow is a regular epi. This is how we say that $c \times c = l \cup l^{\mathrm{op}}$ in a model, which is how we express that the order is total.

2. We take the sketch for graphs with nodes a and n and arrows source, target : $a \to n$.We add a new type r with a cone

$$r \to a \overset{\text{source}}{\underset{\text{target}}{\rightrightarrows}} n$$

and a cocone to express that the composite $r \to a \to n$ (the latter arrow being either source or target) is regular epic. In a model, r is the set of loops and the fact that this maps surjectively to the nodes, implies that there is a loop at each node.

Solutions for Chapter 10

Section 10.2

1. A model of $\mathcal{E}$ is a sketch homomorphisms $\mathcal{E} \to \mathcal{C}$. This assigns to the single node of $\mathcal{E}$ an object E and that is all. If we denote the element of $\mathcal{E}$ by 0, then the functor $M \mapsto M(0)$ is evidently an isomorphism between the objects of $\mathcal{C}$ and those of $\mathbf{Mod}(\mathcal{E},\mathcal{C})$. If $\alpha : M \to N$ is a natural transformation, the $\alpha 0 : M(0) \to N(0)$ is an arrow of $\mathcal{C}$ and is subject to no conditions. Thus means that $\mathrm{Nat}(M,N) = \mathrm{Hom}(M(0),N(0))$ and so the functor is an isomorphism.

2. **Nat** has one such node, denoted 1, as does **Stack**. The disjoint union has two such nodes, but when the coequalizer is formed they are identified to a single node in $\mathcal{S}$. This node, and only this, must become a singleton in any model.

Section 10.3

1. If we have one sort 1, one empty cone with vertex 1 and one cocone

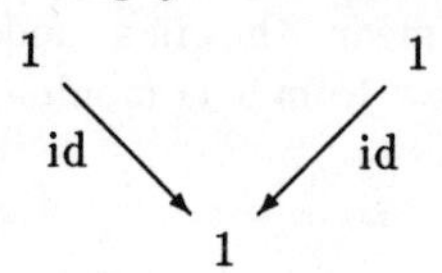

then there is no model in sets because the model must take 1 to the one-element set and that must satisfy that $1 = 1 + 1$, which is impossible. It follows from Exercise 4 of Section 8.6 that this sketch does have a model in the category of monoids.

2. For any $u : s_1 \to s_2$ in $\mathcal{S}$, we have $F(u) : F(s_1) \to F(s_2)$ which gives a commutative square

$$\begin{array}{ccc} M(F(s_1)) & \xrightarrow{M(F(u))} & M(F(s_2)) \\ {\scriptstyle \alpha F(s_1)}\downarrow & & \downarrow{\scriptstyle \alpha F(s_2)} \\ N(F(s_1)) & \xrightarrow[N(F(u))]{} & N(F(s_2)) \end{array}$$

which is the same as

$$\begin{array}{ccc} F^*(M)(s_1) & \xrightarrow{F^*(M)u} & F^*(M)(s_2) \\ {\scriptstyle F^*(\alpha)s_2}\downarrow & & \downarrow{\scriptstyle F^*(N)u} \\ F^*(N)(s_1) & \xrightarrow[F^*(\alpha)s_2]{} & F^*(N)(s_2) \end{array}$$

which is naturality of $F^*(M)$.

3. We have for $\mathrm{id}_{\mathcal{S}} : \mathcal{S} \to \mathcal{S}$ that $(\mathrm{id}_{\mathcal{S}})^*(M) = M \circ \mathrm{id}_{\mathcal{S}} = M$ and for $\alpha : M \to N$, $(\mathrm{id}_{\mathcal{S}})^*(\alpha) = \alpha\,\mathrm{id}_{\mathcal{S}} = \alpha$. Thus $(\mathrm{id}_{\mathcal{S}})^*$ is the identity functor. If $G : \mathcal{R} \to \mathcal{S}$ is another homomorphism of sketches, then $(F \circ G)^*(M) = M \circ F \circ G = G^*(M \circ F) = G^*(F^*(M)) = (G^* \circ F^*)(M)$ and similarly for $\alpha : M \to N$. Thus $(F \circ G)^* = G^* \circ F^*$ which shows that $\mathbf{Mod}_{\mathcal{C}}(-)$ is a contravariant functor.

4. Suppose, say, that σ is a diagram. The other two possibilities are similar. Suppose σ says that

$$s_1 \circ \cdots \circ s_n = t_1 \circ \cdots \circ t_m$$

Then σ is satisfied in $F^*(M)$ if and only if

$$F^*(M)(s_1) \circ \cdots \circ F^*(M)(s_n) = F^*(M)(t_1) \circ \cdots \circ F^*(M)(t_m)$$

which is the same as

$$M(F(s_1)) \circ \cdots \circ M(F(s_n)) = M(F(t_1)) \circ \cdots \circ M(F(t_m))$$

which is the same as the condition that M satisfy $F(\sigma)$.

Solutions for Chapter 11

Section 11.1

1. Let $\mathcal{E}_C$ be the fiber over an object C. If X is an object of $\mathcal{E}_C$, then $F(X) = C$ and $F(\mathrm{id}_X) = \mathrm{id}_C$ by definition of fiber. It follows that id_X is an arrow of $\mathcal{E}_C$. If $f : X \to Y$ and $g : Y \to Z$ are arrows of $\mathcal{E}_C$, then $F(f) = F(g) = \mathrm{id}_C$, so $F(g \circ f) = F(g) \circ F(f) = \mathrm{id}_C \circ \mathrm{id}_C = \mathrm{id}_C$, so $g \circ f$ is an arrow of $\mathcal{E}_C$. Hence $\mathcal{E}_C$ is a subcategory.

2. Let $P : \mathcal{E} \to \mathcal{C}$ be a fibration with cleavage γ. Define $F : \mathcal{C}^{\mathrm{op}} \to \mathbf{Cat}$ by

FF′–1 $F(C)$ is the fiber over C for each object C of $\mathcal{C}$.

FF′–2 For $f : C \to D$ in $\mathcal{C}$ and Y an object of $F(D)$, $Ff(Y)$ is defined to be the domain of the arrow $\gamma(f, Y)$.

FF′–3 For $f : C \to D$ in $\mathcal{C}$ and $u : Y \to Y'$ in $F(D)$, $Ff(u)$ is the unique arrow from $Ff(Y)$ to $Ff(Y')$ given by CA–2 for which

$$\gamma(f, Y') \circ Ff(u) = u \circ \gamma(f, Y)$$

To see that Ff is a functor, note first that $\gamma(f, Y) \circ \mathrm{id}_Y = \mathrm{id}_Y \circ \gamma(f, Y)$ so that $Ff(\mathrm{id}_Y) = \mathrm{id}_{FfY}$. Suppose $u' : Y' \to Y''$. Then

$$\gamma(f, Y'') \circ Ff(u') \circ Ff(u) = u' \circ \gamma(f, Y') \circ Ff(u) = u' \circ u \circ \gamma(f, Y)$$

so by uniqueness $Ff(u' \circ u) = Ff(u') \circ Ff(u)$.

3. Let $P : \mathcal{E} \to \mathcal{C}$ be a fibration. Define $F : \mathcal{C}^{\mathrm{op}} \to \mathbf{Cat}$ as in the preceding problem. If we show that $F(\mathrm{id}_C)$ is the identity on arrows it will have to be the identity on objects because functors preserve source and target. Let C be an object of $\mathcal{C}$ and u an arrow of $F(C)$. Now $F(\mathrm{id}_C)(u)$ is the unique arrow for which $\gamma(\mathrm{id}_C, Y') \circ F(\mathrm{id}_C)(u) = u \circ \gamma(\mathrm{id}_C, Y)$. By SC–1, this requirement becomes $\mathrm{id}_{Y'} \circ F(\mathrm{id}_C)(u) = u \circ \mathrm{id}_Y$. Thus $F(\mathrm{id}_C)(u) = u$ as required.

Now suppose $f : C \to D$ and $g : D \to E$ in $\mathcal{C}$. Let $u : Y \to Y'$ in $F(C)$. Then $F(g \circ f)(u)$ must be the unique arrow for which

$$\gamma(g \circ f, Y') \circ F(g \circ f)(u) = u \circ \gamma(g \circ f, Y)$$

But

$$\begin{aligned} u \circ \gamma(g \circ f, Y) &= u \circ \gamma(g, Y) \circ \gamma(f, Ff(Y)) \\ &= \gamma(g, Y') \circ Fg(u) \circ \gamma(f, Ff(Y)) \\ &= \gamma(g, Y') \circ \gamma(f, Ff(Y')) \circ Fg(Ff(u)) \end{aligned}$$

so by CA–2 and SC–2, $F(g \circ f) = F(g) \circ F(f)$.

4.a. We will also denote the functor by ϕ. It is a fibration if for every element x of $\mathbf{Z}_2$ there is an element u of $\mathbf{Z}_4$ such that $\phi(u) = x$ (that is CA–1) and for every $v \in \mathbf{Z}_4$ and $h \in \mathbf{Z}_2$ such that $x + h = \phi(v) \pmod 2$, there is a unique $w \in \mathbf{Z}_4$ for which $\phi(w) = h$ and $u + w = v \pmod 4$ (which is CA–2). This amounts to saying that ϕ is surjective and every equation $m + y = n \pmod 4$ can be solved uniquely for y, which requires simple case checking (or knowing that $\mathbf{Z}_2$ and $\mathbf{Z}_4$ are groups). An analogous proof works for opfibration.

b. A splitting would be a monoid homomorphism (this follows immediately from SC–1 and SC–2) which would make ϕ a split epimorphism, which it is not by Exercise 2 of Section 2.9.

5.a. P preserves source and target by definition. Let $(h,k) : f \to f'$ and $(h',k') : f' \to f''$. Then

$$P((h',k') \circ (h,k)) = P(h' \circ h, k' \circ k) = k' \circ k = P(h',k') \circ P(h,k)$$

Thus P preserves composition. The identity on an object (A,B) is $(\mathrm{id}_A, \mathrm{id}_B)$, so it follows immediately that P preserves identities.

b. Let $f : C \to D$ in $\mathcal{C}$ and let $k : B \to D$ be an object of $\mathcal{A}$ lying over D. Let $\gamma(f,k)$ be the arrow (u,f) of $\mathcal{A}$ defined in 11.1.5. $P(\gamma(f,k)) = f$ so CA–1 is satisfied. As for CA–2, let $(v,v') : z \to k$ in $\mathcal{A}$ and let h be an arrow of $\mathcal{C}$ such that $f \circ h = v'$. Let w be the unique arrow given by the pullback property for which $u \circ w = v$ and $u' \circ w = h \circ z$. The last equation says that (w,h) is an arrow of $\mathcal{A}$, and $(u,f) \circ (w,h) = (u \circ w, f \circ h) = (v,v')$ as required. The uniqueness property of the pullback means that (w,h) is the only such arrow.

Section 11.2

1. The identity arrow for (x,C) is (x,id_C). Let $(x,f) : (x,C) \to (x',C')$, $(x',f') : (x',C') \to (x'',C'')$ and $(x'',f'') : (x'',C'') \to (x''',C''')$ be arrows of $\mathbf{G}_0(\mathcal{C},F)$. Then $((x'',f'') \circ (x',f')) \circ (x,f) = (x'', f'' \circ f') \circ (x,f) = (x'', (f'' \circ f') \circ f) = (x'', f'' \circ (f' \circ f)) = (x'',f'') \circ ((x',f') \circ (x,f))$ so composition is associative.

2. Let $f : C \to C'$ in $\mathcal{C}$. Let (x,C) be an object of $\mathbf{G}_0(\mathcal{C},F)$ lying over C. Define $\kappa(f,(x,C))$ to be $(x,f) : (x,C) \to (x',C')$ where $x' = Ff(x)$. Now suppose $(x,g) : (x,C) \to (y,C'')$ and suppose $k : C' \to C''$ has the property that $k \circ f = g$. The arrow required by OA–2 is $(x',k) : (x',C') \to (y,C'')$. This is well defined since $Fk(x') = Fk(Ff(x)) = F(k \circ f)(x) = Fg(x)$ which must be y since (y,C'') is the given target of (x,g). Moreover it satisfies OA–2 since $(x',k) \circ (x,f) = (x, k \circ f) = (x,g)$. An arrow satisfying OA–2 must lie over k and have source (x',C') and target (y,C'') so that it can compose with f to give g, so (x',k) is the only possible such arrow.

To see that κ is a splitting, suppose that $(x',f') : (x',C') \to (x'',C'')$. Then $\kappa(f',(x',C')) = (x',f')$ and $(x',f') \circ (x,f) = (x, f' \circ f) = \kappa(f' \circ f, (x,C))$ as required. The verification for identities is even easier.

3. The identity arrow is $(\mathrm{id}_x,\mathrm{id}_C) : (x,C) \to (x,C)$, which is well defined since $F(\mathrm{id}_C)(x) = x$ so $\mathrm{id}_x : F(\mathrm{id}_C)(x) \to x$. Suppose that (u,f) is an arrow from (x,C) to (x',C'), so that $u : Ff(x) \to x'$. Similarly, let $(u',f') : (x',C') \to (x'',C'')$ and $(u'',f'') : (x'',C'') \to (x''',C''')$. Then

$$\begin{aligned} ((u'',f'') \circ (u',f')) \circ (u,f) &= (u'', Ff''(u'), f'' \circ f') \circ (u,f) \\ &= (u'' \circ Ff''(u') \circ F(f'' \circ f')(u), (f'' \circ f') \circ f) \end{aligned}$$

and

$$\begin{aligned} (u'',f'') \circ ((u',f') \circ (u,f)) &= (u'',f'') \circ (u' \circ Ff(u), f' \circ f) \\ &= (u'' \circ Ff''(u' \circ Ff'(u)), f'' \circ (f' \circ f)) \end{aligned}$$

The result follow from the facts that F and Ff'' are functors and composition in C is associative.

4. Let $f : C \to C'$ and let (x,C) lie over C. Define $\kappa(f,(x,C))$ to be

$$(\mathrm{id}_{Ff(x)}, f) : (x,C) \to (Ff(x), C')$$

Let $(u,g) : (x,C) \to (y,C'')$ so that $u : Fg(x) \to y$. Let $k : C' \to C''$ satisfy $k \circ f = g$. Then $(u,k) : (Ff(x),C') \to (y,C'')$ because the domain of u is $Fg(x) = F(k \circ f)(x) = F(k)(Ff(x))$. Furthermore,

$$(u,k) \circ (\mathrm{id}_{Ff(x)}, f) = (u \circ F(\mathrm{id}_{Ff(x)}), k \circ f) = (u,g)$$

as required by OA–2. It follows as in the answer to the second problem that (u,k) is the only possible arrow with this property.

5.

$$\begin{aligned} ((t,m)(t',m'))(t'',m'') &= (t\alpha(m,t'), mm')(t'',m'') \\ &= (t\alpha(m,t')\alpha(mm',t''), mm'm'') \end{aligned}$$

and

$$\begin{aligned} (t,m)((t',m')(t'',m'')) &= (t,m)(t,\alpha(m',t''), m'm'') \\ &= (t\alpha(m, t'\alpha(m',t'')), mm'm'') \end{aligned}$$

However, by MA–4,

$$\alpha(m,t')\alpha(mm',t'') = \alpha(m,t')\alpha(m,\alpha(m',t''))$$

and that is $\alpha(m, t'\alpha(m',t''))$ by MA—2.

6. Let β take an object x of $F(C)$ to (x,C) and an arrow u to (u,id_C). By GC–3, if $v : y \to z$ then

$$(v,\mathrm{id}_C) \circ (u,\mathrm{id}_C) = (v \circ F(\mathrm{id}_C)(u), \mathrm{id}_C \circ \mathrm{id}_C) = (v \circ u, \mathrm{id}_C)$$

so β preserves composition. It is clearly a bijection and preserves identities.

Section 11.3

1. Suppose that $\alpha : F \to G$ is a natural transformation. $\mathbf{G}\alpha$ preserves identities because $\mathbf{G}\alpha(\mathrm{id}_X, \mathrm{id}_C) = \alpha C(\mathrm{id}_X, \mathrm{id}_C) = (\mathrm{id}_{\alpha C(x)}, \mathrm{id}_C)$. Suppose that $(u, f) : (x, C) \to (x', C')$ and $(u', f') : (x', C') \to (x'', C'')$. Then

$$(u', f') \circ (u, f) = (u' \circ Ff'(u), f' \circ f)$$

by GC–3 (Section 11.2). On the other hand,

$$\begin{aligned}
\mathbf{G}\alpha(u', f') \circ \mathbf{G}\alpha(u, f) &= (\alpha C'' u', f') \circ (\alpha C' u, f) \\
&= (\alpha C'' u' \circ Gf'(\alpha C' u), f' \circ f) \\
&= (\alpha C'' u' \circ \alpha C''(Ff'(u)), f' \circ f) \\
&= (\alpha C''(u' \circ Ff'(u)), f' \circ f) \\
&= \mathbf{G}\alpha(u' \circ Ff'(u), f' \circ f)
\end{aligned}$$

where the third equality uses the naturality of α and the fourth uses the fact that $\alpha C''$ is a functor. Thus $\mathbf{G}\alpha$ preserves composition.

2. GR–2 implies that $\mathbf{G}$ preserves identity natural transformations, since any component of an identity transformation is an identity arrow. Let $\alpha : F \to F'$ and $\beta : F' \to F''$ be natural transformations, where F, F' and F'' are functors from $\mathcal{C}$ to **Cat**. Then for $(u, f) : (x, C) \to (x, C')$,

$$\mathbf{G}(\beta \circ \alpha)(u, f) = ((\beta \circ \alpha)C' u, f) = (\beta C'(\alpha C' u), f)$$

and

$$(\mathbf{G}\beta(\mathbf{G}\alpha))(u, f) = \mathbf{G}\beta(\alpha C' u, f) = (\beta C'(\alpha C' u), f)$$

so $\mathbf{G}$ preserves composition.

Section 11.4

1. If $\mathcal{A}$ and $\mathcal{B}$ are monoids, then each has only one object. By WP–1, an object of $\mathcal{A}\,\mathrm{wr}^G\,\mathcal{B}$ is a pair (A, P) where A is the only object of $\mathcal{A}$ and $P : G(A) \to \mathcal{B}$ is a functor. But if $\mathcal{B}$ has only one object and $G(A)$ is discrete, there is only one functor from $G(A)$ to $\mathcal{B}$ – it must take all the objects of $G(A)$ to the only object of $\mathcal{B}$. Hence $\mathcal{A}\,\mathrm{wr}^G\,\mathcal{B}$ has only one object, so is a monoid.

2. Let $\mathcal{A}$ and $\mathcal{B}$ be groups. Since they have only one object each, we can simplify the notation in WP–1 through WP–3 and omit mention of the objects A, A' and A''. The value of G at the only object of $\mathcal{A}$ is a category we will call $\mathcal{G}$. If f is an element of the group $\mathcal{A}$, then Gf is an automorphism of the category $\mathcal{G}$. An object of $\mathcal{A}\,\mathrm{wr}^G\,\mathcal{B}$ is a functor $P : \mathcal{G} \to \mathcal{B}$. An arrow $(f, \lambda) : P \to P'$ consists of an element f of the group $\mathcal{A}$ and a natural transformation $\lambda : P \to P' \circ Gf$. Each component of λ is an element of the group $\mathcal{B}$ so has an inverse; thus λ is an invertible natural transformation. Since f is also a group element, it has an inverse f^{-1}. Let μ be the natural transformation whose component at an object X of $\mathcal{G}$ is $(\lambda(Gf)^{-1}(X))^{-1}$. Then the inverse of the arrow (f,λ) is (F^{-1},μ). To verify this, we calculate

$$(F^{-1},\mu)\circ(f,\lambda) = (f^{-1}\circ f, \mu Gf\circ\lambda)$$

Now $F^{-1}\circ f$ is the identity of $\mathcal{A}$ and for an object X of $\mathcal{G}$,

$$\begin{aligned}(\mu Gf\circ\lambda)X &= \mu(Gf(X))\circ\lambda X\\ &= \lambda((Gf)^{-1}(Gf(X)))^{-1}\circ\lambda X\\ &= (\lambda X)^{-1}\circ\lambda X = \mathrm{id}_X\end{aligned}$$

so (f,λ) is invertible.

Solutions for Chapter 12

Section 12.1

1. By definition, g takes the identity of $F(X)$ to that of M. If $a = (x_1,\ldots,x_n)$ and $b = (y_1,\ldots,y_m)$, then $ab = (x_1,\ldots,x_n,y_1,\ldots,y_m)$ and

$$\begin{aligned}g(ab) &= g(x_1,\ldots,x_n,y_1,\ldots,y_m)\\ &= u(x_1)\cdot\cdots\cdot u(x_n)\cdot u(y_1)\cdot\cdots\cdot u(y_m)\\ &= g(x_1,\ldots,x_n)\cdot g(y_1,\ldots,y_m)\\ &= g(a)\cdot g(b)\end{aligned}$$

2. As we remarked in the text, ηX is a universal element of the functor $\mathrm{Hom}(X,U(-))$. The assumption on γ in the exercise says that it is also a universal element for the same functor. Now Corollary 4.5.12 says there is a unique isomorphism $\phi: E\to F(X)$ for which $\mathrm{Hom}(X,U(\phi))(\gamma) = \eta X$. This translates precisely into the claim of the exercise.

Section 12.2

1. If $S_0\subseteq f^{-1}(T_0)$, let $y\in f(S_0)$. Then there is some $x\in S_0$ such that $f(x) = y$ and since $x\in S_0\subseteq f^{-1}(Y_0)$, $x\in f^{-1}(T_0)$ so that $y = f(x)\in T_0$. Therefore $f(S_0)\subseteq T_0$. Conversely, suppose $f(S_0)\subseteq T_0$. For $x\in S_0$, $f(x)\in F(S_0)$ so that $f(x)\in T_0$, whence $x\in f^{-1}(T_0)$. Therefore $S_0\subseteq f^{-1}(T_0)$.

2. Let $\Sigma\dashv\Delta$. Then for every object C,

$$\begin{aligned}\mathrm{Hom}_{\mathcal{C}}(\Sigma(A,B),C) &\cong \mathrm{Hom}_{\mathcal{C}\times\mathcal{C}}((A,B),\Delta(C))\\ &\cong \mathrm{Hom}_{\mathcal{C}\times\mathcal{C}}((A,B),(C,C))\\ &\cong \mathrm{Hom}_{\mathcal{C}}(A,C)\times\mathrm{Hom}_{\mathcal{C}}(B,C)\end{aligned}$$

which is the mapping condition that determines $A+B$ uniquely.

3. The unit of the adjunction is the map from $\Delta\Pi(A,B) \to (A,B)$ that corresponds under the adjunction to the identity arrow $\Pi(A,B) \to \Pi(A,B)$. Since $\Pi(A,B) = A \times B$, we are looking for an arrow $(A \times B, A \times B) \to (A,B)$. The arrow is $(\mathrm{proj}_1, \mathrm{proj}_2)$ since that is the pair that gives the universal mapping property.

4. There is a sketch for monoids gotten by augmenting that of semigroups given in Section 7.2 by adding a constant of type s and diagrams that force it to be a left and right identity. Call this sketch $\mathcal{M}$. For a set X, let $\mathcal{M}(X)$ denote the sketch gotten from $\mathcal{M}$ by adding to $\mathcal{M}$ the set X of constants. Then the initial algebra for $\mathcal{M}(X)$ is just the free monoid $F(X)$. The reason is that a model of $\mathcal{M}(X)$ is just a model M of $\mathcal{M}$ together with a chosen element $u(x) \in M$ for each $x \in X$, which is just a function $f : X \to U(M)$. Thus an initial model of the sketch is the free monoid generated by X. In 4.7.14, exactly the same construction of adding a set of constants to a sketch and forming the initial model yields the free model on the original sketch generated by that set.

5. This is the content of Proposition 3.1.15.

6. A functor with a left adjoint preserves limits, in particular the terminal object. Since $A \ncong 1$ implies $1 \times A \ncong 1$, $- \times A$ does not preserve the terminal object and hence cannot have a left adjoint.

7. Since it is the arrow that corresponds under adjunction to the identity, its value at B is the unique arrow $\eta B : B \to R_A(B \times A)$ such that the composite

$$B \times A \xrightarrow{\eta B \times \mathrm{id}_A} R_A(B \times A) \times A \xrightarrow{e} B \times A$$

is the identity. In **Set**, $R_A(B \times A)$ is the set of functions from A to $A \times B$, and ηB takes $b \in B$ to the function $a \mapsto (b,a)$.

Section 12.3

1. We use Theorem 12.3.2. We have

$$\mathrm{Hom}_{\mathcal{A}}(LTA, A') \cong \mathrm{Hom}_{\mathcal{B}}(TA, TA') \cong \mathrm{Hom}_{\mathcal{A}}(A, RTA')$$

and

$$\mathrm{Hom}_{\mathcal{B}}(TLB, B') \cong \mathrm{Hom}_{\mathcal{A}}(LB, RB') \cong \mathrm{Hom}_{\mathcal{B}}(B, TRB')$$

2.a. Here and below, we use Hom for $\mathrm{Hom}_{\mathcal{C}}$ and Hom_A for $\mathrm{Hom}_{\mathcal{C}/A}$. If $f : B \to A$ is an object of $\mathcal{C}/A$ and C is an object of $\mathcal{C}$, an arrow $g : f \to P_A(C) = p_2 : C \times A \to A$ has two coordinates, say $g_1 = p_1 \circ g : B \to C$ and $g_2 = p_2 \circ g : B \to A$. It is an arrow in $\mathcal{C}/A$ if and only if $g_2 = f$. Thus there are no conditions on g_1, so $\mathrm{Hom}_A(f, P_A(C)) \cong \mathrm{Hom}(B,C) = \mathrm{Hom}(L_A(f), C)$.

b. This is an application of the preceding exercise.

c.

$$\mathrm{Hom}(B, \Phi(P_A(C))) \cong \mathrm{Hom}(B, [A \to C]) \cong \mathrm{Hom}(B \times A, C)$$
$$\cong \mathrm{Hom}_A(B \times A \to A, C \times A \to A) \cong \mathrm{Hom}_A(P_A(B), P_A(C))$$

d. The hint suggests that we find a natural transformation from the functor $\mathrm{Hom}(-, \Phi(C))$ to $\mathrm{Hom}(-, \Phi(D))$. It is simple to prove that the isomorphism of part (c) is natural in B. Thus we have for any B, the arrow

$$\mathrm{Hom}(B, \Phi(P_A(C))) \cong \mathrm{Hom}(P_A(B), P_A(C))$$
$$\xrightarrow{\mathrm{Hom}(P_A(B), f)} \mathrm{Hom}(P_A(B), P_A(D)) \cong \mathrm{Hom}(B, \Phi(P_A(D)))$$

The fact that these isomorphisms are natural in B implies the existence of the required arrow $\Phi(f)$. The desired commutation is essentially the definition. This part justifies the use of the notation $\Phi(P_A(C))$ for it implies that if $P_A(C) \cong P_A(D)$, then $\Phi(P_A(C)) \cong \Phi(P_A(D))$.

e. The required diagram is given by letting $d = \langle \mathrm{id}_C, f \rangle : C \to C \times A$, $d^0 = \langle \mathrm{proj}_1, \mathrm{proj}_2, \mathrm{proj}_2 \rangle$ and $d^1 = \langle \mathrm{proj}_1, f \circ \mathrm{proj}_1, \mathrm{proj}_2 \rangle : C \times A \to C \times A \times A$. These are readily seen to be arrows in $\mathcal{C}/A$. The equalizer of d^0 and d^1 is

$$\begin{aligned} &\{(c,a) \in C \times A \mid (c,a,a) = (c, f(c), a)\} \\ &= \{(c,a) \in C \times A \mid a = f(c)\} = \{(c, f(c)) \mid c \in C\} \end{aligned}$$

which is the image of d.

f. In the diagram

$$\begin{array}{ccccc} \mathrm{Hom}(B, \Phi(f)) & \longrightarrow & \mathrm{Hom}(B, \Phi(P_A(C))) & \rightrightarrows & \mathrm{Hom}(B, \Phi(P_A L_A P_A(C))) \\ \downarrow & & \downarrow \cong & & \downarrow \cong \\ \mathrm{Hom}_A(P_A(B), f) & \longrightarrow & \mathrm{Hom}_A(P_A(B), P_A(C)) & \rightrightarrows & \mathrm{Hom}_A(P_A(B), P_A L_A P_A(C)) \end{array}$$

both lines are equalizers since both $\mathrm{Hom}(B, -)$ and $\mathrm{Hom}_A(P_A(B), -)$ preserve equalizers. Moreover the middle and right hand vertical arrows are isomorphisms by part (c) and by definition. Hence by uniqueness of equalizers, the left hand vertical arrow is an isomorphism.

g. This is an immediate consequence of the Pointwise Adjointness Theorem 12.3.4.

3. This translates to showing for any $g : A' \to A$ and $h : B \to B'$, that

$$\begin{array}{ccc} \mathrm{Hom}(FA, B) & \xrightarrow{\beta(A,B)} & \mathrm{Hom}(A, UB) \\ {\scriptstyle \mathrm{Hom}(Fg,h)} \downarrow & & \downarrow {\scriptstyle \mathrm{Hom}(g,Uh)} \\ \mathrm{Hom}(FA', B') & \xrightarrow[\beta(A',B')]{} & \mathrm{Hom}(A', UB') \end{array}$$

Applied to an $f \in \mathrm{Hom}(FA, B)$, this requires that we show that

$$U(h \circ f \circ Fg) \circ \eta A' = Uh \circ (Uf \circ \eta A) \circ g$$

But using the functoriality of U and naturality of η, we have

$$U(h \circ f\ ) \circ \eta A' = Uh \circ Uf \circ UFg \circ \eta A' = Uh \circ Uf \circ \eta A \circ g$$

4. The adjointness $F \dashv U$, for $F : \mathcal{C} \to \mathcal{D}$ and $U : \mathcal{D} \to \mathcal{C}$ is equivalent to the natural isomorphism

$$\mathrm{Hom}_{\mathcal{C}}(-, U-) \cong \mathrm{Hom}_{\mathcal{D}}(F-, -) : \mathcal{C}^{\mathrm{op}} \times \mathcal{D} \to \mathbf{Set}$$

But that is the same as the natural isomorphism

$$\mathrm{Hom}_{\mathcal{C}^{\mathrm{op}}}(U^{\mathrm{op}}-, -) \cong \mathrm{Hom}_{\mathcal{D}^{\mathrm{op}}}(-, F^{\mathrm{op}}-) : \mathcal{D} \times \mathcal{C}^{\mathrm{op}} \to \mathbf{Set}$$

and since $(\mathcal{D}^{\mathrm{op}})^{\mathrm{op}} = \mathcal{D}$, the conclusion follows.

Section 12.4

1.a. If $\mathcal{C}/A$ is a slice and $f : B \to A$ is an object of $\mathcal{C}/A$, then there is a functor $F : \mathcal{C}/B \to (\mathcal{C}/A)/f$ that sends $u : C \to B$ to $u : f \circ u \to f$. If $v : D \to B$ is another object of $\mathcal{C}/B$ and $g : u \to v$ is an arrow (which is to say that $v \circ g = u$), then $f \circ v \circ g = f \circ u$ so that we can simply let $F(g : u \to v) = g : F(u) \to F(v)$. There is also a functor $G : (\mathcal{C}/A)/f \to \mathcal{C}/B$ that takes an object $w : g \to f$, whose domain is $g : C \to B$, to $g : C \to B$. Suppose $w' : g' \to f$ is also an arrow of $\mathcal{C}/A$ (hence an object of $(\mathcal{C}/A)/f$) and $x : w' \to w$ is an arrow of $(\mathcal{C}/A)/f$. Then by definition of arrow in a slice category, $w \circ x = w' : g' \to f$ in $\mathcal{C}/A$, hence in $\mathcal{C}$. Thus we can set $G(x : w' \to w) = x : \mathrm{dom}(w') \to \mathrm{dom}(w)$. It is clear that F and G are inverse to each other.

b. Since every slice has products, $\mathcal{C}$ has pullbacks by Proposition 12.4.4. By part (a), so does every slice. Since also every slice has a terminal object ($\mathrm{id} : A \to A$ is automatically terminal in $\mathcal{C}/A$), it follows from Proposition 8.3.7 of Section 8.3 that every slice has finite limits.

c. This follows immediately from Exercise 2 of the preceding section.

Solutions for Chapter 13

Section 13.1

1. In order that h be a homomorphism of algebras, it is required that the diagram

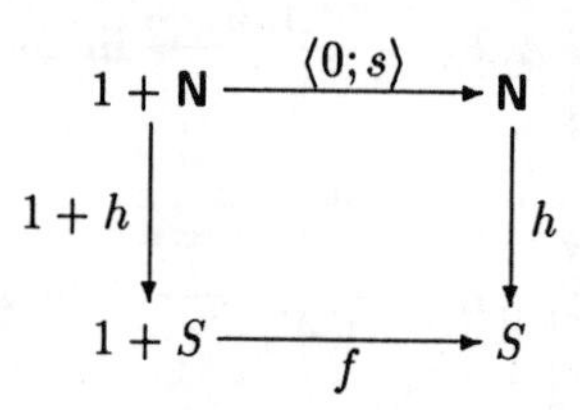

commute. Applied to the element $* \in 1$, this means that $h(0) = f(*)$ and applied to $h(n)$ that $f(h(n)) = h(\mathrm{succ}(n))$. This shows the uniqueness of h and since this is the only condition to be satisfied, it shows that h works.

2. Suppose that 0 is the initial object of some category and that $f : U \rightarrowtail 0$ represents a subobject. Since 0 is initial, there is an arrow $g : 0 \to U$. Since there is only one arrow $0 \to 0$, we have $f \circ g = \mathrm{id}_0$. From this it follows that $f \circ g \circ f = \mathrm{id}_0 \circ f = f = f \circ \mathrm{id}_U$. But since f is monic, it can be cancelled from the *left* of this equation, whence $g \circ f = \mathrm{id}_U$. Thus the subobject represented by $f : U \to 0$ is equivalent to the identity subobject.

3. Two ways of doing this are to make a category with no monomorphisms except the identity and to make a category in which there may be many arrows from the object to itself (and thus it is not initial) but they are all isomorphisms and so equivalent to the identity. For the first, take a category with one object and two distinct arrows 1, and f, with $f = f^2 = f^3 = \cdots$. Then f is not a monomorphism since $f \circ \mathrm{id} = f \circ f$ while $f \neq \mathrm{id}$. For the second, take any group with more than one element viewed as a category with one object.

4. The category associated with S has the elements of S as objects and pairs (x, y) as the unique arrow $y \to x$ when $y \leq x$. Then f takes elements to elements, that is objects to objects and we define f on arrows by $f(x, y) = (f(x), f(y))$ which is well defined since $y \leq x$ implies that $f(y) \leq f(x)$. Since (x, x) is the identity and $f(x, x) = (f(x), f(x))$, we see that f preserves identities. Since $(x, y) \circ (y, z) = (x, z)$ is the composite, it is immediate that f preserves composition as well.

Now an object of $(f : S)$ is an x together with an arrow $f(x) \to x$. Such an arrow exists if and only if $f(x) \leq x$. So the objects of $(f : S)$ can be identified as those elements. If also $f(y) \leq y$, then $y \leq x$ implies that $f(y) \leq f(x)$ and that

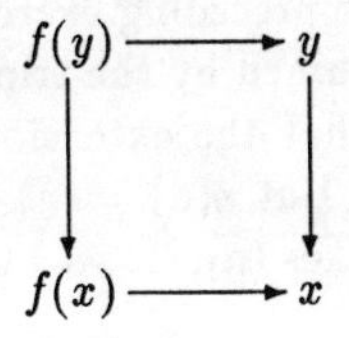

commutes. Thus $(f : S)$ can be identified as the set of all x with $f(x) \leq x$ with the restricted order. If x_0 is initial in $(f : S)$, then $f(y) \leq y$ implies that $x_0 \leq y$. But then $f(x_0) \leq x_0$ implies $f(f(x_0)) \leq f(x_0)$ so that $y = f(x_0)$ is such an element and so $x_0 \leq f(x_0)$. Thus x_0 is fixed and it is clearly the least fixed point.

Section 13.2

1. In effect we claim that if A^* is the set of lists of elements of A (including the empty list $()$), then $\mathrm{rec}(A, B) = A^* \times B$. Let $\mathrm{cons} : A \times A^* \to A^*$ denote the function that adjoins an element of A to the head of a list. Then $r_0(A, B) : B \to A^* \times B$ takes $b \in B$ to $(\langle\rangle, b)$ and $r(A, B) : A \times A^* \to B$ is just $\mathrm{cons} \times \mathrm{id}_B$. Now let $t_0 : B \to X$ and $t : A \times X \to X$ be given. In order that $f : A^* \times B \to X$ make the diagram from 13.2.5 commute, it is necessary that $f(\langle\rangle, b) = t_0(b)$ and that $f(\mathrm{cons}(a, l), b) = t(a, f(l, b))$ for $a \in A$, $l \in A^*$ and $b \in B$. But those conditions define, by induction on the length of a word in A^*, a unique function $f : A^* \times B \to X$ that makes the necessary diagram commute.

2. From the definition of recursive in 13.2.5, we see that when $A = B = 1$, we get $r_0(1,1) : 1 \to \text{rec}(1,1)$ and $r(1,1) : \text{rec}(1,1) \to \text{rec}(1,1)$ (taking rec(1,1) as the product $1 \times \text{rec}(1,1)$) and the diagram of that section describes exactly the universal mapping property of a natural numbers object.

Section 13.3

1. We freely use the Godement rules of Section 4.4 together with the identities satisfied by an adjoint pair given in Section 12.2 to compute

$$\mu \circ T\mu = U\epsilon F \circ UF\eta = U(\epsilon F \circ F\eta) = U(\text{id}_F) = \text{id}_{UF} = \text{id}_T$$

$$\mu \circ \eta T = U\epsilon F \circ \eta UF = (U\epsilon \circ \eta U)F = \text{id}_U F = \text{id}_{UF} = \text{id}_T$$

Now the naturality of ϵ implies that for any $\alpha : R \to S$,

$$\alpha \circ \epsilon R = \epsilon S \circ FU\alpha$$

Applied when $R = FU$, $S = I$ (the identity functor) and $\alpha = \epsilon$, this gives that $\epsilon \circ FU\epsilon = \epsilon \circ \epsilon FU$. We then compute

$$\begin{aligned}\mu \circ \mu T &= U\epsilon F \circ U\epsilon FUF = U(\epsilon \circ \epsilon FU)F = U(\epsilon \circ FU\epsilon)F \\ &= U\epsilon F \circ UFU\epsilon F = \mu \circ T\mu\end{aligned}$$

2. If we can identify the η and μ with those determined by the triple, then the conclusion follows from the preceding exercise. Let us temporarily call the natural transformations determined by the triple $\widehat{\eta}$ and $\widehat{\mu}$. Then $\widehat{\eta}A$ is defined as the unique $A \to A^*$ such that the extension to a monoid homomorphism $F(A) \to F(A)$ is the identity. But $\eta(a) = (a)$ clearly extends to the identity on $F(A)$ since the extension takes $(a_1, \ldots, a_n)$ to $(a_1) \cdots (a_n) = (a_1, \ldots, a_n)$ by definition. Hence $\widehat{\eta} = \eta$.

As for $\widehat{\mu} = U\epsilon F$, that is the function underlying the unique homomorphism from $FUF(A)$ to $F(A)$ that is gotten by extending the identity function from $UF(A)$ to $UF(A)$. This function takes $(a_1, \ldots, a_n)$ to itself and the extension to a homomorphism takes, for example, the two letter string

$$((a_1, \ldots, a_n), (a'_1, \ldots, a'_{n'})) \in FUF(A)$$

to the product (concatenate)

$$(a_1, \ldots, a_n, a'_1, \ldots, a'_{n'}) \in FU(A)$$

Here we see explicitly that μ and $\widehat{\mu}$ agree on two letter strings and it is clear they agree on strings of arbitrary length.

Section 13.4

1. Let us write $f : A \Rightarrow B$ for an arrow in the Kleisli category and $f \bullet g$ for the Kleisli composition. Then for $f : A \Rightarrow B$, we have $f \bullet \eta A = \mu B \circ Tf \circ \eta A = \mu B \circ \eta TB \circ f = \mathrm{id}_B \circ f = f$. For $g : C \Rightarrow A$, we have $\eta A \bullet g = \mu A \circ T\eta A \circ g = \mathrm{id}_A \circ g = g$. In this exercise and many of the later ones we use naturality without comment. Here, for example, the fact that $Tf \circ \eta A = \eta TB \circ f$ is a consequence of the naturality of η.

2. Using the same notation as in the previous problem, we have, for $f : A \Rightarrow B$, $g : B \Rightarrow C$ and $h : C \Rightarrow D$ that $(h \bullet g) \bullet f = \mu D \circ T(\mu C \circ Th \circ g) \circ f = \mu D \circ T\mu D \circ T^2 h \circ Tg \circ f = \mu D \circ \mu TD \circ T^2 h \circ Tg \circ f = \mu D \circ Th \circ \mu C \circ Tg \circ f = h \bullet (g \bullet f)$.

3. Refer to the two solutions above for notation. For an object A, $U\eta A = \mu A \circ T\eta A = \mathrm{id}_{TA} = \mathrm{id}_{UA}$ so U preserves identities. If $f : A \Rightarrow B$ and $g : B \Rightarrow C$ are arrows, then

$$\begin{aligned} Ug \circ Uf &= \mu C \circ Tg \circ \mu B \circ Tf = \mu C \circ \mu TC \circ T^2 g \circ Tf \\ &= \mu C \circ T\mu C \circ T^2 g \circ Tf = \mu C \circ T(\mu C \circ Tg \circ f) \\ &= \mu C \circ T(g \bullet f) = U(g \bullet f) \end{aligned}$$

so that U preserves composition. Thus U is a functor.

As for F, $F\,\mathrm{id}_A = \eta A \circ \mathrm{id}_A = \eta_A$ so F preserves identities. If $f : A \to B$ and $g : B \to C$, then

$$\begin{aligned} Fg \bullet Ff &= \mu C \circ TFg \circ Ff = \mu C \circ T(\eta C \circ g) \circ \eta B \circ f \\ &= \mu C \circ T\eta C \circ Tg \circ \eta B \circ f = \eta C \circ g \circ f = F(g \circ f) \end{aligned}$$

and so F is also a functor.

It is evident that $T = UF$. Also, $\eta A : A \to UFA = TA$ is natural and we let $\epsilon A = \mathrm{id}_{TA} : UFA \Rightarrow A$ in $\mathcal{K}(\mathbf{T})$. Then

$$\epsilon FA \bullet F\eta A = \mu A \circ T(\mathrm{id}_{TA}) \circ \eta TA \circ \eta A = \mu A \circ \eta TA \circ \eta A$$

which is the identity of FA. For the other adjointness, we have for $B = FA$ in $\mathcal{K}(\mathbf{T})$,

$$U\epsilon FA \circ \eta UFA = \mu A \circ T(\mathrm{id}_A) \circ \eta TA = \mu A \circ \eta TA = \mathrm{id}_{TA}$$

This completes the proof.

4. If $g : A' \to A$ and $h : (B, \beta) \to (C, \gamma)$, then what must be shown is that $h \circ f \circ Fg \circ \eta A' = h \circ f \circ \eta A \circ g$. But with $Fg = Tg$, this is just the naturality equation of η.

5.a. Let us take the last point first. Given a monoid structure $\cdot$ and 1 on A, define $\alpha : A^* \to A$ by $\alpha() = 1$, $\alpha(a) = a$ and $\alpha(a_1, a_2, \cdots, a_n) = a_1 \cdot a_2 \cdot \cdots \cdot a_n$. Because of the associativity, it is not necessary to parenthesize that expression. Since $\alpha(a) = a$, the identity $\alpha \circ \eta A = \mathrm{id}_A$ is satisfied. As for the identity $\alpha \circ T\alpha = \alpha \circ \mu A$, let us do this for a list of length two of A^*; the general case follows by an easy induction. So let $l = (a_1, a_2, \cdots, a_n)$ and $l' = (a'_1, a'_2, \cdots a'_{n'})$. If $n = 0$, we have

$$\begin{aligned}\alpha \circ T\alpha(()l') &= \alpha(1, a'_1 \cdot a'_2 \cdot \cdots \cdot a'_{n'}) \\ &= 1 \cdot a'_1 \cdot a'_2 \cdot \cdots \cdot a'_{n'} \\ &= \alpha(l') = \alpha \circ \mu A(()l')\end{aligned}$$

and similarly if $n' = 0$. For $n > 0$ and $n' > 0$,

$$\begin{aligned}\alpha \circ T\alpha(ll') &= \alpha(a_1 \cdot a_2 \cdot \cdots \cdot a_n, a'_1 \cdot a'_2 \cdot \cdots \cdot a'_{n'}) \\ &= a_1 \cdot a_2 \cdot \cdots \cdot a_n \cdot a'_1 \cdot a'_2 \cdot \cdots \cdot a'_{n'} \\ &= \alpha(ll') = \alpha \circ \mu A(ll')\end{aligned}$$

This shows that each monoid gives an algebra structure.

An algebra structure on A assigns to each list $(a_1 a_2 \cdots a_n) \in A^*$ an element $\alpha(a_1 a_2 \cdots a_n) \in A$. In particular, there is a multiplication given by $a \cdot b = \alpha(ab)$. Also let 1 denote $\alpha()$. We must show that $\cdot$ and 1 constitute a monoid structure and that the associated algebra structure is the one we started with. It is already evident that if α is the algebra structure just constructed, then this monoid structure is the original one. We have $\alpha \circ T\alpha(()(a)) = \alpha(1a) = 1 \cdot a$ and $\alpha \circ \mu A(()(a)) = \alpha(a) = a$ so that $1 \cdot a = a$ and similarly $a \cdot 1 = a$. Next $\alpha \circ T\alpha((a),(b,c)) = a \cdot (b \cdot c)$ while $\alpha \circ \mu A((a),(b,c)) = \alpha(abc)$. Similarly, using $((a,b),(c))$ we can show that $(a \cdot b) \cdot c = \alpha(abc)$ and so $\cdot$ is associative. The proof that $a \cdot b \cdot c = \alpha(abc)$ extends by an obvious induction to show that $\alpha(a_1, a_2, \cdots, a_n) = a_1 \cdot a_2 \cdot \cdots \cdot a_n$, which means that the monoid structure determines uniquely the algebra structure.

b. Let (A, α) and (B, β) be algebra structures with corresponding monoid structures that we denote $\cdot$. Let $f : A \to B$. We show that f is a homomorphism of monoids if and only if it is a homomorphism of algebra structures. Suppose f is a homomorphism of algebra structures. Then from $\beta \circ Tf() = \beta() = 1$ and $f \circ \alpha() = f(1)$, we conclude that $f(1) = 1$. From $\beta \circ Tf(a, a') = \beta(f(a)f(a')) = f(a) \cdot f(a')$ and $f \circ \alpha(a, a') = f(a \cdot a')$ we see that f is a monoid homomorphism.

If f is a monoid homomorphism, then

$$\beta \circ Tf() = \beta() = 1 = f(1) = f \circ \alpha()$$

and

$$\beta \circ Tf(a) = \beta(f(a)) = f(\alpha(a)) = f \circ \alpha(a)$$

. If $l = (a_1, a_2, \cdots, a_n)$, then

$$\begin{aligned}\beta \circ Tf(a_1 a_2 \cdots a_l) &= \beta(f(a_1) f(a_2) \cdots f(a_n)) \\ &= f(a_1) \cdot f(a_2) \cdot \cdots \cdot f(a_n) \\ &= f(a_1 \cdot a_2 \cdot \cdots \cdot a_n) \\ &= f \circ \alpha(a_1, a_2, \cdots, a_n)\end{aligned}$$

Section 13.5

1. If A and B are ω-CPOs, let $[A \to B]$ denote the set of monotone functions from A to B that preserve joins of countable chains. Make it a poset by the pointwise ordering, that is $f \le g$ if $f(a) \le g(a)$ for all $a \in A$. We claim it is an

ω-CPO. In fact, if $f_0 \le f_1 \le \cdots$ is a countable increasing chain of such functions, then for each $a \in A$, let $f(a) = \bigvee f_i(a)$. Let us show that f is countably chain complete (it will obviously be the join in that case). If $a_0 \le a_1 \le \cdots$ is a countable increasing sequence with join a we have a double sequence in which every row and every column except perhaps the bottom row consists of a countable increasing sequence and its join.

$$\begin{array}{ccccc} f_0(a_0) & f_0(a_1) & f_0(a_2) & \cdots & f_0(a) \\ f_1(a_0) & f_1(a_1) & f_1(a_2) & \cdots & f_1(a) \\ f_2(a_0) & f_2(a_1) & f_2(a_2) & \cdots & f_2(a) \\ \vdots & \vdots & \vdots & \ddots & \vdots \\ f(a_0) & f(a_1) & f(a_2) & \cdots & f(a) \end{array}$$

For each i, $f(a_i) = \bigvee_j f_j(a_i) \le \bigvee_j f_j(a) = f(a)$ so $f(a)$ is an upper bound for the bottom row. If $b \in B$ were another bound, we would have for each i, $f_i(a) = \bigvee_j f_i(a_j) \le \bigvee_j f(a_j) \le b$ so that $f(a) \le b$. Thus $f(a) = \bigvee f(a_i)$.

Now we must show that if C is any other ω-CPO,

$$\mathrm{Hom}(C, [A \to B]) \cong \mathrm{Hom}(C \times A, B)$$

and this isomorphism is natural in B. If $f : C \times A \to B$ is a function, let $\phi(f) : C \to [A \to B]$ be the function defined by $\phi(f)(c)(a) = f(c, a)$. So far this is just the cartesian closed structure on sets. Now we claim that f is countably chain complete if and only if for all $c \in C$, $\phi(f)(c)$ is countably chain complete *and* $\phi(f)$ is countably chain complete.

Suppose f is countably chain complete. This means that if $c = \bigvee c_i$ and $a = \bigvee a_i$ are sups along increasing chains, then $f(c, a) = \bigvee f(c_i, a_i)$. This can be specialized to the case that all $c_i = c$ to conclude that $f(c, a) = \bigvee f(c, a_i)$ so that $\phi(f)(c)$ is countably chain complete. Similarly, if $c = \bigvee c_i$ is the sup of a countable increasing chain, then the fact that $f(c, a) = \bigvee f(c_i, a)$ implies that $\phi(f)(c) = \bigvee \phi(f)(c_i)$. This shows one direction. By reversing all these implications, one easily shows that if $\phi(f)(c)$ preserves joins along countable chains and $\phi(f)$ also preserves them, then f preserves joins in its variables *separately*. To prove that it preserves them jointly, let $c = \bigvee c_i$ and $a = \bigvee a_i$ and consider the double sequence

$$\begin{array}{ccccc} f(c_0, a_0) & f(c_0, a_1) & f(c_0, a_2) & \cdots & f(c_0, a) \\ f(c_1, a_0) & f(c_1, a_1) & f(c_1, a_2) & \cdots & f(c_1, a) \\ f(c_2, a_0) & f(c_2, a_1) & f(c_2, a_2) & \cdots & f(c_2, a) \\ \vdots & \vdots & \vdots & \ddots & \vdots \\ f(c, a_0) & f(c, a_1) & f(c, a_2) & \cdots & f(c, a) \end{array}$$

in which all rows and all columns are joins of increasing sequences. Clearly, $f(c_i, a_i) \le f(c, a)$ for all i and if b is an upper bound for the $f(c_i, a_i)$, then for all $j \ge i$, $f(c_i, a_j) \le f(c_j, a_j) \le b$, whence $f(c_i, a) = \bigvee_j (c_i, a_j) \le b$. Since this is true for all i, it follows that $f(c, a) \le b$. Hence f preserves joins of countable increasing chains.

2. This is clearly a poset. If $c^0 \le c^1 \le c^2 \le \cdots$ is an increasing chain in $\widehat{P}$, let c^n be the chain $c_0^n \le c_1^n \le c_2^n \le \cdots$. Construct a chain d as follows. Let $d_0 = c_0^0$ and having chosen $d_0 \le d_1 \le \cdots \le d_n$ such that $d_i = c^i_{j(i)}$, choose an integer $j(n+1) > j(n)$ so that $c^n_{j(n)} \le c^{n+1}_{j(n+1)}$. This is always possible since $c^n \le c^{n+1}$. Let $d_{n+1} = c^{n+1}_{j(n+1)}$. An obvious induction shows that $j(n) \ge n$ and so $c_n^i \le d_n$ for $n \ge i$. For a fixed i, the finite set of $n \le i$ does not matter and so $c^i \le d$ for all i. Thus d is an upper bound on the c^i. Suppose $e = e_0 \le e_1 \le e_2 \le \cdots$ is an upper bound on the c^i. Then since $d_n = c^n_{j(n)}$, and $c_n \le e$, it follows that d_n is less than or equal to some term of e, so that $d \le e$, whence $d = e$. Thus $\widehat{P}$ is an ω-CPO. Now suppose that A is an ω-CPO and $f : P \to A$ is an order-preserving map. Define $\widehat{f} : \widehat{P} \to A$ by $\widehat{f}(c) = \bigvee f(c_i)$ for $c = c_0 \le c_1 \le c_2 \le \cdots$. It is easy to see that this preserves the order and therefore the equality. Since the join of a constant sequence is itself, this extends f. Finally, if d is the join of the c^i as constructed above, it easily follows from the fact that the terms of d are a selection of those of the c^i that $\widehat{f}(d) = \bigvee \widehat{f}(c^i)$.

3. Let $\mathcal{C}$ be the category of ω-CPOs and functions that preserve joins along countable increasing chains and let $D : \mathcal{I} \to \mathcal{C}$ be a diagram. Let A be the set that is the limit of the diagram in sets. Then A comes equipped with projections $p_I : A \to DI$ for I an object of $\mathcal{I}$. Say that $a \le a'$ in A if for every object I of $\mathcal{I}$, $p_I(a) \le p_I(a')$. Then A is a poset and the projections preserve order. If $a_0 \le a_1 \le a_2 \le \cdots$ is an increasing chain in A, then for each I, $p_I(a_0) \le p_I(a_1) \le p_I(a_2) \le \cdots$ is an countable increasing chain in DI and has a join a_I. If $\alpha : I \to I'$ is an arrow in $\mathcal{I}$, then

$$D(\alpha)(a_I) = D(\alpha)\left(\bigvee_n (p_I(a_n))\right) = \bigvee_n D(\alpha)(p_I(a_n)) = \bigvee_n p_{I'}(a_n) = a_{I'}$$

so there is a unique $a \in A$ such that $p_I(a) = a_I$. Evidently $a_n \le a$ for each integer n. Also if $a' \in A$ is another upper bound for the $\{a_n\}$ then for each I, $p_I(a_n) \le p_I(a')$ so that $a_I = p_I(a) \le p_I(a')$ and then $a \le a'$. This shows that A is an ω-CPO and it is clear that the projections preserve the joins along countable chains. If B is any poset and $\{f_I\} : B \to D$ is a natural transformation from the constant diagram to D, let $f : B \to A$ be the induced function. Let $b_0 \le b_1 \le b_2 \le \cdots$ be an increasing chain in B with join b. Then for each I, $f_I(b) = \bigvee_n f_I(b_n)$ so that $p_I(f(b)) = \bigvee_n p_I(f(b_n))$. An argument similar to the above shows that this implies $f(b) = \bigvee_n f(b_n)$. The uniqueness of f follows because of the uniqueness of functions into a limit. This completes the proof for limits.

Assuming that the category of posets has colimits, one way to get colimits in the category of ω-CPOs is to form the colimit in the category of posets and apply the completion process of the preceding problem. If A is the colimit and $\widehat{A}$ its completion then any map from A to an ω-CPO extends to a unique map on $\widehat{A}$ that preserves joins along ω-chains. Putting the two universal mapping conditions together gives the result.

4. Since a chain is a directed set, one direction is immediate. For the other direction, let P be an ω-CPO and $D \subseteq P$ be a countable directed set in P. Since D is countable, we can name the elements of D as d_0, d_1, $d_2, \ldots$. Let $c_0 = d_0$ and having chosen $c_0 \leq c_1 \leq c_2 \leq \cdots \leq c_n$ an increasing sequence of elements of D, let $c_{n+1} \in D$ be a common upper bound of d_{n+1} and c_n. A join of the $\{c_n\}$ is clearly an upper bound of D.

Solutions for Chapter 14

Section 14.1

1. Let the pullbacks be

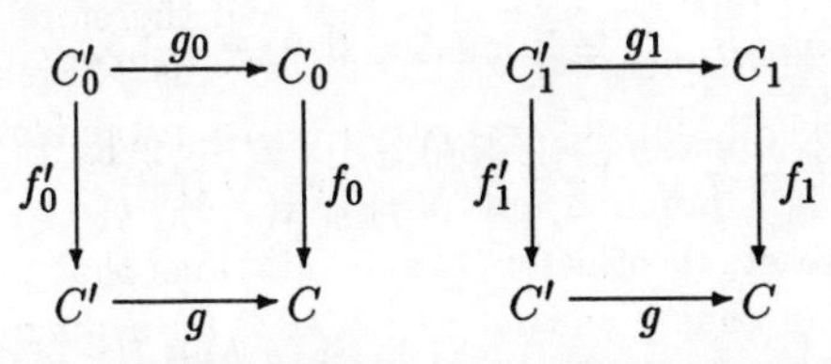

Let $u : C_0 \to C_1$ and $v : C_1 \to C_0$ be the inverse isomorphisms such that $f_0 \circ v = f_1$ and $f_1 \circ u = f_0$. Then $f_1 \circ u \circ g_0 = f_0 \circ g_0 = g \circ f_0'$ so by the universal mapping property of the right hand pullback, there is a unique $u' : C_0' \to C_1'$ such that $g_1 \circ u' = f_1 \circ u$ and $f_1' \circ u' = f_0'$. Similarly, there is a unique $v' : C_1' \to C_0'$ such that $g_0 \circ v' = f_0 \circ v$ and $f_0' \circ v' = f_1'$. Thus f_0' and f_1' belong to the same subobject.

2. For $f : C_0 \rightarrowtail C$, the square

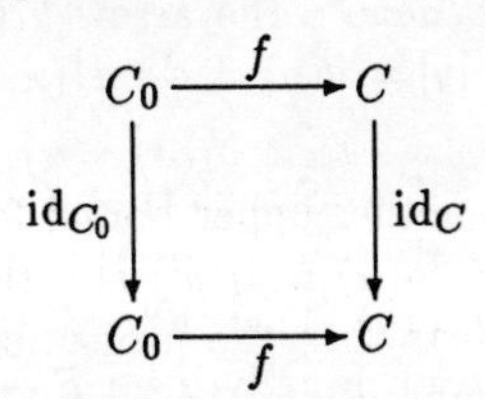

is a pullback. Thus the function induced by the identity of C assigns to each subobject of C the subobject itself (or an equivalent one).

3. If S and T are finite, so is the set of functions between them. In fact if we let $\#S$ denote the number of elements, then $\#[T \to S] = \#(S)^{\#(T)}$. For if $T = \emptyset$, then there is exactly one function from T to S no matter what S is. If $T \neq \emptyset$ and $S = \emptyset$, then there are no functions from T to S. If both are nonempty, we suppose by induction that the conclusion is true for sets with fewer elements than T, let $t \in T$ and $T_0 = T - \{t\}$. A function from T to S is determined by a function from T_0 to S plus an element of S for t to go to. Thus the number of such functions is

$$\#(S)^{\#(T_0)}\#(S) = \#(S)^{\#(T)-1}\#(S) = \#(S)^{\#(T)}$$

which is still finite. Thus the category of finite sets is cartesian closed. The two-element set is also finite so the category of finite sets has a subset classifier.

4. It is not quite sufficient to point out that when S is infinite, the set 2^S of subsets of S is not countable. The point to be made is that $\mathrm{Hom}(1,2^S) \cong \mathrm{Hom}(S,2)$ (where here 2 is the set $\{0,1\}$) and the latter set of functions is not countable, while there is no countable or finite set T for which $\mathrm{Hom}(1,T)$ is not countable or finite. Thus there is no finite or countable set that has the universal mapping property required to be the powerset 2^S.

Section 14.2

1. In the category of sets, $\Omega = 2$ so we must verify that

$$\mathrm{Hom}(B,2^A) \cong \mathrm{Hom}(A \times B, 2) \cong \mathrm{Sub}(A \times B)$$

If $f : B \to 2^A$ is a function, let $\phi(f) : A \times B \to 2$ by $\phi(f)(a,b) = \phi(b)(a)$. If $g : A \times B \to 2$ is a function, let $\psi(g) = \{(a,b) \mid g(a,b) = 1\} \subseteq A \times B$. If $U \subseteq A \times B$ is a subset, then let

$$\rho(U)(b)(a) = \begin{cases} 1 & \text{if } (a,b) \in U \\ 0 & \text{if } (a,b) \notin U \end{cases}$$

It is immediate that $\rho \circ \psi \circ \phi = \mathrm{id}_{\mathrm{Hom}(B,2^A)}$ and similarly for the other two serial composites, from which it is immediate that each of them is invertible, with inverse the composite of the other two.

2. For $x \in S$, let $[x]$ denote the equivalence class containing it. Then $[x] = [y]$ if and only if $(x,y) \in E$. Let $d^0, d^1 : E \to S$ be the two arrows mentioned above and let $d : S \to S/E$ denote the arrow $x \mapsto [x]$. Then for $(x,y) \in E$, $d \circ d^0(x,y) = d(x) = [x] = [y] = d(y) = d \circ d^1(x,y)$ so that d coequalizes d^0 and d^1.

Now suppose that $e^0, e^1 : T \to S$ such that $d \circ e^0 = d \circ e^1$. Then for all $x \in T$, $[e^0(x)] = [e^1(x)]$; equivalently $(e^0(x), e^1(x)) \in E$. Thus we can let $f : T \to E$ by defining $f(x) = (e^0(x), e^1(x))$. Then $d^0 \circ f(x) = d^0(e^0(x), e^1(x)) = e^0(x)$ and similarly $d^1 \circ f(x) = e^1(x)$. Finally if $g : T \to E$ is such that $d^0 \circ g(x) = d^0(e^0(x), e^1(x)) = e^0(x)$ and $d^1 \circ g(x) = e^1(x)$, then $g(x) = (e^0(x), e^1(x)) = f(x)$.

3.a. Suppose that d, e form an equivalence relation. This means that for any monoid N, the pair

$$\mathrm{Hom}(N,d), \mathrm{Hom}(N,e) : \mathrm{Hom}(N,E) \to \mathrm{Hom}(N,M)$$

gives an equivalence relation on the latter. This is another way of saying that

$$\langle \mathrm{Hom}(N,d), \mathrm{Hom}(N,e) \rangle : \mathrm{Hom}(N,E) \to \mathrm{Hom}(N,M) \times \mathrm{Hom}(N,M)$$

is an equivalence relation. Since

$$\mathrm{Hom}(N,M) \times \mathrm{Hom}(N,M) \cong \mathrm{Hom}(N, M \times M)$$

we have that $\langle d,e\rangle : E \to M \times M$ is monic. Since the image of a monoid homomorphism is a submonoid, the image is a submonoid. By letting $N = \mathbf{N}$, whence $\mathrm{Hom}(N,E)$ and $\mathrm{Hom}(N,M)$ are the underlying sets of E and M respectively, we conclude that the image of $\langle d,e\rangle$ is also an equivalence relation on the set underlying M. Thus it is a congruence.

Conversely, let $d,e : E \to M$ have the given property. Then up to isomorphism, we may suppose that $E \subseteq M \times M$ is both a submonoid and an equivalence relation and that d and e are the inclusion followed by the product projections. Then for any monoid N, $\mathrm{Hom}(N,E)$ consists of those pairs of arrows $f,g : N \to E$ such that for all $x \in N$, $(f(x), g(x)) \in E$. Since E is an equivalence relation, it is immediate that this makes $\mathrm{Hom}(N,E)$ into an equivalence relation on $\mathrm{Hom}(N,M)$.

b. Let us suppose that $E \subseteq M \times M$ is a submonoid and simultaneously an equivalence relation. For an $x \in M$, let $[x]$ denote the equivalence class containing x. If $x' \in [x]$ and $y' \in [y]$, then $(x,x') \in E$ and $(y,y') \in E$, whence so is $(x,x')(y,y') = (xy, x'y')$. Thus $x'y' \in [xy]$ which means there an unambiguous multiplication defined on the set M/E of equivalence classes by $[x][y] = [xy]$. The associativity is obvious as is $[1][x] = [x][1] = [x]$. Thus M/E is a monoid and evidently the arrow $p : M \to M/E$, $x \mapsto [x]$ is a monoid homomorphism. $p(x) = p(x')$ if and only if $x' \in [x]$ if and only if $(x,x') \in E$ so E is the kernel pair of p.

4. Let f be the coequalizer of the two arrows $e^0, e^1 : E \to C$ and let the kernel pair be $d^0, d^1 : K \to C$. Since $f \circ e^0 = f \circ e^1$, it follows from the universal mapping property of kernel pairs that there is a unique arrow $e : E \to K$ such that $d^0 \circ e = e^0$ and $d^1 \circ e = e^1$. Now $d^0 \circ f = d^1 \circ f$ is one of the defining properties of kernel pairs. If $g : C \to B$ is an arrow such that $g \circ d^0 = g \circ d^1$, then $g \circ e^0 = g \circ d^0 \circ e = g \circ d^1 \circ e = g \circ e^1$. Since f is the coequalizer of e^0 and e^1, there is a unique $h : D \to B$ such that $h \circ f = g$.

5. Let $S = \{a\}$, $T = \{a,b\}$ and $f : S \to T$ be the inclusion. The kernel pair d^0, d^1 of f is the equality relation on S, that is $d^0 = d^1$, but f is not the coequalizer. For any set U with more than one element and any function $g : S \to U$ there are many functions $h : T \to U$ such that $h \circ f = g$, since $g(b)$ can be any element of U.

6. Let T be a set, $T_0 \subseteq T$ a subset and $f : T_0 \to S$ be an arbitrary function. Define $\widehat{f} : T \to S \cup \{*\}$ by

$$\widehat{f}(x) = \begin{cases} f(x) & \text{if } x \in T_0 \\ * & \text{otherwise} \end{cases}$$

Then $T_0 = \{x \in T \mid \widehat{f}(x) \in S\}$, which is another way of saying that

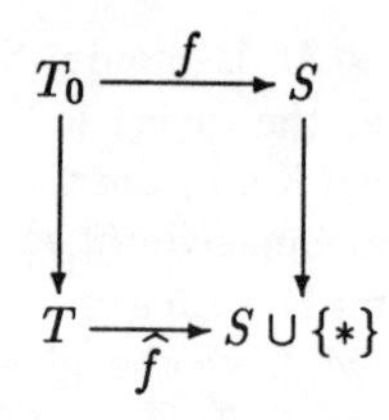

is a pullback. If g is a function with the same property, then $g(x) \in S$ if and only if $\widehat{f}(x)$ is, which means that $g(x) = *$ if and only if $\widehat{f}(x) = *$. Also, if $g(x) \neq *$, then $x \in T_0$ so that $g(x) = f(x) = \widehat{f}(x)$. Thus $g = \widehat{f}$ in all cases.

7. Since each object has a unique arrow to 1, there is a one to one correspondence between subobjects of an object A and partial arrows of A to 1. But partial arrows of A to 1 are in one to one correspondence with arrows $A \to \widetilde{1}$. Thus $\mathrm{Hom}(A, \widetilde{1}) \cong \mathrm{Sub}(A)$ which is the defining property of Ω.

Section 14.4

1. We are thinking of $\mathcal{C} = 0 \rightrightarrows 1$ as a category and 0 and 1 as two objects. The category of graphs is the category of contravariant functors $\mathcal{C} \to \mathbf{Set}$. In particular, the objects 0 and 1 represent functors and the question is which ones. Let us denote the two nonidentity arrows of $\mathcal{C}$ by s and t (mnemonic for 'source' and 'target'). A functor $F : \mathcal{C} \to \mathbf{Set}$ determines and is determined by the two sets $F(1)$ and $F(0)$ and the two arrows $F(s) : F(1) \to F(0)$ and $f(t) : F(1) \to F(0)$. In the case that $F = \mathrm{Hom}(-, 0)$, the two sets are $\mathrm{Hom}(0,0)$ and $\mathrm{Hom}(1,0)$ which are a one-element set and the empty set, respectively, and the functions are the unique functions that exists in that case. Thus the contravariant functor represented by 0 is the graph **1** which we have also called No.

For take the functor represented by 1, the two sets are $\mathrm{Hom}(1,1) = \{\mathrm{id}_1\}$ and $\mathrm{Hom}(0,1) = \{s, t\}$. The two arrows take the element of $\mathrm{Hom}(1,1)$ to s and t respectively. The graph **2** that we have also called No has just one arrow and two nodes which are the source and target of the arrow, respectively, and so is the graph represented by 1.

2. There are two ways of dealing with this question and similar ones. We already have an explicit description of the subobject classifier in this category and we could simply examine the set of nodes and set of arrows and see that their subfunctors have the desired structure. However, there is an easier way, once we know there is a subobject classifier. For the set of subobjects of the graph No is $\mathrm{Hom}(\mathsf{No}, \Omega)$ which, in turn, is isomorphic by the preceding problem, to the set of nodes of Ω. Similarly, the set of subobjects of the graph Ar is $\mathrm{Hom}(\mathsf{Ar}, \Omega)$, which is the set of arrows of Ω.

3. Let $\mathrm{NT}(F, G)$ denote the set of natural transformations between functors F and G. The Yoneda lemma implies that if $[M \to N] = F$, then

$$F(X) \cong \mathrm{NT}(\mathrm{Hom}(-,X), F) \cong \mathrm{NT}(\mathrm{Hom}(-,X), [M \to N])$$
$$\cong \mathrm{NT}(\mathrm{Hom}(-,X) \times M, N)$$

4. We have that

$$\mathrm{Sub}(\mathrm{Hom}(-,X)) \cong \mathrm{NT}(\mathrm{Hom}(-,X),\Omega) \cong \Omega(X)$$

the latter by the Yoneda Lemma.

Section 14.5

1.a. Let $C = \{c \in \mathcal{H} \mid a \wedge c \leq b\}$. Then $a \Rightarrow b = \bigvee_{c \in C} c$. Then

$$a \wedge (a \Rightarrow b) = a \wedge \bigvee_{c \in C} c = \bigvee_{c \in C} (a \wedge c) \leq b$$

since a join of elements less than b is also less than b. Therefore, if $c \leq a \Rightarrow b$ then $a \wedge c \leq b$. On the other hand, if $a \wedge c \leq b$, then $c \in C$ so that $c \leq a \Rightarrow b$.

b. The fact that $a \leq b \wedge c$ if and only if $a \leq b$ and $a \leq c$ means that when the Heyting algebra is considered as a category, $\mathrm{Hom}(a, b \wedge c) = \mathrm{Hom}(a,b) \times \mathrm{Hom}(a,c)$. This means that $b \wedge c$ is the categorical product of b and c. But then $a \wedge c \leq b$ if and only if $c \leq a \Rightarrow b$ means that $\mathrm{Hom}(a \times c, b) = \mathrm{Hom}(c, a \Rightarrow b)$, which means that $a \Rightarrow b$ is the internal hom.

Bibliography

M. Arbib and E. Manes, *Machines in a category: an expository introduction.* SIAM Review **16** (1974), 163–192.

M. Arbib and E. Manes, **Arrows, Structures and Functors: The Categorical Imperative**. Academic Press, 1975.

J. Backus, *The algebra of functional programs: function level reasoning, linear equations, and extended definitions.* In J. Diaz and I. Ramos, eds., **Formalization of Programming Concepts**. Lecture Notes in Computer Science **107** (1981a), 1–43.

J. Backus, *Is computer science based on the wrong fundamental concept of 'program'?* In J. W. deBakker and J. C. van Vliet, eds., **Algorithmic Languages**. North-Holland, 1981b.

E. Bainbridge, P. Freyd, A. Scedrov and P. Scott, *Functional polymorphism.* In G. Huet, ed., **Logical Foundations of Functional Programming**. Addison-Wesley, to appear.

H. Barendregt, **The Lambda Calculus – its Syntax and Semantics**, revised edition. North-Holland, 1984.

M. Barr, *Coequalizers and free triples.* Math. Z. **116** (1970), 307–322.

M. Barr, *Exact categories.* In **Exact Categories and Categories of Sheaves**. Lecture Notes in Mathematics **236**, Springer-Verlag, 1971, 1–120.

M. Barr, *Models of sketches.* Cahiers de Topologie et Géométrie Différentielle Catégorique **27** (1986a), 93–107.

M. Barr, *Representations of categories.* J. Pure Appl. Algebra **41** (1986b), 113–137.

M. Barr, *Fuzzy sets and topos theory.* Canad. Math. Bull. **24** (1986c), 501–508.

M. Barr, *Fixed points in cartesian closed categories.* To appear in the Proceedings of the Conference on Mathematical Foundations of Programming Language Semantics in Boulder, Colo., 1988.

M. Barr, *Models of Horn Theories.* In J. W. Gray and A. Scedrov, eds., **Categories in Computer Science and Logic**. Contemporary Mathematics **92** (1989), 1–7.

M. Barr, C. McLarty and C. Wells, *Variable set theory.* Mimeographed, McGill University, 1985.

M. Barr and C. Wells, **Toposes, Triples and Theories**. Grundlehren der math. Wissenschaften **278,** Springer-Verlag, 1985.

A. Bastiani and C. Ehresmann, *Categories of sketched structures.* Cahiers de Topologie et Géométrie Différentielle **10** (1968), 104–213.

J. L. Bell, **Toposes and Local Set Theories: An Introduction**. Oxford Logic Guides **14**, Oxford University Press, 1988.

J. Bénabou, *Structures algébriques dans les catégories*. Cahiers de Topologie et Géométrie Différentielle **13** (1972), 103–214.

S. Bloom and E. Wagner, *Many-sorted theories and their algebras with some applications to data types*. In M. Nivat and J. C. Reynolds, eds., **Algebraic Methods in Semantics**. Cambridge University Press, 1985.

A. Boileau and A. Joyal, *La logique des topos*. J. Symbolic Logic **46** (1981), 6–16.

R. Brown, P. R. Heath and K. H. Kamps, *Coverings of groupoids and Mayer-Vietoris type sequences*. Proceedings of the Conference on Categorical Topology, Toledo, Ohio, 1983. Heldermann-Verlag, 1984.

M. C. Bunge, *Toposes in logic and logic in toposes*. Topoi **3** (1984), 13–22.

A. Carboni, P. J. Freyd and A. Scedrov, *A categorical approach to realizability and polymorphic types*. In M. Main *et al.*, eds., **Mathematical Foundations of Programming Language Semantics**. Lecture Notes in Computer Science **298**, Springer-Verlag, 1988, 23–42.

J. Cartmell, *Generalized algebraic theories and contextual categories*. Annals Pure Applied Logic **32** (1986), 209–243.

H. G. Chen and J. R. B. Cockett, *Categorical combinators*. Preprint (1989), Department of Computer Science, University of Tennessee, Knoxville, TN.

J. R. B. Cockett, *Locally recursive categories*. Preprint (1987), Department of Computer Science, University of Tennessee, Knoxville, TN.

J. R. B. Cockett, *On the decidability of objects in a locos*. In J. W. Gray and A. Scedrov, eds., **Categories in Computer Science and Logic**. Contemporary Mathematics **92** (1989), 23–46.

T. Coquand, *Categories of embeddings*. Proceedings of the Third Annual Symposium on Logic in Computer Science, Edinburgh, 1988. Computer Society Press and IEEE (1988), 256–263.

M. Coste, *Localization, spectra and sheaf representation*. In M. Fourman, *et al.*, eds., **Applications of Sheaves**. Lecture Notes in Mathematics **753**, Springer-Verlag, 1979, 212–238.

G. Cousineau, P. L. Curien and M. Mauny, *The categorical abstract machine*. In **Functional Programming Languages and Computer Architecture**. Lecture Notes in Computer Science **201**, Springer-Verlag, 1985.

P. L. Curien, **Categorical Combinators, Sequential Algorithms and Functional Programming**. Wiley, 1986.

A. Day, *Filter monads, continuous lattices and closure systems*. Canadian J. Math. **27** (1975), 50–59.

Y. Diers, **Catégories Localizables**. Thèse de doctorat, Université de Paris, 1977.

P. Dybjer, *Category theory and programming language semantics: an overview*. In D. Pitt *et al.*, eds., **Category Theory and Computer Programming**. Lecture Notes in Computer Science **240**, Springer-Verlag, 1986, 165–181.

T. Ehrhard, *A categorical semantics of constructions.* Proceedings of the Third Annual Symposium on Logic in Computer Science, Edinburgh, 1988. Computer Society Press and IEEE (1988), 264–273.

H.-D. Ehrich, *Key extensions of abstract data types, final algebras, and database semantics.* In **Category Theory and Computer Programming**. Lecture Notes in Computer Science **240**, Springer-Verlag, 1986.

H. Ehrig, H.-J. Kreowski, J. Thatcher, E. Wagner and J. Wright, *Parameter passing in algebraic specification languages.* Theoretical Computer Science **28** (1984), 45–81.

H. Ehrig and B. Mahr, **Fundamentals of algebraic specifications I**. Springer-Verlag, 1985.

S. Eilenberg, **Automata, Languages and Machines, Vol. B**. Academic Press, 1976.

S. Eilenberg and S. Mac Lane, **General theory of natural equivalences**. Trans. Amer. Math. Soc. **58** (1945), 231–244.

S. Eilenberg and J. C. Moore, *Adjoint functors and triples.* Illinois J. Math. **9** (1965), 381–398.

M. Fourman, *The logic of topoi.* In J. Barwise, ed., **Handbook of Mathematical Logic**. North-Holland, 1977.

M. Fourman and S. Vickers, *Theories as categories.* In D. Pitt *et al.*, eds., **Category Theory and Computer Programming**. Lecture Notes in Computer Science **240**, Springer-Verlag, 1986, 434–448.

P. Freyd, **Abelian Categories: An Introduction to the Theory of Functors**. Harper and Row, 1964.

P. Freyd, *Aspects of topoi.* Bull. Austral. Math. Soc. **7** (1972), 1–72 and *Corrections. Ibid.* **8** (1973), 467–480.

P. Freyd, POLYNAT *in* PER. In J. W. Gray and A. Scedrov, eds., **Categories in Computer Science and Logic**. Contemporary Mathematics **92** (1989), 67–68.

P. Freyd, J.-Y. Girard, A. Scedrov and P. J. Scott, *Semantic parametricity in polymorphic lambda calculus.* In **Proceedings of the Third Annual Symposium on Logic in Computer Science, Edinburgh, 1988**. Computer Society Press and IEEE (1988), 274–281.

J.-Y. Girard, *Linear Logic.* Theoretical Computer Science **50** (1987), 1-102.

J.-Y. Girard, *Towards a Geometry of Interactions.* In J. W. Gray and A. Scedrov, eds., **Categories in Computer Science and Logic**. Contemporary Mathematics **92** (1989), 69–108.

J.-Y. Girard, P. Taylor and Y. Lafont, **Proofs and Types**. Cambridge University Press, 1989.

R. Godement, **Théorie des faisceaux**. Hermann, 1958.

J. A. Goguen, *Concept representation in natural and artificial languages: axioms, extensions and applications for fuzzy sets.* Int. J. Man-machine Studies **6** (1974), 513–564.

J. A. Goguen, *Abstract errors for abstract data types.* In E. J. Neuhold, ed., **Formal Description of Programming Concepts**. North-Holland, 1978, 491–526.

J. A. Goguen, *What is unification? A categorical view of substitution, equation and solution.* Center for the Study of Language and Information, Report No. CSLI–88–124, 1988.

J. A. Goguen and R. M. Burstall, *A study in the foundations of programming methodology: specifications, institutions, charters and parchments.* In D. Pitt *et al.*, eds., **Category Theory and Computer Programming**. Lecture Notes in Computer Science **240**, Springer-Verlag, 1986, 313–333.

J. Goguen, J.-P. Jouannaud and J. Meseguer, *Operational semantics for order-sorted algebra.* In **Proc. 12th Int. Colloq. on Automata, Languages and Programming, Napflion, Greece**. Lecture Notes in Computer Science **194**, Springer-Verlag, 1985, 221–231.

J. Goguen, J. W. Thatcher and E. G. Wagner, *An initial algebra approach to the specification, correctness and implementation of abstract data types.* In R. Yeh, ed., **Current Trends in Programming Methodology IV**. Prentice-Hall, 1978, 80–149.

J. Goguen, J. W. Thatcher, E. G. Wagner and J. B. Wright, *Initial algebra semantics and continuous algebras.* J. ACM **24** (1977), 68–95.

J. Gray, *Fibred and cofibred categories.* In S. Eilenberg *et al.*, eds., **Proc. La Jolla Conference on Categorical Algebra**. Springer-Verlag, 1966, 21–83.

J. Gray, **Formal Category Theory: Adjointness for 2-Categories**. Lecture Notes in Mathematics **391**, Springer-Verlag, 1974.

J. Gray, *Categorical aspects of data type constructors.* Theoretical Computer Science **50** (1987), 103–135.

J. Gray, *Executable specifications for data-type constructors.* Preprint, University of Illinois, 1988a.

J. Gray, *A categorical treatment of polymorphic operations.* In M. Main *et al.*, eds., **Mathematical Foundations of Programming Language Semantics**. Lecture Notes in Computer Science **298**, Springer-Verlag, 1988b 2–22.

J. Gray, *The category of sketches as a model for algebraic semantics.* In J. W. Gray and A. Scedrov, eds., **Categories in Computer Science and Logic**. Contemporary Mathematics **92** (1989), 109–135.

J. Gray and A. Scedrov, **Categories in Computer Science and Logic**. Contemporary Mathematics **92** (1989).

R. Guitart, *On the geometry of computations.* Cahiers de Topologie et Géométrie Différentielle Catégorique **27** (1986), 107–136.

R. Guitart and C. Lair, *Calcul syntaxique des modèles et calcul des formules internes.* Diagrammes **4** (1980).

R. Guitart and C. Lair, *Existence de diagrammes localement libres 1.* Diagrammes **6** (1981).

R. Guitart and C. Lair, *Existence de diagrammes localement libres 2.* Diagrammes **7** (1982).

T. Hagino, *A typed lambda calculus with categorical type constructors.* In D. Pitt, A. Poigné and D. Rydeheard, eds., **Category Theory and Computer Science**. Lecture Notes in Computer Science **283**, Springer-Verlag, 1987, 140–157.

P. Halmos, **Naive Set Theory**. Van Nostrand, 1960.

J. R. Hindley and J. P. Seldin, **Introduction to Combinators and λ-calculus**. Cambridge University Press, 1986.

J. E. Hopcroft and J. D. Ullman, **Introduction to Automata Theory, Languages and Computation**. Addison-Wesley, 1979.

C. Houghton, *The wreath product of groupoids.* J. London Math. Soc. (2) **10** (1975), 179–188.

G. Huet, *Cartesian closed categories and lambda-calculus.* In G. Cousineau, P.-L. Curien and B. Robinet, eds., **Combinators and Functional Programming Languages**. Lecture Notes in Computer Science **242**, Springer-Verlag, 1986.

J. M. E. Hyland, *The effective topos.* In **The L.E.J. Brouwer Centenary Symposium**. North-Holland, 1982, 165–216.

J. Hyland and A. Pitts, *The theory of constructions: categorical semantics and topos-theoretic models.* In J. W. Gray and A. Scedrov, eds., **Categories in Computer Science and Logic**. Contemporary Mathematics **92** (1989), 137-199.

N. Jacobson, **Basic Algebra I**. W. H. Freeman, 1974.

P. T. Johnstone, **Topos Theory**. Academic Press, 1977.

P. T. Johnstone, J. M. E. Hyland and A. M. Pitts, *Tripos theory,* Proc. Cambridge Phil. Soc. **88** (1980), 205–232.

P. T. Johnstone and R. Paré, eds., **Indexed Categories and their Applications**. Lecture Notes in Mathematics **661**, Springer-Verlag, 1978.

P. T. Johnstone and G. Wraith, *Algebraic theories in toposes.* In P. T. Johnstone and R. Paré, eds., **Indexed Categories and their Applications**. Lecture Notes in Mathematics **661**, Springer-Verlag, 1978, 141–242.

G. M. Kelly, ed. **Category Seminar: Proceedings Sydney Category Theory Seminar 1972/1973**. Lecture Notes in Mathematics **420**, Springer-Verlag, 1974a.

G. M. Kelly, *On clubs and doctrines.* In G. M. Kelly, ed., **Proceedings Sydney Category Theory Seminar 1972/1973**. Lecture Notes in Mathematics **420**, Springer-Verlag, 1974b.

G. M. Kelly, *On the essentially-algebraic theory generated by a sketch.* Bulletin Australian Mathematical Society **26** (1982a), 45–56.

G. M. Kelly, **Basic concepts of enriched category theory**. Cambridge University Press, 1982b.

G. M. Kelly and R. Street, *Review of the elements of 2-categories.* In **Proceedings Sydney Category Theory Seminar 1972/1973**. Lecture Notes in Mathematics **420**, Springer-Verlag, 1974.

H. Kleisli, *Every standard construction is induced by a pair of adjoint functors.* Proc. Amer. Math. Soc. **16** (1965), 544–546.

C. Lair, *Foncteurs d'omission de structures algébriques.* Cahiers de Topologie et Géométrie Différentielle **12** (1971), 447–486.

C. Lair, *Trames et semantiques catégoriques des systèmes de trames.* Diagrammes **18** (1987).

G. Lallement, **Semigroups and Combinatorial Applications.** Wiley, 1979.

F. Lamarche, *A simple model of the theory of constructions.* In J. W. Gray and A. Scedrov, eds., **Categories in Computer Science and Logic.** Contemporary Mathematics **92** (1989), 200–216.

J. Lambek, *Subequalizers.* Canad. Math. Bull. **13** (1970), 337–349.

J. Lambek, *Cartesian closed categories and typed lambda-calculi.* In G. Cousineau, P.-L. Curien and B. Robinet, eds., **Combinators and Functional Programming Languages.** Lecture Notes in Computer Science **242**, Springer-Verlag, 1986, 136–175.

J. Lambek, *Multicategories revisited.* In J. W. Gray and A. Scedrov, eds., **Categories in Computer Science and Logic.** Contemporary Mathematics **92** (1989), 217–240.

J. Lambek and P. Scott, *Aspects of higher order categorical logic.* In J. W. Gray, ed., **Mathematical Applications of Category Theory.** Contemporary Mathematics **30** (1984), 145–174.

J. Lambek and P. Scott, **Introduction to Higher Order Categorical Logic.** Cambridge Studies in Advanced Mathematics **7**. Cambridge University Press, 1986.

F. W. Lawvere, *Functorial semantics of algebraic theories.* Dissertation, Columbia University (1963). Announcement in Proc. Nat. Acad. Sci. **50** (1963), 869–873.

F. W. Lawvere, *The category of categories as a foundation for mathematics.* **Proceedings of the Conference on Categorical Algebra at La Jolla.** Springer-Verlag, 1966.

F. W. Lawvere, *Qualitative distinctions between some toposes of generalized graphs.* In J. W. Gray and A. Scedrov, eds., **Categories in Computer Science and Logic.** Contemporary Mathematics **92** (1989), 261–300.

S. K. Lellahi, *Types abstraits catégoriques: une extension des types abstraits algébriques.* Laboratoire d'Informatique des Systèmes Expérimentaux et leur Modélisation: Research Report 063 (1987).

F. E. J. Linton, *Some aspects of equational categories.* In S. Eilenberg *et al.*, eds., **Proceedings of the Conference on Categorical Algebra at La Jolla.** Springer-Verlag, 1966.

F. E. J. Linton, *An outline of functorial semantics.* Lecture Notes in Mathematics **80**, 7–52. Springer-Verlag, 1969a.

F. E. J. Linton, *Applied functorial semantics.* Lecture Notes in Mathematics **80**, 53–74, Springer-Verlag, 1969b.

S. Mac Lane, **Categories for the Working Mathematician.** Graduate Texts in Mathematics **5**, Springer-Verlag, 1971.

M. Main, A. Melton, M. Mislove and D. Schmidt, eds. **Mathematical Foundations of Programming Language Semantics.** Lecture Notes in Computer Science **298**, Springer-Verlag, 1988.

M. Makkai and R. Paré, **Accessible categories: the foundations of categorical model theory**. Contemporary Mathematics **104** (1990).

M. Makkai and G. Reyes, **First Order Categorical Logic**. Lecture Notes in Mathematics **611**, Springer-Verlag, 1977.

E. Manes, **Algebraic Theories.** Graduate Texts in Mathematics **26**. Springer-Verlag, 1975.

E. Manes, *Weakest preconditions: categorial insights.* In D. Pitt *et al.*, eds., **Category Theory and Computer Programming.** Lecture Notes in Computer Science **240**, Springer-Verlag, 1986, 182–197.

E. Manes and M. Arbib , **Algebraic Approaches to Program Semantics**, Springer-Verlag, 1986.

C. McLarty, *Left exact logic.* J. Pure Applied Algebra **41** (1986), 63–66.

C. McLarty, *Notes toward a new philosophy of logic.* Preprint, Department of Philosophy, Case Western Reserve University (1989).

J. Meseguer and J. Goguen, *Initiality, induction and computability.* In M. Nivat and J. C. Reynolds, eds., **Algebraic Methods in Semantics**. Cambridge University Press, 1985.

J. C. Mitchell and P. J. Scott, *Typed lambda calculus and cartesian closed categories.* In J. W. Gray and A. Scedrov, eds., **Categories in Computer Science and Logic**. Contemporary Mathematics **92** (1989), 301–316.

C. J. Mikkelson, **Lattice Theoretic and Logical Aspects of Elementary Topoi**. Aarhus University Various Publications Series **25** 1976.

W. Nico, *Wreath products and extensions.* Houston J. Math. **9** (1983), 71–99.

F. Oles, *Type algebras, functor categories and block structure.* In M. Nivat and J. C. Reynolds, eds., **Algebraic Methods in Semantics**. Cambridge University Press, 1985.

D. Pitt, *Categories.* In D. Pitt *et al.*, eds., **Category Theory and Computer Programming**. Lecture Notes in Computer Science **240**, Springer-Verlag, (1986), 6–15.

D. Pitt, S. Abramsky, A. Poigné and D. Rydeheard, eds. **Category Theory and Computer Programming.** Lecture Notes in Computer Science **240**, Springer-Verlag, 1986.

D. Pitt, A. Poigné and D. Rydeheard, eds., **Category Theory and Computer Science**. Lecture Notes in Computer Science **283**, Springer-Verlag, 1987.

A. Pitts, *Fuzzy sets do not form a topos.* Fuzzy Sets and Systems **8** (1982), 101–104.

G. D. Plotkin, *Dijkstra's predicate transformers and Smyth's powerdomains.* In D. Bjørner, ed., **Abstract Software Specifications.** Lecture Notes in Computer Science **86**, Springer-Verlag, 1980, 527-553.

A. Poigné, *Elements of categorical reasoning: products and coproducts and some other (co-)limits.* In D. Pitt *et al.*, eds., **Category Theory and Computer Programming**. Lecture Notes in Computer Science **240**, Springer-Verlag, 1986, 16–42.

A. Poigné, *Category theory and logic.* In D. Pitt *et al.*, eds., **Category Theory and Computer Programming**. Lecture Notes in Computer Science **240**, Springer-Verlag, 1986, 103–142.

A. J. Power, *An algebraic formulation for data refinement.* Preprint, Department of Mathematics, Case Western Reserve University, Cleveland, OH 44106, USA.

A. J. Power and C. Wells, *A formalism for the specification of essentially-algebraic structures in 2-categories.* Preprint, Department of Mathematics, Case Western Reserve University, Cleveland, OH 44106, USA.

H. Reichel, **Initial Computability, Algebraic Specifications and Partial Algebras**. Clarendon Press, 1987.

J. C. Reynolds, *Using category theory to design implicit conversions and generic operators.* In N. D. Jones, ed., **Proceedings of the Aarhus Workshop on Semantics-Directed Compiler Generation**. Lecture Notes in Computer Science **94**, Springer-Verlag, 1980,

J. Rhodes and B. Tilson, *The kernel of monoid homomorphisms.* J. Pure Applied Algebra **62** (1989), 227–268.

J. Rhodes and P. Weil, *Decomposition techniques for finite semigroups* I, II. J. Pure Applied Algebra **62** (1989), 269–284.

R. Rosebrugh, *On algebras defined by operations and equations in a topos.* J. Pure Applied Algebra **17** (1980).

G. Rosolini, *Categories and effective computation.* In **Category Theory and Computer Science**. Lecture Notes in Computer Science **283**, Springer-Verlag, 1987.

D. E. Rydeheard and R. M. Burstall, *Monads and theories: a survey for computation.* In M. Nivat and J. C. Reynolds, eds., **Algebraic Methods in Semantics**. Cambridge University Press, 1985.

D. E. Rydeheard and R. M. Burstall, *A categorical unification algorithm.* In D. Pitt *et al.*, eds., **Category Theory and Computer Programming**. Lecture Notes in Computer Science **240**, Springer-Verlag, 1986, 493–505.

D. E. Rydeheard and R. M. Burstall, **Computational Category Theory**. International Series in Computer Science, Prentice Hall, 1988.

D. E. Rydeheard and J. G. Stell, *Foundations of equational deduction: categorical treatment of equational proofs and unification algorithms.* Preprint, 1987.

A. Scedrov, *A guide to polymorphic λ-calculus.* To appear.

D. Schmidt, **Denotational Semantics: A methodology for language development.** Allyn and Bacon, 1986.

D. Scott, *Continuous lattices.* In F. W. Lawvere, ed., **Toposes, Algebraic Geometry and Logic**. Lecture Notes in Mathematics **274**, Springer-Verlag, (1972), 97-136.

D. Scott, *Relating theories of the λ-calculus.* In J. R. Hindley and J. P. Seldin, eds., **To H. B. Curry: Essays on Combinatory Logic, Lambda Calculus and Formalism**. Academic Press, 1980.

D. Scott, *Domains for denotational semantics.* In M. Nielson and E. M. Schmidt, eds., **Automata, Languages and Programming**. Lecture Notes in Computer Science **140**, Springer-Verlag, 1982, 577–613.

R. Sedgewick, **Algorithms**. Addison-Wesley, 1983.

R. Seely, *Natural deduction and the Beck condition.* Z. Math. Logik und Grundlagen Math. **29** (1983), 505–542.

R. Seely, *Locally cartesian closed categories and type theory.* Proc. Cambridge Philos. Soc. **95** (1984), 33–48.

R. Seely, *Modelling computations: A 2-categorical framework.* LICS 1986.

R. Seely, *Categorical semantics for higher order polymorphic lambda calculus.* J. Symbolic Logic **52** (1987), 969–989.

R. Seely, *Linear logic, ∗-autonomous categories and cofree coalgebras.* In J. W. Gray and A. Scedrov, eds., **Categories in Computer Science and Logic**. Contemporary Mathematics **92** (1989), 371–382.

M. B. Smyth, *Power domains and predicate transformers: a topological view.* In J. Diaz, ed., **Automata, Languages and Programming**. Lecture Notes in Computer Science **154**, Springer-Verlag, 1983a, 662–675.

M. B. Smyth, *The largest cartesian closed category of domains.* Theoretical Computer Science **27**, 1983b.

M. B. Smyth and G. D. Plotkin, *The category-theoretic solution of recursive domain equations.* SIAM J. Computing **11** (1983), 761–783.

R. Street, *The algebra of oriented simplexes.* J. Pure Applied Algebra **49** (1987), 283–335.

R. Street and R. Walters, *The comprehensive factorization of a functor.* Bull. Amer. Math. Soc. **79** (1973), 936–941.

J. W. Thatcher, E. G. Wagner and J. B. Wright, *Data type specification: parametrization and the power of specification techniques.* ACM Transactions on Programming Languages and Systems 4 (1982), 711–732.

D. Thérien, *Two-sided wreath product of varieties of finite categories.* To appear in J. Pure Applied Algebra.

B. Tilson, **Categories as algebra: an essential ingredient in the theory of semigroups**. Preprint (1986), to appear in J. Pure Applied Algebra.

H. Volger, *On theories which admit initial structures.* Technical Report MIP–8708, Fakultät für Math. und Informatik, Universität Passau, 1987.

H. Volger, *Model theory of deductive databases.* In Lecture Notes in Computer Science **329**, Springer-Verlag, 1988.

E. G. Wagner, *Algebraic theories, data types and control constructs.* Fundamenta Informatica **9** (1986a), 343–370.

E. G. Wagner, *Categories, data types and imperative languages.* In D. Pitt *et al.*, eds., **Category Theory and Computer Programming**. Lecture Notes in Computer Science **240**, Springer-Verlag, 1986b, 143–162.

E. G. Wagner, *Categorical semantics, or extending data types to include memory.* In H.-J. Kreowski, ed., **Recent Trends in Data Type Specification.** Informatik-Fachberichte **116**, Springer-Verlag, 1986c.

E. G. Wagner, *A categorical treatment of pre- and post-conditions.* Theoretical Computer Science **53**, 1987.

E. G. Wagner, *Semantics of block structured languages with pointers.* In M. Main *et al.*, eds., **Mathematical Foundations of Programming Language Semantics.** Lecture Notes in Computer Science **298**, Springer-Verlag, 1988, 57–84.

E. G. Wagner, S. Bloom and J. W. Thatcher, *Why algebraic theories?* In M. Nivat and J. C. Reynolds, eds., **Algebraic Methods in Semantics.** Cambridge University Press, 1985.

C. Wells, *Some applications of the wreath product construction.* Amer. Math. Monthly **83** (1976), 317–338.

C. Wells, *A Krohn-Rhodes theorem for categories.* J. Algebra **64** (1980), 37–45.

C. Wells, *Wreath product decomposition of categories I.* Acta Sci. Math. Szeged **52** (1988a), 307–319.

C. Wells, *Wreath product decomposition of categories II.* Acta Sci. Math. Szeged **52** (1988b), 321–324.

C. Wells, *A generalization of the concept of sketch.* To appear in Theoretical Computer Science (1990).

C. Wells and M. Barr, *The formal description of data types using sketches.* In M. Main *et al.*, eds., **Mathematical Foundations of Programming Language Semantics.** Lecture Notes in Computer Science **298**, Springer-Verlag, 1988, 490–527.

J. H. Williams, *Notes on the FP style of functional programming.* In J. Darlington, P. Henderson and D. Turner, eds., **Functional Programming and its Applications.** Cambridge University Press, 1982.

G. C. Wraith, *Algebras over theories.* Colloq. Math. **23** (1971), 180–190.

O. Wyler, *Algebraic theories of continuous lattices.* In **Continuous Lattices.** Lecture Notes in Mathematics **871**, Springer-Verlag, 1981, 390–413.

S. N. Zilles, P. Lucas and J. W. Thatcher, *A look at algebraic specifications.* IBM T. J. Watson Research Center Research Report RJ 3568 (#41985), 1982.

Index